T0271393

Contemporary Developments in Finite Fields and Applications

Contemporary Developments in Finite Fields and Applications

Editors

Anne Canteaut
INRIA, France

Gove Effinger
Skidmore College, USA

Sophie Huczynska
University of St Andrews, UK

Daniel Panario
Carleton University, Canada

Leo Storme
Ghent University, Belgium

 World Scientific

NEW JERSEY · LONDON · SINGAPORE · BEIJING · SHANGHAI · HONG KONG · TAIPEI · CHENNAI · TOKYO

Published by

World Scientific Publishing Co. Pte. Ltd.

5 Toh Tuck Link, Singapore 596224

USA office: 27 Warren Street, Suite 401-402, Hackensack, NJ 07601

UK office: 57 Shelton Street, Covent Garden, London WC2H 9HE

British Library Cataloguing-in-Publication Data

A catalogue record for this book is available from the British Library.

CONTEMPORARY DEVELOPMENTS IN FINITE FIELDS AND APPLICATIONS

Copyright © 2016 by World Scientific Publishing Co. Pte. Ltd.

All rights reserved. This book, or parts thereof, may not be reproduced in any form or by any means, electronic or mechanical, including photocopying, recording or any information storage and retrieval system now known or to be invented, without written permission from the publisher.

For photocopying of material in this volume, please pay a copying fee through the Copyright Clearance Center, Inc., 222 Rosewood Drive, Danvers, MA 01923, USA. In this case permission to photocopy is not required from the publisher.

ISBN 978-981-4719-25-4

Printed in Singapore

Preface

The twenty articles in this review volume illustrate the wide variety of areas of mathematics and computer science in which finite fields play a fundamental role. In addition, they illustrate progress which is being made now, in 2016, in these areas, which is why the words contemporary developments are in the title. Those areas include polynomials over finite fields, curves, surfaces and graphs over finite fields, discrete logarithms, coding theory, cryptography, and more.

The bulk of the material contained in this volume is closely related to lectures presented in July 2015, at the 12th International Conference on Finite Fields and Their Applications (Fq12), held in Saratoga Springs, New York. More than 100 mathematicians and computer scientists were present at that conference, and 84 lectures were given over five days. The contributions to this volume were rigorously reviewed, and as a result they represent well the quality of the conference.

I wish to give many thanks to my four fellow editors for their hard work and excellent advice on deciding what belongs in this volume. They are Anne Canteaut (French National Research Institution in Computer Science), Sophie Huczynska (University of St Andrews), Daniel Panario (Carleton University), and Leo Storme (Ghent University). In addition I wish to give special thanks to David Thomson (Carleton University) for providing invaluable technical support in the preparation of this volume. Finally, let me give thanks to Rochelle Kronzek, Acquisitions Editor for World Scientific, for encouraging us to produce this volume and opening the necessary doors for that to happen, and finally to E.H. Chionh, Editor for World Scientific, for seeing the volume through the full process to completion.

Gove Effinger, Skidmore College, Managing Editor

Contents

Contents

Divisibility of L-polynomials for a family of curves

I. Blanco–Chacón

School of Mathematics and Statistics, University College Dublin,
Dublin 4, Ireland `ivan.blanco-chacon@ucd.ie`

R. Chapman

Department of Mathematics, University of Exeter, Exeter, EX4 4QE, UK
`r.j.chapman@exeter.ac.uk`

S. Fordham

School of Mathematics and Statistics, University College Dublin,
Dublin 4, Ireland `stiofain.fordham@ucdconnect.ie`

G. McGuire

School of Mathematics and Statistics, University College Dublin,
Dublin 4, Ireland `gary.mcguire@ucd.ie`

We consider the question of when the L-polynomial of one curve divides the L-polynomial of another curve. A theorem of Tate gives an answer in terms of jacobians. We consider the question in terms of the curves. The last author gave an invited talk at the 12th International Conference on Finite Fields and Their Applications on this topic, and stated two conjectures. In this article we prove one of those conjectures.

1. Introduction

Let p be a prime and $q = p^f$ where f is a positive integer and let $\mathbb{F}_q$ denote the finite field of order q. Let X be a smooth projective variety over $\mathbb{F}_q$, of dimension d. Let $\overline{X} = X(\overline{\mathbb{F}}_q)$ be the corresponding variety over the algebraic closure of $\mathbb{F}_q$ and let $F \colon \overline{X} \to \overline{X}$ be the Frobenius morphism.

1

The zeta function $Z_X(t)$ of X is defined by

$$\log Z_X(t) = \sum_{m \geq 1} \frac{t^m}{m} N_m,$$

where N_m is the cardinality of the set $X(\mathbb{F}_{q^m})$: the points of X with values in $\mathbb{F}_{q^m}$. Via the Weil conjectures (proved by Weil, Dwork, Grothendieck and others), one knows that $Z_X(t)$ is a rational function and may be written in the form

$$Z_X(t) = \frac{P_1(t) \cdots P_{2d-1}(t)}{P_0(t) \cdots P_{2d}(t)}, \tag{1}$$

where each of the $P_i(t) = \det(1 - F^*t; H^i(\overline{X}, \mathbb{Q}_\ell))$ are polynomials with coefficients in $\mathbb{Z}$, where $H^i(\overline{X}, \mathbb{Q}_\ell)$ is the ith ℓ-adic cohomology ($\ell \neq p$) of $\overline{X}$ with coefficients in $\mathbb{Q}_\ell$ and F^* is the map on cohomology induced by F.

In the case that $X = C$ is a curve then the zeta function of C has the form

$$Z_C(t) = \frac{L_C(t)}{(1-t)(1-qt)},$$

and the numerator $L_C(t) = P_1(t)$ is called the L-polynomial of C.

We wish to consider the question of divisibility of L-polynomials. In previous papers [2],[3], we have studied conditions under which the L-polynomial of one curve divides the L-polynomial of another curve. In this article we discuss two divisibility conjectures for specific families of curves, and prove one of them.

2. Two Families of Curves

A hyperelliptic curve X of genus $g > 1$ over $\mathbb{F}_q$ is the projective non-singular model of the affine curve

$$y^2 + Q(x)y = P(x), \qquad P(x), Q(x) \in \mathbb{F}_q[x],$$

where

$$2g + 1 \leq \max\{2 \deg Q(x), \deg P(x)\} \leq 2g + 2.$$

2.1. The C_k Family

For a positive integer k, define the curve C_k over $\mathbb{F}_2$ to be the projective non-singular model of the curve with affine equation

$$y^2 + y = x^{2^k+1} + x.$$

The genus of C_k is 2^{k-1}, the affine model of C_1 is smooth everywhere, and the affine model of C_k for $k > 1$ has one singular point at ∞.

Conjecture 1. *The L-polynomial of C_k is divisible by the L-polynomial of C_1.*

The first six L-polynomials over $\mathbb{F}_2$, computed and factored into irreducible factors over $\mathbb{Z}$ using MAGMA [4] are

C_1: $2t^2 + 2t + 1$

C_2: $(2t^2 + 2t + 1)(2t^2 + 1)$

C_3: $(2t^2 + 2t + 1)(2t^2 - 2t + 1)(4t^4 + 4t^3 + 2t^2 + 2t + 1)$

C_4: $(2t^2 + 2t + 1)^2(2t^2 - 2t + 1)(2t^2 + 1)(16t^8 + 1)$

C_5: $(2t^2 + 2t + 1)^2(2t^2 - 2t + 1)^2(16t^8 - 16t^7 + 8t^6 - 4t^4 + 2t^2 - 2t + 1)$
$\qquad \times (16t^8 + 16t^7 + 8t^6 - 4t^4 + 2t^2 + 2t + 1)^2$

C_6: $(2t^2 - 1)^2(2t^2 + 1)^4(4t^4 - 2t^2 + 1)^3(4t^4 + 2t^2 + 1)^2$
$\qquad \times (2t^2 - 2t + 1)^3(2t^2 + 2t + 1)^3(4t^4 - 4t^3 + 2t^2 - 2t + 1)^2$
$\qquad \times (4t^4 + 4t^3 + 2t^2 + 2t + 1)^3.$

2.2. The E_k Family

For a positive integer k, define the curve E_k over $\mathbb{F}_2$ to be the projective non-singular model of the curve with affine model
$$y^2 + xy = x^{2^k+3} + x.$$
The genus of E_k is $2^{k-1} + 1$ and similar to above, the affine model of E_1 is smooth everywhere, and the affine model of E_k for $k > 1$ has one singular point at ∞.

Conjecture 2. *The L-polynomial of E_k is divisible by the L-polynomial of E_1.*

In an invited talk at the Fq12 conference, the last author spoke about this topic and stated these two conjectures. Conjecture 2 is proposed and discussed in Ahmadi et al.[3]. In this paper we will prove Conjecture 1.

3. Other Approaches

Here we discuss three possible approaches to proving the conjectures. The first two do not seem to work for Conjecture 1, but the third method does work as we will show in this paper. None of these methods appear to work for proving Conjecture 2.

3.1. *Number of Rational Points*

The following theorem was proved in [2].

Theorem 1 (Ahmadi–McGuire). *Let $C(\mathbb{F}_q)$ and $D(\mathbb{F}_q)$ be smooth projective curves such that*

(i) *$C(\mathbb{F}_q)$ and $D(\mathbb{F}_q)$ have the same number of points over infinitely many extensions of $\mathbb{F}_q$.*

(ii) *The L-polynomial of C over $\mathbb{F}_{q^k}$ has no repeated roots, for all $k \geq 1$.*

Then there exists a positive integer s such that the L-polynomial of $D(\mathbb{F}_{q^s})$ is divisible by the L-polynomial of $C(\mathbb{F}_{q^s})$.

The first hypothesis holds for the curves C_k, but the second hypothesis does not. Thus we cannot use Theorem 1 to prove Conjecture 1. To see that the first hypothesis holds, we use the following theorem proved by Lahtonen–McGuire–Ward [11].

Theorem 2. *Let $K = \mathbb{F}_{2^n}$ where n is a non-negative odd integer. Let*
$$Q(x) = \mathrm{Tr}(x^{2^k+1} + x^{2^j+1}), \qquad for\ 0 \leq j < k.$$
Then if $\gcd(k \pm j, n) = 1$, then the number of zeros of Q in K is
$$2^{n-1} + \left(\frac{2}{n}\right) 2^{(n-1)/2},$$
where $(\frac{2}{n})$ is the Jacobi symbol.

If we put $j = 0$ in Theorem 2 then $Q(x) = \mathrm{Tr}(x^{2^k+1} + x)$. It follows that C_1 and C_k have the same number of rational points over $\mathbb{F}_{2^m}$ for any m with $\gcd(k, m) = 1$. Therefore the first hypothesis of Theorem 1 holds.

The C_k curves are supersingular so the L-polynomial of C_1 (which is $2t^2 + 2t + 1$) has repeated roots over some extensions of $\mathbb{F}_2$, something that can also be seen directly. Therefore the second hypothesis of Theorem 1 does not hold.

We remark that the second hypothesis of Theorem 1 *does* hold for the E_k curves, see [3]. However, we are unable to prove that the first hypothesis holds, although it is conjectured that it does.

A similar but different theorem was proved in [3].

Theorem 3 (Ahmadi–McGuire–Rojas-León). *Let* C *and* D *be two smooth projective curves over* $\mathbb{F}_q$. *Assume there exists a positive integer* $k > 1$ *such that*

(i) $\#C(\mathbb{F}_{q^m}) = \#D(\mathbb{F}_{q^m})$ *for every* m *that is not divisible by* k, *and*
(ii) the k-*th powers of the roots of* $L_C(t)$ *are all distinct.*

Then $L_D(t) = q(t^k)\, L_C(t)$ *for some polynomial* $q(t)$ *in* $\mathbb{Z}[t]$.

We cannot use Theorem 3 to prove Conjecture 1, because the first hypothesis does not hold. It is not true that C_1 and C_k have the same number of rational points over $\mathbb{F}_{2^m}$ for any m not divisible by k (or another integer). This can be seen by looking at small examples using a computer algebra package.

It is interesting to compare the first hypothesis in Theorem 3 with the first hypothesis in Theorem 1 (which does hold for C_1 and C_k).

3.2. *The Kani–Rosen theorem*

Let X be an affine variety over a field k with coordinate ring A. Given an action of an algebraic group G on X, one may construct a so-called quotient variety X/G given by $\operatorname{Spec}(A^G)$ where A^G denotes the ring of invariants of A with the induced action of G. If furthermore, G is reductive then A^G is finitely generated so X/G is also an affine variety (A^G will be reduced if A is).

Let G be a finite subgroup of the automorphism group of a curve C and let $\operatorname{Jac}(C)$ denote the Jacobian of C. The Kani–Rosen theorem [9, thm. B] concerns isogenies and idempotents in the rational group algebra $\mathbb{Q}[G]$ and is useful is proving divisibility relations between L-polynomials.

Theorem 4 (Kani–Rosen). *Let* $G \subseteq \operatorname{Aut}(C)$ *be a (finite) subgroup such that* $G = H_1 \cup H_2 \cup \ldots \cup H_r$ *where the subgroups* $H_i \subseteq G$ *satisfy* $H_i \cap H_j = \{1\}$ *when* $i \neq j$. *Then there is an isogeny relation*

$$\operatorname{Jac}(C)^{r-1} \times \operatorname{Jac}(C/G)^g \cong \operatorname{Jac}(C/H_1)^{h_1} \times \ldots \times \operatorname{Jac}(C/H_r)^{h_r},$$

where $g = |G|$ *and* $h_i = |H_i|$.

For any subgroup H of G there is an idempotent

$$\varepsilon_H = \frac{1}{|H|} \sum_{h \in H} h.$$

If G is the Klein 4-group with subgroups H_1, H_2, H_3, we have the idempotent relation

$$\varepsilon_1 + 2\varepsilon_G = \varepsilon_{H_1} + \varepsilon_{H_2} + \varepsilon_{H_3}.$$

Applying the Kani–Rosen theorem we get an isogeny

$$\mathrm{Jac}(C) \times \mathrm{Jac}(C/G)^2 \sim \mathrm{Jac}(C/H_1) \times \mathrm{Jac}(C/H_2) \times \mathrm{Jac}(C/H_3).$$

In order to apply this isogeny to C_k, we need two involutions in the automorphism group of C_k. We want involutions that are defined over $\mathbb{F}_2$. One is the hyperelliptic involution

$$\iota : (x, y) \mapsto (x, y + 1)$$

and the other is the map from [6]

$$\phi : (x, y) \mapsto (x + 1, y + B(x))$$

where $B(x) = x + x^2 + x^4 + x^8 + \cdots + x^{2^{k-1}}$. Then

$$\iota \circ \phi = \phi \circ \iota : (x, y) \mapsto (x + 1, y + 1 + B(x)).$$

Note that ϕ is an involution if and only if $B(1) = 0$ if and only if k is even. When k is odd, ϕ has order 4 and $\phi^2 = \iota$. When k is even, ϕ and ι together generate a Klein 4-group in $\mathrm{Aut}(C_k)$. In fact we have the following.

Proposition 1. *If k is odd then there are no non-hyperelliptic involutions on C_k of the form $(x, y) \mapsto (x + 1, y + B(x))$ where $B(x)$ is a linearised polynomial $B(x) = \sum_{i \geq 0} a_i x^{2^i}$ with $a_i \in \mathbb{F}_2$.*

Proof. Such a $B(x)$ must satisfy $B(1) = 0$ and $B(x)^2 + B(x) = x^{2^k} + x$. The resulting conditions thus imposed on the coefficients a_i mean that $B(x) = \sum_{i=0}^{k-1} x^{2^i}$ but then $B(1) \neq 0$ if k is odd.   □

Remark 1. It follows now from van der Geer and van der Vlugt [6] that this exhausts the subgroup of $\mathrm{Aut}_{\mathbb{F}_2}(C_k)$ fixing the branch points of $C_k \to \mathbb{P}^1$.

Therefore, the first problem in using the Kani–Rosen theorem to prove Conjecture 1 is that we only have the appropriate automorphism group for k even. It therefore appears that for k odd, one cannot use the Kani–Rosen theorem to prove the conjecture, at least not directly.

3.3. *Kleiman–Serre*

The following theorem is well-known in the area.

Theorem 5. *(Kleiman–Serre) If there is a surjective morphism of curves $C \longrightarrow C'$ that is defined over $\mathbb{F}_q$ then $L_{C'}(t)$ divides $L_C(t)$.*

Proof. (Sketch) Given a surjective morphism $f : C \to C'$ one obtains an induced map f^* on the étale cohomology groups that is injective (Kleiman [10, Prop. 1.2.4]). Given the interpretation of the polynomials $P_i(t)$ described in the introduction (equation 1) as determinants via the Weil conjectures, the result follows. □

We will use this result in the next section to prove Conjecture 1.

4. Proof of Conjecture 1

We prove Conjecture 1 using Theorem 5. In fact we will prove something more general: that there is a map from C_k to C_l for any integer l dividing k. Putting $l = 1$ proves Conjecture 1.

Before we construct the morphism we consider a simpler case as motivation. Let A_k denote the smooth projective model of the affine curve defined over $\mathbb{F}_2$

$$y^2 + y = x^{2^k} + x.$$

One can easily verify that the map

$$(x, y) \mapsto (\mathrm{Tr}_{nk/k}(x), y), \qquad \text{for } \mathrm{Tr}_{nk/k}(x) = x + \sum_{i=1}^{n-1} x^{2^{ik}},$$

is a morphism $A_{nk} \to A_k$ for n, k positive integers and $n > 1$.

The similarity between the curves C_k and A_k is apparent, however the morphism above differs quite radically from the one to be described below.

Theorem 6. *Let $k > l$ be integers with l dividing k. Then there is a non-constant morphism $C_k \longrightarrow C_l$ defined over $\mathbb{F}_2$.*

Proof. Write $k = lr$ and set $q = 2^l$. We claim that there is a morphism of the form $x \mapsto f(x)$, $y \mapsto y + g(x)$ from C_k to C_l where f and g are polynomials. For this to be the case it suffices that

$$f(x)^{q+1} + f(x) = x^{q^r+1} + x + g(x)^2 + g(x). \tag{2}$$

Let us take

$$f(x) = \sum_{j=0}^{r-1} x^{q^j}$$

and

$$g(x) = \sum_{j=1}^{r-1} x^{q^j} + \sum_{0 \le i < j \le r-1} \sum_{s=0}^{l-1} x^{2^l(q^i+q^j)}.$$

Then

$$f(x) + f(x)^{q+1} = \sum_{j=0}^{r-1} x^{q^j} + \left(x + \sum_{j=1}^{r-1} x^{q^j} \right) \left(x^{q^r} + \sum_{j=1}^{r-1} x^{q^j} \right)$$

$$= \sum_{j=1}^{r-1} x^{q^j} + x + x^{1+q^r} + \sum_{j=1}^{r-1} x^{1+q^j} + \sum_{j=1}^{r-1} x^{q^r+q^j} + \sum_{j=1}^{r-1} x^{2q^j}$$

and

$$g(x)^2 + g(x) = \sum_{j=1}^{r-1} x^{q^j} + \sum_{j=1}^{r-1} x^{2q^j} + \sum_{0 \le i < j \le r-1} (x^{q^i+q^j} + x^{q(q^i+q^j)})$$

$$= \sum_{j=1}^{r-1} x^{q^j} + \sum_{j=1}^{r-1} x^{2q^j} + \sum_{0 \le i < j \le r-1} x^{q^i+q^j} + \sum_{1 \le i < j \le r} x^{q^i+q^j}$$

$$= \sum_{j=1}^{r-1} x^{q^j} + \sum_{j=1}^{r-1} x^{2q^j} + \sum_{j=1}^{r-1} x^{1+q^j} + \sum_{i=1}^{r-1} x^{q^i+q^r}.$$

Subtracting these gives (2). $\square$

Corollary 1. *Conjecture 1 is true.*

The Corollary follows from Theorem 5 and Theorem 6.

We used Theorem 5 to prove Conjecture 1. We remark that Theorem 5 cannot be used to prove Conjecture 2, because it is shown in [3] that there is no morphism $E_2 \longrightarrow E_1$. Thus a proof of Conjecture 2 will probably use different methods.

As a final remark, we point out where the argument of Theorem 6 breaks down in odd characteristic for the analogous curves

$$C_k^{(p)} : \qquad y^p - y = x^{p^k+1} + x,$$

where p is an odd prime. In the case $k = 2$, $l = 1$, in order to give a morphism of the form $(x, y) \mapsto (f(x), y + g(x))$ from $C_2^{(p)}$ to $C_1^{(p)}$ we need to find polynomials f and g with

$$f(x)^{p+1} + f(x) = x^{p^2+1} + x + g(x)^p - g(x).$$

If we take $f(x) = x + x^p$ by analogy with Theorem 6, then we require

$$x^{p^2+p} + x^{2p} + x^{p+1} + x^p = g(x)^p - g(x),$$

but this is insoluble for polynomial g unless $p = 2$.

Notwithstanding the above, the analogous conjecture for odd p does appear to be true based on computations for small k, p.

Conjecture 3. *Let p be an odd prime. Then the L-polynomial of $C_1^{(p)}$ divides the L-polynomial of $C_k^{(p)}$.*

Acknowledgment I.B.–C., S.F. and G.M. are supported by Science Foundation Ireland grant 13/IA/1914 and are members of the Computational and Adaptive Systems Laboratory (CASL) in University College Dublin. I.B.–C. is a member of the MICINN project MTM2010-17389. S.F. is partially supported by an Irish Department of Education scholarship.

References

[1] J. D. Achter, Fiber products and class groups of hyperelliptic curves, Proc. Amer. Math. Soc. **138** (2010), no. 9, 3159–3161. MR2653940 (2011g:11218)

[2] O. Ahmadi and G. McGuire, Curves over finite fields and linear recurring sequences, Surveys in Combinatorics 2015, Cambridge.

[3] O. Ahmadi, G. McGuire and A. Rojas-León, Decomposing Jacobians of curves over finite fields in the absence of algebraic structure, J. Number Theory **156** (2015), 414–431. MR3360347

[4] W. Bosma, J. Cannon and C. Playoust, The Magma algebra system. I. The user language, J. Symbolic Comput. **24** (1997), no. 3-4, 235–265. MR1484478

[5] S. Farnell and R. Pries, Families of Artin-Schreier curves with Cartier-Manin matrix of constant rank, Linear Algebra Appl. **439** (2013), no. 7, 2158–2166. MR3090462

[6] G. van der Geer and M. van der Vlugt, Reed-Muller codes and supersingular curves. I, Compositio Math. **84** (1992), no. 3, 333–367. MR1189892 (93k:14038)

[7] G. van der Geer and M. van der Vlugt, Fibre products of Artin-Schreier curves and generalized Hamming weights of codes, J. Combin. Theory Ser. A **70** (1995), no. 2, 337–348. MR1329398 (96a:94019)

[8] D. Glass and R. Pries, Hyperelliptic curves with prescribed p-torsion, Manuscripta Math. **117** (2005), no. 3, 299–317. MR2154252 (2006e:14039)

[9] E. Kani and M. Rosen, Idempotent relations and factors of Jacobians, Math. Ann. **284** (1989), no. 2, 307–327. MR1000113 (90h:14057)

[10] S. L. Kleiman, Algebraic cycles and the Weil conjectures, in *Dix exposés sur la cohomologie des schémas*, 359–386, North-Holland, Amsterdam. MR0292838 (45 #1920)

[11] J. Lahtonen, G. McGuire and H. N. Ward, Gold and Kasami-Welch functions, quadratic forms, and bent functions, Adv. Math. Commun. **1** (2007), no. 2, 243–250. MR2306312 (2008d:11137)

[12] R. Lidl and H. Niederreiter, *Finite fields*, Encyclopedia of Mathematics and its Applications, 20, Addison-Wesley, Reading, MA, 1983. MR0746963 (86c:11106)

[13] B. Poonen, Varieties without extra automorphisms. II. Hyperelliptic curves, Math. Res. Lett. **7** (2000), no. 1, 77–82. MR1748289 (2001g:14052b)

[14] J. Scholten and H. J. Zhu, Hyperelliptic curves in characteristic 2, Int. Math. Res. Not. 2002, no. 17, 905–917. MR1899907 (2003d:11089)

Divisibility of exponential sums associated to binomials over $\mathbb{F}_p$

Francis Castro, Raúl Figueroa, Puhua Guan

Department of Mathematics, University of Puerto Rico, Río Piedras Campus {franciscastr, junioyjulio, pguan31}@gmail.com

Jose Ortiz-Ubarri

Department of Computer Science, University of Puerto Rico, S.J., Río Piedras Campus jose.ortiz23@upr.edu

1. Introduction

Exponential sums are applicable to many areas of mathematics and their divisibility is an example of this as they are used to characterize important properties of objects in applied mathematics. Specifically, a lot of effort has been put into the study of the p-adic divisibility of the roots of the L-function associated to the exponential sum. As the value of an exponential sum is equal to the sum of the roots of the associated L-function, any estimate on the roots implies an estimate for the divisibility of the exponential sum. Sometimes, roots of the L-function associated to the exponential sum have the same p-divisibility and when added together, the p-divisibility of the exponential sum increases. In general, there are good estimates for the divisibility of exponential sums as illustrated by [1–3, 5, 9, 10, 14].

We compute the exact divisibility of exponential sums associated to $F(X) = X^{d_1} + bX^{d_2}$ over $\mathbb{F}_p$, where $d_1 - d_2 = \frac{p-1}{m}, m \in \{2,3\}$, and $p > 2$ is a prime number. In particular, a correspondence is established between the $b \in \mathbb{F}_p^*$ such that $F(X) = X^{d_1} + bX^{d_2}$ is a permutation of $\mathbb{F}_p$ and the roots of some identified polynomial. Motivated by our results, we state a conjecture relating divisibility of exponential sums to permutation polynomials of $\mathbb{F}_p$.

2. Preliminaries

Given an integer j, we consider the integers j_i with $0 \leq j_i < p$ such $j = \sum_{i=0}^{r} j_i p^i$. Then, the p-weight of j is defined as $\sigma_p(j) = \sum_{i=0}^{r} j_i$. From now on, we assume that a polynomial $F(X) = \sum_{i=1}^{N} a_i X^{d_i}$ is a nonconstant polynomial of degree less than $p - 1$. In this paper we consider p to be odd.

Let $\mathbb{Q}_p$ be the p-adic field with ring of integers $\mathbb{Z}_p$. Let $\mathcal{T}$ denote the Teichmüller representatives of $\mathbb{F}_p$ in $\mathbb{Q}_p$. Denote by ξ a primitive p-th root of unity in $\overline{\mathbb{Q}_p}$. Define $\theta = 1 - \xi$ and denote by ν_θ the valuation over θ. Note that $\nu_\theta(p) = p - 1$ and $v_p(x) = \frac{\nu_\theta(x)}{p-1}$.

Let $\phi : \mathbb{F}_p \to \mathbb{Q}(\xi)$ be a nontrivial additive character. The exponential sum associated to $F(X) = \sum_{i=1}^{N} a_i X^{d_i}$ is defined as follows:

$$S_p(F) = \sum_{x \in \mathbb{F}_p} \phi(F(x)).$$

Note that if the exact p-divisibility of the exponential sum $\sum_{x \in \mathbb{F}_p} \phi(F(x))$ is a real number, then $S_p(F)$ will not be divisible by an arbitrary power of p and therefore $S_p(F) \neq 0$. The next theorem gives a lower bound for the valuation of an exponential sum with respect to θ.

Theorem 1 ([10]). *Let* $F(X) = \sum_{i=1}^{N} a_i X^{d_i}$, $a_i \neq 0$. *If* $S_p(F)$ *is the exponential sum*

$$S_p(F) = \sum_{x \in \mathbb{F}_p} \phi(F(x)), \tag{1}$$

then $\nu_\theta(S_p(F)) \geq \mu_p(d_1, \ldots, d_N)$, *where*

$$\mu_p(d_1, \ldots, d_N) = \min_{(j_1, \ldots, j_N)} \left\{ \sum_{i=1}^{N} j_i \mid 0 \leq j_i < p \right\},$$

for $(j_1, \ldots, j_N) \neq (0, \ldots, 0)$ *a solution to the modular equation*

$$d_1 j_1 + d_2 j_2 + \ldots + d_N j_N \equiv 0 \bmod p - 1. \tag{2}$$

Remark 1. If $F(X) = aX^{d_1}$ is a monomial, then $\mu_p(d_1) = \frac{p-1}{\gcd(p-1, d_1)}$ for $a \neq 0$. If $F(X)$ is not a monomial, we can assume that $d_1 > d_2$. We have that $\mu_p(d_1, \ldots, d_N) \leq p - 1 + d_2 - d_1 < p - 1$ since $(d_2, p - 1 - d_1, 0, \ldots, 0)$ is a solution of the modular equation (2). In [7], the author proved the following: If F is not a monomial over $\mathbb{F}_p (p \geq 5)$, then $\mu_p(d_1, \ldots, d_N) \leq \frac{p-1}{2}$.

Following the notation in [10], we expand the exponential sum $S_p(F)$:

$$S_p(F) = \sum_{j_1=0}^{p-1} \cdots \sum_{j_N=0}^{p-1} \left[\prod_{i=1}^{N} c(j_i)\right] \left[\sum_{t\in\mathcal{T}} t^{d_1 j_1 + \cdots + d_N j_N}\right] \left[\prod_{i=1}^{N} a_i'^{j_i}\right], \quad (3)$$

where a_i''s are the Teichmüller representatives of the coefficients a_i of F, and $c(j_i)$ is defined in Lemma 1 below. Each solution $(j_1, \cdots, j_N)$ of (2) is associated to a term T in the above sum with

$$\nu_\theta(T) = \nu_\theta\left(\left[\prod_{i=1}^{N} c(j_i)\right] \left[\sum_{t\in\mathcal{T}} t^{d_1 j_1 + \cdots + d_N j_N}\right] \left[\prod_{i=1}^{N} a_i'^{j_i}\right]\right) \quad (4)$$

$$= \sum_{i=1}^{N} j_i.$$

Sometimes there is not equality on the valuation of $S_p(F)$ because there could be more than one solution $(j_1, \ldots, j_N)$ providing the minimum value for $\sum_{i=1}^{N} j_i$, for example, when the associated terms are similar some of them could add to produce higher powers of θ dividing the exponential sum.

From now on, we call any solution $(j_1, \cdots, j_N)$ of (2) that has $\nu_\theta(T) = \mu_p(d_1, \ldots, d_N)$ of minimum value a *minimal solution*. We need to use the following lemma together with Stickelberger's Theorem to compute the exact divisibility.

Lemma 1 ([3]). *There is a unique polynomial* $C(X) = \sum_{j=0}^{p-1} c(j) X^j \in \mathbb{Q}_p(\xi)[X]$ *of degree* $p-1$ *such that*

$$C(t) = \xi^t, \quad \text{for all } t \in \mathcal{T}.$$

Moreover, the coefficients of $C(X)$ *satisfy*

$$c(0) = 1$$

$$(p-1)c(p-1) = -p$$

$$(p-1)c(j) = g(j) \quad \text{for } 0 < j < p-1,$$

where $g(j)$ *is the Gauss sum,*

$$g(j) = \sum_{t\in\mathcal{T}^*} t^{-j} \xi^t. \quad (5)$$

Theorem 2 (Stickelberger [8]). *For* $0 \le j < p-1$,

$$\frac{g(j)j!}{\theta^{\sigma_p(j)}} \equiv -1 \bmod \theta. \quad (6)$$

Before we state a theorem of Rogers ([13]) we need to define $u_p(F)$. Let $u_p(F)$ be the smallest positive integer k such that $\sum_{x\in\mathbb{F}_p} F(x)^k \neq 0$ in $\mathbb{F}_p$, where $F(X)$ is a polynomial over $\mathbb{F}_p$.

Theorem 3 ([13]). *A polynomial* $F(X)$ *over* $\mathbb{F}_p$ *is a permutation polynomial over* $\mathbb{F}_p$ *if and only if* $u_p(F) > \frac{p-1}{2}$.

3. Divisibility of Exponential Sums

In this section we compute the exact divisibility of some exponential sums over $\mathbb{F}_p$. We study the divisibility of exponential sums of the type $S_p(aX^{d_1} + bX^{d_2})$, where $d_1 - d_2 = \frac{p-1}{m}, m \in \{2,3\}$. The modular equation associated to $S_p(aX^{d_1} + bX^{d_2})$ has more than one minimal solution, hence the coefficients of $F(X)$ can contribute to the divisibility of $S_p(F)$. For the two families considered in this paper we prove that: $\nu_\theta(S_p(d_1, d_2)) > \frac{p-1}{2} \Leftrightarrow F(X) = aX^{d_1} + bX^{d_2}$ is a permutation of $\mathbb{F}_p$. These results motivated us to state a conjecture relating divisibility of exponential sums over $\mathbb{F}_p$ with permutation polynomials of $\mathbb{F}_p$.

In the following lemma, we compute the exact divisibility of $S_p(F)$.

Lemma 2. *Suppose* $ab \neq 0$, $d_1 - d_2 = \dfrac{p-1}{2}$ *and* $\gcd(d_1, d_2) = 1$. *If*

- $d_2 \equiv 1 \bmod 2$ *and* $P_1(a,b) = \sum_{l \equiv 1 \bmod 2} \binom{\frac{p-1}{2}}{l} a^l b^{\frac{p-1}{2}-l}$ *then*

$$\nu_\theta(S_p(aX^{d_1} + bX^{d_2})) \begin{cases} = \frac{p-1}{2} & \text{if } P_1(a,b) \not\equiv 0 \bmod p \\ > \frac{p-1}{2} & \text{otherwise.} \end{cases}$$

- $d_2 \equiv 0 \bmod 2$ *and* $P_0(a,b) = \sum_{l \equiv 0 \bmod 2} \binom{\frac{p-1}{2}}{l} a^l b^{\frac{p-1}{2}-l}$, *then*

$$\nu_\theta(S_p(aX^{d_1} + bX^{d_2})) \begin{cases} = \frac{p-1}{2} & \text{if } P_0(a,b) \not\equiv 0 \bmod p \\ > \frac{p-1}{2} & \text{otherwise.} \end{cases}$$

Proof. In this case the modular equation that we need to consider is

$$d_1 i + d_2 j \equiv 0 \bmod p - 1 \leftrightarrow d_2(i+j) + (\frac{p-1}{2})i \equiv 0 \bmod p - 1. \quad (7)$$

The last congruence implies that $(\frac{p-1}{2})$ divides $d_2(i+j)$. Using that $\gcd(d_1, d_2) = 1$, we obtain that $(\frac{p-1}{2})$ divides $i + j$. Note that if $\gcd(\frac{p-1}{2}, d_2) > 1$, then $\gcd(d_1, d_2) > 1$. Therefore $\mu_p(d_1, d_2) = \frac{p-1}{2}$ since $\frac{p-1}{2}$ divides $\mu_p(d_1, d_2)$ and $\mu_p(d_1, d_2) < p - 1$. We are going to compute the minimal solutions of (7) whenever $d_2 \equiv 1 \bmod 2$. The other case follows in a similar way. In this case we have two subcases whether $d_1 - d_2$ is even or odd. Note that $(l, \frac{p-1}{2} - l)$ is a minimal solution of the modular equation (7) for $l \equiv 1 \bmod 2$ and $1 \leq l \leq \frac{p-1}{2}$. Note that if $\frac{p-1}{2}$ is odd, then $l = 1, \ldots, \frac{p-1}{2}$. Now suppose that $\frac{p-1}{2}$ is odd. The p-divisibility of $S_p(aX^{d_1} + bX^{d_2})$ is controlled by

$$(p-1) \sum_{l=1}^{\frac{p+1}{4}} a^{2l-1} b^{\frac{p+1-4l}{2}} c(2l-1) c\left(\frac{p+1-4l}{2}\right). \quad (8)$$

Now we apply Stickelberger's theorem to a term of (8):

$$\frac{p-1}{\theta^{\frac{p-1}{2}}}\left(a^{2l-1}b^{\frac{p+1-4l}{2}}c(2l-1)c(\frac{p+1-4l}{2})\right) =$$

$$\frac{p-1}{\theta^{\frac{p-1}{2}}}\left(a^{2l-1}b^{\frac{p+1-4l}{2}}(\frac{g(2l-1)}{p-1})(\frac{g(\frac{p+1-4l}{2})}{p-1})\right) \equiv -\frac{a^{2l-1}b^{\frac{p+1-4l}{2}}}{(2l-1)!(\frac{p+1-4l}{2})!} \bmod \theta.$$

Hence

$$\frac{p-1}{\theta^{\frac{p-1}{2}}}\sum_{l=1}^{\frac{p+1}{4}}a^{2l-1}b^{\frac{p+1-4l}{2}}c(2l-1)c\left(\frac{p+1-4l}{2}\right)$$

$$\equiv -\sum_{l=1}^{\frac{p+1}{4}}\frac{a^{2l-1}b^{\frac{p+1-4l}{2}}}{(2l-1)!(\frac{p+1-4l}{2})!} \bmod \theta.$$

If

$$\sum_{l=1}^{\frac{p+1}{4}}\frac{a^{2l-1}b^{\frac{p+1-4l}{2}}}{(2l-1)!(\frac{p+1-4l}{2})!} \equiv 0 \bmod \theta$$

then

$$\sum_{l=1}^{\frac{p+1}{4}}\frac{a^{2l-1}b^{\frac{p+1-4l}{2}}}{(2l-1)!(\frac{p+1-4l}{2})!} \equiv 0 \bmod p$$

since a, b are congruent to an integer modulo p. Now we multiply the last equation by $(\frac{p-1}{2})!$ to obtain

$$\sum_{l=1}^{\frac{p+1}{4}}\binom{\frac{p-1}{2}}{2l-1}a^{2l-1}b^{\frac{p+1-4l}{2}} \equiv 0 \bmod p.$$

This completes the proof for the case $\frac{p-1}{2}$ is odd.

If $\frac{p-1}{2}$ is even, then the minimal solutions of modular equation (7) are $(l, \frac{p-1}{2}-l)$, where $l = 1, \ldots, \frac{p-1}{2}-1$. Then p-divisibility of $S_p(aX^{d_1}+bX^{d_2})$ is controlled by

$$(p-1)\sum_{l=1}^{\frac{p-1}{4}}a^{2l-1}b^{\frac{p+1-4l}{2}}c(2l-1)c\left(\frac{p+1-4l}{2}\right). \qquad (9)$$

Applying Stickelberger's theorem to a term of the summation (9):

$$\frac{p-1}{\theta^{\frac{p-1}{2}}}\left(a^{2l-1}b^{\frac{p+1-4l}{2}}c(2l-1)c(\frac{p+1-4l}{2})\right) \equiv \frac{a^{2l-1}b^{\frac{p+1-4l}{2}}}{(p-1)(2l-1)!(\frac{p+1-4l}{2})!} \bmod \theta.$$

Note that

$$\frac{p-1}{\theta^{\frac{p-1}{2}}} \sum_{l=1}^{\frac{p-1}{4}} a^{2l-1} b^{\frac{p+1-4l}{2}} c(2l-1)c(\frac{p+1-4l}{2}) \equiv - \sum_{l=1}^{\frac{p-1}{4}} \frac{a^{2l-1} b^{\frac{p+1-4l}{2}}}{(2l-1)!(\frac{p+1-4l}{2})!} \mod \theta.$$

If

$$\sum_{l=1}^{\frac{p-1}{4}} \frac{a^{2l-1} b^{\frac{p+1-4l}{2}}}{(2l-1)!(\frac{p+1-4l}{2})!} \equiv 0 \mod \theta$$

then

$$\sum_{l=1}^{\frac{p-1}{4}} \frac{a^{2l-1} b^{\frac{p+1-4l}{2}}}{(2l-1)!(\frac{p+1-4l}{2})!} \equiv 0 \mod p$$

since a, b are congruent to an integer modulo p. Now we multiply the last equation by $(\frac{p-1}{2})!$ to obtain

$$\sum_{l=1}^{\frac{p-1}{4}} \binom{\frac{p-1}{2}}{2l-1} a^{2l-1} b^{\frac{p+1-4l}{2}} \equiv 0 \mod p$$

controls the divisibility of $S_p(d_1, d_2)$ in this case.

$\square$

Remark 2. In Lemma 2 when $d_1 \equiv 0 \mod 2$, we have that $P_0(1,0) = 0$. Note this case is not included in the Lemma 2. Note 0 is a simple root of $P_0(1,b)$. The degree of $P_0(1,b)$ is equal to $\frac{p-1}{2}$.

The following theorem describes completely the divisibility of the exponential sums $S_p(aX^{d_1} + bX^{d_2})$ satisfying $d_1 - d_2 = \dfrac{p-1}{2}$.

Theorem 4. *With the notation of Lemma 2.*

- *Let d_2 be an odd natural number. Any root α of the polynomial*

$$P_1(b) = \sum_{l \equiv 1 \bmod 2} \binom{\frac{p-1}{2}}{l} b^{\frac{p-1}{2}-l}$$

  satisfies that $\alpha^2 - 1$ is a quadratic residue.
- *Let d_2 be an even natural number. Any root α of the polynomial*

$$P_0(b) = \sum_{l \equiv 0 \bmod 2} \binom{\frac{p-1}{2}}{l} b^{\frac{p-1}{2}-l}$$

  satisfies that $\alpha^2 - 1$ is a quadratic residue.

•

$$\nu_\theta(S_p(X^{d_1}+bX^{d_2})) = \begin{cases} \frac{p-1}{2} \iff b^2-1 \text{ is not a quadratic residue} \\ \quad\quad\quad \text{of } \mathbb{F}_p \text{ or } b = \pm 1. \\ \infty \iff b^2-1 \text{ is a quadratic residue} \\ \quad\quad\quad \text{of } \mathbb{F}_p. \end{cases}$$

Proof. If $X^{d_1} + \alpha X^{d_2}$ is a permutation of $\mathbb{F}_p$, then α is a root of $P_j(b)$ since $S_p(d_1, d_2) = 0$ for $j \in \{0, 1\}$. Hence, the number of permutation polynomials of the type $X^{d_1} + bX^{d_2}$ is less than or equal to the degree of $P_j(b)$. In [4], Carlitz proved that $X^{d_1} + bX^{d_2}$ is a permutation of $\mathbb{F}_p$ if and only if $b^2 \neq 1$ and $b^2 - 1$ is a quadratic residue of $\mathbb{F}_p$. In [11], Mullen-Niederreiter proved that the number of b's in $\mathbb{F}_p^*$ such that $X^{d_1} + bX^{d_2}$ is a permutation of $\mathbb{F}_p$ is equal to $\frac{p-3}{2}$. The degree of $P_1(b)$ is equal to $\frac{p-3}{2}$ which coincides with the number of permutation polynomials of the type $X^{d_1} + bX^{d_2}$ with $b \neq 0$. Note that the number roots $\alpha \neq 0$ of $P_0(b)$ is less than or equal to $\frac{p-3}{2}$ (equivalently after simplification $\frac{P(b)}{b}$ has degree $\frac{p-3}{2}$ see Remark 2) coinciding with the number of permutation polynomials of the type $X^{d_1} + bX^{d_2}$ with $b \neq 0$. Hence the number of permutations of the type $X^{d_1} + bX^{d_2}$ with $b \neq 0$ coincides with roots of the polynomial P_j for $j = 0, 1$. In particular, the polynomials split completely over $\mathbb{F}_p$. We have that $\nu_\theta(S_p(X^{d_1} + bX^{d_2})) = \frac{p-1}{2}$ if $P_j(b) \neq 0$ and $\nu_\theta(S_p(X^{d_1} + bX^{d_2})) = \infty$ if $P_j(b) = 0$. $\square$

Remark 3. Note that $P_j(\alpha) = 0$ if and only if $P_j(\alpha^{-1}) = 0$ for $\alpha \neq 0$.

Example 1. Let $p = 43$. In this case

$$P_1(b) = 42 + 5b^2 + 35b^4 + 2b^6 + 29b^8 + 13b^{10} + b^{12} + 35b^{14} + 33b^{16} + 3b^{18} + 22b^{20}$$
$$= 22\,(b+6)\,(b+15)\,(b+37)\,(b+38)\,(b+24)\,(b+29)$$
$$(b+27)\,(b+39)\,(b+28)\,(b+22)\,(b+5)\,(b+10)$$
$$(b+19)\,(b+31)\,(b+21)\,(b+33)\,(b+12)\,(b+16)\,(b+14)\,(b+4)$$

Now we consider the divisibility of exponential sums associated to $X^{d_1} + bX^{d_2}$ with $d_1 - d_2 = \dfrac{p-1}{3}$.

Theorem 5. *Let* $p \equiv 1 \bmod 3$ *be a odd prime,* $d_1 > d_2$ *positive integers,* $d_1 - d_2 = \dfrac{p-1}{3}$, $P(b) = \displaystyle\sum_{i \equiv -d_2 \bmod 3} \binom{\frac{p-1}{3}}{i} b^{\frac{p-1}{3}-i}$ *and* $b \neq 0$. *If* $\gcd(d_1, d_2) = 1$, *then*

•

$$\nu_\theta(S_p(X^{d_1} + bX^{d_2})) = \begin{cases} \frac{p-1}{3} & \text{if } P(b) \neq 0 \\ = \infty & \text{otherwise.} \end{cases}$$

• $P(b) = 0$ *if and only if $X^{d_1} + bX^{d_2}$ is a permutation of $\mathbb{F}_p$.*

Proof. The modular equation associated to $X^{d_1} + bX^{d_2}$ is

$$d_1 i + d_2 j \equiv 0 \bmod p - 1. \tag{10}$$

The minimal solutions of $d_1 i + d_2 j \equiv 0 \bmod p - 1$ satisfy $i + j = \frac{p-1}{3}$, where $i \equiv -d_2 \bmod 3 (-d_2 \in \{0, 1, 2\})$. Therefore

$$P(b) = \sum_{i \equiv -d_2 \bmod 3} \binom{\frac{p-1}{3}}{i} b^{\frac{p-1}{3} - i}$$

controls the p-divisibility of $S_p(F)$. If $P(b) \neq 0$, then

$$\nu_\theta(S_p(X^{d_1} + bX^{d_2})) = \frac{p-1}{3}.$$

If $P(b) = 0$, then

$$\nu_\theta(S_p(X^{d_1} + bX^{d_2})) > \frac{p-1}{3}.$$

Note that the coefficient of X^{p-1} in the expansion $(X^{d_1} + bX^{d_2})^t \bmod (X^p - X)$ is the sum of terms of type $\binom{t}{i}$, where $(i, t-i)$ is a solution of (10). This implies $u_p(F) > \frac{p-1}{3}$. Since the solutions of (10) are divisible by $\frac{p-1}{3}$, then $u_p(F) \geq 2(\frac{p-1}{3}) > \frac{p-1}{2}$. Using Theorem 3, we have $F(X)$ is a permutation of $\mathbb{F}_p$.

$\square$

Example 2. We consider the case when $d_1 = \frac{p+2}{3}$ and $d_2 = 1$. In this case we have three subcases: $\frac{p-1}{3} \bmod 3 \in \{0, 1, 2\}$.

Case 1. $\frac{p-1}{3} \equiv 2 \bmod 3$. In this case we have that

$$P(b) = \binom{\frac{p-1}{3}}{2} b^{\frac{p-1}{3} - 2} + \binom{\frac{p-1}{3}}{5} b^{\frac{p-1}{3} - 5} + \cdots + \binom{\frac{p-1}{3}}{\frac{p-1}{3} - 3} b^3 + 1.$$

Note that $P(X) \in \mathbb{F}_p[X^3]$. In particular for $p = 43, d_1 = 14, d_2 = 1$, we have

$$P(b) = 1 + 20\,b^3 + 36\,b^6 + 24\,b^9 + 5\,b^{12}$$
$$= 5\,(b + 37)\,(b + 28)\,(b + 39)\,(b + 19)\,(b^6 + b^3 + 34)\,(b + 42)\,(b + 7).$$

Hence $F(X) = X^{14} + bX$ is a permutation of $\mathbb{F}_{43}$ for $b \in \{1, 4, 6, 15, 24, 36\}$.

Case 2. $\frac{p-1}{3} \equiv 1 \mod 3$. In this case we have that

$$P(b) = \binom{\frac{p-1}{3}}{2} b^{\frac{p-1}{3}-2} + \binom{\frac{p-1}{3}}{5} b^{\frac{p-1}{3}-5} + \cdots + \binom{\frac{p-1}{3}}{\frac{p-1}{3}-5} b^5 + \binom{\frac{p-1}{3}}{\frac{p-1}{3}-2} b^2.$$

Note that $P(X) = X^2 P'(X)$, where $P'(0) \neq 0$ and $P'(X) \in \mathbb{F}_p[X^3]$. In particular, for $p = 67, d_1 = 23, d_2 = 1$, we have

$$\begin{aligned}
P(b) &= 30\, b^2 + 3\, b^5 + 46\, b^8 + 56\, b^{11} + 46\, b^{14} + 3\, b^{17} + 30\, b^{20} \\
&= 30\, (b+12)\, (b^3 + 38)\, (b+28)\, (b+15)\, (b+9)\, (b+31)\, b^2\, (b+33) \\
&\quad (b+65)\, (b^3+30)\, (b+8)\, (b+19)\, (b+42)\, (b+13)\, (b+60).
\end{aligned}$$

Hence $F(X) = X^{23} + bX$ is a permutation of $\mathbb{F}_{67}$ for $b \in$ $\{2, 7, 25, 34, 26, 39, 48, 52, 54, 55, 58, 59\}$.

Case 3. $\frac{p-1}{3} \equiv 0 \mod 3$. In this case we have that

$$P(b) = \binom{\frac{p-1}{3}}{2} b^{\frac{p-1}{3}-2} + \binom{\frac{p-1}{3}}{5} b^{\frac{p-1}{3}-5} + \cdots + \binom{\frac{p-1}{3}}{\frac{p-1}{3}-4} b^4 + \binom{\frac{p-1}{3}}{\frac{p-1}{3}-1} b.$$

Note that $P(X) = X P'(X)$, where $P'(0) \neq 0$ and $P'(X) \in \mathbb{F}_p[X^3]$. In particular, for $p = 37, d_1 = 13, d_2 = 1$, we have

$$\begin{aligned}
P(b) &= 12\, b + 14\, b^4 + 15\, b^7 + 29\, b^{10} \\
&= 29\, (b+28)\, (b+21)\, (b+25)\, b\, (b^6 + b^3 + 15).
\end{aligned}$$

Hence $F(X) = X^{13} + bX$ is a permutation of $\mathbb{F}_{37}$ for $b \in \{9, 12, 16\}$.

Conjecture.

$$\nu_\theta(S_p(X^{d_1} + bX^{d_2})) > \frac{p-1}{2} \leftrightarrow F(X) = X^{d_1} + bX^{d_2} \text{ is a permutation of } \mathbb{F}_p.$$

Remark 4. The conjecture is false for field extensions of $\mathbb{F}_p$. We have that

- $\nu_2\left(\sum_{x \in \mathbb{F}_{64}} (-1)^{Tr(x^{17}+x)} \right) = 4 > \nu_2(64^{1/2}) = 3$ but $F(X) = X^{17} + X$
  is not a permutation polynomial of $\mathbb{F}_{64}$.
- $\nu_3\left(\sum_{x \in \mathbb{F}_{3^8}} e^{2\pi i Tr(x^{10}+x)/3} \right) = 6 > 4$ but $F(X) = X^{10} + X$ is not a
  permutation polynomial of $\mathbb{F}_{3^8}$.

In [12], we compute the number of permutation polynomials considered in Theorem 5 for $p \leq 2683$.

4. Conclusion

In this paper we study the divisibility of exponential sums associated to binomials. In [6], the authors compute the exact p-divisibility for exponential sums associated to binomials, but with the degree less than or equal to $\sqrt{p-1}$. The two families considered in this paper are not included in [6]. We reduce the determination of when $F(X) = X^{d_1} + bX^{d_2}$ ($d_1 - d_2 = \frac{p-1}{m}, m \in \{2,3\}$) is permutation of $\mathbb{F}_p$ to the problem of finding solutions of a very specific polynomial. Hence the next step will be the study of the roots of these polynomials.

Acknowledgment   The authors thank the referee for his/her careful reading of the paper as well as giving many helpful comments and corrections. His/Her comments and suggestions improved the presentation of this work.

References

[1] A. Adolphson and S. Sperber. p-adic estimates for exponential sums and the theorem of Chevalley-Warning. *Ann. Sci. École Norm. Sup. (4)*, 20(4):545–556, 1987.

[2] A. Adolphson and S. Sperber. Hasse invariants and mod p solutions of A-hypergeometric systems. *J. Number Theory*, 142:183–210, 2014.

[3] J. Ax. Zeros of polynomials over finite fields. *Amer. J. Math.*, 86:255–261, 1964.

[4] L. Carlitz. Some theorems on permutation polynomials. *Bull. Amer. Math. Soc.*, 68:120–122, 1962.

[5] F. Castro and F. N. Castro-Velez. Improvement to Moreno-Moreno's theorems. *Finite Fields Appl.*, 18(6):1207–1216, 2012.

[6] F. N. Castro, P. Guan, and R. Figueroa. p-adic valuation of exponential sums in one variable associated to binomials. *Uniform Distribution Theory*, (in press).

[7] X. Liu. A problem related to the divisibility of exponential sums. *http://arxiv.org/pdf/1502.07281.pdf*.

[8] C. Moreno. *Algebraic curves over finite fields*, volume 97 of *Cambridge Tracts in Mathematics*. Cambridge University Press, Cambridge, 1991.

[9] O. Moreno and C. J. Moreno. Improvements of the Chevalley-Warning and the Ax-Katz theorems. *Amer. J. Math.*, 117(1):241–244, 1995.

[10] O. Moreno, K. Shum, F. N. Castro, and P. V. Kumar. Tight bounds for Chevalley-Warning-Ax-Katz type estimates, with improved applications. *Proc. London Math. Soc. (3)*, 88(3):545–564, 2004.

[11] G. Mullen and H. Niederreiter. The structure of a group of permutation polynomials. *J. Austral. Math. Soc. Ser. A*, 38(2):164–170, 1985.

[12] J. Ortiz-Ubarri. Counting permutation polynomials over $\mathbb{F}_p$. *http://ccom.uprrp.edu/ jortiz/cpp/*.

[13] L. J. Rogers. Note on functions proper to represent a substitution of a prime number of letters. *Messenger Math.*, 21:44–47, 1981.

[14] S. Sperber. On the p-adic theory of exponential sums. *Amer. J. Math.*, 108(2):255–296, 1986.

Dickson polynomials that are involutions

Pascale Charpin

INRIA-Paris, 2 rue Simone Iff 75012, Paris, France,
Pascale.Charpin@inria.fr

Sihem Mesnager

University of Paris VIII and University of Paris XIII, LAGA, CNRS and
Télécom ParisTech, Paris, France, smesnager@univ-paris8.fr

Sumanta Sarkar

TCS Innovation Labs, Hyderabad 500081, India, Sumanta@atc.tcs.com

Dickson polynomials which are permutations are interesting combinatorial objects and well studied. In this paper, we describe Dickson polynomials of the first kind in $\mathbb{F}_2[x]$ that are involutions over finite fields of characteristic 2. Such description is obtained using modular arithmetic's tools. We give results related to the cardinality and the number of fixed points (in the context of cryptographic application) of this corpus. We also present infinite classes of Dickson involutions. We study Dickson involutions which have a minimal set of fixed points.

1. Introduction

We start with the question related to cryptography: can decryption algorithm be the same as the encryption algorithm? The answer is yes, and in fact, the classic example of this kind of cryptosystem is Enigma. The advantage of having the same encryption and decryption algorithm is that the same implementation of the encryption algorithm works for the decryption also, and hence reduces the implementation cost. Suppose a uniformly chosen permutation $E : X \to X$ is applied as the encryption, where X is the message space, with the additional property that $E(E(x)) = x$ for all $x \in X$, i.e., $E^{-1} = E$. Then E serves for both the encryption and decryption.

Let $\mathbb{F}_{2^n}$ be the finite field of 2^n elements. If the polynomial $F(x)$ defined over $\mathbb{F}_{2^n}$ induces a permutation of $\mathbb{F}_{2^n}$, then $F(x)$ is called a permutation polynomial. Permutations are invertible functions, i.e., for a permutation polynomial $F(x)$ there exists a unique polynomial $F'(x)$ such that $F' \circ F(x) = F \circ F'(x) = x$, for all x. The polynomial $F'(x)$ is called the compositional inverse of F and is generally denoted by F^{-1}. Permutation polynomial is well known for its application in cryptography, coding theory, combinatorial design, etc. For instance, in block ciphers, a permutation F is used as an S-box to build the confusion layer during the encryption process, while in the decryption the inverse of F is required.

A permutation polynomial $F(x)$ for which $F \circ F(x) = x$ is called *involution*, and from the above discussions, it is clear that involution property is important in applications. Dickson polynomials form an important class of permutation polynomials. We would like to refer to the book of Lidl, Mullen and Turnwald [8], where the work on Dickson polynomials, and its developments are presented. Our results are widely derived from those of [8, Chapter 2-3]. Moreover, the Dickson permutations that decompose in cycles of same length are generally studied in [11]; more applications are presented in [12]. Our purpose is to describe precisely this corpus in the case of cycles of length 2, for such permutations over any finite field of characteristic 2. Some proofs are given for clarity; our aim is to propose a clear understanding in order to use easily Dickson involutions.

The Dickson polynomials have been extensively investigated in recent years under different contexts (see for instance [2, 5, 6, 9, 10, 13]). In this paper we treat *Dickson polynomials of the first kind* defined on a finite field of order 2^n.

Definition 1. The Dickson polynomial of the first kind of degree k in indeterminate x and with parameter $a \in \mathbb{F}_{2^n}^*$ is defined by

$$D_k(x,a) = \sum_{i=0}^{\lfloor k/2 \rfloor} \frac{k}{k-i} \binom{k-i}{i} a^k x^{k-2i}, \ k \geq 2 \qquad (1)$$

where $\lfloor k/2 \rfloor$ denotes the largest integer less than or equal to $k/2$.

We treat here the polynomials $D_k(x,1)$ that we will denote by $D_k(x)$ throughout this paper. The set of k such that D_k is a permutation of a given finite field $\mathbb{F}_{2^n}$ is well-known.

Our aim, in this paper, is the study of such polynomials $D_k(x)$ which induce an involution of any fixed finite field. After some preliminaries, in Section 2, the characterization of Dickson involutions is presented in Section 3 (see Theorem 3). Section 4 is devoted to the study of the corpus of involutions. We notably show that it consists in equivalence classes of size 4 and we compute the number of such classes. We also exhibit two infinite classes of Dickson involutions (Theorem 5). In Section 5, we study the fixed points of Dickson involutions. Our study reveals that they generally have a high number of fixed points. We propose lower bounds on this number. We give a precise description of the set of fixed points of Dickson involutions and study the case where this set has a minimal size, when $n = 2m$ with m even. We prove that such minimal set is equal to $\mathbb{F}_{2^{\frac{m}{2}}}$ and we characterize the Dickson involutions which have such set of fixed points (Section 5.3). At the end we give some numerical results.

2. Basic properties

Here we introduce some useful properties, on the Dickson polynomials of $\mathbb{F}_2[x]$. Note that they are known in many different contexts. Dickson polynomial $D_k \in \mathbb{F}_2[x]$ are recursively defined by

$$\begin{aligned} &D_0(x) = 0 \text{ and } D_1(x) = x; \\ &D_{i+2}(x) = xD_{i+1}(x) + D_i(x). \end{aligned} \tag{2}$$

Using this definition it is easy to prove the next properties which we use in the sequel.

Proposition 1. *The polynomials defined by (2) satisfy:*

- $\deg(D_i) = i$,
- $D_{2i}(x) = (D_i(x))^2$,
- $D_{ij}(x) = D_i(D_j(x))$,
- $D_i(x + x^{-1}) = x^i + x^{-i}$,

for all x, for any integer $i, j > 0$.

In this paper, we identify a polynomial on $\mathbb{F}_{2^n}$ (for some n) with its corresponding mapping $x \mapsto F(x)$ from $\mathbb{F}_{2^n}$ to $\mathbb{F}_{2^n}$; F is a permutation when this mapping is bijective. Concerning the Dickson polynomials, we have the following fundamental result.

Theorem 1. [8, Theorem 3.2] *The Dickson polynomial $D_k \in \mathbb{F}_2[x]$ is a permutation on $\mathbb{F}_{2^n}$ if and only if $\gcd(k, 2^{2n} - 1) = 1$.*

Some permutations are *involutive* and are then called *involutions*.

Definition 2. We say that F is an involution on $\mathbb{F}_{2^n}$ when it satisfies

$$F \circ F(x) = x, \quad \text{for all } x \in \mathbb{F}_{2^n}.$$

Note that an involution is equal to its compositional inverse. We will use later the *Jacobi symbols*. We now give its definition and some of its basic properties. Recall that an integer a is a *quadratic residue modulo a prime* p if and only if there is an integer u such that $a \equiv u^2 \pmod{p}$.

Definition 3. Let P be an odd integer, $P > 2$, and $P = p_1^{a_1} p_2^{a_2} \ldots p_k^{a_k}$ be the decomposition of P in prime factors. Let a be any integer. The Jacobi symbol of a is

$$Jac(a, P) = \left(\frac{a}{P} \right) = \left(\frac{a}{p_1} \right)^{a_1} \cdots \left(\frac{a}{p_k} \right)^{a_k},$$

where $\left(\frac{a}{p_i} \right)$, called a *Legendre symbol*, is as follows defined : it is equal to 0 if p_i divides a; otherwise we have:

$$\left(\frac{a}{p_i} \right) = \begin{cases} 1 & \text{if } a \text{ is a quadratic residue modulo } p_i, \\ & \text{i.e., there is } k > 0 \text{ such that } a \equiv k^2 \pmod{p_i}; \\ -1 & \text{if } a \text{ is not a quadratic residue modulo } p_i. \end{cases}$$

And we have these well-known formula on the values $Jac(a, P)$ where a and b are any integer.

$$\left(\frac{ab}{P} \right) = \left(\frac{a}{P} \right) \left(\frac{b}{P} \right). \tag{3}$$

$$\left(\frac{-1}{P} \right) = (-1)^{\frac{P-1}{2}} \quad \text{and} \quad \left(\frac{2}{P} \right) = (-1)^{\frac{P^2-1}{8}}. \tag{4}$$

$$a \equiv b \pmod{P} \implies \left(\frac{a}{P} \right) = \left(\frac{b}{P} \right). \tag{5}$$

Note that if $Jac(a, P) = -1$ then a is a quadratic nonresidue modulo P.

3. Dickson polynomials which induce involutions

A priori, the Dickson permutations cannot be involutive since they are obtained recursively. However they are permutations on a specific finite field. Let $n = 2m$ and k such that $\gcd(k, 2^n - 1) = 1$. Then D_k permutes $\mathbb{F}_{2^m}$. What happens when we compute $D_k \circ D_k$?

From Proposition 1, we have $D_k(D_k(x)) = D_{k^2}(x)$. Thus D_k is an involution on $\mathbb{F}_{2^m}$ if and only if $D_{k^2}(x) \equiv x \pmod{x^{2^m} + x}$. For instance, for $m = 2\ell$,

$$D_{2^\ell} : x \mapsto x^{2^\ell} \text{ is an involution on } \mathbb{F}_{2^m}.$$

We are going to describe the set of k such that D_k is an involution on $\mathbb{F}_{2^m}$, for any fixed m. The proofs of Theorem 2 and Corollary 1 below are derived from the results of [8, Chapters 2-3]. We give these proofs for clarity.

Lemma 1. *For all $x \in \mathbb{F}_{2^m}$ there is $\gamma \in \mathbb{F}_{2^n}^*$, $n = 2m$, such that $x = \gamma + \gamma^{-1}$. Moreover such a γ satisfies either $\gamma^{2^m-1} = 1$ or $\gamma^{2^m+1} = 1$.*

Proof. We denote by Tr the absolute trace on $\mathbb{F}_{2^n}$. For any $x \in \mathbb{F}_{2^m}$, there is γ such that $x = \gamma + \gamma^{-1}$ if and only if the equation $\gamma^2 + \gamma x + 1 = 0$ has a solution in $\mathbb{F}_{2^n}^*$. And this is equivalent to $Tr(1/x) = 0$ which is satisfied for all $x \in \mathbb{F}_{2^m}$.

We have $\gamma + \gamma^{-1}$ in $\mathbb{F}_{2^m}$ if and only if

$$\left(\gamma + \frac{1}{\gamma}\right)^{2^m} = \gamma + \frac{1}{\gamma}, \text{ or equivalently } (\gamma^{2^m} + \gamma)(\gamma^{2^m+1} + 1) = 0,$$

completing the proof. $\qquad\square$

Theorem 2. *Let k, ℓ be two nonzero integers. Then*

$$D_k(x) \equiv D_\ell(x) \pmod{x^{2^m} + x}$$

if and only if

$$k \equiv \ell \pmod{2^m - 1} \text{ or } k \equiv -\ell \pmod{2^m - 1}$$
and
$$k \equiv \ell \pmod{2^m + 1} \text{ or } k \equiv -\ell \pmod{2^m + 1}.$$

Proof. Let us suppose that $D_k(x) \equiv D_\ell(x)$ for any $x \in \mathbb{F}_{2^m}$. From Lemma 1, this is to say that for any $x = \gamma + \gamma^{-1}$

$$D_k(x) = \gamma^k + \left(\frac{1}{\gamma}\right)^k \equiv D_\ell(x) = \gamma^\ell + \left(\frac{1}{\gamma}\right)^\ell, \tag{6}$$

applying Proposition 1, where $\gamma \in \mathbb{F}_{2^n}^*$ such that $\gamma^{2^m-1} = 1$ or $\gamma^{2^m+1} = 1$. Now (6) can be written

$$\gamma^\ell(\gamma^{2k} + 1) + \gamma^k(\gamma^{2\ell} + 1) = 0 \Leftrightarrow (\gamma^\ell + \gamma^k)(\gamma^{\ell+k} + 1) = 0.$$

Thus (6) is equivalent to $\gamma^{\ell+k} = 1$ or $\gamma^{k-\ell} = 1$. Let α be a primitive root of $\mathbb{F}_{2^n}$. We know that we have to consider two forms for γ: $\gamma = \alpha^{s(2^m-1)}$ and $\gamma = \alpha^{t(2^m+1)}$ for some s, t. Then (6) holds for any γ is if and only if the following two conditions are satisfied:

- if $\gamma = \alpha^{s(2^m-1)}$ then $k \pm \ell \equiv 0 \pmod{2^m + 1}$;
- if $\gamma = \alpha^{t(2^m+1)}$ then $k \pm \ell \equiv 0 \pmod{2^m - 1}$.

□

Now we can describe that cases where $D_k(x) \equiv x \pmod{x^{2^m} + x}$. The next corollary can be viewed as an instance of [8, Theorem 3.8].

Corollary 1. *Let $n = 2m$. Let us define*

$$K_n = \{ \, k \mid 1 \le k \le 2^n - 1, \; D_k(x) \equiv x \pmod{x^{2^m} + x} \, \}.$$

Then $K_n = \{ \, 1, \; 2^m, \; 2^n - 2^m - 1, \; 2^n - 2 \, \}$.

Proof. Applying Theorem 2 to the case $\ell = 1$, k must be a solution of one of the four systems of congruences modulo $2^n - 1$:

(i) $k \equiv 1 \pmod{2^m - 1}$ and $k \equiv 1 \pmod{2^m + 1}$
(ii) $k \equiv 1 \pmod{2^m - 1}$ and $k \equiv -1 \pmod{2^m + 1}$
(iii) $k \equiv -1 \pmod{2^m - 1}$ and $k \equiv 1 \pmod{2^m + 1}$
(iv) $k \equiv -1 \pmod{2^m - 1}$ and $k \equiv -1 \pmod{2^m + 1}$.

If $k < 2^m - 1$ then $k = 1$ (case (i)). Moreover $k = 2^m$ is a solution of (ii). We now assume that $2^m < k$.

Now (i) implies that $(2^m - 1)$ and $(2^m + 1)$ divide $k - 1$. Since $(2^m - 1)$ and $(2^m + 1)$ are odd and coprime, only $k = 2^n$ is a solution of (i) and $2^n \equiv 1$ modulo $(2^n - 1)$. Similarly, (iv) implies that $2^n - 1$ divides $k + 1$ so that only $k = 2^n - 2$ is a solution of (iv).

The congruence (ii) implies that $(2^m - 1)$ divides $k - 1$ and $(2^m + 1)$ divides $k + 1$. Thus there is b such that

$$k = b(2^m - 1) + 1 = b(2^m + 1) - 2b + 1, \; i.e., \; k + 1 \equiv -2b + 2 \pmod{2^m + 1},$$

implying $b \equiv 1 \pmod{2^m + 1}$ so that $b = 1$. Further $k = 2^m$.

The congruence (iii) implies that $(2^m - 1)$ divides $k + 1$ and $(2^m + 1)$ divides $k - 1$. Thus there is b such that

$$k = b(2^m + 1) + 1 = b(2^m - 1) + 2b + 1, \; i.e., \; k + 1 \equiv 2b + 2 \pmod{2^m - 1},$$

implying $b \equiv -1 \pmod{2^m - 1}$ so that $b = 2^m - 2$. Further

$$k = (2^m - 2)(2^m + 1) + 1 = 2^{2m} - 2^m + 1.$$

□

Now we are able to describe the set of Dickson involutions. We first need to eliminate the elements of K_n which are not quadratic residues modulo $2^n - 1$.

Lemma 2. *Let $n = 2m$. Then*

- *$2^n - 2^m - 1$ and $2^n - 2$ are quadratic nonresidues modulo $2^n - 1$.*
- *2^m is a quadratic residue modulo $2^n - 1$ if and only if m is even and the square roots are $2^{m/2} S_n$ where S_n are the square roots of 1 modulo $2^n - 1$.*

Proof. Since $2^n - 1 \equiv 3 \pmod 4$, $2^{n-1} - 1$ is an odd integer. Thus we have

$$\left(\frac{-1}{2^n - 1} \right) = (-1)^{\frac{(2^n - 1) - 1}{2}} = (-1)^{2^{n-1} - 1} = -1.$$

On the other hand, one has that $2^n - 1 \equiv 7 \pmod 8$ for every $n \geq 3$ which implies that $\frac{(2^n - 1)^2 - 1}{8}$ is an even integer. Thus

$$\left(\frac{2}{2^n - 1} \right) = (-1)^{\frac{(2^n - 1)^2 - 1}{8}} = 1.$$

by applying (4). Thus $2^m - 2$ is a quadratic nonresidue modulo $2^n - 1$, using (5), since $-1 \equiv 2^m - 2 \pmod{2^n - 1}$. That implies also

$$\left(\frac{2^n - 2^m - 1}{2^n - 1} \right) = \left(\frac{-2^m}{2^n - 1} \right) = \left(\frac{-1}{2^n - 1} \right) \left(\frac{2}{2^n - 1} \right)^m = -1,$$

since $-2^m \equiv 2^n - 2^m - 1 \pmod{2^n - 1}$. Thus $2^n - 2^m - 1$ is a quadratic nonresidue modulo $2^n - 1$.

Secondly, since n is even, 3 divides $2^n - 1$. Now, 2 is a quadratic nonresidue modulo 3 that is $\left(\frac{2}{3} \right) = -1$ which implies that $\left(\frac{2^m}{3} \right) = (-1)^m = 1$ if and only if m is even. Therefore, according to the preceding result, if m is odd, 2^m is a quadratic nonresidue modulo $2^n - 1$. If m is even, 2^m is clearly a quadratic residue since $(2^{m/2})^2 \equiv 2^m$ modulo $2^n - 1$. Now, set

$$S_n = \{\, u \mid 1 \leq u \leq 2^n - 2, \ u^2 \equiv 1 \pmod{2^n - 1} \,\}. \qquad (7)$$

Note that the map $k \mapsto 2^{m/2} k$ is a one-to-one map from the set S_n of all square roots of 1 modulo $2^n - 1$ to the set of all square roots of 2^m modulo $2^n - 1$. $\qquad \square$

Theorem 3. *Consider the Dickson polynomials D_k, $1 \leq k \leq 2^n - 1$, $n = 2m$ with $m \geq 2$. Let S_n be defined by (7). Then D_k is an involution on $\mathbb{F}_{2^m}$ if and only if*

- $k \in S_n$, *when* m *is odd;*
- $k \in S_n \cup 2^{m/2} S_n$ *if* m *is even.*

Proof. We will always consider $D_k(x)$ (mod $x^{2^m} + x$). This polynomial induces an involution if and only if

$$D_k \circ D_k(x) = D_{k^2}(x) = x, \quad \text{for all } x \in \mathbb{F}_{2^m}.$$

From Corollary 1, this is to say that $k^2 \in \{ 1, \ 2^m, \ 2^n - 2^m - 1, \ 2^n - 2 \}$ where k^2 is computed modulo $2^n - 1$. According to Lemma 2, that is equivalent to $k \in S_n$ if m is odd and $k \in S_n \cup 2^{m/2} S_n$ if m is even. $\quad\square$

Remark 1. For all $u \in S_n$, $u^2 \equiv 1$ (mod $2^n - 1$) implies $\gcd(u, 2^n - 1) = 1$. Therefore, we have, for even m, $\gcd(2^{m/2} u, 2^n - 1) = 1$. This is to say that the hypothesis $\gcd(k, 2^n - 1) = 1$ is not necessary in the previous theorem.

4. The set of Dickson involutions

We consider involutions of $\mathbb{F}_{2^m}$ and $n = 2m$ in all this section. There are some immediate observations. Let S_n be defined by (7) and

$$K_n = \{ 1, \ 2^m, \ 2^n - 2^m - 1, \ 2^n - 2 \} \equiv \{\pm 1, \pm 2^m\} \pmod{2^n - 1}$$

Note that K_n is a multiplicative subgroup of S_n. Define an equivalence relation over S_n:

$$s_1 \sim s_2 \quad \text{if and only if} \quad \frac{s_1}{s_2} \in K_n. \tag{8}$$

Lemma 3. *Denote by* $\sigma(s)$ *the class of* $s \in S_n$, *according to the relation (8). Then* $\sigma(s) = \{s, -s, 2^m s, -2^m s\}$ *and we have*

$$D_t(x) \equiv D_s(x) \pmod{x^{2^m} + x}, \quad \text{for any } t \in \sigma(s).$$

Proof. One simply observe that if s and t belong to the same class then D_s and D_t induce the same permutation on $\mathbb{F}_{2^m}$. Indeed, in this case there exists $k \in K_n$ such that $t = ks$. Therefore, according to Corollary 1 and to Proposition 1, we have for all $x \in \mathbb{F}_{2^m}$

$$D_t(x) = D_{sk}(x) = D_s \circ D_k(x) = D_s \circ D_1(x) = D_s(x) \pmod{x^{2^m} + x}.$$

$\square$

Hence each class different from the class of 1 leads to a different non-trivial involution. We will give the exact number of such classes for a fixed n in the next subsection. The Dickson polynomials are closed with respect to composition of polynomials. Moreover the commutativity of integers imply that this composition is commutative. Thus for two non trivial involutions D_s and D_t of $\mathbb{F}_{2^m}$ we have

$$(D_s \circ D_t)^{-1} = D_t^{-1} \circ D_s^{-1} = D_t \circ D_s = D_{ts} = D_{st},$$

proving that D_{st} is an involution too. Now, if s and t are in the same class then $D_{st} = D_{s^2}$, from Lemma 3, where s is a square root of 1 so that $D_{st}(x) \equiv x \pmod{x^{2^m} + x}$. Now suppose that $t \notin \sigma(s)$. If m is even and $t \in \sigma(2^{m/2}s)$ then $D_{st} = (D_{s^2})^{2^{m/2}}$ so that $D_{st}(x) \equiv x^{m/2} \pmod{x^{2^m} + x}$. Otherwise, in other cases, there are more than 4 classes and $st \pmod{2^n - 1} = r$ where r is not in the classes $\{s, t, 2^{m/2}, 2^{m/2}s\}$ (see Example 2 later). We summarize these results with the next lemma.

Lemma 4. *If D_s and D_t are two Dickson involutions of $\mathbb{F}_{2^m}$ then $D_s \circ D_t = D_{st}$ is an involution too. Moreover:*

- *If $t \in \sigma(s)$ then $st \pmod{2^n - 1} \in \sigma(1)$.*
- *If $t = 2^{m/2}$ (m even) then $st \in \sigma(2^{m/2}s)$.*
- *If $t \in \sigma(2^{m/2}s)$ (m even) then $st \pmod{2^n - 1} \in \sigma(2^{m/2})$.*
- *Otherwise, and assuming that s, t are two representatives of non trivial classes we get $st \pmod{2^n - 1} = r$ where r is in another nontrivial class.*

Remark 2. Note that the three first assertions are in fact equivalences since $st \in \sigma(u)$ is equivalent to $t \in \sigma(us^{-1}) \pmod{2^n - 1} = \sigma(us)$ (s being a quadratic residue of 1 modulo $2^n - 1$, its inverse modulo $2^n - 1$ is itself).

Example 1. n=6, m=3: $K_6 = \{1, 8, 55, 62\} = S_6$. For any $k \in K_6$, $D_k(x) = x$ modulo $(x^8 + x)$. For example

$$D_{55}(x) = x + x^{33} + x^9 + x^{41} + x^{49} + x^5 + x^{37} + x^{53} + x^7 + x^{39} + x^{55}$$

$$\equiv x \pmod{x^8 + x}.$$

n=8, m=4: $K_8 = \{1, 16, 239, 254\}$ while $S_8 = \{1, 16, 86, 101, 154, 169, 239, 254\}$. Note that $-86 = 169$, $86 * 16 = 101$ and $86 * (-16) = 154$. Here the non trivial involutions are the $D_k(x) \pmod{x^{16} + x}$ with k in $(S_8 \cup 4S_8) \setminus K_8$. Thus, according to (8), we get three such D_k which are the representatives of the three classes:

$$k \in \{4, 64, 191, 251\} \cup \{86, 101, 154, 169\} \cup \{89, 149, 106, 166\}$$

For instance
$$D_4(x) = x^4 \pmod{x^{16} + x}$$
$$D_{86}(x) = x^2 + x^6 + x^{10} + x^{18} + x^{22} + x^{34} + x^{38} + x^{42} + x^{86}$$
$$+ x^{74} + x^{82} + x^{66} + x^{70}$$
$$= x^2 + x^3 + x^4 + x^8 + x^{12} + x^{11} + x^{14} \pmod{x^{16} + x}.$$
while, with $89 \equiv 86 * 4 \pmod{255}$
$$D_{89}(x) = (D_{86}(x))^4 = x^2 + x^3 + x^{12} + x^8 + x^{11} + x + x^{14} \pmod{x^{16} + x}.$$
n=10, m=5: $K_{10} = \{\ 1, 32, 991, 1022\}$ while
$$S_{10} = \{\ 1, 32, 340, 373, 650, 683, 991, 1022\}.$$
Here we have a unique non trivial involution over $\mathbb{F}_{2^5}$: D_{340}.

4.1. *The number of Dickson involutions*

We now compute the number of Dickson polynomials which induce involutions on $\mathbb{F}_{2^m}$. To this end, we begin with a technical result.

Lemma 5. *The number of quadratic residues of* 1 *modulo* $2^n - 1$ *is equal to* 2^τ *where τ is the number of the prime factors in the prime decomposition of* $2^n - 1$.

Proof. Given a positive integer p, let us denote $\rho(p)$ the number of square roots of unity modulo p, that is, the number of solutions of the congruence equation : $x^2 \equiv 1 \pmod{p}$. Let us show that
$$\rho(pq) = \rho(p)\rho(q), \quad \text{when } p \text{ and } q \text{ are coprime.} \tag{9}$$
To this end, note that according to Chinese's Theorem, $\mathbb{Z}/(pq)\mathbb{Z}$ is isomorphic to $\mathbb{Z}/p\mathbb{Z} \times \mathbb{Z}/q\mathbb{Z}$ via the isomorphism
$$\psi \,:\, x \in \mathbb{Z}/(pq)\mathbb{Z} \mapsto (x \pmod{p}, x \pmod{q}).$$
By construction, in $\mathbb{Z}/p\mathbb{Z} \times \mathbb{Z}/q\mathbb{Z}$, $(a, b)^2 = (c, d)$ is equivalent to $a^2 = c$ and $b^2 = d$ so that $\psi(x^2) = (x^2 \pmod{p}, x^2 \pmod{q})$, proving (9).

Now, one has $\rho(p^\alpha) = 2$ for any odd prime number p and positive integer α. Indeed, suppose that $x^2 \equiv 1 \pmod{p^\alpha}$. Then
$$x^2 - 1 = (x + 1)(x - 1) \equiv 0 \pmod{p^\alpha}$$
and this is equivalent to
$$x + 1 \equiv 0 \pmod{p^\alpha} \text{ or } x - 1 \equiv 0 \pmod{p^\alpha},$$
that is $x \equiv \pm 1 \mod p^\alpha$. Since $2^n - 1$ is an odd number, we can write
$$2^n - 1 = \prod_{i=1}^{\tau} p_i^{\alpha_i}, \quad p_i \text{ is a prime factor,}$$
and the α_i's are positive integers. Then $\rho(2^n - 1) = \prod_{i=1}^{\tau} \rho(p_i^{\alpha_i}) = 2^\tau$. $\quad\square$

Then we have the following result on the number of Dickson polynomials that are involutions. Recall that for any such involution D_k there are four elements $k' \in \sigma(k)$ providing the same involution (see Lemma 3). Thus, the number of Dickson involutions is the number of such classes $\sigma(.)$. The class of 1 is said *trivial* since $D_1(x) = x$.

Theorem 4. *Let m be a positive integer such that $m > 1$ and set $n = 2m$. Let τ be the number of prime factors in the decomposition of $2^n - 1$. Then the number of (non trivial) Dickson polynomials over $\mathbb{F}_{2^m}$ which are involutions is equal to*

$$2^{\tau-2} - 1 \text{ if } m \text{ is odd and } 2^{\tau-1} - 1 \text{ if } m \text{ is even.}$$

Proof. Suppose that m is odd. According to Theorem 3, D_k is an involution if and only if $k \in S_n$, that is k is a quadratic residue of 1 modulo $2^n - 1$. The number of such k is equal to 2^τ by Lemma 5 . Now, according to (8) and Lemma 3, the number of pairwise different Dickson polynomial is equal to $2^\tau/4 = 2^{\tau-2}$ since K_n is of cardinality 4. Suppose that m is even. One can repeat again similar arguments as those of the odd case except that, in the even case, D_k is an involution if and only $k \in S \cup 2^{m/2}S$. It means that we have to replace 2^τ by $2^{\tau+1}$ in the preceding calculation. We then conclude by excluding the class of 1. $\qquad\square$

Remark 3. When m is odd, non trivial Dickson involutions exist for $\tau > 2$. Since $n = 2m$, $2^n - 1 = (2^m - 1)(2^m + 1)$. If $2^m - 1$ is prime and $2^m + 1$ is a power of 3 then $\tau = 2$, implying that non trivial Dickson involutions do not exist. It is the case for $n = 4, 6$. We prove below that this never holds for $m > 3$.

Proposition 2. *Let τ be the number of Dickson involutions on $\mathbb{F}_{2^m}$. Then $\tau \geq 3$ for any $m > 3$, i.e., for any $m > 3$ non trivial Dickson involutions do exist.*

Proof. We assume that $m > 3$. When m is even we have

$$2^n - 1 = (2^m + 1)(2^{m/2} - 1)(2^{m/2} + 1)$$

implying $\tau \geq 3$. From now on m is odd. Thus 3 divides $2^m + 1$ and we have $2^m + 1 \equiv 3m \pmod 9$ because

$$2^m + 1 = (3-1)^m + 1$$
$$= 3^m + \binom{m}{1}3^{m-1}(-1) + \ldots + \binom{m}{m-1}3(-1)^{m-1}.$$

If 3 does not divide m then $2^m + 1 = 3\ell$ where ℓ is coprime with 3 and with $2^m - 1$. In this case, we conclude that $2^n - 1$ has at least three prime divisors.

Now suppose that $m = 3^i p$, for some odd p which is coprime with 3. Then 9 divides $2^m + 1$. If $p > 1$ then $2^3 - 1$ and $2^p - 1$, which are coprime, both divide $2^m - 1$ providing at least three prime divisors of $2^n - 1$. If $p = 1$ then $i \geq 2$ so that $2^9 - 1$ divides $2^m - 1$. Since $2^9 = 17 \times 73$, we get again three prime divisors of $2^n - 1$. □

Example 2.

$n = 4, 6$ We seen above that $\tau = 2$. There is no Dickson polynomials which are involutions except $D_1(x) = x$.

$n = 8$, $m = 4$, $2^n - 1 = 255 = 17 \times 5 \times 3$, $\tau = 3$. The number of nontrivial Dickson polynomials which are involutions is equal to $2^{\tau - 1} - 1 = 3$.

$n = 10$, $m = 5$, $2^n - 1 = 1023 = 31 \times 11 \times 3$, $\tau = 3$. The number of nontrivial Dickson polynomials which are involutions is equal to $2^{\tau - 2} - 1 = 1$.

$n = 12$, $m = 6$, $2^n - 1 = 4095 = 3^2 \times 5 \times 7 \times 13$, $\tau = 4$. The number of Dickson polynomials which are involutions is equal to $2^{\tau - 1} = 8$ if we include the trivial class $\sigma(1)$. A set of representatives of these classes is:

$$\{\, 1, 181, 1574, 1756 \,\} \cup \{\, 8, 1448, 307, 1763\} \subset S_{12} \cup 8 * S_{12}.$$

Note that $181 * 1574 = -1756$ modulo 4095.

4.2. *Dickson involutions of very high degree*

Our previous results show that it is generally easy to get a Dickson involution on $\mathbb{F}_{2^m}$ by computation. However, for a very high m it could be difficult to obtain such involution which is neither trivial nor equal to $x^{m/2}$ (m even). Also, it could be convenient to have the use of specific k such that D_k is an involution which is not trivial. To explain that this is possible, we exhibit below several infinite class of Dickson involutions.

Theorem 5. *Let* $n = 2m$, $m \geq 4$. *Assume that* $\gcd(m, 3) = 1$ *and set* $n = 3r + e$, $e \in \{1, 2\}$. *Then the polynomial*

$$D_k(x) \pmod{x^{2^m} + x}, \quad \text{where } k = \begin{cases} (2^n - 1)/3 + 1 & \text{if } e = 2 \\ (2^n - 1)/3 - 1 & \text{if } e = 1 \end{cases},$$

is an involution of $\mathbb{F}_{2^m}$, *such that* $k \notin \sigma(1)$; *moreover for even* m, $k \notin \sigma(2^{m/2})$. *Consequently, the Dickson monomial* $D_k(x) = x^k$ *is also an involution.*

Proof. We first prove that $k^2 = 1$ modulo $2^n - 1$:

$$k^2 - 1 = \left(\frac{2^n - 1}{3} \pm 1\right)^2 - 1 = \left(\frac{2^n - 1}{3}\right)^2 \pm \frac{2(2^n - 1)}{3}$$

$$= (2^n - 1)\left(\frac{2^n - 1 \pm 6}{9}\right)$$

with $2^n = 2^e(2^3 + 1 - 1)^r \equiv 2^e(-1)^r \pmod 9$. Let $a = 2^n - 1 \pm 6 \pmod 9$. We have to examine when a is zero modulo 9.

Note that, by hypothesis, $e \neq 0$ since a cannot be 0 modulo 9 when $e = 0$. When $e = 2$ (r even), we get $2^n \equiv 4 \pmod 9$; thus $a \in \{0, -3\}$. The value -3, obtained from $a \equiv 4 - 1 - 6 \pmod 9$ is not suitable. Hence, for $e = 2$ we take $a \equiv 4 - 1 + 6 \pmod 9$, that is $k = (2^n - 1)/3 + 1$. Similarly, if $e = 1$ (r odd) then $2^n \equiv -2 \pmod 9$ and $a \in \{3, 0\}$. In this case, only $k = (2^n - 1)/3 - 1$ is suitable.

Now, k is given by its binary expansion. With this representation, the *weight of* k, say $w(k)$, is the number of terms of this expression. Here we have $w(k) = m$ for $e = 1$ and $w(k) = m + 1$ for $e = 2$. Recall that the class of 1 is $K_n = \{\pm 1, \pm 2^m\}$. Clearly the corresponding weights are $\{1, n - 1\}$ and this holds for the class of $2^{m/2}$ too (for even m). This completes the proof. $\qquad\square$

Example 3. The first pairs (n, k), obtained by Theorem 5 are: $(8, 86)$, $(10, 340)$, $(14, 5462)$ and $(16, 21844)$. According to Theorem 3, we have also $(8, 2^2 * 86)$, and $(16, 2^4 * 21844)$.

Theorem 6. *Let* ℓ *be any even integer* ($\ell > 0$). *Set* $n = \ell(2^\ell + 1)$ *and* $m = n/2$. *Then the polynomial*

$$D_k(x), \quad (\mathrm{mod}\ x^{2^m} + x) \quad \text{where } k = \sum_{i=1}^{2^\ell} 2^{i\ell},$$

is an involution of $\mathbb{F}_{2^m}$, *such that* $k \notin \sigma(1)$; *moreover for even* m, $k \notin \sigma(2^{m/2})$. *Consequently, the Dickson monomial* $D_k(x) = x^k$ *is also an involution.*

Proof. First, it is easy to prove that $k^2 = 1$ modulo $2^n - 1$, since

$$2^n - 1 = (2^\ell - 1)(2^{\ell 2^\ell} + 2^{\ell(2^\ell - 1)} + \cdots + 2^\ell + 1) = (2^\ell - 1)(k + 1).$$

Hence we have

$$k^2 - 1 = \left(\frac{2^n - 1}{2^\ell - 1} - 1\right)^2 - 1 = \left(\frac{2^n - 1}{2^\ell - 1}\right)^2 - 2\frac{2^n - 1}{2^\ell - 1}$$

$$= (2^n - 1)\left(\frac{2^n - 1 - 2(2^\ell - 1)}{(2^\ell - 1)^2}\right) = (2^n - 1)\left(\frac{k - 1}{2^\ell - 1}\right).$$

But $k = \sum_{i=1}^{2^\ell} \left((2^\ell - 1) + 1\right)^i \equiv 1 \pmod{2^\ell - 1}$. Hence $2^n - 1$ divides $k^2 - 1$.

Now, k is given by its binary expansion. With this representation, the *weight of k* is $w(k) = 2^\ell = (n - \ell)/\ell$ where $\ell \geq 2$. As in the proof of Theorem 5, this completes the proof. □

Remark 4. The first value of n, in Theorem 6, is $n = 10$, for $\ell = 2$. In this case we check that $k = 340$, as explained in Example 1. This value is also obtained by Theorem 5 but the two classes are different. Note that in Theorem 6 the condition $\gcd(m, 3) = 1$ is not necessary.

5. Fixed points of the Dickson involutions

A *fixed point* of any polynomial $P(x)$ is an element ρ such that $P(\rho) = \rho$. In [14], an empirical study on the number of fixed points of involutions and general permutations were made. Permutations are important building blocks in block ciphers, and for a secure design purpose S-boxes are chosen with very good cryptographic properties. The observation in [14] was that the number of fixed points is correlated to the cryptographic properties like nonlinearity and the maximum XOR entry table. Precisely, lower is the number of fixed points better is the value of nonlinearity and the maximum XOR entry table. Thus the authors proposed to choose permutation S-boxes with a few fixed points.

5.1. *General description*

The number of fixed points of polynomials $D_k(x)$ on $\mathbb{F}_{2^m}$ is computed in [8] as a function of m and k. We give this result in our context with a sketch of proof.

Theorem 7.[8, Theorem 3.34] *Denote by $\mathcal{F}(k, m)$ the set of fixed points of the Dickson polynomial D_k (over $\mathbb{F}_{2^m}$). Then the cardinality of $\mathcal{F}(k, m)$ is*

$$|\mathcal{F}(k, m)| = \frac{1}{2}\left(\gcd(2^m + 1, k + 1) + \gcd(2^m - 1, k + 1)\right.$$
$$\left. + \gcd(2^m + 1, k - 1) + \gcd(2^m - 1, k - 1)\right) - 1. \quad (10)$$

Sketch of Proof. According to Lemma 1, any $x \in \mathbb{F}_{2^m}$ can be written

$$x = \gamma + \frac{1}{\gamma}, \ \gamma \in \mathbb{F}_{2^n}^*, \ \gamma^{2^m+1} = 1 \ \text{ or } \ \gamma^{2^m-1} = 1 \tag{11}$$

Thus $D_k(x) = D_k(\gamma + \gamma^{-1}) = \gamma^k + \gamma^{-k}$. Further, x is a fixed point of D_k if and only if

$$\gamma^k + \gamma^{-k} = \gamma + \gamma^{-1}, \ i.e., \ (\gamma^{k+1} - 1)(\gamma^{k-1} - 1) = 0.$$

Note that γ and γ^{-1} provide the same x. $\diamond$

The proof of Theorem 7 gives explicitly the fixed points of D_k : x is written as in (11) and either $\gamma^{k-1} = 1$ or $\gamma^{k+1} = 1$. We can be more precise in the case where D_k is an involution. Set

$$\begin{aligned} r_1 = \gcd(2^m - 1, k - 1), \ & r_2 = \gcd(2^m - 1, k + 1), \\ s_1 = \gcd(2^m + 1, k - 1), \ & s_2 = \gcd(2^m + 1, k + 1), \end{aligned} \tag{12}$$

providing $|\mathcal{F}(k, m)| = (r_1 + r_2 + s_1 + s_2)/2 - 1$. From Theorem 3 and its proof, we must distinguish two cases:

- $k \in S_n$ where we have $k^2 \equiv 1 \pmod{2^n - 1}$, i.e., $2^n - 1$ divides $k^2 - 1$;
- for even m, let $k \in 2^{m/2} S_n$; in this case, we have $k^2 = 2^m s^2$ where $s \in S_n$. Hence $k^2 \equiv 2^m \pmod{2^n - 1}$ and then

$$k^2 - 1 = (k+1)(k-1) \equiv 2^m - 1 \pmod{2^n - 1}.$$

These properties will be used respectively in Corollaries 2 and 3.

Corollary 2. *Consider any involution D_k on $\mathbb{F}_{2^m}$ as described by Theorem 3 with $k \in S_n$. We assume that D_k is not the identity modulo $(x^{2^m} + x)$, i.e., $k \notin \{\pm 1, \ \pm 2^m\}$. Let α be a primitive root of $\mathbb{F}_{2^n}$, $n = 2m$. Then $2^m - 1 = r_1 r_2$ and $2^m + 1 = s_1 s_2$. Moreover $\mathcal{F}(k, m)$ is the set of $\gamma + \gamma^{-1}$ where*

$$\gamma \in \left\{ \begin{array}{l} \alpha^{i r_2 (2^m+1)}, \ 0 \leq i \leq (r_1 - 1)/2, \ \alpha^{j r_1 (2^m+1)}, \ 0 \leq j \leq (r_2 - 1)/2 \\ \alpha^{\ell s_2 (2^m-1)}, \ 0 \leq \ell \leq (s_1 - 1)/2, \ \alpha^{t s_1 (2^m-1)}, \ 0 \leq t \leq (s_2 - 1)/2 \end{array} \right\}.$$

Proof. Since $2^n - 1$ divides $k^2 - 1$, we get directly from the definition (see (12)): $2^m - 1 = r_1 r_2$, $2^m + 1 = s_1 s_2$ and

$$\gcd(2^n - 1, k - 1) = r_1 s_1, \ \gcd(2^n - 1, k + 1) = r_2 s_2.$$

Now we have two cases:

- If $\gamma^{2^m-1} = 1$ then $\gamma = \alpha^{u(2^m+1)} = \alpha^{us_1 s_2}$ for some u. If $\gamma^{k-1} = 1$ then $u = ir_2$, $0 \leq i \leq r_1 - 1$. If $\gamma^{k+1} = 1$ then $u = jr_1$, $0 \leq j \leq r_2 - 1$.
- If $\gamma^{2^m+1} = 1$ then $\gamma = \alpha^{u(2^m-1)} = \alpha^{ur_1 r_2}$ for some u. If $\gamma^{k-1} = 1$ then $u = \ell s_2$, $0 \leq \ell \leq s_1 - 1$. If $\gamma^{k+1} = 1$ then $u = ts_1$, $0 \leq t \leq s_2 - 1$.

We observe that $\alpha^{(r_1-i)r_2 s_1 s_2} = \alpha^{-ir_2 s_1 s_2}$ and this holds for the three other kinds of α^e above, completing the proof. □

Remark 5. Define M_n, $n = 2m$, the set of factors of $2^n - 1$, as follows:

$$M_n = \{p_i^{e_i} \mid i = 1, \ldots s\} \text{ with } 2^n - 1 = \prod_{i=1}^{s} p_i^{e_i}, \ p_i \text{ is prime.}$$

Clearly, for any i, $p_i^{e_i}$ is a factor either of $2^m - 1$ or of $2^m + 1$. Such a factor divides either $k - 1$ or $k + 1$ when $2^n - 1$ divides $k^2 - 1$. Thus, all elements of M_n are involved in the computation of $|\mathcal{F}(k, m)|$ when $k \in S_n$.

Remark 6. It is to be noted that in Theorem 3.36 of [8], a lower bound on the number of fixed points of Dickson polynomials which are permutations has been given. For characteristic 2, this lower bound equals 2. Throughout our results it appears that such a lower bound is higher for Dickson involutions.

We now include the case where $k \in 2^{m/2} S_n$. We notably give a count of the fixed points of the Dickson involution D_k, $k \in \sigma(2^{m/2})$.

Corollary 3. *Let $n = 2m$ where m is even. Consider any involution D_k on $\mathbb{F}_{2^m}$ such that $k \in 2^{\frac{m}{2}} S_n$. Then $2^m - 1$ divides $k^2 - 1$ and we have $2^m - 1 = r_1 r_2$,*

$$s_1 = s_2 = 1 \text{ so that } |\mathcal{F}(k, m)| = \frac{r_1 + r_2}{2}.$$

In particular the cardinality of the set of fixed points of D_k, $k \in \sigma(2^{\frac{m}{2}})$, equals $2^{\frac{m}{2}}$.

Moreover, $\mathcal{F}(k, m)$ is the set of $\gamma + \gamma^{-1}$ where

$$\gamma \in \left\{ \alpha^{ir_2(2^m+1)}, \ 0 \leq i \leq (r_1 - 1)/2 \right\} \cup \left\{ \alpha^{jr_1(2^m+1)}, \ 0 \leq j \leq (r_2 - 1)/2 \right\}$$

(where α is a primitive root of $\mathbb{F}_{2^n}$ and s_i, r_i are given by (12)).

Proof. Recall that the number of fixed points of D_k is:

$$(r_1 + r_2 + s_1 + s_2)/2 - 1,$$

where r_1, r_2, s_1 and s_2 are from (12).

As noticed at the beginning of this section, $2^m - 1$ divides $k^2 - 1$ for $k \in 2^{\frac{m}{2}} S_n$. More precisely $k^2 \equiv 2^m \pmod{2^n - 1}$, i.e.,

$$(k-1)(k+1) = \ell(2^n - 1) + 2^m - 1, \text{ for some } \ell > 0.$$

Thus $\gcd(2^m+1, k^2-1) = \gcd(2^m+1, 2^m-1) = 1$, proving that $s_1 = s_2 = 1$. Since $\gcd(2^m - 1, k^2 - 1) = 2^m - 1$, $2^m - 1 = r_1 r_2$. Further

$$|\mathcal{F}(k, m)| = (r_1 + r_2 + 2)/2 - 1 = (r_1 + r_2)/2.$$

If $k \in \sigma(2^{\frac{m}{2}})$ then $k = 2^{\frac{m}{2}} s$ with $s \in K_n$ (see Lemma 3). Note $\mathcal{F}(k, m) = \mathcal{F}(k', m)$ for any element of $k' \in \sigma(k)$. Thus, we can assume that $k = 2^{\frac{m}{2}}$. Since $k^2 - 1 = 2^m - 1$, we have directly $r_1 = 2^{\frac{m}{2}} - 1$ and $r_2 = 2^{\frac{m}{2}} + 1$. Then

$$|\mathcal{F}(2^{\frac{m}{2}}, m)| = \frac{2^{\frac{m}{2}} - 1 + 2^{\frac{m}{2}} + 1}{2} = 2^{\frac{m}{2}}.$$

To compute the set $\mathcal{F}(k, m)$ we proceed as previously, for Corollary 2:

- If $\gamma^{2^m-1} = 1$ then $\gamma = \alpha^{u(2^m+1)}$ for some u. If $\gamma^{k-1} = 1$ then $u = ir_2, 0 \le i \le r_1 - 1$. If $\gamma^{k+1} = 1$ then $u = jr_1, 0 \le j \le r_2 - 1$.
- If $\gamma^{2^m+1} = 1$ then $\gamma = \alpha^{u(2^m-1)}$ for some u. But $\gcd(k^2 - 1, 2^m + 1) = 1$ implies that $\gamma^{k-1} = 1$ (resp. $\gamma^{k+1} = 1$) if and only if $2^m + 1$ divides u. Thus $\gamma = \alpha^{(2^m+1)(2^m-1)} = 1$.

$\square$

Example 4. Let $n = 12$ and $k = 181$. We have

$$N = 2^{12} - 1 = 63 \times 65 = (9 \times 7) \times (5 \times 13) \text{ and}$$

$$180 = 4 \times 5 \times 9, \quad 182 = 2 \times 7 \times 13.$$

With notation of Corollary 2, $r_1 = 9$, $r_2 = 7$, $s_1 = 5$ and $s_2 = 13$. Thus the cardinality of $\mathcal{F}(181, 12)$ equals $(7 + 13 + 9 + 5)/2 - 1 = 16$. and, α being a primitive root of $\mathbb{F}_{2^{12}}$,

$$\mathcal{F}(181, 12) = \left\{ \alpha^d + \alpha^{-d}, d \in \left\{ 0, \frac{N}{9}, \dots, \frac{4N}{9}, \frac{N}{7}, \dots, \frac{3N}{7}, \frac{N}{5}, \frac{2N}{5}, \frac{N}{13}, \dots, \frac{6N}{13} \right\} \right\}.$$

Now we take $k = 2^3 \times 181 \equiv 1448 \pmod{2^{12} - 1}$. We have $2^6 - 1 = r_1 r_2$ with $r_1 = 1$ and $r_2 = 63$. Moreover $s_1 = s_2 = 1$. Then the cardinality of $\mathcal{F}(1448, 12)$ equals $64/2 = 32$. According to Corollary 3, we have

$$\mathcal{F}(1448, 12) = \left\{ \alpha^d + \alpha^{-d}, \ d \in \{0, 65, 2 * 65, \dots, 31 * 65\} \right\}.$$

5.2. Bounds on the number of fixed points

Below we give the proof that for even m, the minimum value of $|\mathcal{F}(k,m)|$ is $2^{\frac{m}{2}}$.

Theorem 8. *Let I be the index set such that for all $k \in I$, D_k is an involution on $\mathbb{F}_{2^m}$. Then for even m,*

$$\min_{k \in I} |\mathcal{F}(k,m)| = |\mathcal{F}(2^{\frac{m}{2}}, m)| = 2^{\frac{m}{2}}.$$

Proof. Let $n = 2m$. Since m is even, $k \in S_n$ or $k \in 2^{\frac{m}{2}} S_n$. We treat the proof in two cases.

Case 1: $k \in 2^{\frac{m}{2}} S_n$.

Following Corollary 3, the minimum value of $|\mathcal{F}(k,m)|$ is the minimum value that $(r_1 + r_2)/2$ can have with the constraint $r_1 r_2 = 2^m - 1$. So now we have to deal with an optimization problem. To deal with this we consider the following related optimization problem over positive real numbers.

Minimize $f(x,y) = x + y$

Subject to :

$$xy = 2^m - 1,$$

where x, y are positive real numbers.

Using *Lagrange multiplier method* [1, Section 3.1.3], we get that the minimum value of $f(x,y)$ is $2\sqrt{2^m - 1}$ which is obtained at $x = \sqrt{2^m - 1}$ and $y = \sqrt{2^m - 1}$.

So if r_1 and r_2 were real numbers then the minimum value of $(r_1 + r_2)$ would have been $2\sqrt{2^m - 1}$. Note that the closest integer to $\sqrt{2^m - 1}$ is $2^{\frac{m}{2}}$ and $2 \cdot 2^{\frac{m}{2}} = (2^{\frac{m}{2}} + 1) + (2^{\frac{m}{2}} - 1)$. If $k = 2^{\frac{m}{2}}$, then $r_1 = 2^{\frac{m}{2}} + 1$ and $r_2 = 2^{\frac{m}{2}} - 1$ for which $r_1 r_2 = 2^m - 1$, and hence the minimum value of $(r_1 + r_2)/2$ is $2^{\frac{m}{2}}$.

Case 2: $k \in S_n$.

In this case $k^2 - 1 \equiv 0 \pmod{2^n - 1}$. Therefore, $2^m - 1 = r_1 r_2$ and $2^m + 1 = s_1 s_2$. Now we have to consider the optimization problem:

Minimize $(r_1 + r_2 + s_1 + s_2)/2 - 1$

Subject to :

$$r_1 r_2 = 2^m - 1,$$
$$s_1 s_2 = 2^m + 1.$$

More precisely, we need to minimize the value $(r_1 + r_2)$ subject to $r_1 r_2 = 2^m - 1$, and $(s_1 + s_2)$ subject to $s_1 s_2 = 2^m + 1$. It is clear that one of s_1

and s_2 is greater than 1, *i.e.*, $(s_1 + s_2) > 2$. We already have seen in the previous case that the minimum value of $r_1 + r_2$ is $2 \cdot 2^{\frac{m}{2}}$. Therefore, the minimum value of $(r_1 + r_2 + s_1 + s_2)/2 - 1$ is greater than $2^{\frac{m}{2}}$.

Thus from the two cases the proof is clear.   □

For odd m, the exact minimum value of $|\mathcal{F}(k, m)|$ is not clear. However, we can derive a lower bound on the minimum value of $|\mathcal{F}(k, m)|$ that we present below. Note that if m is odd, then $3 | (2^m + 1)$.

Theorem 9. *Let I be the index set such that for all $k \in I$, D_k is an involution on $\mathbb{F}_{2^m}$, where m is odd. Then*

(1) $\min\limits_{k \in I} |\mathcal{F}(k, m)| > 2^{m-1} + \lceil \sqrt{2^m + 1} \rceil - 1$, *when $2^m - 1$ is prime,*

(2) $\min\limits_{k \in I} |\mathcal{F}(k, m)| > \lceil (\sqrt{2^m - 1} + \sqrt{2^m + 1}) \rceil - 1$, *when $2^m - 1$ is composite,*

Proof. Since m is odd, $k \in S_n$. Consider r_1, r_2, s_1, s_2 as they are given in (12). Since $2^n - 1$ divides $k^2 - 1$, we have $r_1 r_2 = 2^m - 1$ and $s_1 s_2 = 2^m + 1$. To find the minimum value of $|\mathcal{F}(k, m)|$, we need to minimize the value of $(r_1 + r_2 + s_1 + s_2)/2 - 1$ subject to the constraints $r_1 r_2 = 2^m - 1$ and $s_1 s_2 = 2^m + 1$. Note that we can consider this optimization problem over the set of real numbers, that will give us a lower bound of the minimum value of $(r_1 + r_2 + s_1 + s_2)/2 - 1$ Therefore we consider the following optimization problem:

Minimize $(r_1 + r_2 + s_1 + s_2)/2 - 1$
Subject to :

$$r_1 r_2 = 2^m - 1,$$
$$s_1 s_2 = 2^m + 1.$$

First we prove *(i)*, i.e., when $2^m - 1$ is prime. In this case, $r_1 + r_2 = 2^m$. If s_1 and s_2 were real valued, the minimum of $s_1 + s_2$ would have been $2\sqrt{2^m + 1}$ (using the Lagrange Multiplier method), and the minimum value of $(r_1 + r_2 + s_1 + s_2)/2 - 1$ would be $2^{m-1} + \sqrt{2^m + 1} - 1$. Since r_1, r_2, s_1, s_2 are all integers, thus the actual minimum value is greater than $2^{m-1} + \lceil \sqrt{2^m + 1} \rceil - 1$.

For *(ii)*, we have that $2^m - 1$ is composite. Then considering all of r_1, r_2, s_1, s_2 as real numbers, and using the Lagrange Multiplier method, we get the minimum value of $(r_1 + r_2 + s_1 + s_2)/2 - 1$ is $\sqrt{2^m - 1} + \sqrt{2^m + 1} - 1$. Since r_1, r_2, s_1, s_2 are all integers, we have that the actual minimum value is greater than $\lceil (\sqrt{2^m - 1} + \sqrt{2^m + 1}) \rceil - 1$.   □

Remark 7. Theorem 8 and 9 clearly says that the Dickson involutions $D_k(x)$ have high number of fixed points jeopardizing their use as S-boxes in block ciphers. However the structure of the set of fixed points is of interest allowing to have easy methods to reduce its size. A special example is studied in the next section.

Let us see how good these bounds are. In Table 1, we compare the lower bound obtained in Theorem 9 and the exact minimum value of $|\mathcal{F}(k,m)|$ for some odd m.

m	Lower bound of $|\mathcal{F}(k,m)|$ from Theorem 9	Exact minimum value of $|\mathcal{F}(k,m)|$
5	21	22
7	75	86
9	45	62
11	90	398

5.3. *Minimal sets of fixed points*

In this section m is even and we consider Dickson involutions D_k over $\mathbb{F}_{2^n}$ such that $k \in 2^{m/2} S_n$ where S_n, defined by (7), is the set of $1 \leq u \leq 2^n - 2$ such that $u^2 \equiv 1 \pmod{2^n - 1}$. From Theorem 8, we know that the minimal size of $\mathcal{F}(k,m)$ is $2^{m/2}$. It is exactly $2^{m/2}$ for those $k \in \sigma(2^{\frac{m}{2}})$ and we will point out that other k satisfy this property. We study these special cases now. And first we have to precise that in this case the set $\mathcal{F}(k,m)$ is the subfield of order $2^{\frac{m}{2}}$.

Proposition 3. *Let $n = 2m = 4t$ and k is such that D_k is an involution. Then $|\mathcal{F}(k,m)| = 2^t$ if and only if $k = 2^t s$ with $s^2 \equiv 1 \pmod{2^n - 1}$ and either (i) or (ii) holds:*

(i) $r_1 = 2^t - 1$ *and* $r_2 = 2^t + 1$;
(ii) $r_1 = 2^t + 1$ *and* $r_2 = 2^t - 1$,

where r_1, r_2 are as given in (12).

Proof. It is clear that if $k = 2^t s$ and (i) or (ii) holds, then

$$|\mathcal{F}(k,m)| = \frac{r_1 + r_2}{2} = 2^t.$$

Next assume that $|\mathcal{F}(k,m)| = 2^t$. From Case 2 of Theorem 8, we get that if $k \in S_n$, then $|\mathcal{F}(k,m)| > 2^t$. Therefore, it is necessary to have $k = 2^t s$, where $s \in S_n$. Moreover, from Corollary 3 we have $r_1 r_2 = 2^m - 1$ and $r_1 + r_2 = 2^{t+1}$. Thus r_1 and r_2 are the integer roots of the equation

$$X^2 - 2^{t+1}X + 2^m - 1 = 0 \iff (X - 2^t)^2 = 1.$$

Hence $(r_1, r_2) = (2^t - 1, 2^t + 1)$ or $(r_1, r_2) = (2^t + 1, 2^t - 1)$, completing the proof. $\qquad\square$

Remark 8. We could give a more general result following Corollary 3 by studying the roots of $X^2 + 2EX + 2^m - 1 = 0$, where $E = |\mathcal{F}(k,m)|$. One can say that E is suitable if and only if $D = E^2 - (2^m - 1)$ is a square. If it is we get r_1 and r_2. Further, the problem is to know if such k exists.

Lemma 6. *Let $n = 2m = 4t$, $k = 2^t s$ with $s^2 \equiv 1 \pmod{2^n - 1}$. Then $\mathcal{F}(k,m) = \mathbb{F}_{2^t}$ if and only if either (i) or (ii) (of Proposition 3) holds.*

Proof. We apply Corollary 3 assuming that (i) holds. Let $x = \gamma + \gamma^{-1}$ be a fixed point of D_k. Note that $\gamma^{2^m - 1} = 1$ which means $\gamma \in \mathbb{F}_{2^m}$. The cardinality of $\mathcal{F}(k,m)$ is

$$\frac{r_1 + r_2}{2} = \frac{2^t + 1 + 2^t - 1}{2} = 2^t.$$

If $\gamma = \alpha^{ir_2(2^m + 1)}$ then $\gamma^{2^t - 1} = 1$ so that $\gamma \in \mathbb{F}_{2^t}$. If $\gamma = \alpha^{jr_1(2^m + 1)}$ then $\gamma^{2^t + 1} = 1$ so that $\gamma^{-1} = \gamma^{2^t}$. Hence $\gamma \mapsto \gamma + \gamma^{-1}$ is a function from $\mathbb{F}_{2^m}$ to $\in \mathbb{F}_{2^t}$. Thus in both cases we have $x \in \mathbb{F}_{2^t}$. The proof is completed since $|\mathcal{F}(k,m)| = 2^t$. The proof assuming that (ii) holds is similar, by exchanging r_1 and r_2.

Conversely, if $\mathcal{F}(k,m) = \mathbb{F}_{2^t}$ then $|\mathcal{F}(k,m)| = 2^t$ and we apply Proposition 3. $\qquad\square$

Now, considering $k = 2^t s$ we want to identify those s such that D_k is an involution with fixed points set $\mathbb{F}_{2^t}$. In the following we avoid the trivial cases $s \in \sigma(1)$. The case $k = 2^t$ was treated before (Corollary 3).

Theorem 10. *Let $n = 2m = 4t$. Let $k \equiv 2^t s \pmod{2^n - 1}$, $s \notin \sigma(1)$. Then $s^2 \equiv 1 \pmod{2^n - 1}$ and $\mathcal{F}(k,m) = \mathbb{F}_{2^t}$ if and only if either (a) or (b) holds:*

(a) $s - 1 = (2^m - 1)L$ *where* $0 < L < 2^m$ *and* $L^2 - L \equiv 0 \pmod{2^m + 1}$;
(b) $s + 1 = (2^m - 1)L$ *where* $0 < L < 2^m$ *and* $L^2 + L \equiv 0 \pmod{2^m + 1}$.

Moreover if s satisfies (a) then there is $s' \in \sigma(s)$ which satisfies (b) and vice versa.

Proof. Assume that (a) holds. Then $s + 1 = (2^m - 1)L + 2$ and

$$s^2 - 1 = (2^m - 1)L(s + 1),$$

where

$$L(s + 1) = L^2(2^m - 1) + 2L = L^2(2^m + 1 - 2) + 2L$$
$$\equiv -2(L^2 - L) \equiv 0 \pmod{2^m + 1}. \tag{13}$$

Therefore $s^2 \equiv 1 \pmod{2^n - 1}$. Also, we have

$$k - 1 = 2^t(2^m - 1)L + 2^t - 1 \equiv 2^t - 1 \pmod{2^m - 1}$$

and

$$k + 1 = 2^t(2^m - 1)L + 2^t + 1 \equiv 2^t + 1 \pmod{2^m - 1},$$

that is, we are in the case (i) of Proposition 3. Thus $\mathcal{F}(k, m) = \mathbb{F}_{2^t}$ from Lemma 6. Similarly, if (b) holds then $s - 1 = (2^m - 1)L - 2$ and

$$L((2^m - 1)L - 2) = L^2(2^m + 1 - 2) - 2L \equiv -2(L^2 + L) \pmod{2^m + 1}.$$

We prove Proposition 3),(ii), by computing $k \pm 1$ using $s = (2^m - 1)L - 1$. To conclude this part of proof, note that if s satisfies (a) then

$$(-s) + 1 = -((2^m - 1)L + 1) + 1 = (2^m - 1)R \quad \text{where } R = (-L).$$

Then $R^2 + R = L^2 - L \equiv 0 \pmod{2^m + 1}$ showing that $-s$ (which is in $\sigma(s)$) satisfies (b).

Conversely, assume that $s^2 \equiv 1 \pmod{2^n - 1}$ and $\mathcal{F}(k, m) = \mathbb{F}_{2^t}$. From Lemma 6, either (i) or (ii) holds. Assume that (i) holds; so $r_1 = \gcd(2^m - 1, k - 1) = 2^t - 1$. This implies

$$k - 1 = 2^t s - 1 = (2^t - 1)s + (s - 1) \quad \text{where } (2^t - 1) \text{ divides } s - 1.$$

Similarly, $k + 1 = (2^t + 1)s - (s - 1)$ where $(2^t + 1)$ divides $s - 1$, since $r_2 = \gcd(2^m - 1, k + 1) = 2^t + 1$. Hence $2^m - 1$ divides $s - 1$. Moreover we have $s - 1 < 2^n - 2$. Then $s - 1 = (2^m - 1)L$ where $L < 2^m$. Now we have $L(s+1) \equiv 0 \pmod{2^m + 1}$, because $s^2 - 1 = (2^m - 1)L(s+1)$. This implies $L^2 - L \equiv 0 \pmod{2^m + 1}$, since as shows (13) $L(s + 1) \equiv -2(L^2 - L)$ $\pmod{2^m + 1}$. The proof is similar (symmetric) if we suppose that (ii) holds. □

Remark 9. By Theorem 10 we have a partial characterization of those $k \in 2^t S_n$ whose set of fixed points equals 2^t. Clearly we have a good algorithm by computing $(2^m - 1)L$ such that

$$L^2 \pm L - R(2^m + 1) = 0, \quad \text{has integer solutions } L.$$

If there are solutions, one is negative and then is not suitable. The main question is : for which R the integer $1 + 4R(2^m + 1)$ is a square?

Example 5. We give here the (non trivial) involutions D_k which are such that $k \in 2^t S_n$ and $\mathcal{F}(k, m) = \mathbb{F}_{2^t}$ $(m = 2t)$, for $m \leq 14$. For $2 \leq t \leq 6$ we found zero or one such k:

n=12, t=3: $k = 307 = 2^3 s$ where $s + 1 = 1574 + 1 = 63 * 25$.
n=20, t=5: $k = 51118 = 2^5 s$ where $s + 1 = (2^{10} - 1) * 450$.
n=24, t=6: $k = 6908201 = 2^6 s$ where $s + 1 = (2^{12} - 1) * 1445$.

For $n = 28$ we found three classes whose representatives are $k = 2^7 s$ where $s = (2^{14} - 1)L + 1$, $L \in \{1131, 6555, 7685\}$.

Several observations arise, since our numerical results.

If $2^m + 1$ is prime then $2^m + 1$ must divide L or $s + 1$ (resp. $s - 1$) which is impossible unless $s = 2^m$. Indeed $L < 2^m$ and $s + 1 = (2^m - 1)L + 2$ in case (a) (resp. $s - 1 = (2^m - 1)L - 2$ in case (b)). When $s + 1 \equiv -2L + 2$ (mod $2^m + 1$) $2^m + 1$ divides $s + 1$ only if $L = 1$, that is $s = 2^m$. Respectively $s - 1 \equiv -2L - 2$ (mod $2^m + 1$) and $2^m + 1$ cannot divide $s - 1$. This explains why we have no result for $m = 4$ and $m = 8$, two cases where $2^m + 1$ is prime.

Proposition 4. *Let $n = 2m = 4t$. If D_k is such that $\mathcal{F}(k, m) = \mathbb{F}_{2^t}$, $k = 2^t s$ with $s \in S_n$, then $|\mathcal{F}(s, m)| > 2^{m-1}$.*

Proof. Recall that r_i and s_i, $i = 1, 2$, are given by (12), replacing k by s. From Theorem 10, we have either $r_1 = 2^m - 1$ or $r_2 = 2^m - 1$ and, respectively, $r_2 = 1$ or $r_1 = 1$. According to Theorem 7 and Corollary 2, we get

$$|\mathcal{F}(s, m)| = \frac{(2^m - 1) + 1}{2} + \frac{s_1 + s_2}{2} - 1 = 2^{m-1} + \frac{s_1 + s_2}{2} - 1$$

where $s_1 s_2 = 2^m + 1$. $\qquad\square$

6. Numerical results

In Table 6, we present some numerical results related to number of equivalence classes of Dickson involutions and the respective number of fixed points. Notation is as follows: m is the dimension of the field, $\mathcal{N}_m$ is the total number of Dickson involutions over $\mathbb{F}_{2^m}$, k is such that D_k is a Dickson involution on $\mathbb{F}_{2^m}$, where k represents one equivalence class as described in (8), and $|\mathcal{F}(k,m)|$ is the corresponding number of fixed points.

From Table 6, note that for odd m, the minimum value of $|\mathcal{F}(k,m)|$ is much larger than $2^{\frac{m}{2}}$, considering $2^{\frac{m}{2}}$ as a real value.

m	$\mathcal{N}_m$	k	$|\mathcal{F}(k,m)|$	m	$\mathcal{N}_m$	k	$|\mathcal{F}(k,m)|$
4	16	1	16	7	8	1	128
		4	4			5333	86
		86	12				
		89	8				
5	8	1	32				
		340	22				
6	32	1	64	8	32	1	256
		8	8			16	16
		181	16			9011	44
		307	8			12851	156
		1448	32			17749	28
		1574	40			21589	172
		1756	40			30584	144
		1763	32			30599	128

Table 1.6 - Dickson involutions of 1st kind over $\mathbb{F}_{2^m}$, $4 \leq m \leq 8$, and their fixed points.

Some Dickson involutions. We do not give the polynomials D_k where k is in the classes of 1 and of $2^{m/2}$. Polynomials are given modulo $(x^{2^m} + x)$. Involutions on $\mathbb{F}_{2^m}$ for $m = 3, 4$ are given in Example 1. Recall that $D_{2^{m/2}s}(x) = (D_s(x))^{2^{m/2}}$ where we compute only D_s. When $m = 5$ we have only one non trivial involution:

$$D_{340}(x) = x^4 + x^5 + x^6 + x^8 + x^9 + x^{16} + x^{17} + x^{21} + x^{24} + x^{28} + x^{30}.$$

Let $m = 6$. We have $\{181, 1574, 1756\}$ from S_{12} and $\{1448, 307, 1763\}$ in

$2^3 S_{12}$, respectively.

$$D_{181}(x) = x + x^2 + x^4 + x^6 + x^7 + x^9 + x^{10} + x^{12} + x^{13} + x^{33} + x^{39}$$
$$+ x^{41} + x^{45} + x^{49} + x^{55}.$$

$$D_{1574}(x) = x^2 + x^8 + x^9 + x^{10} + x^{12} + x^{13} + x^{32} + x^{33} + x^{41} + x^{45}$$
$$+ x^{48} + x^{49} + x^{56} + x^{60} + x^{62}.$$

$$D_{1756}(x) = x^4 + x^6 + x^7 + x^8 + x^{32} + x^{39} + x^{48} + x^{55} + x^{56} + x^{60} + x^{62}.$$

For $m = 7$ we have only one non trivial involution:

$$D_{5333}(x) = x + x^2 + x^3 + x^4 + x^8 + x^{10} + x^{11} + x^{16} + x^{18} + x^{19} + x^{20}$$
$$+ x^{32} + x^{34} + x^{35} + x^{36} + x^{40} + x^{42} + x^{43} + x^{64} + x^{67} + x^{75}$$
$$+ x^{83} + x^{96} + x^{112} + x^{120} + x^{124} + x^{126}.$$

7. Conclusion

In this paper we have characterized Dickson polynomials of the first kind, with the constant $a = 1$, that are involutions. We studied some properties of Dickson involutions over $\mathbb{F}_{2^m}$, for a fixed m, trying to have a clear description of this corpus. In particular, we show that its size increases with the number of prime divisors of $2^n - 1$. We have studied the set of fixed points and noticed that the number of fixed points of a Dickson involution is generally *high*. Nevertheless, we have obtained a precise description of such points. In particular, we have described the set of Dickson involutions whose set of fixed point is $\mathbb{F}_{2^t}$ $(t = \frac{n}{4})$. Removing many of those fixed points in such a way that keeps the involution property intact will give involution with a few fixed points. Actually, several techniques exist to reduce the number of fixed points [3, 4].

Acknowledgment

The third author did this work while he was at the Centre of Excellence in Cryptology at Indian Statistical Institute, Kolkata.

References

[1] D. P. Bertsekas. Nonlinear Programming (Second ed.), 1999, Cambridge, MA.: Athena Scientific.
[2] P. Charpin and G. Gong. Hyperbent functions, Kloosterman sums and Dickson polynomials. *IEEE Transactions on Information Theory*, Vol. 54, No. 9, pages 4230-4238, September 2008.

[3] P. Charpin, S. Mesnager and S. Sarkar. On involutions of finite fields, *Proceedings of 2015 IEEE International Symposium on Information Theory, ISIT 2015*, Hong-Kong, 2015.

[4] P. Charpin, S. Mesnager and S. Sarkar. Involutions over the Galois field $\mathbb{F}_{2^n}$, *IEEE Transactions on Information Theory*. To appear, 2016.

[5] J. F. Dillon. Geometry, codes and difference sets: exceptional connections. In *Codes and designs (Columbus, OH, 2000)*, volume 10 of *Ohio State Univ. Math. Res. Inst. Publ.*, pages 73–85. de Gruyter, Berlin, 2002.

[6] J. F. Dillon and H. Dobbertin. New cyclic difference sets with Singer parameters. *Finite Fields and their Applications*, vol. 10, n. 3, pages 342-389, 2004.

[7] X.-D. Hou, G.L. Mullen, J.A. Sellers and J.L. Yucas. Reversed Dickson polynomials over finite fields, *Finite Fields Appl.*, 15 (2009), pages. 748-773.

[8] R. Lidl, G.L. Mullen and G. Turnwald, *Dickson Polynomials*, Pitman Monographs in Pure and Applied Mathematics, Vol. 65, Addison-Wesley, Reading, MA 1993.

[9] S. Mesnager, Bent and Hyper-bent functions in polynomial form and their link with some exponential sums and Dickson Polynomials. *IEEE Transactions on Information Theory*, Vol 57, No 9, pages 5996-6009, 2011.

[10] S. Mesnager, Semi-bent functions from Dillon and Niho exponents, Kloosterman sums and Dickson polynomials. *IEEE Transactions on Information Theory*, Vol 57, No 11, pages 7443-7458, 2011.

[11] I.M. Rubio, G.L. Mullen, C. Corrada and F. N. Castro, Dickson permutation polynomials that decompose in cycles of the same length, *Finite fields and applications*, 229239, *Contemp. Math.*, 461, Amer. Math. Soc., Providence, RI, 2008.

[12] A. Sakzad, M.-R. Sadeghi and D. Panario, Cycle structure of permutation functions over finite fields and their applications, *Advances in Math. Communications* 6, 2012.

[13] G. Wu, N. Li, T. Helleseth and Y. Zhang, Some classes of monomial complete permutation polynomials over finite fields of characteristic two, *Finite Fields and Their Applications*, Volume 28, July 2014, Pages 148-165.

[14] A.M. Youssef, S.E. Tavares and H.M. Heys, A new class of substitution-permutation networks, Proceedings of *selected Areas in Cryptography, SAC-96*, pp 132-147.

Constructing elliptic curves and curves of genus 2 over finite fields

Kirsten Eisenträger

The Pennsylvania State University, University Park, PA, USA

eisentra@math.psu.edu

1. Introduction

In cryptography, the security of discrete-log-based systems depends on the the largest prime factor of the group order. Groups of points on elliptic curves and Jacobians of hyperelliptic curves of low genus can be used in these systems. Hence it is desirable to be able to construct curves over finite fields such that the resulting group order is prime. The problem of constructing elliptic curves with a given number of points has been studied extensively. The standard approach is to compute the Hilbert class polynomial for a quadratic imaginary number field. The running time of the best known algorithms is $\tilde{O}(|D|)$, where D is the discriminant of the quadratic imaginary field. In this paper we present the Chinese Remainder Theorem (CRT) algorithm by Agashe-Lauter-Venkatesan [ALV04] for constructing elliptic curves with a prescribed number of points and describe the improvements to it given in [BBEL08].

The constructions for elliptic curves can be generalized to curves of genus 2. Almost all approaches for constructing genus 2 curves rely on computing the Igusa class polynomials of quartic CM fields. In this paper we present the CRT algorithm for generating genus 2 curves defined over finite fields with a prescribed number of points on their Jacobian given in [EL10]. The algorithm in [EL10] first determines the Igusa class polynomials modulo p for certain small primes p. Then the Igusa class polynomials are computed using the Chinese Remainder Theorem and a bound on the denominators of the coefficients. We will also describe some improvements to this algorithm that were given in [BGL11] and [LR13]. As for elliptic curves, the CRT method for genus 2 curves is one of several known methods for constructing curves and we will briefly describe some of the other approaches as well.

2. Generating elliptic curves with a prescribed number of points

In this section we give an algorithm for the following problem (under certain conditions): Given a prime number ℓ and a positive integer N, construct an elliptic curve E_0 defined over $\mathbb{F}_\ell$ such that $\#E_0(\mathbb{F}_\ell) = N$.

2.1. *Properties of elliptic curves over finite fields*

Let ℓ be a prime and let E be an elliptic curve defined over the finite field $\mathbb{F}_\ell$ with ℓ elements. Denote by $\operatorname{End}(E)$ its ring of endomorphisms that are defined over $\overline{\mathbb{F}}_\ell$.

For any such curve E, the number of points on the curve that are defined over $\mathbb{F}_\ell$ lie in the Hasse-Weil interval:

$$\#E(\mathbb{F}_\ell) \in [\ell + 1 - 2\sqrt{\ell}, \ell + 1 + 2\sqrt{\ell}].$$

The curve E has a special endomorphism π, called the *Frobenius endomorphism*, which is given by $\pi : (x, y) \mapsto (x^\ell, y^\ell)$. Its characteristic polynomial is $P(X) = X^2 - tX + \ell$; t is referred to as the trace of Frobenius. It is customary to associate the Frobenius endomorphism with a root of its characteristic polynomial. So since π satisfies the above quadratic equation, we identify it with an element of $\mathbb{Q}(\sqrt{D})$ where $D = t^2 - 4\ell$. When the trace of π is nonzero modulo ℓ, we call E ordinary and in this case $\operatorname{End}(E)$ is an order in $\mathbb{Q}(\sqrt{D})$ and $D < 0$. The number of points on E over $\mathbb{F}_\ell$ is determined by the trace of Frobenius, $\#E(\mathbb{F}_\ell) = \ell + 1 - t$.

Possible Group Orders. Let E be an elliptic curve E defined over $\mathbb{F}_\ell$ whose endomorphism ring is an order in a quadratic imaginary number field $K = \mathbb{Q}(\sqrt{D})$, with K not equal to $\mathbb{Q}(i)$ or $\mathbb{Q}(\zeta_3)$. Then there are only two possibilities for $\#E(\mathbb{F}_\ell)$, which can be seen as follows: as before, let $P(x) = X^2 - tX + \ell$ be the characteristic polynomial of the Frobenius endomorphism of E. Let $\pi, \overline{\pi}$ be the zeros of $P(x)$. Then $\pi\overline{\pi} = \ell$, so π and $\overline{\pi}$ have norm ℓ, and (π) and $(\overline{\pi})$ are the only ideals of $\mathcal{O}_K$ of norm ℓ. Hence the only elements of norm ℓ in $\mathcal{O}_K$ are $\pi, \overline{\pi}, -\pi, -\overline{\pi}$. The characteristic polynomial of a curve E' whose Frobenius endomorphism corresponds to $-\pi$ or $-\overline{\pi}$ is then $X^2 + tX + \ell$, and for such a curve E' we have $\#E'(\mathbb{F}_\ell) = \ell + 1 + t$. So the only possible group orders for a curve whose endomorphism ring is an order in K are $\ell + 1 - t$ or $\ell + 1 + t$. The curves E and E' are quadratic twists of each other.

2.2. *Complex Multiplication (CM) method*

Now we can give an outline of an algorithm for our problem:

Problem 1. *Given a prime $\ell > 3$ and a positive integer N with*

$$N \in [\ell + 1 - 2\sqrt{\ell}, \ell + 1 + 2\sqrt{\ell}],$$

construct an elliptic curve $E_0/\mathbb{F}_\ell$ with $\#E_0(\mathbb{F}_\ell) = N$.

Here is a sketch of the standard algorithm for finding such a curve E_0, which is called the Complex Multiplication (CM) method:

Algorithm 1. CM method (see [AM93])
Input: N, ℓ as above.

(1) Let π be the Frobenius endomorphism of the desired curve E_0, $t = \ell + 1 - N$ its trace, and $D = t^2 - 4\ell$. Assume $t \not\equiv 0$ modulo ℓ. Let $K := \mathbb{Q}(\sqrt{D})$.
(2) Compute the *Hilbert class polynomial* $H_D(X)$ of K.
(3) Let j_0 be a root of H_D modulo ℓ. Take the curve $E_0/\mathbb{F}_\ell$ with j-*invariant* $j(E_0) = j_0$.
(4) Compute $\#E_0(\mathbb{F}_\ell)$. If $\#E_0(\mathbb{F}_\ell) \neq N$, then a quadratic twist of E_0 has N points over $\mathbb{F}_\ell$ and is the desired curve.

Remark 1. In step (3) of the above algorithm we reconstruct an elliptic curve E_0 from its j-invariant j_0. The j-invariant of an elliptic curve of the form $E : y^2 = x^3 + Ax + b$ is $j(E) = \frac{256 \cdot 27 \cdot A^3}{4A^3 + 27B^2}$. Two elliptic curves E, E' defined over a field F have the same j-invariant if and only if they are isomorphic over the algebraic closure of F. Assuming $j_0 \neq 0, 1728$ and $\ell \neq 2, 3$, an elliptic curve over $\mathbb{F}_\ell$ with j-invariant j_0 is given by the Weierstrass equation $E_0 : y^2 = x^3 + 3kx + 2k$, where $k = \frac{j_0}{1728 - j_0}$.

The endomorphism ring of E_0 is an order in K. By the discussion in the previous section that means that E_0 has either $\ell + 1 - t$ or $\ell + 1 + t$ points, so either E_0 or a twist of E_0 is the desired curve.

The above algorithm shows that constructing elliptic curves with a prescribed number of points reduces to computing the Hilbert class polynomial of the imaginary quadratic number field $\mathbb{Q}(\sqrt{D})$. We will now define the Hilbert class polynomial and give an algorithm for computing it.

2.3. Complex multiplication and Hilbert class polynomials

Let K be a field of characteristic zero, E an elliptic curve over K, and denote by $\mathrm{End}(E)$ its ring of endomorphisms defined over $\overline{K}$. We need the notion of complex multiplication to define the Hilbert class polynomial $H_D(X)$ of K.

For most elliptic curves in characteristic zero we have $\mathrm{End}(E) = \mathbb{Z}$, and the only endomorphisms are the "multiplication-by-m" maps. The CM case is the case where $\mathrm{End}(E)$ is strictly larger than $\mathbb{Z}$.

Definition 1. The elliptic curve E has *complex multiplication (CM) by* $\mathcal{O}$ if $\mathrm{End}(E)$ is an order $\mathcal{O}$ in $\mathbb{Q}(\sqrt{D})$ with $D < 0$.

Example 1. The elliptic curve $E : y^2 = x^3 - x$ defined over $\mathbb{C}$ has endomorphism ring $\mathrm{End}(E)$ which is strictly larger than $\mathbb{Z}$ since it contains the map

$$\phi : (x, y) \mapsto (-x, iy).$$

It is easy to check that $\phi \circ \phi$ is $[-1] : P \mapsto -P$. So E has CM by $\mathbb{Z}[i]$.

Now we are ready to define the Hilbert class polynomial.

Definition 2. Let $K = \mathbb{Q}(\sqrt{D})$ with $D < 0$, and let $\mathcal{O}_K$ be its ring of integers. The Hilbert class polynomial $H_D(X)$ of K is the polynomial whose roots are exactly the j-invariants of elliptic curves over $\mathbb{C}$ with CM by $\mathcal{O}_K$.

$$H_D(X) = \prod_{j(E) \in \mathrm{Ell}(D)} (X - j(E)).$$

Here $\mathrm{Ell}(D) = \{j(E) : \mathrm{End}(E) \cong \mathcal{O}_K\}$ is the finite set of j-invariants of elliptic curves over $\mathbb{C}$ with CM by $\mathcal{O}_K$.

From the theory of complex multiplication we have the following theorem:

Theorem 1.*[Sil94, Theorem II.6.1] The polynomial $H_D(X)$ has integer coefficients.*

The approach for finding the desired curve E_0 will be to compute H_D and then reduce H_D modulo ℓ to find curves over $\mathbb{F}_\ell$ whose endomorphism ring is $\mathcal{O}_K$.

2.4. *Algorithms for computing Hilbert class polynomials*

We have seen that to construct elliptic curves with a given number of points, it is enough to compute the Hilbert class polynomial H_D. This problem has been studied extensively, and there are several methods known for computing Hilbert class polynomials: the complex analytic method by Atkin and Morain [AM93] computes $H_D(X)$ by listing all binary quadratic forms corresponding to elliptic curves with CM by $\mathcal{O}_K$. It then evaluates the j-functions as a floating-point integer with sufficient precision, takes the product, and rounds the coefficients to the nearest integer. The p-adic approach (see Couveignes-Henocq [CH02], and Bröker [Brö08]) uses p-adic lifting to approximate the roots and recognize the coefficients of the polynomial. The Chinese Remainder Theorem method by Agashe-Lauter-Venkatesan [ALV04] and the modified CRT method by Belding-Bröker-Enge-Lauter [BBEL08] compute $H_D(X)$ modulo p for sufficiently many small primes p and then use the Chinese Remainder Theorem to compute the integer coefficients of the polynomial $H_D(X)$. The running time for all of the algorithms is exponential in $\log(|D|)$, and the complexity analysis and comparison in [BBEL08, Section 5.4] show that the fastest version of the CRT algorithm has expected run time $\tilde{O}(|D|)$, and that the run time of the complex analytic algorithm and the p-adic method is also $\tilde{O}(|D|)$.

We now give an overview of the CRT method described in [ALV04] and the improvements made in [BBEL08]. For simplicity of notation we only deal with the case where D is the discriminant of the maximal order in $\mathbb{Q}(\sqrt{D})$.

Algorithm 2. CRT algorithm for computing $H_D(X)$[ALV04, Section 3.1]

Input: $D < 0$, discriminant of the maximal order in the quadratic imaginary number field $\mathbb{Q}(\sqrt{D})$.

(1) Compute an upper bound B on the coefficients of $H_D(X)$.
(2) Compute $H_D(X)$ modulo small primes $p_1, \ldots, p_n$ which split completely in the Hilbert class field of K and such that $\prod_i p_i > B$.
(3) Use the Chinese Remainder Theorem to find the coefficients of $H_D(X)$.

The point of the algorithm is that we can compute $H_D(X)$ mod p for suitably chosen small primes p directly without knowing $H_D(X)$. To compute $H_D(X)$ mod p we use the following proposition.

Proposition 1.[ALV04, Proposition 4.1] *Let $D < 0$ be the discriminant of the maximal order $\mathcal{O}_K$ of $K := \mathbb{Q}(\sqrt{D})$. Let p be a rational prime such* *that $4p = t^2 - Du^2$ for some integers u and t. Let $\mathrm{Ell}'(D)$ be the set of $\overline{\mathbb{F}}_p$ isomorphism classes of elliptic curves over $\mathbb{F}_p$ with $\mathrm{End}(E) = \mathcal{O}_K$. Then*

$$H_D(X) \pmod{p} = \prod_{[E'] \in \mathrm{Ell}'(D)} (X - j(E')).$$

In particular, $H_D(X)$ mod p splits completely into linear factors.

When D is the discriminant of the maximal order and p is a prime of the form $4p = t^2 - D$ we can compute $\mathrm{Ell}'(D)$ as follows:

Proposition 2. *[ALV04, Proposition 4.2] Suppose p is a prime and $t \neq 0$ is an integer such that $4p = t^2 - D$. Let E' be an elliptic curve over $\mathbb{F}_p$. Then $[E'] \in \mathrm{Ell}'(D)$ if and only if $\#E'(\mathbb{F}_p)$ is either $p + 1 - t$ or $p + 1 + t$.*

This leads us to the following algorithm for computing all factors of $H_D(X)$ mod p:

Algorithm 3.

(1) For each $j \in \mathbb{F}_p$, create an elliptic curve E' over $\mathbb{F}_p$ with $j(E') = j$.
(2) If $\#E'(\mathbb{F}_p)$ is either $p + 1 - t$ or $p + 1 + t$, then $H_D(X) \pmod{p}$ has a factor of $(X - j(E'))$.

Remark 2. The first step in the above algorithm, creating an elliptic curve with a given j-invariant, is easy, as described in Remark 1. For the second step we can use the well known point counting algorithm by Schoof.

2.5. *Improvements*

The algorithm in [ALV04] for computing H_D mod p required an exhaustive search through all possible j-invariants to find all curves E with $\mathrm{End}(E) = \mathcal{O}_K$. The improvement by Belding-Bröker-Enge-Lauter [BBEL08] finds one curve $E/\mathbb{F}_p$ with $\mathrm{End}(E) = \mathcal{O}_K$ and then obtains the others via the action of the class group $\mathrm{Cl}(\mathcal{O}_K)$ on the set $\mathrm{Ell}'(D)$.

To compute the action of the class group, they compute ℓ_i-isogenous curves, where the ℓ_i are the norms of ideals whose classes generate the class group $\mathrm{Cl}(\mathcal{O}_K)$. One way this can be done is by computing the modular polynomials $\Phi_{\ell_i}(X, Y)$, and using the fact that if $j_0 \in \mathbb{F}_p$ is the j-invariant of some curve E with endomorphism ring $\mathcal{O}$, then the roots in $\mathbb{F}_p$ of $\Phi_{\ell_i}(X, j_0)$ correspond to curves, ℓ_i-isogenous to E, with endomorphism ring $\mathcal{O}$ (see [BBEL08, Section 3.1]). Under GRH, the algorithm in [BBEL08] has expected runtime $O(|D|(\log|D|)^{7+o(1)})$.

3. Generating genus 2 curves with a prescribed number of points

We now generalize the elliptic curve construction to generate genus 2 curves with a given number of points. The points on a curve of genus 2 do not form a group, so we use as group elements the points on the *Jacobian* of the curve.

3.1. *Background and definitions*

Let $p > 2$ be a prime, and let C be a curve of genus 2 defined over the finite field $\mathbb{F}_p$ with p elements. A genus 2 curve is hyperelliptic, so C has a model of the form $y^2 = f(x)$ with $f(x) \in \mathbb{F}_p[x]$ a polynomial of degree 5 or 6 without repeated roots (in $\overline{\mathbb{F}}_p$). By a point on C we mean a pair (x, y) with coefficients in $\overline{\mathbb{F}}_p$ that satisfies $y^2 = f(x)$, or a "point at infinity". (If f has degree 5, there is one point at infinity, if f has degree 6 there are two points at infinity.) A *divisor* D of C is a finite formal sum of the form $D = \sum m_i P_i$, where the m_i are integers and the P_i are points of C; the degree of D is $\sum m_i$. Let h be a rational function on C. We associate with h its divisor $(h) = \sum m_i P_i$, where the P_i are the zeros and poles of h (in $\overline{\mathbb{F}}_p$) with multiplicities m_i. A divisor of such a nonzero function (h) is called principal, and it is well known that a principal divisor has degree 0. The divisors of degree 0 form a group $D_0(C)$, and the principal divisors form a subgroup $\mathrm{Princ}(C)$ of $D_0(C)$. The *Jacobian* of C is the quotient group

$$\mathrm{Jac}(C) = D_0(C)/\mathrm{Princ}(C).$$

Example 2. Let $f(x)$ be a monic polynomial of degree 5 over a field K of cardinality at least 5. Let $\lambda_0, \lambda_1, \lambda_2$ be distinct elements of K, not equal to 0 or 1. Then

$$C : y^2 = x(x - 1)(x - \lambda_0)(x - \lambda_1)(x - \lambda_2)$$

is a curve of genus 2 that has one point at infinity ∞.

It follows from the Riemann-Roch theorem that elements of the Jacobian of a genus 2 curve C can be represented by divisors of the form

$$D = \sum_{i=1}^{r} P_i - r \cdot \infty \text{ with } P_i \in C \text{ and } r \leq 2.$$

Let C be a curve of genus 2 over $\mathbb{F}_p$. Let $P(X)$ denote the characteristic polynomial of the Frobenius endomorphism π on $\mathrm{Jac}(C)$. The polynomial

$P(X)$ is monic of degree 4 and has integer coefficients. Its roots $a_1, \ldots, a_4$ have absolute value $p^{1/2}$.

Again, the characteristic polynomial $P(X)$ determines the number of points on C. It also determines the number of points on its Jacobian $\mathrm{Jac}(C)$. For any $m \geq 1$, we have $\#C(\mathbb{F}_{p^m}) = 1 - \sum_{i=1}^{4} a_i^m + p^m$, $\# \mathrm{Jac}(C)(\mathbb{F}_{p^m}) = \prod_{i=1}^{4}(1 - a_i^m)$, and

$$\# \mathrm{Jac}(C)(\mathbb{F}_p) = \frac{1}{2}\#C(\mathbb{F}_{p^2}) + \frac{1}{2}\#C(\mathbb{F}_p)^2 - p.$$

3.2. *Complex multiplication of genus 2 curves and quartic CM fields*

As in the elliptic curve case, we will work with curves whose Jacobians have endomorphism rings that are orders in CM fields. For elliptic curves the associated CM fields were quadratic imaginary fields. For genus 2 curves the associated fields have degree 4 over $\mathbb{Q}$.

Definition 3. A curve C of genus 2 has CM if the endomorphism ring of its Jacobian $\mathrm{End}(\mathrm{Jac}(C))$ is an order in a quartic CM field K, i.e. if $K = K_0(\sqrt{d})$ with K_0 real quadratic and $d \in K_0$ totally negative. Here $d \in K_0$ is totally negative if for both embeddings σ_1, σ_2 of K_0 into $\mathbb{R}$, we have $\sigma_i(d) < 0$ $(i = 1, 2)$.

3.3. *Outline of algorithm*

Our algorithm solves the following problem under certain conditions.

Problem: Given (ℓ, N_1, N_2) with ℓ prime, and N_1, N_2 positive integers, find a genus 2 curve C over the prime field $\mathbb{F}_\ell$ such that $\#C(\mathbb{F}_\ell) = N_1$ and $\#C(\mathbb{F}_{\ell^2}) = N_2$.

Approach: Given a triple (ℓ, N_1, N_2) we generate a curve as follows.

As in the case of the elliptic curves, the first step is to find the quartic CM field K such that $\mathrm{End}(\mathrm{Jac}(C)) \subseteq K$. To do this we find quartic polynomial satisfied by Frobenius. We write

$$N_1 = \ell + 1 - s_1, \quad N_2 = \ell^2 + 1 + 2s_2 - s_1^2.$$

and solve for s_1, s_2. The quartic CM field K is then generated over $\mathbb{Q}$ by a root of $X^4 - s_1 X^3 + s_2 X^2 - \ell s_1 X + \ell^2$.

Restrictions: We assume that $\gcd(s_2, \ell) = 1$, which forces the Jacobian of C to be ordinary. We also restrict to *primitive* quartic CM fields. A quartic CM field is not primitive iff it is Galois over $\mathbb{Q}$ and biquadratic. We further assume that K does not contain a cyclotomic field.

This leads to the following approach for constructing genus 2 curves:

Algorithm 4. Constructing genus 2 curves (see [EL10, Section 3])
Given (ℓ, N_1, N_2) with restrictions as above:

(1) Compute the quartic CM field K.
(2) Compute the Igusa class polynomials H_1, H_2, H_3 of K.
(3) From a triple of roots modulo ℓ of H_1, H_2, H_3, construct a genus 2 curve over $\mathbb{F}_\ell$ using the algorithms of Mestre [Mes91] and Cardona-Quer [CQ05].
(4) Test whether the curve generated has the correct number of points on the Jacobian. A curve C with $\#C(\mathbb{F}_\ell) = N_1$ and $\#C(\mathbb{F}_{\ell^2}) = N_2$ will have $\#J(C)(\mathbb{F}_\ell) = N = (N_1^2 + N_2)/2 - \ell$. If the curve does not have the required number of points on the Jacobian, a twist of the curve may be used. In the case where 4 group orders are possible for the pair (ℓ, K), a different triple of invariants may be tried until the desired group order is obtained.

Remark 3. Let C be a curve of genus 2 defined over $\mathbb{F}_\ell$ with CM by the maximal order of the primitive quartic CM field K, and assume that $K \neq \mathbb{Q}(\zeta_5)$. In [EL10, Proposition 4] it is shown that there 2 or 4 possibilities for the group order $\# \mathrm{Jac}(C)(\mathbb{F}_\ell)$. If there are four possible group orders and the constructed curve C does not have the right number of points on its Jacobian, then taking a twist of C may not be sufficient to get the desired curve and a different triple of invariants may be necessary.

Next we will define the Igusa invariants and the Igusa class polynomials of a genus 2 curve. Then we will describe an algorithm for computing them.

3.4. *Igusa invariants and Igusa class polynomials*

Using classical invariant theory over a field of characteristic zero, Clebsch defined a triple of invariants of a binary sextic f defining a genus 2 curve $y^2 = f(x)$. Bolza showed how those invariants could also be expressed in terms of theta functions on the period matrix associated to the Jacobian variety and its canonical polarization over $\mathbb{C}$. Igusa showed how these invariants could be extended to work in arbitrary characteristic [Igu67, p. 848], and so the invariants are often referred to as Igusa or Clebsch-Bolza-Igusa invariants.

In [Igu60, p. 620] Igusa defined invariants I_2, I_4, I_6, I_{10} and gave a bijection between isomorphism classes of genus 2 curves over K and points $(I_2 : I_4 : I_6 : I_{10})$ in weighted projective space with $I_{10} \neq 0$.

The ring of rational functions of the coarse moduli space for hyperelliptic curves of genus 2 is generated by the absolute Igusa invariants, which can be defined as:

$$ i_1 := \frac{I_2^5}{I_{10}}, \quad i_2 := \frac{I_2^3 I_4}{I_{10}}, \quad i_3 := \frac{I_2^2 I_6}{I_{10}}. $$

(See [vW99, p. 313].) This choice of generators is not unique. The absolute Igusa invariants can be viewed as a generalization of the j-invariant for elliptic curves in the following sense: if i_1 is non-zero and the characteristic is not 2 or 3, these invariants agree for two genus 2 curves exactly when the two curves are isomorphic over an algebraically closed field. The moduli space of genus 2 curves is 3-dimensional and so three invariants are needed to specify a curve up to isomorphism over an algebraically closed field.

Given a primitive quartic CM field K, let $\mathcal{A}$ be a system of representatives for the set of isomorphism classes of principally polarized abelian varieties over $\mathbb{C}$ having complex multiplication by $\mathcal{O}_K$. For each abelian variety $A \in \mathcal{A}$, let $(i_1(A), i_2(A), i_3(A))$ be the absolute Igusa invariants of A. Then the *Igusa class polynomials* H_j, for $j = 1, 2, 3$, are defined to be

$$ H_j := \prod_{A \in \mathcal{A}} (X - i_j(A)). $$

Remark 4. The Igusa class polynomials are polynomials with coefficients in $\mathbb{Q}$. While the Hilbert class polynomial has integer coefficients, the Igusa class polynomials can have denominators. In order to run the algorithm in Theorem 2 below for computing them, we need to know which primes occur in the denominator and to which power. In [GL07], Goren and Lauter proved a bound on the primes which appear in the factorization of the denominators. In [GL12], they also gave a bound on the powers to which those primes appear. This bound, however, is not sharp. Independently, Bruinier and Yang formulated a conjecture that would give a way to compute the denominators of the Igusa class polynomials precisely [BY06]. Yang [Yan10, Yan13] proved this conjecture under certain assumptions on the quartic number field K, and Lauter-Viray [LV15] proved the conjecture for all primitive quartic CM fields. Using [LV15] we can compute a bound on the denominators of the coefficients of H_1, H_2, H_3.

3.5. *Algorithm for computing Igusa class polynomials*

We have seen that in order to construct genus 2 curves with a prescribed number of points on the Jacobian, it suffices to compute the Igusa class polynomials of the quartic CM field K. As in the elliptic curve case, there are several known approaches for doing this. The complex analytic approach, which is the analogue for genus two curves of the Atkin-Morain CM method for elliptic curves, computes the Igusa class polynomials of a quartic CM field K by evaluating the modular invariants of all the abelian varieties of dimension 2 with CM by $\mathcal{O}_K$. The CM algorithm was first implemented by Spallek[Spa94], van Wamelen [vW99], and Weng [Wen03].

In this paper we present the CRT method for computing the Igusa class polynomials of a primitive quartic CM field K given in [EL10]. This approach is a generalization of the CRT method for elliptic curves [ALV04]. It first computes the Igusa class polynomials modulo p for sufficiently many small primes p. The Igusa class polynomials are then found using the Chinese Remainder Theorem and a bound on the denominators of the coefficients.

Theorem 2.[*EL10, Theorem 1*] *Given a primitive quartic CM field K, with totally real subfield K_0, the following algorithm finds the Igusa class polynomials of K:*

Algorithm 5. Computing Igusa class polynomials H_1, H_2, H_3 for K

(1) Produce a collection S of small rational primes $p \in S$ satisfying:

    (i) p splits completely in K and splits completely into principal ideals in K^*, the reflex of K.

    (ii) Let B be the set of all primes of bad reduction for the genus 2 curves with CM by K. Then $S \cap B = \emptyset$.

    (iii) $\prod_{p \in S} p > c$, where c is a constant related to the size of the coefficients and denominators of the Igusa class polynomials.

(2) Form the class polynomials H_1, H_2, H_3 modulo p for each $p \in S$. Let $H_{j,p}(X) := H_j(X) \mod p$. Then $H_{j,p}(X) = \prod_{C \in T_p}(X - i_j(C))$, where T_p is the collection of $\overline{\mathbb{F}}_p$-isomorphism classes of genus 2 curves over $\mathbb{F}_p$ whose Jacobian has endomorphism ring isomorphic to $\mathcal{O}_K$.

(3) *Chinese Remainder Step.* Form $H_j(X)$ from $\{H_{j,p}\}_{p \in S}$ ($j = 1, 2, 3$).

To compute the Igusa class polynomials modulo p for a prime p

satisfying the conditions above we do the following (see [EL10, Section 5.3]):

Algorithm 6. Computing Igusa class polynomials modulo p

(1) Loop through all possible triples of Igusa invariants (i_1, i_2, i_3). For each triple, construct a curve C over $\mathbb{F}_p$ with $(i_1, i_2, i_3) = (i_1(C), i_2(C), i_3(C))$ using Mestre's algorithm.

(2) For each curve C constructed, check whether $\mathrm{End}(\mathrm{Jac}(C)) = \mathcal{O}_K$.

3.6. *Improvements for computing Igusa class polynomials*

The first improvement to the CRT method was given by Lauter and Freeman [FL08] who gave a more efficient algorithm to check whether $\mathrm{End}(\mathrm{Jac}(C)) = \mathcal{O}_K$. Checking this condition was needed in Step (2) of Algorithm 6 above for computing the Igusa class polynomials modulo p.

A further improvement to computing the Igusa class polynomials modulo p was given in [BGL11]: instead of looping through all possible triples of Igusa invariants and checking for each curve C whether $\mathrm{End}(\mathrm{Jac}(C)) = \mathcal{O}_K$, as in Algorithm 6 above, the approach by Bröker-Gruenewald-Lauter finds *one* curve whose Jacobian has endomorphism ring equal to $\mathcal{O}_K$ (a so called *maximal curve*). The other curves in the isogeny class are then found by using computable $(3, 3)$ isogenies.

Another improvement to this approach was proposed by Lauter and Robert in [LR13]. They showed that it was not necessary to find a maximal curve first. Instead they first find a curve C whose endomorphism ring is only an order in $\mathcal{O}_K$ and then give a probabilistic algorithm for "going up" to a maximal curve. Heuristically this improves the running time from p^3 per prime p to $p^{3/2}$ per prime p. Lauter-Robert give a mostly-heuristic analysis of their algorithm and state that at best their algorithm has quasiquadratic complexity in the discriminant of the quartic number field.

Acknowledgments   The author was partially supported by National Science Foundation grant DMS-1056703.

References

[ALV04]   Amod Agashe, Kristin Lauter, and Ramarathnam Venkatesan. Constructing elliptic curves with a known number of points over a prime field. In *High primes and misdemeanours: lectures in honour of the*

60th birthday of Hugh Cowie Williams, volume 41 of *Fields Inst. Commun.*, pages 1–17. Amer. Math. Soc., Providence, RI, 2004.

[AM93]   A. O. L. Atkin and F. Morain. Elliptic curves and primality proving. *Math. Comp.*, 61(203):29–68, 1993.

[BBEL08] Juliana Belding, Reinier Bröker, Andreas Enge, and Kristin Lauter. Computing Hilbert class polynomials. In *Algorithmic number theory*, volume 5011 of *Lecture Notes in Comput. Sci.*, pages 282–295. Springer, Berlin, 2008.

[BGL11]   Reinier Bröker, David Gruenewald, and Kristin Lauter. Explicit CM theory for level 2-structures on abelian surfaces. *Algebra Number Theory*, 5(4):495–528, 2011.

[Brö08]   Reinier Bröker. A *p*-adic algorithm to compute the Hilbert class polynomial. *Math. Comp.*, 77(264):2417–2435, 2008.

[BY06]   Jan Hendrik Bruinier and Tonghai Yang. CM-values of Hilbert modular functions. *Invent. Math.*, 163(2):229–288, 2006.

[CH02]   Jean-Marc Couveignes and Thierry Henocq. Action of modular correspondences around CM points. In *Algorithmic number theory (Sydney, 2002)*, volume 2369 of *Lecture Notes in Comput. Sci.*, pages 234–243. Springer, Berlin, 2002.

[CQ05]   Gabriel Cardona and Jordi Quer. Field of moduli and field of definition for curves of genus 2. In *Computational aspects of algebraic curves*, volume 13 of *Lecture Notes Ser. Comput.*, pages 71–83. World Sci. Publ., Hackensack, NJ, 2005.

[EL10]   Kirsten Eisenträger and Kristin Lauter. A CRT algorithm for constructing genus 2 curves over finite fields. In *Arithmetics, geometry, and coding theory (AGCT 2005)*, volume 21 of *Sémin. Congr.*, pages 161–176. Soc. Math. France, Paris, 2010.

[FL08]   David Freeman and Kristin Lauter. Computing endomorphism rings of Jacobians of genus 2 curves over finite fields. In *Algebraic geometry and its applications*, volume 5 of *Ser. Number Theory Appl.*, pages 29–66. World Sci. Publ., Hackensack, NJ, 2008.

[GL07]   Eyal Z. Goren and Kristin E. Lauter. Class invariants for quartic CM fields. *Ann. Inst. Fourier (Grenoble)*, 57(2):457–480, 2007.

[GL12]   Eyal Z. Goren and Kristin E. Lauter. Genus 2 curves with complex multiplication. *Int. Math. Res. Not. IMRN*, (5):1068–1142, 2012.

[Igu60]   Jun-ichi Igusa. Arithmetic variety of moduli for genus two. *Ann. of Math. (2)*, 72:612–649, 1960.

[Igu67]   Jun-ichi Igusa. Modular forms and projective invariants. *Amer. J. Math.*, 89:817–855, 1967.

[LR13]   Kristin E. Lauter and Damien Robert. Improved CRT algorithm for class polynomials in genus 2. In *ANTS X—Proceedings of the Tenth Algorithmic Number Theory Symposium*, volume 1 of *Open Book Ser.*, pages 437–461. Math. Sci. Publ., Berkeley, CA, 2013.

[LV15]   Kristin Lauter and Bianca Viray. An arithmetic intersection formula for denominators of Igusa class polynomials. *Amer. J. Math.*, 137(2):497–533, 2015.

[Mes91]   Jean-François Mestre. Construction de courbes de genre 2 à partir de leurs modules. In *Effective methods in algebraic geometry (Castiglioncello, 1990)*, volume 94 of *Progr. Math.*, pages 313–334. Birkhäuser Boston, Boston, MA, 1991.

[Sil94]   Joseph H. Silverman. *Advanced topics in the arithmetic of elliptic curves*, volume 151 of *Graduate Texts in Mathematics*. Springer-Verlag, New York, 1994.

[Spa94]   Anne-Monika Spallek. Kurven vom Geschlecht 2 und ihre Anwendung in Public-Key-Kryptosystemen. Ph.D. Thesis. Universität Gesamthochschule Essen, 1994.

[vW99]   Paul van Wamelen. Examples of genus two CM curves defined over the rationals. *Math. Comp.*, 68(225):307–320, 1999.

[Wen03]   Annegret Weng. Constructing hyperelliptic curves of genus 2 suitable for cryptography. *Math. Comp.*, 72(241):435–458 (electronic), 2003.

[Yan10]   Tonghai Yang. An arithmetic intersection formula on Hilbert modular surfaces. *Amer. J. Math.*, 132(5):1275–1309, 2010.

[Yan13]   Tonghai Yang. Arithmetic intersection on a Hilbert modular surface and the Faltings height. *Asian J. Math.*, 17(2):335–381, 2013.

A family of plane curves with two or more Galois points in positive characteristic

Satoru Fukasawa

Department of Mathematical Sciences, Faculty of Science, Yamagata University, Kojirakawa-machi 1-4-12, Yamagata 990-8560, JAPAN

s.fukasawa@sci.kj.yamagata-u.ac.jp

1. Introduction

In this study, we present new examples of plane curves with two or more Galois points as a family, and we describe the number of Galois points for these curves using finite fields.

Let K be an algebraically closed field of characteristic $p \geq 0$, and let $C \subset \mathbb{P}^2_K$ be an irreducible plane curve of degree $d \geq 3$. H. Yoshihara introduced the notion of the *Galois point* in 1996 (see [9, 14]). If the function field extension $K(C)/K(\mathbb{P}^1)$ induced by the projection $\pi_P : C \dashrightarrow \mathbb{P}^1$ from a point $P \in \mathbb{P}^2$ is Galois, then the point P is said to be Galois with respect to C. When a Galois point P is contained in $\mathbb{P}^2 \setminus C$, we call P an outer Galois point. The number of outer Galois points for C is denoted by $\delta'(C)$, which would be interesting to determine. For example, there are applications of the distribution of Galois points to finite geometry (see [6]). When C is smooth, $\delta'(C)$ is completely determined (see [4, 8, 9, 14]); however, determining $\delta'(C)$ for (singular) curves C in general is difficult. For example, curves C satisfying $\delta'(C) > 1$ are very rare [15].

In this study, we present new examples. Let $p > 0$, $e \geq 1$, and:

$$g_1(x) := x^{p^e} + \alpha_{e-1}x^{p^{e-1}} + \cdots + \alpha_1 x^p + \alpha_0 x,$$
$$g_2(y) := y^{p^e} + \beta_{e-1}y^{p^{e-1}} + \cdots + \beta_1 y^p + \beta_0 y,$$

where $\alpha_{e-1}, \ldots, \alpha_0, \beta_{e-1}, \ldots, \beta_0 \in K$ and $\alpha_0\beta_0 \neq 0$. Assume that $\ell \geq 2$ is an integer such that ℓ is not divisible by p, ℓ divides $p^e - 1$, and ℓ divides $p^i - 1$, if $\alpha_i \neq 0$ or $\beta_i \neq 0$. We consider the curve $C \subset \mathbb{P}^2$ of degree $p^e \ell$ (the projective closure of the affine curve) defined by:

$$g_1(x)^\ell + \lambda g_2(y)^\ell + \mu = 0, \tag{I}$$

where $\lambda, \mu \in K \setminus \{0\}$. Our primary theorem describes the number of Galois points in terms of finite fields, as follows.

Theorem 1. *Let the characteristic* $p > 0$, $q = p^e$, $C \subset \mathbb{P}^2$ *be the plane curve given by equation* (I), *and let* $\alpha \in K$ *satisfy* $\alpha^{q\ell} + \lambda = 0$.

    (a) *The curve* C *is irreducible and has exactly* ℓ *singular points.*

    (b) *The genus of the smooth model satisfies* $p_g(C) \geq \frac{q\ell(q(\ell-1)-2)}{2} + 1$. *Furthermore, the equality holds if and only if* $g_1(\alpha X) - \alpha^q g_2(X)$ *is zero as a polynomial.*

    (c) $\delta'(C) \geq 2$.

    (d) *If* ℓ *is odd,* $\ell - 1$ *is not divisible by* p, *and* $g_1(\alpha X) - \alpha^q g_2(X) = 0$, *then* $\delta'(C) = 2$.

    (e) *Let* $\ell = 2$, $g_1(X) = g_2(X)$, $\mathbb{F}_{q_0} := \mathbb{F}_q \cap (\bigcap_{\{i>0:\alpha_i \neq 0\}} \mathbb{F}_{p^i})$, *and* $\lambda \in \mathbb{F}_{q_0}$. *If* $\alpha \in \mathbb{F}_{q_0}$ *(resp.,* $\alpha \notin \mathbb{F}_{q_0}$*), then* $\delta'(C) \geq q_0 - 1$ *(resp.,* $\delta'(C) \geq q_0 + 1$*). Furthermore, if* $q_0 = q$, *then the equality holds.*

Assertion (e) is proved by using projective geometry over a finite field $\mathbb{F}_q$. It is interesting that the set of Galois points coincides with the set of $\mathbb{F}_q$-rational points on the line $Z = 0$ but not on C if $q_0 = q$ (see Lemmas 2, 5, and 6).

As another aspect, our curves have non-trivial automorphism groups. For example, a well-known curve studied by Subrao belongs to our family.

Example 1. Taking $p > 2$, $\ell = 2$, $g_1(x) = x^q - x$, $g_2(y) = y^q - y$ and $\lambda = -1$ and introducing the variable change $X = x + y$ and $Y = x - y$, we obtain the equation:

$$(X^q - X)(Y^q - Y) + \mu = 0$$

studied by D. Subrao [12],[7, Example 11.89]. This is known to be an example of an irreducible (ordinary) curve whose automorphism group exceeds the Hurwitz bound under specific assumptions. If $q = p$, R. C. Valentini and M. L. Madan [13] showed that the automorphism group is a semidirect product of an elementary abelian p-group of order p^2 with a dihedral group of order $2(p - 1)$. If K is a non-Archimedean valued field and $|\mu| < 1$, G. Cornelissen, F. Kato, and A. Kontogeorgis [1] determined the automorphism group.

For Galois points in $p = 2$, see [5].

Remark 1. The curve with equation (I) and $\ell = 1$ was studied in [3].

2. Preliminaries

We introduce a system $(X : Y : Z)$ of homogeneous coordinates on $\mathbb{P}^2$ with local coordinates $x = X/Z$, $y = Y/Z$ for the affine piece $Z \neq 0$. For an irreducible plane curve $C \subset \mathbb{P}^2$, the singular locus of C is denoted by $\mathrm{Sing}(C)$, the normalization of C by $\pi : \hat{C} \to C$, and the composition map $\pi_P \circ \pi : \hat{C} \to \mathbb{P}^1$ by $\hat{\pi}_P$. For a point $R \in C \setminus \mathrm{Sing}(C)$, $T_R C \subset \mathbb{P}^2$ is the (projective) tangent line at R. For a projective line $L \subset \mathbb{P}^2$ and a point $R \in C \cap L$, $I_R(C, L)$ is the intersection multiplicity of C and L at R. The line passing through points P and R is denoted by $\overline{PR}$ when $P \neq R$. If $\hat{R} \in \hat{C}$, the ramification index of $\hat{\pi}_P$ at $\hat{R}$ is denoted by $e_{\hat{R}}$. If $R \in C \setminus \mathrm{Sing}(C)$ and $\pi(\hat{R}) = R$, then we use the same symbol e_R for $e_{\hat{R}}$, by abuse of terminology. We note the following elementary fact.

Fact 0.1. Let $P \in \mathbb{P}^2 \setminus C$, $\hat{R} \in \hat{C}$, and $\pi(\hat{R}) = R$. Let $h = 0$ be a local equation for the line $\overline{PR}$ in a neighborhood of R. For π_P,

$$e_{\hat{R}} = \mathrm{ord}_{\hat{R}}(\pi^* h).$$

In particular, if R is smooth, then $e_R = I_R(C, \overline{PR})$.

The following fact is useful (see [11, III. 7.2]).

Fact 0.2. Let C, C' be smooth curves, $\theta : C \to C'$ a Galois covering of degree d, and $R, R' \in C$.

  (a) If $\theta(R) = \theta(R')$, then $e_R = e_{R'}$.
  (b) The index e_R divides the degree d.

Note that the polynomial $g(X) = X^{p^e} + \alpha_{e-1} X^{p^{e-1}} + \cdots + \alpha_1 X^p + \alpha_0 X \in K[X]$ has the property:

$$g(X + Y) = g(X) + g(Y) \quad (\in K[X, Y]).$$

3. Proof of assertions (a), (b), and (c) in Theorem 1

Let $p > 0$, $\ell \geq 2$ an integer not divisible by p, and $g(x) = x^{p^e} + \alpha_{e-1} x^{p^{e-1}} + \cdots + \alpha_1 x^p + \alpha_0 x$, where $\alpha_{e-1}, \ldots, \alpha_0 \in K$ and $\alpha_0 \neq 0$. We consider the plane curve $C \subset \mathbb{P}^2$ (the projective closure of the affine curve) defined by:

$$f(x, y) := g(x)^\ell + h(y) = 0,$$

where $h(y)$ is a polynomial of degree $p^e \ell$. Assume that ℓ divides $p^e - 1$, and $p^i - 1$, if $\alpha_i \neq 0$. For a constant $a \in K$ with $g(a) = 0$, the linear transformation defined by $(X : Y : Z) \mapsto (X + aZ : Y : Z)$ is denoted by σ_a. Let $q = p^e$, $P = (1 : 0 : 0)$, and $K_P := \{\sigma_a : g(a) = 0\}$. Then, K_P is a subgroup of $\mathrm{Aut}(\mathbb{P}^2)$ of order q. We take a primitive ℓ-th root ζ of unity. Let τ be the linear transformation given by $(X : Y : Z) \mapsto (\zeta X : Y : Z)$, and let $G := K_P\langle\tau\rangle$. Since $g(\zeta a) = \zeta g(a) = 0$ if $g(a) = 0$, $\langle\tau\rangle K_P = K_P\langle\tau\rangle$. This indicates that G is a subgroup of $\mathrm{Aut}(\mathbb{P}^2)$ of order $q\ell$. Let L be a line passing through P. L is given by $cY + dZ = 0$ for some $c, d \in K$. Since $\sigma_a(x : -d : c) = (x + ac : -d : c)$ and $\tau(x : -d : c) = (\zeta x : -d : c)$, $\xi(L) = L$ for any $\xi \in G$. Since $g(x + a) = g(x) + g(a) = g(x)$ for any a with $g(a) = 0$ and $g(\zeta x) = \zeta g(x)$, $\xi(C) = C$ for any $\xi \in G$. The group G will be the Galois group G_P at P, if it can be proved that C is irreducible, because G yields $q\ell$ automorphisms of C that commute with the projection $\pi_P : C \to \mathbb{P}^1$.

Proposition 1. *Let $C \subset \mathbb{P}^2$ be given by $f(x, y) = g(x)^\ell + h(y) = 0$. If $h(y)$ is a separable polynomial of degree $q\ell$, then C is irreducible. Furthermore, the genus of the smooth model satisfies $p_g(C) \geq \frac{q\ell(q(\ell-1)-2)}{2} + 1$.*

Proof. Let C_0 be an irreducible component of C with degree d_0 and the genus of the smooth model $p_g(C_0)$. Since $f_x = \ell g(x)^{\ell-1}$ and $f_y = h_y$, C is smooth in the affine plane $Z \neq 0$. In this affine plane, C_0 is smooth and does not intersect another irreducible component. Let $F(X, Y, Z) = Z^{q\ell}f(X/Z, Y/Z)$, $G(X, Z) = Z^q g(X/Z)$, and $H(Y, Z) = Z^{q\ell}h(Y/Z)$. For $\xi \in K$ with $h(\xi) = 0$, the line defined by $Y - \xi Z = 0$ is denoted by L_ξ. The scheme $C \cap L_\xi$ is given by $F(X, \xi Z, Z) = G(X, Z)^\ell = 0$. Since $G(X, Z)$ is a separable polynomial of degree q, the multiplicity of the set $C \cap L_\xi$ is ℓ at each point, and the set $C \cap L_\xi$ consists of exactly q points. Therefore, the multiplicity of the set $C_0 \cap L_\xi$ is ℓ at each point, and the set $C_0 \cap L_\xi$ consists of exactly d_0/ℓ points. We consider the projection $\pi_P : C_0 \to \mathbb{P}^1$ from P. Since each point of $C_0 \cap L_\xi$ is a tame ramification with index ℓ,

$$2p_g(C_0) - 2 \geq d_0(-2) + q\ell \times (d_0/\ell)(\ell - 1)$$

by the Riemann–Hurwitz formula. Then, $2p_g(C_0) - 2 \geq d_0(q(\ell-1)-2)$. On the other hand, by the genus formula, $2p_g(C_0) - 2 \leq (d_0 - 3)d_0$. Therefore, $d_0 \geq q(\ell-1) + 1$. We find that any irreducible component of C is of degree at least $(q\ell)/2 + 1$, since the degree of C is $q\ell$, C is irreducible. According to the Riemann–Hurwitz formula again, $p_g(C) \geq \frac{q\ell(q(\ell-1)-2)}{2} + 1$. $\square$

We consider the plane curves with equation (I) in the Introduction.

Points $(1 : 0 : 0)$ and $(0 : 1 : 0)$ are outer Galois, by the argument before Proposition 1, and $\delta'(C) \geq 2$.

Remark 2. The Galois group G at $(1 : 0 : 0)$ or $(0 : 1 : 0)$ is isomorphic to $(\mathbb{Z}/p\mathbb{Z})^{\oplus e} \rtimes \langle \zeta \rangle$, since we have the split exact sequence of groups (see [2, Theorem 2]):

$$0 \to (\mathbb{Z}/p\mathbb{Z})^{\oplus e} \to G \to \langle \zeta \rangle \to 1.$$

Proposition 2. *Let* $C \subset \mathbb{P}^2$ *be given by* $f(x, y) := g_1(x)^\ell + \lambda g_2(y)^\ell + \mu = 0$, *and let* $\alpha \in K$ *satisfy* $\alpha^{q\ell} + \lambda = 0$.

(a) *$\mathrm{Sing}(C) = \{(\zeta^i \alpha : 1 : 0) : 0 \leq i \leq \ell - 1\}$. In particular, C has exactly ℓ singular points.*

(b) *Assume that $g_1(\alpha X) - \alpha^q g_2(X)$ is zero as a polynomial. Then, $\pi^{-1}(Q)$ consists of exactly q points, and there exist q distinct tangent directions at Q, for any $Q \in \mathrm{Sing}(C)$. In this case, $p_g(C) = \frac{q\ell(q(\ell-1)-2)}{2} + 1$.*

(c) *Assume that $g_1(\alpha X) - \alpha^q g_2(X) \neq 0$. The multiplicity at Q is at most $q - 1$, and $\pi^{-1}(Q)$ consists of at most q/v points for some v a power of p. In this case, the genus of the smooth model satisfies $p_g(C) \geq \frac{q\ell(q(\ell-1)-1)}{2} + 1$.*

Proof. We prove (a). Let $F(X, Y, Z) = Z^{q\ell} f(X/Z, Y/Z)$, $G_1(X, Z) = Z^q g_1(X/Z)$, and $G_2(Y, Z) = Z^q g_2(Y/Z)$.

$$\frac{\partial F}{\partial X} = \ell G_1(X, Z)^{\ell-1} \alpha_0 Z^{q-1}, \quad \frac{\partial F}{\partial Y} = \lambda \ell G_2(Y, Z)^{\ell-1} \beta_0 Z^{q-1},$$

$$\frac{\partial F}{\partial Z} = \ell(q-1) G_1(X, Z)^{\ell-1} \alpha_0 X Z^{q-2} + \lambda \ell(q-1) G_2(Y, Z)^{\ell-1} \beta_0 Y Z^{q-2}.$$

Therefore, the singular locus is given by $Z = X^{q\ell} + \lambda Y^{q\ell} = 0$. We have $\mathrm{Sing}(C) = \{(\zeta^i \alpha : 1 : 0) : 0 \leq i \leq \ell - 1\}$, which consists of exactly ℓ points.

We prove (b). $\mathrm{Sing}(C) = \{(\zeta^i \alpha : 1 : 0) : 0 \leq i \leq \ell - 1\}$ by (a). Let $Q = (\alpha : 1 : 0)$. We consider the projection π_Q from Q. If we take $t := x - \alpha y$, then the projection π_Q is given by $\pi_Q(x : y : 1) = (t : 1)$, and we have a field extension $K(t, y)/K(t)$ with the equation $\hat{f}(t, y) := g_1(t + \alpha y)^\ell + \lambda g_2(y)^\ell + \mu = 0$. Note that:

$$g_1(t + \alpha y)^\ell = (g_1(t) + g_1(\alpha y))^\ell = \sum_{i=0}^{\ell} \binom{\ell}{i} g_1(t)^{\ell-i} g_1(\alpha y)^i,$$

$$\hat{f}(t, y) = \sum_{i=0}^{\ell-1} \binom{\ell}{i} g_1(t)^{\ell-i} g_1(\alpha y)^i + g_1(\alpha y)^\ell + \lambda g_2(y)^\ell + \mu,$$

and

$$g_1(\alpha y)^\ell + \lambda g_2(y)^\ell = \prod_{0 \le i \le \ell-1} (g_1(\alpha y) - \zeta^i \alpha^q g_2(y)).$$

Since $g_1(\alpha y) - \zeta^i \alpha^q g_2(y) = (\alpha^q - \zeta^i \alpha^q) y^q + ($ lower terms $)$, and $\alpha^q - \zeta^i \alpha^q \ne 0$ if $0 < i \le \ell - 1$, the degree of $\hat{f}(t, y)$ as a polynomial over $K(t)$ is larger than $q(\ell - 1)$ if and only if $g_1(\alpha y) - \alpha^q g_2(y) \ne 0$.

Assume that $g_1(\alpha X) - \alpha^q g_2(X) = 0$. Then, $g_1(\alpha X)^\ell + \lambda g_2(X)^\ell = 0$ and:

$$\sum_{i=0}^{\ell-1} \binom{\ell}{i} g_1(t)^{\ell-1-i} g_1(\alpha y)^i + \frac{\mu}{g_1(t)} = 0.$$

According to [11, III. 1. 14], for each solution t_0 of $g_1(t) = 0$, $\hat{\pi}_Q^{-1}((t_0 : 1))$ consists of a single point $\hat{Q}_{t_0} \in \hat{C}$ with $e_{\hat{Q}_{t_0}} = q(\ell - 1)$ for the projection $\hat{\pi}_Q$. Therefore, $\pi^{-1}(Q)$ consists of exactly q points, and there exist exactly q distinct tangent directions. We consider the projection π_P from $P = (1 : 0 : 0)$. The set $\hat{\pi}_P^{-1}(\pi_P(Q)) = \pi^{-1}(C \cap \{Z = 0\})$ consists of exactly $q\ell$ points. Therefore, $\hat{\pi}_P$ is not ramified at any point in $\pi^{-1}(C \cap \{Z = 0\})$. By the Riemann–Hurwitz formula, $p_g(C) = \frac{q\ell(q(\ell-1)-2)}{2} + 1$.

Assuming that $g_1(\alpha X) - \alpha^q g_2(X) \ne 0$, the degree of the projection π_Q is at least $q(\ell - 1) + 1$; this implies that the multiplicity at Q is at most $q - 1$ and that the cardinality of $\pi^{-1}(C \cap \{Z = 0\})$ is strictly less than $q\ell$. When we consider the projection π_P from $P = (1 : 0 : 0)$, there exists a ramification point in $\pi^{-1}(C \cap \{Z = 0\})$; let v be the index at this point. By Fact 0.2(a)(b), v is the index for any point in $\pi^{-1}(C \cap \{Z = 0\})$, and v divides $p^e \ell$. Since $C \cap \{Z = 0\}$ consists of exactly ℓ points, v divides q, and $\pi^{-1}(Q)$ consists of q/v points. By the Riemann–Hurwitz formula,

$$2p_g(C) - 2 \ge q\ell(-2) + q\ell \times \frac{q\ell}{\ell}(\ell - 1) + \frac{q\ell}{v} \times v.$$

Then, $p_g(C) \ge \frac{q\ell(q(\ell-1)-1)}{2} + 1$. $\qquad\qquad\square$

Example 2. There exist curves whose genera attain the lower bound, as in assertion (b) for each q and ℓ. Let $g_1(X) = g_2(X)$, and $\lambda = -1$. Then, $\alpha = 1 \in K$ satisfies $\alpha^{q\ell} + \lambda = 1 - 1 = 0$, and $g_1(\alpha X) - \alpha^q g_2(X) = g_1(X) - g_1(X) = 0$.

4. Proof of assertion (d) in Theorem 1

Assume that ℓ is odd, $\ell - 1$ is not divisible by p, and $g_1(\alpha X) - \alpha^q g_2(X) = 0$. Then, $p > 2$ and $p_g(C) = \frac{q\ell(q(\ell-1)-2)}{2} + 1$. Moreover, $m \ge 3$, if m

divides $q\ell$. We prove $\delta'(C) = 2$. For a smooth point R, we call R a flex if $I_R(C, T_R C) \geq 3$.

Lemma 1. *Let* $R = (x_0 : y_0 : 1) \in C \setminus \mathrm{Sing}(C)$. *Then,* R *is a flex if and only if* $g_1(x_0) = 0$ *or* $g_2(y_0) = 0$. *In this case,* $I_R(C, T_R C) = \ell$ *and* $T_R C$ *passes through* $(1 : 0 : 0)$ *or* $(0 : 1 : 0)$.

Proof. We consider the Hessian matrix:

$$H(f) = \begin{pmatrix} f_{xx} & f_{xy} & f_x \\ f_{xy} & f_{yy} & f_y \\ f_x & f_y & 0 \end{pmatrix}$$

$$= \begin{pmatrix} \ell(\ell-1)g_1(x)^{\ell-2}\alpha_0^2 & 0 & \ell g_1(x)^{\ell-1}\alpha_0 \\ 0 & \lambda\ell(\ell-1)g_2(y)^{\ell-2}\beta_0^2 & \lambda\ell g_2(y)^{\ell-1}\beta_0 \\ \ell g_1(x)^{\ell-1}\alpha_0 & \lambda\ell g_2(y)^{\ell-1}\beta_0 & 0 \end{pmatrix}.$$

Then, $\det H(f) = -\lambda\alpha_0^2\beta_0^2\ell^3(\ell-1)g_1(x)^{\ell-2}g_2(y)^{\ell-2}(\lambda g_2(y)^\ell + g_1(x)^\ell)$. It is well known that $R \in C \setminus \mathrm{Sing}(C)$ is a flex if and only if $\det H(f)(R) = 0$ (see, for example,[10, I.1.5]). Since by assumption, $\ell(\ell-1)$ is not divisible by p, R is a flex if and only if $g_1(x_0) = 0$ or $g_2(y_0) = 0$.

If $g_1(x_0) = 0$, then $g_2(y_0)^\ell + \mu = 0$, by the defining equation. Since $f(x, y_0) = g_1(x)^\ell$, $T_R C$ is defined by $Y - y_0 Z$ and $I_R(C, T_R C) = \ell$. This line passes through $(1 : 0 : 0)$. $\square$

Proof of Theorem 1(d). Assume by contradiction that a point P' in the affine plane $Z \neq 0$ is outer Galois. Let Q be a singular point. Since $\pi^{-1}(Q)$ consists of exactly q points by Proposition 2(b), $\hat{\pi}_{P'}$ is ramified at some point in $\pi^{-1}(Q)$ if and only if P' lies on the line given by some tangent direction at Q. In this case, P' is ramified at one point in $\pi^{-1}(Q)$ and unramified at the other $q-1$ points. By Fact 0.2(a), this is a contradiction. Therefore, we may assume that $\hat{\pi}_{P'}$ is unramified at each point in $\pi^{-1}(Q)$ for any singular point Q. Note that $\hat{\pi}_{P'}$ is ramified at some point $\hat{R} \in \hat{C}$, by the Riemann–Hurwitz formula. Now, $R = \pi(\hat{R}) \in C \setminus \mathrm{Sing}(C)$. By Fact 0.2(b) and the assumption on the degree, $I_R(C, T_R C) \geq 3$. According to Lemma 1, $I_R(C, T_R C) = \ell$. By the Riemann–Hurwitz formula again, there exist at least three points R_1, R_2, R_3 such that $T_{R_i}C \ni P'$ for each i, and $T_{R_i}C \neq T_{R_j}C$, if $i \neq j$. By Lemma 1, P' must be $(1 : 0 : 0)$ or $(0 : 1 : 0)$; this is a contradiction.

Assume that $P' \in \{Z = 0\}$ is outer Galois. Since $\pi^{-1}(C \cap \{Z = 0\})$ consists of exactly $q\ell$ points by Proposition 2(b), the projection $\hat{\pi}_{P'}$ is unramified at such points. Note that $\hat{\pi}_{P'}$ is ramified at some point $\hat{R} \in \hat{C}$, by the Riemann–Hurwitz formula. Now, $R = \pi(\hat{R}) \in C \setminus \mathrm{Sing}(C)$. By Fact 0.2(b) and the assumption on the degree, $I_R(C, T_R C) \geq 3$. By Lemma 1, P' must be $(1:0:0)$ or $(0:1:0)$. $\qquad\square$

5. Proof of assertion (e) in Theorem 1

Let $\ell = 2$, $g_1(X) = g_2(X) = g(X) := X^{p^e} + \alpha_{e-1} X^{p^{e-1}} + \cdots + \alpha_0 X$, and $\mathbb{F}_{q_0} := \mathbb{F}_q \cap (\bigcap_{\{i > 0 : \alpha_i \neq 0\}} \mathbb{F}_{p^i})$. Then, $p > 2$, since p does not divide $\ell = 2$. We consider the curve C defined by:

$$g(x)^2 + \lambda g(y)^2 + \mu = 0,$$

where $\lambda \in \mathbb{F}_{q_0}$. Note that $g(\gamma x) = \gamma g(x)$ if $\gamma \in \mathbb{F}_{q_0}$, and $\alpha^2 = -\lambda$, since $\alpha^{2q} = -\lambda = -\lambda^q$. The following Lemma implies the former assertion of (e).

Lemma 2. *Let $\gamma \in \mathbb{F}_{q_0}$ and $P' = (\gamma : 1 : 0)$. If $\gamma^2 + \lambda \neq 0$, then P' is outer Galois. In particular, $\delta'(C) \geq q_0 - 1$, and $\delta'(C) \geq q_0 + 1$, if $\alpha \notin \mathbb{F}_{q_0}$.*

Proof. The projection $\pi_{P'}$ from P' is given by $\pi_{P'}(x : y : 1) = (x - \gamma y : 1)$. Let $t := x - \gamma y$; then, we have the field extension $K(t, y)/K(t)$ given by $\hat{f}(t, y) := g(t + \gamma y)^2 + \lambda g(y)^2 + \mu = 0$. Since $g(\gamma x) = \gamma g(x)$ and $\gamma^2 + \lambda \neq 0$,

$$
\begin{aligned}
\hat{f}(t, y) &= (g(t) + \gamma g(y))^2 + \lambda g(y)^2 + \mu \\
&= (\gamma^2 + \lambda) g(y)^2 + 2\gamma g(t) g(y) + g(t)^2 + \mu \\
&= (\gamma^2 + \lambda) g(y) \left(g(y) + \frac{2\gamma}{\gamma^2 + \lambda} g(t) \right) + g(t)^2 + \mu.
\end{aligned}
$$

Let $\eta := \frac{2\gamma}{\gamma^2 + \lambda} \in \mathbb{F}_{q_0}$. Note that $\eta^{q_0} = \eta$ and $g(\eta \cdot x) = \eta \cdot g(x)$. The set $\{y + \beta : g(\beta) = 0\} \cup \{-y - \eta t + \beta : g(\beta) = 0\} \subset K(t, y)$ consists of all roots of $\hat{f}(t, y) \in K(t)[y]$; therefore, P' is outer Galois.

Since the number of elements $\gamma \in \mathbb{F}_{q_0}$ with $\gamma^2 + \lambda = 0$ is at most two, $\delta'(C) \geq q_0 + 1 - 2 = q_0 - 1$. If there exists $\gamma \in \mathbb{F}_{q_0}$ with $\gamma^2 + \lambda = 0$, then $\gamma = \pm\alpha$ and hence, $\alpha \in \mathbb{F}_{q_0}$. Therefore, $\delta'(C) \geq q_0 + 1$ if $\alpha \notin \mathbb{F}_{q_0}$. $\qquad\square$

Next, we consider the latter assertion of (e). Using the Hessian matrix (see the proof of Lemma 1), we have the following.

Lemma 3. *The curve C has no flex in the affine plane $Z \neq 0$.*

Assume that $\mathbb{F}_q = \mathbb{F}_{q_0}$; then,

$$(x^q + \alpha_0 x)^2 + \lambda(y^q + \alpha_0 y)^2 + \mu = 0.$$

Introducing the variable change $Z \mapsto (1/\sqrt[q-1]{-\alpha_0})Z$, we can assume that $\alpha_0 = -1$. In this case, we prove that all outer Galois points are $\mathbb{F}_q$-rational points on the line $Z = 0$.

We determine special, multiple tangent lines.

Lemma 4. *Let $R \in C \setminus \mathrm{Sing}(C)$. If $T_R C$ has q distinct contact points in C, then $T_R C$ intersects the line $Z = 0$ at an $\mathbb{F}_q$-rational point.*

Proof. The Gauss map $\gamma : C \dashrightarrow (\mathbb{P}^2)^* \cong \mathbb{P}^2$, which sends a smooth point R to the tangent line $T_R C$ at R, is given by

$$(\partial F/\partial X : \partial F/\partial Y : \partial F/\partial Z)$$
$$= (-2(x^q - x) : -2\lambda(y^q - y) : 2x(x^q - x) + 2\lambda y(y^q - y))$$
$$= (x^q - x : \lambda(y^q - y) : -x(x^q - x) - \lambda y(y^q - y)).$$

Let $R_i = (x_i : y_i : 1)$ for $i = 1, \ldots, q$, let $R_i \neq R_j$ and let $T_{R_i} C = T_{R_j} C$ for any i, j. Assume that $x_1^q - x_1 \neq 0$. Then, $x_i^q - x_i \neq 0$. Since $T_{R_1} C = T_{R_i} C$, $(y_1^q - y_1)/(x_1^q - x_1) = (y_i^q - y_i)/(x_i^q - x_i)$ and

$$\frac{\mu}{(x_1^q - x_1)^2} = -1 - \lambda \left(\frac{y_1^q - y_1}{x_1^q - x_1} \right)^2 = \frac{\mu}{(x_i^q - x_i)^2}$$

by the defining equation, $x_1^q - x_1 = \pm(x_i^q - x_i)$ and $y_1^q - y_1 = \pm(y_i^q - y_i)$.

Assume that there exists a value $i \neq 1$ such that $x_1^q - x_1 = x_i^q - x_i$. We can assume $i = 2$. Then, $y_1^q - y_1 = y_2^q - y_2$, and there exist $\beta_x, \beta_y \in \mathbb{F}_q$ such that $x_2 = x_1 + \beta_x$ and $y_2 = y_1 + \beta_y$. If $\beta_y = 0$, then $\beta_x = 0$, by the condition $x_1 + \lambda y_1(y_1^q - y_1)/(x_1^q - x_1) = x_2 + \lambda y_2(y_2^q - y_2)/(x_2^q - x_2)$. Therefore, $\beta_y \neq 0$. Using the condition $x_1 + \lambda y_1(y_1^q - y_1)/(x_1^q - x_1) = x_2 + \lambda y_2(y_2^q - y_2)/(x_2^q - x_2)$ yields $\lambda(y_1^q - y_1)/(x_1^q - x_1) = -\beta_x/\beta_y$. Then, $T_{R_1} C$ is defined by:

$$X - \frac{\beta_x}{\beta_y} Y + \left(-x_1 + y_1 \frac{\beta_x}{\beta_y} \right) Z = 0,$$

and meets the line $Z = 0$ at the point $(\beta_x : \beta_y : 0)$, which is $\mathbb{F}_q$-rational.

Assume that $x_1^q - x_1 = -(x_i^q - x_i)$ for any i. Then, there exists $\alpha_i \in \mathbb{F}_q$ such that $x_i = -x_1 + \alpha_i$. Then, $x_3 - x_2 = (-x_1 + \alpha_3) - (-x_1 + \alpha_2) = \alpha_3 - \alpha_2 \in \mathbb{F}_q$. We can reduce to the above case.

If $y_1^q + y_1 \neq 0$, then we have the same assertion, similar to the above discussion. $\qquad\square$

Lemma 5. *Assume that $\alpha \notin \mathbb{F}_q$. Then, any outer Galois point is an $\mathbb{F}_q$-rational point on the line $Z = 0$. In particular, $\delta'(C) \leq q + 1$.*

Proof. Let $P' = (\gamma : 1 : 0)$. We consider the projection from P'. The projection $\pi_{P'}$ from P' is given by $\pi_{P'}(x : y : 1) = (x - \gamma y : 1)$. Let $t := x - \gamma y$. Then, we have the field extension $K(t, y)/K(t)$ given by $\hat{f}(t, y) := ((t + \gamma y)^q - (t + \gamma y))^2 + \lambda(y^q - y)^2 + \mu = 0$. We have

$$\hat{f}(t, y) = ((t^q - t) + (\gamma^q y^q - \gamma y))^2 + \lambda(y^q - y)^2 + \mu$$
$$= (\gamma^{2q} + \lambda)y^{2q} - 2(\gamma^{q+1} + \lambda)y^{q+1} + (\text{ lower terms }).$$

Note that $\gamma^{2q} + \lambda = 0$ if and only if $\gamma^2 + \lambda = 0$, since $\lambda \in \mathbb{F}_q$. If $\gamma^2 + \lambda \neq 0$, then the degree of $\hat{f}(t, y)$ is $2q$. However, if $\gamma^2 + \lambda = 0$, then $\gamma^2 = \alpha^2$, and the degree is $q + 1$, since $\gamma^{q+1} + \lambda = \alpha^{q+1} - \alpha^2 = \alpha(\alpha^q - \alpha) \neq 0$. In addition,

$$\frac{dt}{dy} = \frac{\gamma((t^q - t) + (\gamma^q y^q - \gamma y)) + \lambda(y^q - y)}{(t^q - t) + (\gamma^q y^q - \gamma y)}.$$

We consider the condition $\hat{f}(t, y) = dt/dy = 0$. Taking $t^q - t + (\gamma^q y^q - \gamma y) = -(\lambda/\gamma)(y^q - y)$, we have $(\lambda^2/\gamma^2 + \lambda)(y^q - y)^2 + \mu = 0$. If $\lambda \neq -\gamma^2$, we have a solution (t_0, y_0) satisfying $\hat{f}(t_0, y_0) = (dt/dy)(t_0, y_0) = 0$. This implies that $\pi_{P'}$ is ramified at some point R in the affine plane $Z \neq 0$. If $\lambda = -\gamma^2$ (this implies that P' is singular), then $\hat{f}(t, y)$ is of degree $q + 1$, and we do not have a solution (t_0, y_0) satisfying $\hat{f}(t_0, y_0) = (dt/dy)(t_0, y_0) = 0$. This implies that $\pi_{P'}$ is not ramified at any point R in the affine plane $Z \neq 0$.

Let $O \in \mathbb{P}^2 \setminus C$ be a point in the affine plane $Z \neq 0$, and let Q be a singular point. Then, the line $\overline{OQ}$ passes through exactly $q + 1$ smooth points with multiplicity one except for Q, by the above discussion. Since the cardinality of $\pi^{-1}(Q)$ is smaller than $q - 1$ by Proposition 2(c), the projection $\hat{\pi}_O$ is ramified at some point in $\pi^{-1}(Q)$. By Fact 0.2(a), O is not Galois. Therefore, all Galois points lie on the line $Z = 0$.

Let $P' = (\gamma : 1 : 0)$ be outer Galois. By Lemma 3, $I_R(C, T_R C) = 2$. By Fact 0.2(a), $T_R C$ contains exactly q contact points. By Lemma 4, the point given by $T_R C \cap \{Z = 0\}$ is $\mathbb{F}_q$-rational. Therefore, P' is $\mathbb{F}_q$-rational. $\square$

Lemma 6. *Assume that $\alpha \in \mathbb{F}_q$. Any outer Galois point is an $\mathbb{F}_q$-rational point on the line $Z = 0$. In particular, $\delta'(C) \leq q - 1$.*

Proof. First, we prove that points on the affine plane $Z \neq 0$ are not outer Galois. Assume by contradiction that P' in the affine plane $Z \neq 0$ is outer Galois. Similar to the proof of assertion (d), we can assume that $\hat{\pi}_{P'}$ is unramified at each point in $\pi^{-1}(Q)$ for any singular point Q. Note that $\hat{\pi}_{P'}$ is ramified at some point $\hat{R} \in \hat{C}$, by the Riemann–Hurwitz formula. Now, $R = \pi(\hat{R}) \in C \setminus \text{Sing}(C)$. By Lemma 3, $I_R(C, T_R C) = 2$. By Fact

0.2(a), $T_R C$ contains exactly q contact points. According to the Riemann–Hurwitz formula again, considering $p_g(C)$, there exist $d = 2q$ tangent lines containing P' and q contact points. However, by Lemma 4, there exist at most $q + 1$ such lines. Since $q + 1 < 2q = d$, this is a contradiction.

Let $P' \in \{Z = 0\}$ be outer Galois. Similar to the proof of assertion (d), $\pi_{P'}$ is ramified at some point R in the affine plane $Z \neq 0$. By Lemma 3, $I_R(C, T_R C) = 2$. By Fact 0.2(a), $T_R C$ contains exactly q contact points. By Lemma 4, the point given by $T_R C \cap \{Z = 0\}$ is $\mathbb{F}_q$-rational. Therefore, P' is $\mathbb{F}_q$-rational.

By Proposition 2(a), $\mathrm{Sing}(C) = \{(\pm \alpha : 1 : 0)\}$. The two singular points are $\mathbb{F}_q$-rational. Therefore, $\delta'(C) \leq q + 1 - 2 = q - 1$. $\qquad\square$

As a consequence of Lemmas 5 and 6, we have the latter assertion of (e).

Acknowledgment   The author was partially supported by JSPS KAK-ENHI Grant Numbers 22740001, 25800002.

References

[1] G. Cornelissen, F. Kato and A. Kontogeorgis, Discontinuous groups in positive characteristic and automorphisms of Mumford curves, Math. Ann. **320** (2001), 55–85.

[2] S. Fukasawa, On the number of Galois points for a plane curve in positive characteristic, II, Geom. Dedicata **127** (2007), 131–137.

[3] S. Fukasawa, Classification of plane curves with infinitely many Galois points, J. Math. Soc. Japan **63** (2011), 195–209.

[4] S. Fukasawa, Complete determination of the number of Galois points for a smooth plane curve, Rend. Sem. Mat. Univ. Padova **129** (2013), 93–113.

[5] S. Fukasawa, Galois points for a plane curve in characteristic two, J. Pure Appl. Algebra **218** (2014), 343–353.

[6] S. Fukasawa, M. Homma and S. J. Kim, Rational curves with many rational points over a finite field, in "Arithmetic, Geometry, Cryptography and Coding Theory," Contemp. Math. **574**, Amer. Math. Soc., Providence, RI, 2012, pp. 37–48.

[7] J. W. P. Hirschfeld, G. Korchmáros and F. Torres, *Algebraic curves over a finite field*, Princeton Ser. Appl. Math., Princeton Univ. Press, Princeton, 2008.

[8] M. Homma, Galois points for a Hermitian curve, Comm. Algebra **34** (2006), 4503–4511.

[9] K. Miura and H. Yoshihara, Field theory for function fields of plane quartic curves, J. Algebra **226** (2000), 283–294.

[10] I. R. Shafarevich, *Basic algebraic geometry, 1. Varieties in projective space*, Second edition, Springer-Verlag, Berlin, 1994.

[11] H. Stichtenoth, *Algebraic function fields and codes*, Universitext, Springer-Verlag, Berlin, 1993.

[12] D. Subrao, The p-rank of Artin-Schreier curves, Manuscripta Math. **16** (1975), 169–193.

[13] R. C. Valentini and M. L. Madan, A Hauptsatz of L. E. Dickson and Artin-Schreier extensions, J. Reine Angew. Math. **318** (1980), 156–177.

[14] H. Yoshihara, Function field theory of plane curves by dual curves, J. Algebra **239** (2001), 340–355.

[15] H. Yoshihara and S. Fukasawa, List of problems, available at: http: //hyoshihara.web.fc2.com/openquestion.html

Permutation polynomials of $\mathbb{F}_{q^2}$ of the form $aX + X^{r(q-1)+1}$

Xiang-dong Hou

Department of Mathematics and Statistics, University of South Florida, Tampa, FL 33620, USA xhou@usf.edu

Let q be a prime power, $2 \le r \le q$, and $f = aX + X^{r(q-1)+1} \in \mathbb{F}_{q^2}[X]$, where $a \ne 0$. The conditions on r, q, a that are necessary and sufficient for f to be a permutation polynomial (PP) of $\mathbb{F}_{q^2}$ are not known. (Such conditions are known under an additional assumption that $a^{q+1} = 1$.) In this paper, we prove the following: (i) If f is a PP of $\mathbb{F}_{q^2}$, then $\gcd(r, q+1) > 1$ and $(-a)^{(q+1)/\gcd(r,q+1)} \ne 1$. (ii) For a fixed $r > 2$ and subject to the conditions that $q + 1 \equiv 0 \pmod{r}$ and $a^{q+1} \ne 1$, there are only finitely many (q, a) for which f is a PP of $\mathbb{F}_{q^2}$. Combining (i) and (ii) confirms a recent conjecture regarding the type of permutation binomial considered here.

1. Introduction

Let $\mathbb{F}_q$ denote the finite field with q elements. A polynomial $f \in \mathbb{F}_q[X]$ is called a *permutation polynomial* (PP) of $\mathbb{F}_q$ if it induces a permutation of $\mathbb{F}_q$. In general, it is difficult to predict the permutation property of a polynomial from its algebraic appearance; even for binomials, the question remains challenging. Historical accounts and contemporary reviews of permutation polynomials can be found in a few book chapters and recent survey papers; see [1],[2],[7, Ch. 7],[8, Ch. 8].

In this paper, we are concerned with the following question. Let $f = aX + X^{r(q-1)+1} \in \mathbb{F}_{q^2}[X]$, where $2 \le r \le q$ and $a \in \mathbb{F}_{q^2}^*$. When is f a PP of $\mathbb{F}_{q^2}$? Our interest and curiosity in this question were inspired and elevated by several recent results. Under the assumption that $a^{q+1} = 1$, the answer to question was provided by Zieve [9].

Theorem 1.[9] *Assume that $a^{q+1} = 1$. Then f is a PP of $\mathbb{F}_{q^2}$ if and only if $(-a)^{(q+1)/\gcd(r,q+1)} \ne 1$ and $\gcd(r - 1, q + 1) = 1$.*

The proof of Theorem 1 given in [9] uses degree one rational functions over $\mathbb{F}_{q^2}$ which permute the $(q+1)$-th roots of unity in $\mathbb{F}_{q^2}$. For that proof, the assumption that $a^{q+1} = 1$ is essential. In the general situation, that is, without the assumption that $a^{q+1} = 1$, the question has been answered for $r = 2, 3, 5, 7$ in [3, 5, 6]. It turns out that for $r = 3, 5, 7$ and $a \in \mathbb{F}_{q^2}^*$ with $a^{q+1} \neq 1$, there are numerous but *finitely many* (q, a) for which f is a PP of $\mathbb{F}_{q^2}$. A conjecture has been formulated based on this observation:

Conjecture 1. [2, 6] *Let $r > 2$ be a fixed prime. Under the assumption that $a^{q+1} \neq 1$ $(a \in \mathbb{F}_{q^2}^*)$, there are only finitely many (q, a) for which f is a PP of $\mathbb{F}_{q^2}$.*

Remark 1. Conjecture 1 is false for $r = 2$. It follows from [3, Theorems A and B] that $a\mathrm{X} + \mathrm{X}^{2(q-1)+1}$ $(a \in \mathbb{F}_{q^2}^*)$ is a PP of $\mathbb{F}_{q^2}$ if and only if q is odd and $(-a)^{(q+1)/2} = -1$ or 3.

In the present paper, we confirm Conjecture 1 by proving the following.

Theorem 2. *Assume that f is a PP of $\mathbb{F}_{q^2}$. Then $\gcd(r, q+1) > 1$ and $(-a)^{(q+1)/\gcd(r,q+1)} \neq 1$. In particular, if r is a prime, then $q + 1 \equiv 0 \pmod{r}$ and $(-a)^{(q+1)/r} \neq 1$.*

Theorem 3. *Let $r > 2$ be fixed. Assume that $q \geq r$, $q + 1 \equiv 0 \pmod{r}$ and $a^{q+1} \neq 1$ $(a \in \mathbb{F}_{q^2}^*)$. Subject to these conditions, there are only finitely many (q, a) for which f is a PP of $\mathbb{F}_{q^2}$.*

It is well known that f is a PP of $\mathbb{F}_{q^2}$ if and only 0 is the unique root of f in $\mathbb{F}_{q^2}$ and

$$\sum_{x \in \mathbb{F}_{q^2}} f(x)^s = 0 \qquad \text{for all } 1 \leq s \leq q^2 - 2.$$

The starting point of the paper is the computation of the power sums $\sum_{x \in \mathbb{F}_{q^2}} f(x)^s$, which is carried out in Section 2. Theorem 2, proved in Section 3, is a straightforward consequence of the computation in Section 2. The proof Theorem 3 is rather involved; it is given in Section 4 preceded by a description of the proof strategy.

2. The Power Sums

Throughout the paper, we always assume that $f = a\mathrm{X} + \mathrm{X}^{r(q-1)+1} \in \mathbb{F}_{q^2}[\mathrm{X}]$, where $2 \leq r \leq q$ and $a \neq 0$. All other conditions are considered additional. The following fact is obvious.

Fact 0.1. 0 is the only root of f in $\mathbb{F}_{q^2}$ if and only if $(-a)^{(q+1)/\gcd(r,q+1)} \neq 1$.

For $0 \leq \alpha, \beta \leq q-1$ with $(\alpha, \beta) \neq (0,0)$, we have

$$\sum_{x \in \mathbb{F}_{q^2}} f(x)^{\alpha+\beta q} = \sum_{x \in \mathbb{F}_{q^2}^*} x^{\alpha+\beta q}(a + x^{r(q-1)})^{\alpha+\beta q}$$

$$= \sum_{x \in \mathbb{F}_{q^2}^*} x^{\alpha+\beta q} \sum_{i,j} \binom{\alpha}{i}\binom{\beta}{j} a^{\alpha+\beta q-i-jq} x^{r(q-1)(i+jq)}$$

$$= a^{\alpha+\beta q} \sum_{i,j} \binom{\alpha}{i}\binom{\beta}{j} a^{-i-jq} \sum_{x \in \mathbb{F}_{q^2}^*} x^{\alpha+\beta q+r(q-1)(i-j)}.$$

In the above, the inner sum is 0 unless $\alpha + \beta q \equiv 1 \pmod{q-1}$, i.e., $0 \leq \alpha \leq q-1$ and $\beta = q-1-\alpha$, in which case,

$$\sum_{x \in \mathbb{F}_{q^2}} f(x)^{\alpha+(q-1-\alpha)q}$$

$$= a^{(\alpha+1)(1-q)} \sum_{i,j} \binom{\alpha}{i}\binom{q-1-\alpha}{j} a^{-i-jq} \sum_{x \in \mathbb{F}_{q^2}^*} x^{(q-1)[-\alpha-1+r(i-j)]}$$

$$= -a^{(\alpha+1)(1-q)} \sum_{-\alpha-1+r(i-j)\equiv 0 \,(\mathrm{mod}\, q+1)} \binom{\alpha}{i}\binom{q-1-\alpha}{j} a^{-i-jq}.$$

As i runs over the interval $[0, \alpha]$ and j over the interval $[0, q-1-\alpha]$, the range of $-\alpha-1+r(i-j)$ is

$$I_\alpha := \bigl[r\alpha - \alpha - 1 - r(q-1),\; r\alpha - \alpha - 1\bigr]. \tag{1}$$

Thus

$$\sum_{x \in \mathbb{F}_{q^2}} f(x)^{\alpha+(q-1-\alpha)q} = -a^{(\alpha+1)(1-q)} S_q(\alpha, a), \tag{2}$$

where

$$S_q(\alpha, a) = \sum_{-\alpha-1+r(i-j)\in I_\alpha \cap (q+1)\mathbb{Z}} \binom{\alpha}{i}\binom{q-1-\alpha}{j} a^{-i-jq}. \tag{3}$$

Lemma 1. *Assume that $a^{q+1} = 1$ but $(-a)^{(q+1)/d} \neq 1$, where $d = \gcd(r, q+1)$. Then, for $0 \leq \alpha \leq q-1$,*

$$S_q(\alpha, a) = \begin{cases} a^{-\alpha-1} & \text{if } (\alpha+1)(r-1) \equiv 0 \pmod{q+1}, \\ 0 & \text{otherwise.} \end{cases} \tag{4}$$

Proof. We only have to consider the case $\alpha + 1 \equiv 0 \pmod{d}$. Let $k_0 \in \mathbb{Z}$ be the smallest such that $\alpha + 1 + k_0(q + 1) \in r\mathbb{Z}$ and

$$k_0(q + 1) \geq r\alpha - \alpha - 1 - r(q - 1). \tag{5}$$

Then

$$\left(k_0 - \frac{r}{d}\right)(q + 1) < r\alpha - \alpha - 1 - r(q - 1). \tag{6}$$

We derive from (5) and (6) that

$$\begin{cases} \left(k_0 + (d - 2)\dfrac{r}{d}\right)(q + 1) < r\alpha - \alpha - 1, \\ \left(k_0 + d\dfrac{r}{d}\right)(q + 1) > r\alpha - \alpha - 1. \end{cases}$$

Therefore

$$\left[\alpha + 1 + \left(I_\alpha \cap (q + 1)\mathbb{Z}\right)\right] \cap r\mathbb{Z} = \alpha + 1 + k_0(q + 1) + \frac{r(q + 1)}{d}L, \tag{7}$$

where $L = \{0, \ldots, d - 1\}$ or $\{0, \ldots, d - 2\}$.

If $L = \{0, \ldots, d - 2\}$, then

$$\left(k_0 + (d - 1)\frac{r}{d}\right)(q + 1) > r\alpha - \alpha - 1,$$

which is equivalent to

$$\alpha + 1 + k_0(q + 1) + \frac{d - 1}{d}(q + 1)r \geq r\alpha + r. \tag{8}$$

Also note that (6) is equivalent to

$$\alpha + 1 + k_0(q + 1) - \frac{1}{d}(q + 1)r \leq r\alpha - rq. \tag{9}$$

Taking the difference of (8) and (9), we see that the equal sign holds in both (8) and (9). Hence

$$\alpha + 1 + k_0(q + 1) = \left(\alpha + 1 - \frac{d - 1}{d}(q + 1)\right)r, \tag{10}$$

which further implies that $(\alpha + 1)(r - 1) \equiv 0 \pmod{q + 1}$. On the other hand, if $(\alpha + 1)(r - 1) \equiv 0 \pmod{q + 1}$ and k_0 is given by (10), reversing the above argument gives $L = \{0, \ldots, d - 2\}$. Thus we have proved that $L = \{0, \ldots, d - 2\}$ if and only if $(\alpha + 1)(r - 1) \equiv 0 \pmod{q + 1}$.

Now we have
$$S_q(\alpha, a)$$

$$= \sum_{-\alpha-1+r(i-j)\in I_\alpha \cap (q+1)\mathbb{Z}} \binom{\alpha}{i}\binom{q-1-\alpha}{j} a^{-i+j} \qquad \text{(since } a^{q+1}=1)$$

$$= \sum_{i-j\in\frac{1}{r}(\alpha+1+k_0(q+1))+\frac{q+1}{d}L} \binom{\alpha}{i}\binom{q-1-\alpha}{j} a^{-i+j} \qquad \text{(by (7))}$$

$$= \sum_{\alpha-i+j\in\alpha-\frac{1}{r}(\alpha+1+k_0(q+1))-\frac{q+1}{d}L} \binom{\alpha}{\alpha-i}\binom{q-1-\alpha}{j} a^{-i+j}$$

$$= \sum_{l\in L} a^{-\frac{1}{r}(\alpha+1+k_0(q+1))-\frac{q+1}{d}l} \binom{q-1}{\alpha-\frac{1}{r}(\alpha+1+k_0(q+1))-\frac{q+1}{d}l}$$

$$= \sum_{l\in L} a^{-\frac{1}{r}(\alpha+1+k_0(q+1))-\frac{q+1}{d}l}(-1)^{\alpha-\frac{1}{r}(\alpha+1+k_0(q+1))-\frac{q+1}{d}l}$$

$$= (-1)^\alpha(-a)^{-\frac{1}{r}(\alpha+1+k_0(q+1))}\sum_{l\in L}(-a)^{-\frac{q+1}{d}l}.$$

(For the next-to-last step, note that $\alpha - \frac{1}{r}(\alpha+1+k_0(q+1)) - \frac{q+1}{d}l \in \alpha-\frac{1}{r}(\alpha+1+I_\alpha) = [0, q-1]$ by (7) and (1).) If $(\alpha+1)(r-1) \not\equiv 0 \pmod{q+1}$, then $L = \{0,\ldots,d-1\}$ and $\sum_{l\in L}(-a)^{-\frac{q+1}{d}l} = 0$. If $(\alpha+1)(r-1) \equiv 0 \pmod{q+1}$, then $L = \{0,\ldots,d-2\}$ and hence

$$S_q(\alpha, a) = (-1)^{\alpha+1}(-a)^{-\frac{1}{r}(\alpha+1+k_0(q+1))}(-a)^{-\frac{q+1}{d}(d-1)}$$

$$= (-1)^{\alpha+1}(-a)^{-(\alpha+1-\frac{d-1}{d}(q+1))-\frac{q+1}{d}(d-1)} \qquad \text{(by (10))}$$

$$= a^{-\alpha-1}.$$

$\square$

Remark 2. Note that $\gcd(r-1, q+1) = 1$ if and only if there is no $0 \le \alpha \le q-1$ such that $(\alpha+1)(r-1) \equiv 0 \pmod{q+1}$. Hence Lemma 1 and Fact 0.1 provide an alternate proof of Theorem 1.

Lemma 2. *Assume that* $q+1 \equiv 0 \pmod r$, $0 \le \alpha \le q-1$, $\alpha+1 \equiv 0 \pmod r$ *and* $q \ge (r-1)(\alpha+2)$. *Then*

$$S_q(\alpha, a) = (-a)^{\frac{\alpha+1}{r}q}\Big(\sum_{l=0}^{r-1} v^l\Big)g_\alpha(v), \qquad (11)$$

where $v = (-a)^{-\frac{q+1}{r}q}$ *and*

$$g_\alpha(X) = (X-1)\sum_{l=0}^{r-1} X^l \sum_{i=1}^{\alpha}(-1)^i\binom{\alpha}{i}\binom{i+\alpha-\frac{\alpha+1}{r}+\frac{l}{r}}{\alpha}\frac{X^{ri}-1}{X^r-1} + (-1)^\alpha.$$

$$(12)$$

Proof. Since $q \geq (r-1)(\alpha+2)$, it is easy to see that $I_\alpha \cap (q+1)\mathbb{Z} = \{-(r-1)(q+1), \ldots, -(q+1), 0\}$. Therefore,

$$
S_\alpha(\alpha, a)
$$

$$
= \sum_{-\alpha-1+r(i-j)\in I_\alpha \cap (q+1)\mathbb{Z}} \binom{\alpha}{i}\binom{q-1-\alpha}{j} a^{-i-jq}
$$

$$
= \sum_{l=0}^{r-1} \sum_{-\alpha-1+r(i-j)=-l(q+1)} \binom{\alpha}{i}\binom{q-1-\alpha}{j} a^{-i-jq}
$$

$$
= \sum_{l=0}^{r-1} \sum_{i=0}^{\alpha} \binom{\alpha}{i}\binom{q-1-\alpha}{\frac{1}{r}(l(q+1)-\alpha-1)+i} a^{-i-[\frac{1}{r}(l(q+1)-\alpha-1)+i]q}
$$

$$
= a^{\frac{\alpha+1}{r}q} \sum_{l=0}^{r-1} \sum_{i=0}^{\alpha} \binom{\alpha}{i}\binom{-1-\alpha}{\frac{1}{r}(l(q+1)-\alpha-1)+i} a^{-\frac{\alpha+1}{r}q(l+ri)}
$$

$$
(\text{since } \frac{1}{r}(l(q+1)-\alpha-1)+i \leq q-1)
$$

$$
= a^{\frac{\alpha+1}{r}q} \sum_{l=0}^{r-1} \sum_{i=0}^{\alpha} \binom{\alpha}{i}(-1)^{\frac{1}{r}(l(q+1)-\alpha-1)+i}\binom{\alpha+\frac{1}{r}(l(q+1)-\alpha-1)+i}{\alpha}
$$

$$
\cdot a^{-\frac{q+1}{r}q(l+ri)}
$$

$$
= (-a)^{\frac{\alpha+1}{r}q} \sum_{l=0}^{r-1} \sum_{i=0}^{\alpha} (-1)^i \binom{\alpha}{i}\binom{i+\alpha-\frac{\alpha+1}{r}+\frac{l}{r}}{\alpha} v^{l+ri}.
$$

$$(13)$$

We also have

$$
\sum_{l=0}^{r-1} \sum_{i=0}^{\alpha} (-1)^i \binom{\alpha}{i}\binom{i+\alpha-\frac{\alpha+1}{r}+\frac{l}{r}}{\alpha} v^l
$$

$$
= \sum_{l=0}^{r-1} v^l \sum_{i=0}^{\alpha} \binom{\alpha}{\alpha-i}(-1)^{\frac{1}{r}(l(q+1)-\alpha-1)}\binom{-1-\alpha}{\frac{1}{r}(l(q+1)-\alpha-1)+i}
$$

$$
= \sum_{l=0}^{r-1} v^l (-1)^{\frac{1}{r}(l(q+1)-\alpha-1)}\binom{-1}{\alpha+\frac{1}{r}(l(q+1)-\alpha-1)}
$$

$$
= (-1)^\alpha \sum_{l=0}^{r-1} v^l.
$$

$$(14)$$

Combining (13) and (14) gives

$$S_q(\alpha, a)$$

$$= (-a)^{\frac{\alpha+1}{r}} q \sum_{l=0}^{r-1} \sum_{i=0}^{\alpha} (-1)^i \binom{\alpha}{i} \binom{i + \alpha - \frac{\alpha+1}{r} + \frac{l}{r}}{\alpha} (v^{l+ri} - v^l + v^l)$$

$$= (-a)^{\frac{\alpha+1}{r}} q$$

$$\cdot \left[\sum_{l=0}^{r-1} v^l \sum_{i=0}^{\alpha} (-1)^i \binom{\alpha}{i} \binom{i + \alpha - \frac{\alpha+1}{r} + \frac{l}{r}}{\alpha} (v^{ri} - 1) + (-1)^\alpha \sum_{l=0}^{r-1} v^l \right]$$

$$= (-a)^{\frac{\alpha+1}{r}} q \left(\sum_{l=0}^{r-1} v^l \right)$$

$$\cdot \left[(v-1) \sum_{l=0}^{r-1} v^l \sum_{i=1}^{\alpha} (-1)^i \binom{\alpha}{i} \binom{i + \alpha - \frac{\alpha+1}{r} + \frac{l}{r}}{\alpha} \frac{v^{ri} - 1}{v^r - 1} + (-1)^\alpha \right]$$

$$= (-a)^{\frac{\alpha+1}{r}} q \left(\sum_{l=0}^{r-1} v^l \right) g_\alpha(v).$$

$\square$

In (12), we may treat $g_\alpha(X)$ as a polynomial over $\mathbb{Z}[1/r]$. The reason is that for all $i \in \mathbb{Z}$ and $j \in \mathbb{N}$, $\binom{i/r}{j}$ is a p-adic integer for all primes $p \nmid r$, and hence $\binom{i/r}{j} \in \mathbb{Z}[1/r]$. We have

$$g_\alpha(0) = -\sum_{i=1}^{\alpha} (-1)^i \binom{\alpha}{i} \binom{i + \alpha - \frac{\alpha+1}{r}}{\alpha} + (-1)^\alpha$$

$$= -\sum_{i=1}^{\alpha} \binom{\alpha}{i} \binom{-\alpha - 1}{i - \frac{\alpha+1}{r}} (-1)^{\frac{\alpha+1}{r}} + (-1)^\alpha$$

$$= -(-1)^{\frac{\alpha+1}{r}} \sum_{i=0}^{\alpha} \binom{\alpha}{\alpha - i} \binom{-\alpha - 1}{i - \frac{\alpha+1}{r}} + (-1)^\alpha$$

$$= -(-1)^{\frac{\alpha+1}{r}} \binom{-1}{\alpha - \frac{\alpha+1}{r}} + (-1)^\alpha$$

$$= -(-1)^{\frac{\alpha+1}{r}} (-1)^{\alpha - \frac{\alpha+1}{r}} + (-1)^\alpha = 0.$$

Hence

$$h_\alpha(X) := X^{-1} g_\alpha(X) \in (\mathbb{Z}[1/r])[X]. \tag{15}$$

3. Proof of Theorem 2

Given that f is a PP of $\mathbb{F}_{q^2}$, we show that $\gcd(r, q+1) > 1$. Assume to the contrary that $\gcd(r, q+1) = 1$. Then there exists a unique $j_0 \in \{0, 1, \dots, q\}$ such that $-1 - rj_0 \equiv 0 \pmod{q+1}$. Note that $j_0 \neq q$ since otherwise $r \equiv 1 \pmod{q+1}$, which would imply that $r \geq q + 2$, a contradiction. By (3),

$$S_q(0, a) = \sum_{-1-rj \equiv 0 \,(\mathrm{mod}\, q+1)} \binom{q-1}{j} a^{-jq}$$

$$= \binom{q-1}{j_0} a^{-j_0 q} = (-1)^{j_0} a^{-j_0 q} \neq 0,$$

which is a contradiction.

4. Proof of Theorem 3

Fix $r \geq 3$ and assume that $f = a\mathsf{X} + \mathsf{X}^{r(q-1)+1}$ is a PP of $\mathbb{F}_{q^2}$, where $q \geq r$, $q + 1 \equiv 0 \pmod{r}$ and $a^{q+1} \neq 1$ ($a \in \mathbb{F}_{q^2}^*$). Our goal is to show that there are only finitely many possibilities for q. The strategy and the outline of the proof are the following.

1. We show that $\gcd_{\mathbb{Q}[\mathsf{X}]}\{h_\alpha : \alpha > 0, \ \alpha \equiv -1 \pmod{r}\} = 1$, where h_α is defined in (15). It follows that there are only finitely many possibilities for $p := \mathrm{char}\,\mathbb{F}_q$.
2. If for a certain characteristic p, there are infinitely many possibilities for q, then we must have $p \in \{2, 3, 5\}$.
3. For each $p \in \{2, 3, 5\}$, we prove that there are only finitely many possibilities for q.

4.1. *Finiteness of the possibilities of the characteristic*

Let $\tau > 1$ be a prime power with $\tau \equiv -1 \pmod{r}$ and $\mathrm{char}\,\mathbb{F}_\tau = t$. Set $k = (\tau + 1)/r$. Note that the polynomial $g_\alpha(\mathsf{X})$ in (12) can be treated as a polynomial in $\mathbb{F}_t[\mathsf{X}]$. For Lemmas 3 and 5, we only require that $r \geq 2$. Lemma 4 requires that $r \geq m$ and Lemmas 6 – 8 require that $r \geq 3$.

Lemma 3. *Assume that $x \neq 0$ is a common root of g_τ, g_{τ^3} and g_{τ^4-2} in some extension of $\mathbb{F}_t$. Then, $x \in \mathbb{F}_{\tau^2}^*$, $x^r \in \mathbb{F}_\tau^* \setminus \{1\}$, and*

$$x^r = \frac{x(rx + 1 - r)}{(r+1)x - r}, \tag{16}$$

where $(r+1)x - r \neq 0$.

Proof. All polynomials considered in this proof are in characteristic t.

1° Let $\alpha = \tau^{2m+1}$, where $m \geq 0$ is an integer. Note that $\alpha \equiv -1$ (mod r) and that (12) gives

$$- g_\alpha = (X-1)\frac{X^{\alpha r} - 1}{X^r - 1} \sum_{l=0}^{r-1} \binom{2\alpha - \frac{\alpha+1}{r} + \frac{l}{r}}{\alpha} X^l + 1. \tag{17}$$

Let $k' = (\alpha+1)/r$. In the ring $\mathbb{Z}_t$ of t-adic integers,

$$\frac{1}{r} = \frac{k'}{1+\alpha} \equiv k'(1-\alpha) \pmod{\alpha^2}.$$

Note that $k' \equiv \frac{1}{r} \equiv k \pmod{t}$. For $l \in \mathbb{Z}$,

$$2\alpha - \frac{\alpha+1}{r} + \frac{l}{r} \equiv 2\alpha - k' + lk'(1-\alpha) \pmod{\alpha^2}$$
$$= k'(l-1) + \alpha(2 - k'l).$$

When $1 \leq l \leq r-1$, we have

$$\binom{2\alpha - \frac{\alpha+1}{r} + \frac{l}{r}}{\alpha} = \binom{k'(l-1) + \alpha(2 - k'l)}{\alpha}$$
$$= \binom{2 - k'l}{1} \qquad \text{(since } 0 \leq k'(l-1) < \alpha)$$
$$= 2 - k'l = 2 - kl.$$

When $l = 0$,

$$\binom{2\alpha - \frac{\alpha+1}{r}}{\alpha} = \binom{2\alpha - k'}{\alpha} = \binom{\alpha - k' + \alpha}{\alpha} = 1.$$

Therefore,

$$- g_\alpha$$
$$= (X-1)\frac{X^{\alpha r} - 1}{X^r - 1}\left[1 + \sum_{l=1}^{r-1}(2 - kl)X^l\right] + 1$$
$$= (X^r - 1)^{\alpha-1}(X-1)\left[-1 + \sum_{l=0}^{r-1}(2 - kl)X^l\right] + 1$$
$$= (X^r - 1)^{\alpha-1}(X-1)$$
$$\cdot \left[-1 + 2\frac{X^r - 1}{X - 1} - kX\left(rX^{r-1}(X-1)^{-1} - (X^r - 1)(X-1)^{-2}\right)\right] + 1$$
$$= (X^r - 1)^{\alpha-1}\left[\left(1 + \frac{kX}{X - 1}\right)(X^r - 1) - X\right] + 1. \tag{18}$$

(In the last step, we used the relation $kr \equiv 1 \pmod{t}$.) Since $g_\tau(x) = 0 = g_{\tau^3}(x)$, it follows from (18) that

$$(x^r - 1)^{\tau - 1} = (x^r - 1)^{\tau^3 - 1}.$$

Therefore $(x^r - 1)^\tau = (x^r - 1)^{\tau^3}$, which gives $x^{r\tau^2} = x^r$; that is, $x^r \in \mathbb{F}_{\tau^2}$. Clearly, $x^r \neq 1$. (Otherwise, by (18), we have $-g_\tau(x) = 1 \neq 0$, which is a contradiction.) Using (18), the condition $g_\tau(x) = 0$ becomes

$$-(x^r - 1)^{\tau - 1}x(x - 1) + (x^r - 1)^\tau\left[(1 + k)x - 1\right] + x - 1 = 0. \tag{19}$$

Treating (19) as a quadratic equation in x with coefficients in $\mathbb{F}_{\tau^2}$, we conclude that $x \in \mathbb{F}_{\tau^4}$.

    $2°$ Let $\beta = \tau^4 - 2$. (Note that $\beta \equiv -1 \pmod{r}$.) Let $k'' = (\beta + 1)/r$. In $\mathbb{Z}_t$,

$$\frac{1}{r} = \frac{k''}{\tau^4 - 1} \equiv -k'' \pmod{\tau^4},$$

and $k'' \equiv -\frac{1}{r} \equiv -k \pmod{t}$. For $1 \leq i \leq \alpha$ and $0 \leq l \leq r - 1$, we have

$$i + \beta - \frac{\beta + 1}{r} + \frac{l}{r} \equiv i - 2 - k'' - lk'' \pmod{\tau^4}.$$

Hence

$$\binom{i + \beta - \frac{\beta+1}{r} + \frac{l}{r}}{\beta} = \binom{i - 2 - k''(1 + l)}{\beta}$$

$$= \begin{cases} -1 & \text{if } i - 2 - k''(1 + l) \equiv -1 \pmod{\tau^4}, \\ 1 & \text{if } i - 2 - k''(1 + l) \equiv -2 \pmod{\tau^4}, \\ 0 & \text{otherwise.} \end{cases} \tag{20}$$

(In the last step, we used the fact that the base t digits of β are $(t - 2, t - 1, \ldots, t - 1)$.) Since

$$i - 2 - k''(1 + l) \leq \tau^4 - 4 - k'' < \tau^4 - 4$$

and

$$i - 2 - k''(1 + l) \geq -1 - k''r = -1 - (\beta + 1) = -\tau^4,$$

we see that for $c = -1$ and -2, $i - 2 - k''(1 + l) \equiv c \pmod{\tau^4}$ if and only if $i - 2 - k''(1 + l) = c$. Therefore we can rewrite (20) as

$$\binom{i + \beta - \frac{\beta+1}{r} + \frac{l}{r}}{\beta} = \begin{cases} -1 & \text{if } i - 2 - k''(1 + l) = -1, \\ 1 & \text{if } i - 2 - k''(1 + l) = -2, \\ 0 & \text{otherwise.} \end{cases} \tag{21}$$

By (12) and (21), we have

$$0 = -g_\beta(x)$$

$$= -(x-1)\sum_{l=0}^{r-1} x^l \left[(-1)^{k''(1+l)} \binom{\beta}{k''(1+l)+1} \frac{x^{r(k''(1+l)+1)}-1}{x^r-1} \right.$$

$$\left. + (-1)^{k''(1+l)} \binom{\beta}{k''(1+l)} \frac{x^{rk''(1+l)}-1}{x^r-1} \right] + 1$$

$$= -(x-1)\sum_{l=0}^{r-1} x^l (-1)^{k''(1+l)} \binom{\tau^4-2}{k''(1+l)+1} + 1. \tag{22}$$

(For the last step, note that $x^{rk''} = x^{\beta+1} = x^{\tau^4-1} = 1$ since $x \in \mathbb{F}_{\tau^4}^*$.) Note that

$$k''(1+l)+1 \begin{cases} < \tau^4 & \text{if } 0 \le l < r-1, \\ = \tau^4 & \text{if } l = r-1. \end{cases}$$

Hence

$$\binom{\tau^4-2}{k''(1+l)+1}$$
$$= \begin{cases} \binom{-2}{k''(1+l)+1} = (-1)^{k''(1+l)+1}\big(k''(1+l)+2\big) & \text{if } 0 \le l < r-1, \\ 0 & \text{if } l = r-1. \end{cases}$$

Thus (22) becomes

$$0 = (x-1)\sum_{l=0}^{r-2} (k''(1+l)+2)x^l + 1$$

$$= (x-1)\left[-x^{r-1} + \sum_{l=0}^{r-1}(2-k(1+l))x^l \right] + 1 \quad \text{(since } k'' \equiv -k \pmod{t}\text{)}$$

$$= (x-1)\left[-x^{r-1} + (2-k)\frac{x^r-1}{x-1} - kx\big(rx^{r-1}(x-1)^{-1} \right.$$
$$\left. - (x^r-1)(x-1)^{-2}\big) \right] + 1$$

$$= -x^r + x^{r-1} + (2-k)(x^r-1) - x^r + \frac{kx(x^r-1)}{x-1} + 1$$

$$= \frac{k}{x-1}(x^r-1) + x^{r-1} - 1.$$

$$\tag{23}$$

Multiplying (23) by x gives

$$0 = \left(1 + \frac{kx}{x-1}\right)(x^r - 1) - x + 1. \tag{24}$$

On the other hand, by (18),

$$0 = -g_\tau(x) = (x^r - 1)^{\tau-1}\left[\left(1 + \frac{kx}{x-1}\right)(x^r - 1) - x\right] + 1. \tag{25}$$

Combining (24) and (25) gives $(x^r - 1)^{\tau-1} = 1$, which implies that $x^r \in \mathbb{F}_\tau$. It follows from (19) that $x \in \mathbb{F}_{\tau^2}$.

Finally, by (24), we have

$$x^r\left[(r+1)x - r\right] = x(rx + 1 - r).$$

If $(r+1)x - r = 0$, then $rx + 1 - r = 0$; the two equations imply that $x = 1$, which is a contradiction. Hence we have $(r+1)x - r \neq 0$ and (16) follows. $\qquad\square$

Lemma 4. *Let* $x \in \overline{\mathbb{F}}_t$ *(the algebraic closure of* $\mathbb{F}_t$*) be such that* $x^r \in \mathbb{F}_\tau^* \setminus \{1\}$. *Let* m *be a positive integer, and assume that* $r \geq m$ *and* $k := (\tau+1)/r > 2(m-1)$. *Then*

$$g_{m-1+m\tau}(x) = (x-1) \sum_{\substack{0 \leq i_0 \leq m-1 \\ 0 \leq i_1 \leq m}} (-1)^{i_0+i_1} \binom{m-1}{i_0}\binom{m}{i_1} \frac{x^{r(i_0+i_1)} - 1}{x^r - 1}$$

$$\cdot \left[\sum_{l=0}^{r-1} \binom{i_0 + m - 1 + (l-m)k}{m-1}\binom{i_1 + m - lk}{m} x^l\right.$$

$$\left. - \sum_{l=0}^{m-1} \binom{i_0 + m - 1 + (l-m)k}{m-1}\binom{i_1 + m - lk}{m-1} x^l\right] - 1. \tag{26}$$

Proof. Let $\alpha = m - 1 + m\tau$. For any integer $i \geq 0$, we have $\binom{\alpha}{i} = 0$ unless $i = i_0 + i_1\tau$, where $0 \leq i_0 \leq m - 1$ and $0 \leq i_1 \leq m$, in which case,

$$\binom{\alpha}{i} = \binom{m-1}{i_0}\binom{m}{i_1}.$$

Thus by (12),

$$g_\alpha(x) = (x-1) \sum_{l=0}^{r-1} x^l \sum_{\substack{0 \leq i_0 \leq m-1 \\ 0 \leq i_1 \leq m}} (-1)^{i_0+i_1} \binom{m-1}{i_0}\binom{m}{i_1}$$

$$\cdot \binom{i_0 + i_1\tau + \alpha - \frac{\alpha+1}{r} + \frac{l}{r}}{\alpha} \frac{x^{r(i_0+i_1)} - 1}{x^r - 1} - 1. \tag{27}$$

We have

$$i_0 + i_1\tau + \alpha - \frac{\alpha+1}{r} + \frac{l}{r}$$
$$\equiv i_0 + i_1\tau + m - 1 + m\tau - mk + lk(1-\tau) \pmod{\tau^2} \tag{28}$$
$$= (i_0 + m - 1 + (l-m)k) + (i_1 + m - lk)\tau.$$

When $0 \le l \le m - 1$,

$$\begin{cases} i_0 + m - 1 + (l-m)k + \tau \le 2m - 2 - k + \tau < \tau, \\ i_0 + m - 1 + (l-m)k + \tau \ge m - 1 - mk + \tau \\ \qquad\qquad = \tau + m - 1 - \dfrac{m}{r}(\tau+1) \ge 0. \end{cases} \tag{29}$$

When $m \le l \le r - 1$,

$$0 \le i_0 + m - 1 + (l-m)k \le 2m - 2 + (r-1-m)k = 2m - 2 - (m+1)k + \tau + 1 < \tau. \tag{30}$$

(To see the last inequality, consider the cases $m = 1$ and $m > 1$ separately; note that when $m > 1$, $k > 2(m-1) \ge 2$.) Therefore,

$$\binom{i_0 + i_1\tau + \alpha - \frac{\alpha+1}{r} + \frac{l}{r}}{\alpha}$$
$$= \binom{(i_0 + m - 1 + (l-m)k) + (i_1 + m - lk)\tau}{m-1 \qquad + \qquad m\tau}$$

(by (28) and the fact that $m < \tau$, which is easy to see)

$$= \begin{cases} \dbinom{i_0 + m - 1 + (l-m)k}{m-1}\dbinom{i_1 + m - 1 - lk}{m} & \text{if } 0 \le l \le m-1, \\[3mm] \dbinom{i_0 + m - 1 + (l-m)k}{m-1}\dbinom{i_1 + m - lk}{m} & \text{if } m \le l \le r-1, \end{cases}$$

(by (29) and (30)).
$$\tag{31}$$

Combining (27) and (31) gives

$$g_\alpha(x) = (x-1) \sum_{\substack{0 \leq i_0 \leq m-1 \\ 0 \leq i_1 \leq m}} (-1)^{i_0+i_1} \binom{m-1}{i_0}\binom{m}{i_1} \frac{x^{r(i_0+i_1)}-1}{x^r-1}$$

$$\cdot \left[\sum_{l=0}^{m-1} \binom{i_0+m-1+(l-m)k}{m-1}\binom{i_1+m-1-lk}{m} x^l \right.$$

$$+ \left. \sum_{l=m}^{r-1} \binom{i_0+m-1+(l-m)k}{m-1}\binom{i_1+m-lk}{m} x^l \right] - 1$$

$$= (x-1) \sum_{\substack{0 \leq i_0 \leq m-1 \\ 0 \leq i_1 \leq m}} (-1)^{i_0+i_1} \binom{m-1}{i_0}\binom{m}{i_1} \frac{x^{r(i_0+i_1)}-1}{x^r-1}$$

$$\cdot \left[\sum_{l=0}^{r-1} \binom{i_0+m-1+(l-m)k}{m-1}\binom{i_1+m-lk}{m} x^l \right.$$

$$- \left. \sum_{l=0}^{m-1} \binom{i_0+m-1+(l-m)k}{m-1}\binom{i_1+m-1-lk}{m-1} x^l \right] - 1.$$

$\square$

Lemma 5. *Let $x \in \overline{\mathbb{F}}_t$ be such that $x^r \in \mathbb{F}_\tau^* \setminus \{1\}$. Assume that $k := (\tau+1)/r > 2$. Then*

$$g_{1+\tau+\tau^3}(x)$$

$$= (x-1) \sum_{0 \leq i_0,i_1,i_3 \leq 1} (-1)^{i_0+i_1+i_3} \frac{x^{r(i_0+i_1+i_3)}-1}{x^r-1} \tag{32}$$

$$\cdot \left[\sum_{l=0}^{r-1} (i_0+1+(l-2)k)(i_1+1-(l-1)k)(i_3+1-lk)x^l \right.$$

$$\left. - (i_0+1-2k)(i_1+i_3+k+1) - (i_0+1-k)(i_3+1-k)x \right] - 1.$$

Proof. Let $\alpha = 1+\tau+\tau^3$. For any integer $i \geq 0$, we have $\binom{\alpha}{i} = 0$ unless $i = i_0 + i_1\tau + i_3\tau^3$, where $0 \leq i_0, i_1, i_3 \leq 1$, in which case,

$$\binom{\alpha}{i} = 1.$$

Thus (12) gives

$$g_\alpha(x) = (x-1) \sum_{l=0}^{r-1} x^l \sum_{0 \le i_0, i_1, i_3 \le 1} (-1)^{i_0+i_1+i_3}$$

$$\cdot \binom{i_0 + i_1\tau + i_3\tau^3 + \alpha - \frac{\alpha+1}{r} + \frac{l}{r}}{\alpha} \frac{x^{r(i_0+i_1+i_3)} - 1}{x^r - 1} - 1. \tag{33}$$

We have

$$i_0 + i_1\tau + i_3\tau^3 + \alpha - \frac{\alpha+1}{r} + \frac{l}{r}$$

$$\equiv i_0 + i_1\tau + i_3\tau^3 + 1 + \tau + \tau^3 - k(2 - \tau + \tau^2) + lk(1 - \tau + \tau^2 - \tau^3)$$

$$(\bmod \ \tau^4)$$

$$= (i_0 + 1 + (l-2)k) + (i_1 + 1 + k(l-1)(\tau - 1))\tau + (i_3 + 1 - lk)\tau^3.$$

When $0 \le l \le 1$,

$$i_0 + 1 + (l-2)k + \tau \le 2 - k + \tau < \tau,$$

and

$$i_0 + 1 + (l-2)k + \tau \ge 1 - 2k + \tau = \tau + 1 - \frac{2}{r}(\tau+1) \ge 0.$$

Thus

$$\binom{i_0 + i_1\tau + i_3\tau^3 + \alpha - \frac{\alpha+1}{r} + \frac{l}{r}}{\alpha}$$

$$= (i_0 + 1 + (l-2)k) \binom{i_1 + k(l-1)(\tau-1)\tau + (i_3 + 1 - lk)\tau^3}{\tau} + \frac{(i_3 + 1 - lk)\tau^3}{\tau^3}. \tag{34}$$

When $l = 0$,

$$\begin{cases} i_1 + k(l-1)(\tau-1) + \tau^2 \ge -k(\tau-1) + \tau^2 = -\frac{1}{r}(\tau+1)(\tau-1) + \tau^2 \ge 0, \\ i_1 + k(l-1)(\tau-1) + \tau^2 \le 1 - k(\tau-1) + \tau^2 < \tau^2. \end{cases}$$

When $l = 1$,

$$\begin{cases} i_1 + k(l-1)(\tau-1) \ge 0, \\ i_1 + k(l-1)(\tau-1) \le 1 < \tau^2. \end{cases}$$

Hence

$$\binom{i_1 + k(l-1)(\tau-1)\tau + (i_3 + 1 - lk)\tau^3}{\tau} + \frac{(i_3 + 1 - lk)\tau^3}{\tau^3} = \begin{cases} (i_1 + k)i_3 & \text{if } l = 0, \\ i_1(i_3 + 1 - k) & \text{if } l = 1. \end{cases} \tag{35}$$

When $2 \le l \le r - 1$,

$$\begin{cases} i_0 + 1 + (l-2)k \ge 0, \\ i_0 + 1 + (l-2)k \le 2 + (r-3)k = 2 - 3k + \tau + 1 < \tau, \end{cases}$$

and

$$\begin{cases} i_1 + 1 + k(l-1)(\tau-1) \ge 0, \\ i_1 + 1 + k(l-1)(\tau-1) \le 2 + k(r-3)(\tau-1) \le 2 - 3 + kr(\tau-1) \\ \qquad\qquad\qquad\qquad\qquad\qquad < (\tau+1)(\tau-1) < \tau^2. \end{cases}$$

Hence

$$\binom{i_0 + i_1\tau + i_3\tau^3 + \alpha - \frac{\alpha+1}{r} + \frac{l}{r}}{\alpha} \tag{36}$$
$$= \big(i_0 + 1 + (l-2)k\big)\big(i_1 + 1 - (l-1)k\big)(i_3 + 1 - lk).$$

Gathering (34) – (36), we have

$$\binom{i_0 + i_1\tau + i_3\tau^3 + \alpha - \frac{\alpha+1}{r} + \frac{l}{r}}{\alpha}$$

$$= \big(i_0 + 1 + (l-2)k\big) \cdot \begin{cases} (i_1 + k)i_3 & \text{if } l = 0, \\ i_1(i_3 + 1 - k) & \text{if } l = 1, \\ (i_1 + 1 - (l-1)k)(i_3 + 1 - lk) & \text{if } 2 \le l \le r - 1. \end{cases} \tag{37}$$

Now by (33) and (37),

$$g_\alpha(x) = (x-1) \sum_{0 \le i_0, i_1, i_3 \le 1} (-1)^{i_0 + i_1 + i_3} \frac{x^{r(i_0 + i_1 + i_3)} - 1}{x^r - 1}$$
$$\cdot \Big[(i_0 + 1 - 2k)(i_1 + k)i_3 + (i_0 + 1 - k)i_1(i_3 + 1 - k)x$$
$$+ \sum_{l=2}^{r-1} (i_0 + 1 + (l-2)k)(i_1 + 1 - (l-1)k)(i_3 + 1 - lk)x^l \Big] - 1$$

$$= (x-1) \sum_{0 \le i_0, i_1, i_3 \le 1} (-1)^{i_0 + i_1 + i_3} \frac{x^{r(i_0 + i_1 + i_3)} - 1}{x^r - 1}$$
$$\cdot \Big[\sum_{l=0}^{r-1} (i_0 + 1 + (l-2)k)(i_1 + 1 - (l-1)k)(i_3 + 1 - lk)x^l$$
$$- (i_0 + 1 - 2k)(i_1 + i_3 + k + 1) - (i_0 + 1 - k)(i_3 + 1 - k)x \Big] - 1.$$

$\square$

Recall from (15) that for positive integers α with $\alpha \equiv -1 \pmod{r}$ we have $h_\alpha = \mathbf{X}^{-1} g_\alpha \in (\mathbb{Z}[1/r])[\mathbf{X}]$.

Lemma 6. *Let $r \geq 3$. Assume that τ is a prime such that $\tau \equiv -1 \pmod{r}$, $k := (\tau+1)/r > 2$ and $\tau \nmid \rho(r)$, where $\rho(r)$ is a nonzero integer, depending only on r, to be defined in (47). Then*

$$\gcd_{\mathbb{F}_\tau[\mathbf{X}]}(h_\tau, h_{\tau^3}, h_{\tau^4-2}, h_{1+\tau+\tau^3}, h_{2\tau+1}) = 1. \tag{38}$$

Proof. $1°$ We claim that $h_\tau(0) \neq 0$ in $\mathbb{F}_\tau$.

Using the first step of (18), we have

$$-g'_\tau(0) = 1 + (-1)(2-k) = -1 + k.$$

Since $k = (\tau+1)/r$, we have $1 < k < \tau$. Thus $g'_\tau(0) \neq 0$ and hence $h_\tau(0) \neq 0$.

$2°$ Assume to the contrary that $h_\tau, h_{\tau^3}, h_{\tau^4-2}, h_{1+\tau+\tau^3}, h_{2\tau+1}$ have a common root $x \in \overline{\mathbb{F}}_\tau$. By $1°$, $x \neq 0$. By Lemma 16, $x^r \in \mathbb{F}_\tau^* \setminus \{1\}$, $(r+1)x - r \neq 0$, and (16) holds.

By Lemma 5, we have (32). We need to rewrite (32) in a form suitable for further computation. Let $y = x^r$. For any integer $k \geq 0$, define

$$s_k = \sum_{l=0}^{r-1} l^k \mathbf{X}^l \in \mathbb{Z}[\mathbf{X}]. \tag{39}$$

(Note that s_k also depends on r. Since r is fixed, it is suppressed in the notation.) The polynomial s_k satisfies the recursive relation

$$\begin{cases} s_0 = \dfrac{\mathbf{X}^r - 1}{\mathbf{X} - 1}, \\ s_k = \dfrac{1}{\mathbf{X} - 1}\left[(-1)^k \displaystyle\sum_{j=0}^{k-1}(-1)^j \binom{k}{j} s_j - (-1)^k + (r-1)^k \mathbf{X}^r\right], \quad k > 0. \end{cases} \tag{40}$$

We can rewrite (32) as

$$\begin{aligned} &g_{1+\tau+\tau^3}(x) \\ &= (x-1) \sum_{0 \leq i_0, i_1, i_3 \leq 1} (-1)^{i_0+i_1+i_3} \frac{y^{i_0+i_1+i_3} - 1}{y - 1} \\ &\quad \cdot \big[a_0 s_0(x) + a_1 s_1(x) + a_2 s_2(x) + a_3 s_3(x) \\ &\quad - (i_0 + 1 - 2k)(i_1 + i_3 + k + 1) - (i_0 + 1 - k)(i_3 + 1 - k)x\big] - 1, \end{aligned} \tag{41}$$

where
$$\begin{cases} a_0 = 1 - k - 2k^2 + i_0 + ki_0 + i_1 - 2ki_1 + i_0i_1 + i_3 - ki_3 - 2k^2i_3 \\ \quad\quad + i_0i_3 + ki_0i_3 + i_1i_3 - 2ki_1i_3 + i_0i_1i_3, \\ a_1 = -k + 4k^2 + 2k^3 - 2ki_0 - k^2i_0 + 2k^2i_1 - ki_0i_1 \\ \quad\quad + 3k^2i_3 - ki_0i_3 + ki_1i_3, \\ a_2 = -k^2 - 3k^3 + k^2i_0 - k^2i_1 - k^2i_3, \\ a_3 = k^3. \end{cases}$$
(42)

Combining (41) and (42) allows us to express $g_{1+\tau+\tau^3}(x)$ as a rational function in x and y, and, with the substitutions $k = 1/r$ and
$$y = \frac{x(rx + 1 - r)}{(r+1)x - r},$$
that expression becomes
$$g_{1+\tau+\tau^3}(x) = \frac{x(x-1)^2}{\left((r+1)x - r\right)^3} G_{1+\tau+\tau^3}(x), \tag{43}$$
where
$$\begin{aligned} G_{1+\tau+\tau^3}(\mathrm{X}) \\ = \left(4r^3 - 3r + 1\right)\mathrm{X}^4 + \left(-9r^3 + 6r^2 + 5r - 1\right)\mathrm{X}^3 + \left(6r^3 - 5r^2 - 7r - 1\right)\mathrm{X}^2 \\ + \left(-r^3 - 2r^2 + 11r - 7\right)\mathrm{X} + r^2 - 3r + 2. \end{aligned}$$
(44)

In the same way, and starting from (16) with $m = 2$, we find that
$$2g_{1+2\tau}(x) = \frac{x(x-1)^3}{\left((r+1)x - r\right)^3} G_{1+2\tau}(x), \tag{45}$$
where
$$\begin{aligned} G_{1+2\tau}(\mathrm{X}) = \left(12r^3 + 2r^2 - 8r + 2\right)\mathrm{X}^3 + \left(-28r^3 + 31r^2 + 16r - 7\right)\mathrm{X}^2 \\ + \left(20r^3 - 44r^2 + 14r + 6\right)\mathrm{X} - 4r^3 + 11r^2 - 6r - 1. \end{aligned}$$
(46)

The resultant of $G_{1+\tau+\tau^3}$ and $G_{1+2\tau}$ is
$$\begin{aligned} \rho(r) &:= R(G_{1+\tau+\tau^3}, G_{1+2\tau}) \\ &= -(r-2)(r-1)^2(r+1)(2r-1) \\ &\quad \cdot (2616r^{10} - 4994r^9 + 212r^8 - 21785r^7 + 58174r^6 \\ &\quad - 30241r^5 + 2258r^4 - 30963r^3 + 12562r^2 - 433r - 78). \end{aligned}$$
(47)

Since $r \geq 3$, we have $\rho(r) \neq 0$ in $\mathbb{Z}$. (It is easy to show that $\rho(r)$, as a polynomial in r, has no integer roots ≥ 3.) Therefore we may choose τ such that $\tau \nmid \rho(r)$. However, since x is a common root of $g_{1+\tau+\tau^3}$ and $g_{1+2\tau}$, it is a common root of $G_{1+\tau+\tau^3}$ and $G_{1+2\tau}$, which contradicts the fact that $R(G_{1+\tau+\tau^3}, G_{1+2\tau}) \not\equiv 0 \pmod{\tau}$. $\square$

Lemma 7. *Assume $r \geq 3$. We have*

$$\gcd_{\mathbb{Q}[X]}\{h_\alpha : \alpha > 0,\ \alpha \equiv -1 \pmod{r}\} = 1.$$

Proof. Assume the contrary; that is, there exists a primitive polynomial $d \in \mathbb{Z}[X]$ with $\deg d > 0$ such that $d \mid h_\alpha$ in $\mathbb{Q}[X]$ for all $\alpha > 0$ with $\alpha \equiv -1$ $\pmod{r}$. Choose a prime τ satisfying the conditions in Lemma 6 such that τ does not divide the leading coefficient of d, that is, $\deg_{\mathbb{F}_\tau[X]} d = \deg_{\mathbb{Q}[X]} d$. For all $\alpha > 0$ with $\alpha \equiv -1 \pmod{r}$, since $h_\alpha \in (\mathbb{Z}[1/r])[X]$, we have $d \mid h_\alpha$ in $\mathbb{F}_\tau[X]$. This a contradiction to (38). $\qquad\square$

Lemma 8. *Fix $r \geq 3$, and assume that f is a PP of $\mathbb{F}_{q^2}$, where $q \geq r$, $q + 1 \equiv 0 \pmod{r}$ and $a^{q+1} \neq 1$ ($a \in \mathbb{F}_{q^2}^*$). Then there are only finitely many possibilities for $p := \operatorname{char} \mathbb{F}_q$.*

Proof. By Lemma 7, there exist $0 < \alpha_1 < \cdots < \alpha_m$ with $\alpha_i \equiv -1$ $\pmod{r}$, $1 \leq r \leq m$, such that

$$\gcd_{\mathbb{Q}[X]}(h_{\alpha_1}, \ldots, h_{\alpha_m}) = 1.$$

Therefore, there exist $a_1, \ldots, a_m \in \mathbb{Z}[X]$ such that

$$a_1 h_{\alpha_1} + \cdots + a_m h_{\alpha_m} = A \in \mathbb{Z} \setminus \{0\}. \tag{48}$$

If $q < (r-1)(\alpha_m + 2)$, there are only finitely many possibilities for p. If $q \geq (r-1)(\alpha_m + 2)$, by Lemma 2, $h_{\alpha_1}, \ldots, h_{\alpha_m}$ have a common root in $\overline{\mathbb{F}}_p$. It follows from (48) that $p \mid A$. Hence there are only finitely many possibilities for p. $\qquad\square$

4.2. *Finiteness of the possibilities of q in a given characteristic*

Lemma 9. *Fix $r \geq 3$, and let p be a prime such that there is a power τ of p with $\tau \equiv -1 \pmod{r}$. If f is a PP of $\mathbb{F}_{q^2}$, where $\operatorname{char} \mathbb{F}_q = p$, $q \equiv -1$ $\pmod{r}$, $q \geq \tau^4$ and $a^{q+1} \neq 1$ ($a \in \mathbb{F}_{q^2}^*$), then $p \in \{2, 3, 5\}$.*

Proof. Since f is a PP of $\mathbb{F}_{q^2}$ and $q \geq \tau^4$, it follows that g_α, $\alpha \in \{\tau, 1 + 2\tau, 2 + 3\tau, 3 + 4\tau, 1 + \tau + \tau^3, -2 + \tau^4\}$, have a common root $x \in \overline{\mathbb{F}}_p \setminus \{0\}$ with $x^r \in \mathbb{F}_\tau^* \setminus \{1\}$. By Lemma 3, $(r+1)x - r \neq 0$, and (16) holds. The equations (43) and (45) remain valid. Moreover, $g_{2+3\tau}(x)$ and $g_{3+4\tau}(x)$ can be computed in a way similar to the computation of $g_{1+2\tau}(x)$. To sum up, we have

$$g_{1+\tau+\tau^3}(x) = \frac{x(x-1)^2}{\left((r+1)x - r\right)^3} G_{1+\tau+\tau^3}(x), \tag{49}$$

$$2g_{1+2\tau}(x) = \frac{x(x-1)^3}{\left((r+1)x - r\right)^3} G_{1+2\tau}(x), \tag{50}$$

$$12g_{2+3\tau}(x) = \frac{(x-1)^2}{\left((r+1)x - r\right)^5} G_{2+3\tau}(x), \tag{51}$$

$$144g_{3+4\tau}(x) = \frac{(x-1)^2}{\left((r+1)x - r\right)^7} G_{3+4\tau}(x), \tag{52}$$

where

$$
\begin{aligned}
&G_{1+\tau+\tau^3}(\mathbf{X}) \\
&= \left(4r^3 - 3r + 1\right)\mathbf{X}^4 + \left(-9r^3 + 6r^2 + 5r - 1\right)\mathbf{X}^3 + \left(6r^3 - 5r^2 - 7r - 1\right)\mathbf{X}^2 \\
&\quad + \left(-r^3 - 2r^2 + 11r - 7\right)\mathbf{X} + r^2 - 3r + 2,
\end{aligned}
\tag{53}
$$

$$
\begin{aligned}
G_{1+2\tau}(\mathbf{X}) &= \left(12r^3 + 2r^2 - 8r + 2\right)\mathbf{X}^3 + \left(-28r^3 + 31r^2 + 16r - 7\right)\mathbf{X}^2 \\
&\quad + \left(20r^3 - 44r^2 + 14r + 6\right)\mathbf{X} - 4r^3 + 11r^2 - 6r - 1,
\end{aligned}
\tag{54}
$$

$$
\begin{aligned}
&G_{2+3\tau}(\mathbf{X}) \\
&= 2\left(480r^5 - 56r^4 - 612r^3 + 574r^2 - 210r + 4\right)\mathbf{X}^8 \\
&\quad + 2\left(-2904r^5 + 2390r^4 + 2655r^3 - 4033r^2 + 2004r - 238\right)\mathbf{X}^7 \\
&\quad + 2\left(7464r^5 - 11148r^4 - 1230r^3 + 9711r^2 - 6477r + 1266\right)\mathbf{X}^6 \\
&\quad + 2\left(-10572r^5 + 22426r^4 - 9387r^3 - 8690r^2 + 9177r - 2414\right)\mathbf{X}^5 \\
&\quad + 2\left(8940r^5 - 24118r^4 + 19116r^3 - 502r^2 - 5736r + 1958\right)\mathbf{X}^4 \\
&\quad + 2\left(-4560r^5 + 14658r^4 - 15783r^3 + 5571r^2 + 1002r - 762\right)\mathbf{X}^3 \\
&\quad + 2\left(1344r^5 - 4928r^4 + 6354r^3 - 3305r^2 + 375r + 142\right)\mathbf{X}^2 \\
&\quad + 2\left(-204r^5 + 830r^4 - 1197r^3 + 728r^2 - 147r - 10\right)\mathbf{X} \\
&\quad + 2\left(12r^5 - 54r^4 + 84r^3 - 54r^2 + 12r\right),
\end{aligned}
\tag{55}
$$

$$G_{3+4\tau}(\mathbf{X})$$
$$= \big(226800r^7 - 211140r^6 - 188340r^5 + 388485r^4 - 255456r^3 + 74058r^2$$
$$- 6204r + 597\big)\mathbf{X}^{12} + \big(-2030400r^7 + 3135420r^6 + 355372r^5 - 3641401r^4$$
$$+ 3175096r^3 - 1217746r^2 + 175660r - 7873\big)\mathbf{X}^{11} + \big(8169120r^7$$
$$- 17438796r^6 + 6486732r^5 + 12248029r^4 - 15562968r^3 + 7606042r^2$$
$$- 1549092r + 94501\big)\mathbf{X}^{10} + \big(-19474560r^7 + 52594740r^6 - 39521644r^5$$
$$- 13998585r^4 + 38790944r^3 - 24354306r^2 + 6413804r - 556953\big)\mathbf{X}^9$$
$$+ \big(30522960r^7 - 98911656r^6 + 106243584r^5 - 18404742r^4 - 50668800r^3$$
$$+ 44318196r^2 - 14546448r + 1649658\big)\mathbf{X}^8 + \big(-32977152r^7 + 123769656r^6$$
$$- 168070536r^5 + 81291294r^4 + 25818720r^3 - 46698708r^2 + 19296552r$$
$$- 2678946\big)\mathbf{X}^7 + \big(25026624r^7 - 106072632r^6 + 170852376r^5 - 120540462r^4$$
$$+ 17400096r^3 + 26245236r^2 - 15232440r + 2516178\big)\mathbf{X}^6 + \big(-13329792r^7$$
$$+ 62623944r^6 - 115027224r^5 + 100925334r^4 - 36600336r^3 - 4219764r^2$$
$$+ 6945288r - 1415658\big)\mathbf{X}^5 + \big(4888080r^7 - 25116660r^6 + 51237324r^5$$
$$- 52014495r^4 + 25779696r^3 - 3616638r^2 - 1606668r + 481329\big)\mathbf{X}^4$$
$$+ \big(-1183680r^7 + 6598476r^6 - 14687844r^5 + 16566267r^4 - 9691800r^3$$
$$+ 2414310r^2 + 74028r - 96093\big)\mathbf{X}^3 + \big(175392r^7 - 1059516r^6 + 2548316r^5$$
$$- 3117839r^4 + 2021816r^3 - 621326r^2 + 43484r + 10249\big)\mathbf{X}^2 + \big(-13824r^7$$
$$+ 91332r^6 - 236988r^5 + 310403r^4 - 215952r^3 + 73910r^2 - 8436r - 445\big)\mathbf{X}$$
$$+ 432r^7 - 3168r^6 + 8872r^5 - 12288r^4 + 8944r^3 - 3264r^2 + 472r.$$

$$(56)$$

Assume to the contrary that $p \notin \{2, 3, 5\}$.

1° We claim that $r \neq 1, 2$ in $\mathbb{F}_p$.
If $r = 1$ in $\mathbb{F}_p$, by (53) and (54),

$$G_{1+\tau+\tau^3}(\mathbf{X}) = \mathbf{X}(1 - 7\mathbf{X} + \mathbf{X}^2 + 2\mathbf{X}^3),$$
$$G_{1+2\tau}(\mathbf{X}) = 2(1 - 2\mathbf{X} + 6\mathbf{X}^2 + 4\mathbf{X}^3).$$

We have

$$A \cdot (1 - 7\mathbf{X} + \mathbf{X}^2 + 2\mathbf{X}^3) + B \cdot (1 - 2\mathbf{X} + 6\mathbf{X}^2 + 4\mathbf{X}^3) = 3 \cdot 31,$$

where

$$A = 4(-5 + 26\mathbf{X} + 17\mathbf{X}^2), \qquad B = 113 - 18\mathbf{X} - 34\mathbf{X}^2.$$

When $p \neq 31$, this contradicts the fact that x is a common root of $G_{1+\tau+\tau^3}$ and $G_{1+2\tau}$. When $p = 31$, we find that

$$\gcd_{\mathbb{F}_{31}[\mathsf{X}]}\left(1 - 7\mathsf{X} + \mathsf{X}^2 + 2\mathsf{X}^3, \; G_{2+3\tau}(\mathsf{X})\right) = 1,$$

which is also a contradiction.

Similarly, if $r = 2$ in $\mathbb{F}_p$, we have

$$G_{1+\tau+\tau^3}(\mathsf{X}) = \mathsf{X}(\mathsf{X} - 1)(3\mathsf{X} - 1)(9\mathsf{X} - 1),$$
$$G_{1+2\tau}(\mathsf{X}) = 1 + 18\mathsf{X} - 75\mathsf{X}^2 + 90\mathsf{X}^3,$$

and

$$A \cdot (3\mathsf{X} - 1)(9\mathsf{X} - 1) + B \cdot (1 + 18\mathsf{X} - 75\mathsf{X}^2 + 90\mathsf{X}^3) = 2 \cdot 89,$$

where

$$A = 101 - 210\mathsf{X} - 120\mathsf{X}^2, \qquad B = 77 + 36\mathsf{X}.$$

When $p \neq 89$, we have a contradiction. When $p = 89$, we find that

$$\gcd_{\mathbb{F}_{89}[\mathsf{X}]}\left((3\mathsf{X} - 1)(9\mathsf{X} - 1), \; G_{2+3\tau}(\mathsf{X})\right) = 1,$$

which is again a contradiction.

$2°$ We computed the resultant $R(G_\alpha, G_\beta)$ for all $\alpha, \beta \in \{1+\tau+\tau^3, 1+2\tau, 2+3\tau, 3+4\tau\}$, $\alpha \neq \beta$. The following selected results are to be used in the next step.

$R(G_{1+\tau+\tau^3}, G_{1+2\tau})$
$= -(r - 2)(r - 1)^2(r + 1)(2r - 1)(2616r^{10} - 4994r^9 + 212r^8 - 21785r^7$
$\quad + 58174r^6 - 30241r^5 + 2258r^4 - 30963r^3 + 12562r^2 - 433r - 78),$

$R(G_{1+\tau+\tau^3}, G_{2+3\tau})$
$= -432(r - 2)(r - 1)^2(904200192r^{24} - 26766761472r^{23} + 331824091392r^{22}$
$\quad - 2431984416768r^{21} + 12259306904112r^{20} - 46191235200768r^{19}$
$\quad + 136361704467312r^{18} - 323448347180178r^{17} + 622072535670810r^{16}$
$\quad - 963022456113483r^{15} + 1163874205192488r^{14} - 1011383557230121r^{13}$
$\quad + 458542507975098r^{12} + 246769336119765r^{11} - 714942562087848r^{10}$
$\quad + 750787901224431r^9 - 493424131713792r^8 + 213252937500746r^7$
$\quad - 55526043586236r^6 + 4871506549056r^5 + 1905998384136r^4$
$\quad - 727704486624r^3 + 99138790224r^2 - 4740664936r - 30723984),$

$$R(G_{1+2\tau}, G_{2+3\tau})$$
$$= 64(r-1)^2 \big(3034644480r^{24} - 32529595392r^{23} + 657287337600r^{22}$$
$$- 8299978411200r^{21} + 56251922766208r^{20} - 235334307782820r^{19}$$
$$+ 667348751791160r^{18} - 1357351029780397r^{17} + 2048898884042996r^{16}$$
$$- 2339185451094502r^{15} + 2030190962218068r^{14} - 1317041479048822r^{13}$$
$$+ 596228225529184r^{12} - 146983940173022r^{11} - 9832362841412r^{10}$$
$$+ 18243146047288r^9 - 3480285918736r^8 - 419376654818r^7$$
$$+ 134266774188r^6 + 4006393734r^5 - 501750288r^4 + 168774858r^3$$
$$+ 13728780r^2 - 279387r - 26676\big),$$

$$R(G_{2+3\tau}, G_{3+4\tau}) = -2166612408926208(r-1)^4 r F(r),$$

where $F(r)$ is a polynomial of degree 61 in r with integer coefficients; see [4]. Let $R'(G_\alpha, G_\beta)$ denote the expression obtained from $R(G_\alpha, G_\beta)$ with the factors r, $r-1$ and $r-2$ removed. For $\alpha_1, \beta_1, \alpha_2, \beta_2 \in \{1+\tau+\tau^3, 1+2\tau, 2+3\tau, 3+4\tau\}$, we treat $R'(G_{\alpha_1}, G_{\beta_1})$ and $R'(G_{\alpha_2}, G_{\beta_2})$ as polynomials in r with integer coefficients, whose resultant $R(R'(G_{\alpha_1}, G_{\beta_1}), R'(G_{\alpha_2}, G_{\beta_2}))$ is thus an integer. We computed

$$R\big(R'(G_{1+\tau+\tau^3}, G_{1+2\tau}), R'(G_{1+\tau+\tau^3}, G_{2+3\tau})\big),$$
$$R\big(R'(G_{1+\tau+\tau^3}, G_{1+2\tau}), R'(G_{1+2\tau}, G_{2+3\tau})\big),$$
$$R\big(R'(G_{1+\tau+\tau^3}, G_{1+2\tau}), R'(G_{2+3\tau}, G_{3+4\tau})\big),$$

each of which is a very large integer. However, the gcd of the above three integers is $2 \cdot 3^2 \cdot 5$, which is nonzero in $\mathbb{F}_p$. Thus we have a contradiction. $\square$

Remark. In fact, we computed

$$R(R'(G_{\alpha_1}, G_{\beta_1}), R'(G_{\alpha_2}, G_{\beta_2})) \tag{57}$$

for all $\alpha_1, \beta_1, \alpha_2, \beta_2 \in \{1 + \tau + \tau^3, 1 + 2\tau, 2 + 3\tau, 3 + 4\tau\}$ with $\alpha_1 \neq \beta_1$, $\alpha_2 \neq \beta_2$ and $\{\alpha_1, \beta_1\} \neq \{\alpha_2, \beta_2\}$. It turned out that the gcd of all such integers is $2 \cdot 3^2 \cdot 5$. Therefore, consideration of all the resultants in (57) still cannot eliminate the possibilities of $p = 2, 3, 5$.

The following corollary is immediate from Lemma 9

Corollary 1. *Fix $r \geq 3$ and a prime p with $p \notin \{2, 3, 5\}$. Then there are only finitely many q with char $\mathbb{F}_q = p$ and $q \equiv -1 \pmod{r}$ for which f is a PP of $\mathbb{F}_{q^2}$, where $a^{q+1} \neq 1$ $(a \in \mathbb{F}_{q^2}^*)$.*

4.3. *The cases* $p = 2, 3, 5$

Lemma 10. *Fix* $r \geq 3$, *and let* $p \in \{2, 3, 5\}$. *Fix a power* τ *(> 2) of* p *such that* $\tau \equiv -1 \pmod{r}$, *assuming that such a power exists. Let* q *be a power of* p *with* $q \equiv -1 \pmod{r}$. *If* $q \geq r\tau^4$, *then* f *is not a PP of* $\mathbb{F}_{q^2}$, *where* $a^{q+1} \neq 1$ $(a \in \mathbb{F}_{q^2}^*)$.

Proof. Assume to the contrary that f is a PP of $\mathbb{F}_{q^2}$. It follows from Lemmas 2 and 3 that g_α, $\alpha \in \{\tau, 1+2\tau, 2+3\tau, 3+4\tau, 1+\tau+\tau^3, -2+\tau^4\}$, have a common root $x \in \overline{\mathbb{F}}_p \setminus \{0\}$ with $x^r \in \mathbb{F}_\tau^* \setminus \{1\}$ and $(r+1)x - r \neq 0$ that satisfies (16).

    Case 1. Assume $p = 2$. Clearly, $r = 1$ in $\mathbb{F}_2$. It follows that

$$G_{1+\tau+\tau^3}(\mathsf{X}) = \mathsf{X}(1 + \mathsf{X} + \mathsf{X}^2).$$

Since $p = 2$, $G_{1+2\tau}$, $G_{2+3\tau}$ and $G_{3+4\tau}$ in (53) – (55) are all 0, due to the fact that they arise from $2g_{1+2\tau}(x)$, $12g_{2+3\tau}(x)$ and $144g_{3+4\tau}(x)$, respectively. It is possible to compute new polynomials arising in a similar way from $g_{1+2\tau}(x)$, $g_{2+3\tau}(x)$ and $g_{3+4\tau}(x)$. The result from $g_{1+2\tau}(x)$ is still 0. The result from $g_{2+3\tau}(x)$, whose computation is detailed below, will be useful.

    Let $k = (\tau + 1)/r$ and $y = x^r$. By (26) (with $m = 3$),

$$
\begin{aligned}
g_{2+3\tau}(x) = (x - 1) &\sum_{\substack{0 \leq i_0 \leq 2 \\ 0 \leq i_1 \leq 3}} (-1)^{i_0 + i_1} \binom{2}{i_0}\binom{3}{i_1} \frac{y^{i_0 + i_1} - 1}{y - 1} \\
&\cdot \left[\sum_{l=0}^{r-1} \binom{i_0 + 2 + (l-3)k}{2}\binom{i_1 + 2 - lk}{3} x^l \right. \\
&\left. - \sum_{l=0}^{2} \binom{i_0 + 2 + (l-3)k}{2}\binom{i_1 + 2 - lk}{2} x^l \right] - 1.
\end{aligned}
\tag{58}
$$

    Let F be a finite degree unramified extension of $\mathbb{Q}_2$ such that the finite field $\mathcal{O}_F/2\mathcal{O}_F$ contains x, where $\mathcal{O}_F$ is the ring of integers of F. Let $\mathfrak{x} \in \mathcal{O}_F$ be a lift of x, and set $\mathfrak{y} = \mathfrak{x}^r$. Then

$$
\mathfrak{y} = \frac{\mathfrak{x}(r\mathfrak{x} + 1 - r)}{(r+1)\mathfrak{x} - r} \pmod{2}.
\tag{59}
$$

Let $\mathfrak{g} \in \mathcal{O}_F$ denote the expression obtained from (58) with x and y replaced by $\mathfrak{x}$ and $\mathfrak{y}$, respectively. Then $\mathfrak{g}$ can be computed in a manner similar to

(41). We have

$$\mathfrak{g} = (\mathfrak{r} - 1) \sum_{\substack{0 \le i_0 \le 2 \\ 0 \le i_1 \le 3}} (-1)^{i_0 + i_1} \binom{2}{i_0} \binom{3}{i_1} \frac{\mathfrak{y}^{i_0 + i_1} - 1}{\mathfrak{y} - 1}$$

$$\cdot \Big[a_0 s_0(\mathfrak{r}) + a_1 s_1(\mathfrak{r}) + a_2 s_2(\mathfrak{r}) + a_3 s_3(\mathfrak{r}) + a_4 s_4(\mathfrak{r}) + a_5 s_5(\mathfrak{r}) \quad (60)$$

$$- \sum_{l=0}^{2} \binom{i_0 + 2 + (l-3)k}{2} \binom{i_1 + 2 - lk}{2} \mathfrak{r}^l \Big] - 1,$$

where s_i is given by (39) and (40), and

$$a_0 = \frac{i_0^2 i_1^3}{12} + \frac{i_0^2 i_1^2}{2} + \frac{11 i_0^2 i_1}{12} + \frac{i_0^2}{2} - \frac{1}{2} i_0 i_1^3 k + \frac{i_0 i_1^3}{4} - 3 i_0 i_1^2 k + \frac{3 i_0 i_1^2}{2} - \frac{11 i_0 i_1 k}{2}$$

$$+ \frac{11 i_0 i_1}{4} - 3 i_0 k + \frac{3 i_0}{2} + \frac{3 i_1^3 k^2}{4} - \frac{3 i_1^3 k}{4} + \frac{i_1^3}{6} + \frac{9 i_1^2 k^2}{2} - \frac{9 i_1^2 k}{2} + i_1^2$$

$$+ \frac{33 i_1 k^2}{4} - \frac{33 i_1 k}{4} + \frac{11 i_1}{6} + \frac{9 k^2}{2} - \frac{9k}{2} + 1,$$

$$a_1 = -\frac{1}{4} i_0^2 i_1^2 k - i_0^2 i_1 k - \frac{11 i_0^2 k}{12} + \frac{1}{6} i_0 i_1^3 k + \frac{3}{2} i_0 i_1^2 k^2 + \frac{1}{4} i_0 i_1^2 k + 6 i_0 i_1 k^2$$

$$- \frac{7 i_0 i_1 k}{6} + \frac{11 i_0 k^2}{2} - \frac{7 i_0 k}{4} - \frac{1}{2} i_1^3 k^2 + \frac{i_1^3 k}{4} - \frac{9 i_1^2 k^3}{4} - \frac{3 i_1^2 k^2}{4} + i_1^2 k$$

$$- 9 i_1 k^3 + \frac{7 i_1 k^2}{2} + \frac{3 i_1 k}{4} - \frac{33 k^3}{4} + \frac{21 k^2}{4} - \frac{k}{3},$$

$$a_2 = \frac{1}{4} i_0^2 i_1 k^2 + \frac{i_0^2 k^2}{2} - \frac{1}{2} i_0 i_1^2 k^2 - \frac{3}{2} i_0 i_1 k^3 - \frac{5}{4} i_0 i_1 k^2 - 3 i_0 k^3 - \frac{i_0 k^2}{3} + \frac{i_1^3 k^2}{12}$$

$$+ \frac{3 i_1^2 k^3}{2} - \frac{i_1^2 k^2}{4} + \frac{9 i_1 k^4}{4} + \frac{15 i_1 k^3}{4} - \frac{19 i_1 k^2}{12} + \frac{9 k^4}{2} + k^3 - \frac{5 k^2}{4},$$

$$a_3 = -\frac{i_0^2 k^3}{12} + \frac{1}{2} i_0 i_1 k^3 + \frac{i_0 k^4}{2} + \frac{3 i_0 k^3}{4} - \frac{i_1^3 k^3}{4} - \frac{3 i_1 k^4}{2} - \frac{i_1 k^3}{4} - \frac{3 k^5}{4}$$

$$- \frac{9 k^4}{4} + \frac{5 k^3}{12},$$

$$a_4 = -\frac{i_0 k^4}{6} + \frac{i_1 k^4}{4} + \frac{k^5}{2} + \frac{k^4}{4},$$

$$a_5 = -\frac{k^5}{12}.$$

$$(61)$$

Applying (40) and (61) to (60) gives

$$\mathfrak{g} = \frac{1}{12 (\mathfrak{r} - 1)^5} H(k, r, \mathfrak{r}, \mathfrak{y}), \qquad (62)$$

where $H(k, r, \mathfrak{x}, \mathfrak{y})$ is a polynomial in $k, r, \mathfrak{x}, \mathfrak{y}$ with integer coefficients. (The expression $H(k, r, \mathfrak{x}, \mathfrak{y})$ is too lengthy to be included here but can be easily generated with computer assistance.) Since $\tau > 2$, we have $2^2 \mid \tau$, and hence $k = (\tau + 1)/r \equiv 1/r \pmod{2^2}$. Thus, the binomial coefficients in (58) remain the same modulo 2 when k is replaced by $1/r$. Therefore,

$$
\begin{aligned}
\mathfrak{g} &\equiv \frac{1}{12(\mathfrak{x} - 1)^5} H(1/r, r, \mathfrak{x}, \mathfrak{y}) \pmod 2 \\
&= \frac{1}{r^5(\mathfrak{x} - 1)^5} H_1(r, \mathfrak{x}, \mathfrak{y}),
\end{aligned}
\tag{63}
$$

where $H_1(r, \mathfrak{x}, \mathfrak{y}) = (r^5/12)H(1/r, r, \mathfrak{x}, \mathfrak{y})$ is a polynomial in $\mathfrak{x}$ and $\mathfrak{y}$ whose coefficients are of the form $\frac{1}{6}C(r)$ with $C \in \mathbb{Z}[X]$ and $\frac{1}{6}C(\mathbb{Z}) \subset \mathbb{Z}$. Making the substitution (59) in (63) gives

$$
\mathfrak{g} \equiv \frac{(\mathfrak{x} - 1)^2}{\left((r+1)\mathfrak{x} - r\right)^5} L(\mathfrak{x}) \pmod 2,
\tag{64}
$$

where

$$
\begin{aligned}
L(X) = \frac{1}{6}\Big[&(480r^5 - 56r^4 - 612r^3 + 574r^2 - 210r + 4)X^8 \\
&+ (-2904r^5 + 2390r^4 + 2655r^3 - 4033r^2 + 2004r - 238)X^7 \\
&+ (7464r^5 - 11148r^4 - 1230r^3 + 9711r^2 - 6477r + 1266)X^6 \\
&+ (-10572r^5 + 22426r^4 - 9387r^3 - 8690r^2 + 9177r - 2414)X^5 \\
&+ (8940r^5 - 24118r^4 + 19116r^3 - 502r^2 - 5736r + 1958)X^4 \\
&+ (-4560r^5 + 14658r^4 - 15783r^3 + 5571r^2 + 1002r - 762)X^3 \\
&+ (1344r^5 - 4928r^4 + 6354r^3 - 3305r^2 + 375r + 142)X^2 \\
&+ (-204r^5 + 830r^4 - 1197r^3 + 728r^2 - 147r - 10)X \\
&+ 12r^5 - 54r^4 + 84r^3 - 54r^2 + 12r \Big].
\end{aligned}
$$

The coefficients of L are also of the form $\frac{1}{6}C(r)$ with $C \in \mathbb{Z}[X]$ and $\frac{1}{6}C(\mathbb{Z}) \subset \mathbb{Z}$. These coefficients, modulo 2, only depend on r modulo 4. The reduction of (64) in $\mathcal{O}_F/2\mathcal{O}_F$ is

$$
g_{2+3\tau}(x) = \frac{(x-1)^2}{\left((r+1)x - r\right)^5} L(x).
\tag{65}
$$

(In fact, $L = \frac{1}{12}G_{2+3\tau}$ in $\mathbb{Q}[X]$, where $G_{2+3\tau}$ is given in (55). For $p = 2$, (65) implies (51) but not the converse.) We find that

$$
L(X) = \begin{cases} X^2(1 + X + X^2 + X^4 + X^5) & \text{if } r \equiv 1 \pmod 4, \\ X^4 & \text{if } r \equiv -1 \pmod 4, \end{cases}
$$

and we always have

$$\gcd_{\mathbb{F}_2[X]}(1 + X + X^2, L) = 1.$$

This is a contradiction since $G_{1+\tau+\tau^3}(X) = X(1 + X + X^2)$ and $L(X)$ have a nonzero common root x.

Case 2. Assume $p = 3$. First assume $r \equiv 1 \pmod 3$. We have

$$G_{1+\tau+\tau^3}(X) = X - X^2 + X^3 - X^4 = X(1 - X)(X^2 + 1).$$

It follows that $x^2 = -1$. Then $x^r = \pm 1$ or $\pm x$. By (16) we also have

$$x^r = \frac{x^2}{-x - 1} = \frac{1}{x + 1}.$$

If $x^r = \pm 1$, then $x + 1 = \pm 1$, and hence $x \in \mathbb{F}_3$. Thus $x^2 \neq -1$, which is a contradiction. If $x^r = \pm x$, then $x + 1 = \mp x$, which forces $x = 1$, which is also a contradiction.

Now assume $r \equiv -1 \pmod 3$. We have $G_{1+\tau+\tau^3}(X) = -X + X^2 = X(X - 1)$. This is impossible since $x \neq 0, 1$.

Case 3. Assume $p = 5$.

Case 3.1. Assume $r \equiv 1 \pmod 5$. We have

$$G_{1+\tau+\tau^3}(X) = X(1 + 3X + X^2 + 2X^3),$$
$$\frac{1}{2}G_{1+2\tau}(X) = X(1 + 2X + 3X^2),$$

and

$$\gcd_{\mathbb{F}_5[X]}(1 + 3X + X^2 + 2X^3,\ 1 + 2X + 3X^2) = 1.$$

Hence we have a contradiction.

Case 3.2. Assume $r \equiv 2 \pmod 5$. We have

$$\frac{1}{2}G_{1+2\tau}(X) = 4 + 3X,$$
$$G_{2+3\tau}(X) = X(4 + 4X + 3X^2 + X^4 + X^5 + 3X^7),$$

and

$$\gcd_{\mathbb{F}_5[X]}(4 + 3X,\ 4 + 4X + 3X^2 + X^4 + X^5 + 3X^7) = 1.$$

Hence we have a contradiction. (Note. In this case, $G_{1+\tau+\tau^3}(X) = X(1 - X)(4 + 3X)$, which is a multiple of $G_{1+2\tau}(X)$.)

Case 3.3. Assume $r \equiv -1 \pmod 5$. We have $\frac{1}{2}G_{1+2\tau}(\mathtt{X}) = \mathtt{X}(\mathtt{X} - 2)$. Thus $x = 2$. However, by (16),

$$x^r = \frac{x(rx + 1 - r)}{\big((r+1)x - r\big)} = 0,$$

which is a contradiction.

Case 3.4. Assume $r \equiv -2 \pmod 5$. We have

$$G_{1+\tau+\tau^3}(\mathtt{X}) = 2 + \mathtt{X},$$
$$\frac{1}{2}G_{1+2\tau}(\mathtt{X}) = 2 + 2\mathtt{X} + 4\mathtt{X}^2,$$

and

$$\gcd_{\mathbb{F}_5[\mathtt{X}]}(2 + \mathtt{X},\ 2 + 2\mathtt{X} + 4\mathtt{X}^2) = 1.$$

Hence we have a contradiction. $\qquad\square$

In conclusion, Theorem 3 follows from Lemma 8, Corollary 1 and Lemma 10.

References

[1] X. Hou, *A survey of permutation binomials and trinomials over finite fields*, Proceedings of the 11th International Conference on Finite Fields and Their Applications, Magdeburg, Germany, 2013, Contemporary Mathematics **632**, 177 – 191, 2015.

[2] X. Hou, *Permutation polynomials over finite fields — a survey of recent advances*, Finite Field Appl. Finite Fields Appl., **32** (2015), 82 – 119.

[3] X. Hou, *Determination of a type of permutation trinomials over finite fields, II*, Finite Fields Appl. **35** (2015), 16 – 35.

[4] X. Hou, *Permutation Polynomials of* $\mathbb{F}_{q^2}$ *of the form* $a\mathtt{X} + \mathtt{X}^{r(q-1)+1}$, arXiv:1510.00437, 2015.

[5] X. Hou and S. D. Lappano, *Determination of a type of permutation binomials over finite fields*, J. Number Theory 147 (2015) 14 – 23.

[6] S. D. Lappano, *A note regarding permutation binomials over* $\mathbb{F}_{q^2}$, Finite Fields Appl. **34** (2015), 153 – 160.

[7] R. Lidl and H. Niederreiter, *Finite Fields*, 2nd ed., Cambridge Univ. Press, Cambridge, 1997.

[8] G. L. Mullen and D. Panario, *Handbook of Finite Fields*, Taylor & Francis, Boca Raton, 2013.

[9] M. E. Zieve, *Permutation polynomials on* $\mathbb{F}_q$ *induced from bijective Rédei functions on subgroups of the multiplicative group of* $\mathbb{F}_q$, arXiv:1310.0776, 2013.

Character sums and generating sets

Ming-Deh A. Huang, Lian Liu

University of Southern California, Los Angeles, CA, USA

mdhuang, lianliu@usc.edu

1. Introduction

The problem of finding small generating sets for a given group is a relaxation of finding primitive elements in finite fields. In computational algebra, it is often desired to find small generating sets for given groups. For example, in the index calculus method for solving the discrete logarithm problem, one is interested in finding reasonably small generating sets over which many relations can be found [1, 6]. Generating sets are also involved in explicit construction of expander graphs [3, 9], which has been extensively studied and applied in many areas including computational complexity theory, coding theory and communication networks [2, 4].

An important case that has been extensively studied in the past few decades is the problem of finding small generating sets for the multiplicative group of a given finite field [3, 9, 11]. A fundamental result of Chung [3] states that if $\sqrt{p} > n - 1$, the set $x + \mathbb{F}_p := \{x + t : t \in \mathbb{F}_p\}$ forms a generating set for $\mathbb{F}_{p^d}^{\times} \simeq (\mathbb{F}_p[x]/f)^{\times}$, where p is a prime number, and $f \in \mathbb{F}_p[x]$ is an irreducible polynomial of degree d. This result was critical to recent advances in the discrete-log problem and the problem of finding primitive elements for finite fields in small characteristics [5, 6].

Chung's proof revealed the relationship between the character sum over a subset of group elements and their generated subgroup. Informally speaking, a good generating set should have a relatively small character sum (in absolute value) compared with its cardinality.

In this paper, we generalize Chung's idea to all finite abelian groups. We demonstrate that given an arbitrary finite abelian group G together with a subset S of its elements, S generates G if for every nontrival multiplicative character defined on G, the absolute value of the character sum over S is smaller than $|S|$.

As an illustration for this result, we study multiplicative groups of algebras of the form $\mathbb{F}_p[x]/f^e$, where $f \in \mathbb{F}_p[x]$ is an irreducible polynomial of degree d, and $e \geq 1$ is an integer. When $e = 1$, this case reduces to Chung's situation. Clearly, when $e > 1$, the algebra is no longer a field, and we show that its multiplicative group is no longer cyclic, which makes this situation interesting to explore. We begin with two basic types of generating sets for $(\mathbb{F}_p[x]/f^e)^\times$: if $\sqrt{p} > de - 1$, we can simply use $\{t - x : t \in \mathbb{F}_p\}$ as a generating set for this group; otherwise, if $p \geq e$ holds, then we may use $\{\pi(t) - x : t \in \mathbb{F}_p/f\}$ as a generating set, where π is an embedding of $\mathbb{F}_p[x]/f$ into $\mathbb{F}_p[x]/f^e$ which can be computed efficiently. For both cases, we also show upper bounds on the length of the product needed to present an element in the group. We then propose an algorithm for constructing small generating sets based on these two types of sets. The size of the generating set returned by our algorithm depends on the choice of the parameters, p, d and e. For the special case when $p \geq e$ are fixed and d is a perfect power of a fixed number, the size of the resulting generating set is proven to be $p^{O(\log d)}$.

In order to study the empirical performance of our construction, we ran experiments in order to see whether the size of the generating set can be substantially reduced in practice, where a practically optimal generating set is constructed by drawing random subsets from our original construction. From the experimental results, we observe that this randomized construction also gives generating sets of size $p^{O(\log d)}$. But in many cases, the constant scalar for the exponent, $\log d$, is reduced by approximately a half, which means that a square root number of random elements in our original construction would usually be sufficient as a generating set. Whether this is true in general remains an interesting open question.

As an application, we show the explicit construction of a new family of regular directed expander graphs based on $(\mathbb{F}_p[x]/f^e)^\times$ and its generating sets. In particular, our construction yields a graph with $(p^d - 1)p^{d(e-1)}$ vertices. The graph achieves a spectral gap of at least $\gamma \in (0, 1)$ if the degree $k = p^c$ satisfies $p \geq e$, $c|d$ and $(1 - \gamma)\sqrt{k} \geq de/c - 1$.

2. Background

2.1. *Dirichlet characters for finite abelian groups*

Given G any finite abelian group, by a Dirichlet characters χ of G, we mean a group homomorphism $\chi : G \to \mathbb{C}^\times$. All possible Dirichlet characters of G form a set of size $|G|$, which is denoted by $X(G)$.

For any G a nontrivial finite abelian group, by the structure theorem, we can decompose it into a direct sum of cyclic groups:

$$G \simeq \bigoplus_{i=1}^{m} \mathbb{Z}/d_i\mathbb{Z}$$

where $d_i > 1$ are integers and $m \geq 1$. Fix a decomposition of G, an element $g \in G$ can be decomposed into $g \simeq g_1 \oplus \ldots \oplus g_m$ according to the decomposition. Let χ be a Dirichlet character of G, then for each g,

$$\chi : g \to \prod_{i=1}^{m} \omega_{d_i}^{g_i}$$

where ω_d stands for a dth root of unity.

There are d_i different choices of ω_{d_i} for each i, so there are $\prod_{i=1}^{m} d_i = |G|$ possible characters, and $X(G)$ is precisely the set of all such characters. When $\omega_1 = \ldots = \omega_m = 1$, χ is said to be *trivial*, i.e., $\chi(g) = 1$ for all $g \in G$. Otherwise, χ is said to be *nontrivial*.

Lemma 1. *Given a finite abelian group G with $|G| = n$, the set of complex vectors $\{[\chi(g_1), \ldots, \chi(g_n)]^\top : \chi \in X(G)\}$ form an orthogonal basis for $\mathbb{C}^n$.*

Proof. Fix a decomposition of G, and suppose two vectors u, v are given by

$$u = \begin{bmatrix} \chi_1(g_1) \\ \vdots \\ \chi_1(g_n) \end{bmatrix} = \begin{bmatrix} \omega_{d_1}^0 \ldots \omega_{d_m}^0 \\ \vdots \\ \omega_{d_1}^{d_1-1} \ldots \omega_{d_m}^{d_m-1} \end{bmatrix}$$

$$v = \begin{bmatrix} \chi_2(g_1) \\ \vdots \\ \chi_2(g_n) \end{bmatrix} = \begin{bmatrix} \theta_{d_1}^0 \ldots \theta_{d_m}^0 \\ \vdots \\ \theta_{d_1}^{d_1-1} \ldots \theta_{d_m}^{d_m-1} \end{bmatrix}$$

where $\chi_1 \neq \chi_2 \in X(G)$ and $\omega_{d_i}, \theta_{d_i}$ are d_i-th roots of unity. Verify the inner product between u and itself, and there we have

$$\langle u, u \rangle = \sum_{i_1=0}^{d_1-1} \cdots \sum_{i_m=0}^{d_m-1} \overline{\left(\prod_{j=1}^{m} \omega_{d_j}^{i_j} \right)} \left(\prod_{j=1}^{m} \omega_{d_j}^{i_j} \right)$$

$$= \sum_{i_1=0}^{d_1-1} \cdots \sum_{i_m=0}^{d_m-1} \prod_{j=1}^{m} \left(\overline{\omega}_{d_j} \omega_{d_j} \right)^{i_j}$$

$$= \sum_{i_1=0}^{d_1-1} \cdots \sum_{i_m=0}^{d_m-1} \prod_{j=1}^{m} 1$$

$$= n > 0$$

If $\chi_1 \neq \chi_2$, there must be some k such that $\omega_{d_k} \neq \theta_{d_k}$. We assume without loss of generality that $\omega_{d_1} \neq \theta_{d_1}$. Thus, $\overline{\omega}_{d_1} \theta_{d_1} = \phi_{d_1}$, where ϕ_{d_1} is an nontrivial d_1-th root of unity. Then the inner product between u and v is given by

$$\langle u, v \rangle = \sum_{i_1=0}^{d_1-1} \cdots \sum_{i_m=0}^{d_m-1} \overline{\left(\prod_{j=1}^{m} \omega_{d_j}^{i_j} \right)} \left(\prod_{j=1}^{m} \theta_{d_j}^{i_j} \right)$$

$$= \left(\sum_{i_1=0}^{d_1-1} \left(\overline{\omega}_{d_1} \theta_{d_1} \right)^{i_1} \right) \left(\sum_{i_2=0}^{d_2-1} \cdots \sum_{i_m=0}^{d_m-1} \overline{\left(\prod_{j=2}^{m} \omega_{d_j}^{i_j} \right)} \left(\prod_{j=2}^{m} \theta_{d_j}^{i_j} \right) \right)$$

$$= \left(\sum_{i_1=0}^{d_1-1} \left(\phi_{d_1} \right)^{i_1} \right) \left(\sum_{i_2=0}^{d_2-1} \cdots \sum_{i_m=0}^{d_m-1} \overline{\left(\prod_{j=2}^{m} \omega_{d_j}^{i_j} \right)} \left(\prod_{j=2}^{m} \theta_{d_j}^{i_j} \right) \right)$$

$$= 0 \cdot \left(\sum_{i_2=0}^{d_2-1} \cdots \sum_{i_m=0}^{d_m-1} \overline{\left(\prod_{j=2}^{m} \omega_{d_j}^{i_j} \right)} \left(\prod_{j=2}^{m} \theta_{d_j}^{i_j} \right) \right)$$

$$= 0$$

$\square$

2.2. Cayley graphs, character sums and generating sets

Given G a finite abelian group and $S \subseteq G$ a subset of elements, the corresponding Cayley graph, denoted by $\mathcal{G}(G, S)$, is constructed as follows:

- For each element $g \in G$, create a vertex g in $\mathcal{G}(G, S)$;
- Create a directed edge $g_i \to g_j$ in $\mathcal{G}(G, S)$ if and only if $g_i s = g_j$ for some $s \in S$.

Cayley graphs are regular graphs, meaning that each vertex has the same in and out degree. Each graph is associated with an adjacency matrix M, where each entry $M_{i,j} = 1$ if $g_i \to g_j$ is an edge, and 0 otherwise. One may verify that for Cayley graphs, the eigenvalues can be written as a character sum, as stated in Lemma 2.

Lemma 2. *Given G a finite abelian group and $S \subseteq G$ a subset of elements. Let M denote the adjacency matrix of $\mathcal{G}(G,S)$, then the set of eigenvalues of M are $\{\sum_{s \in S} \chi(s) : \chi \in X(G)\}$.*

Proof. The eigenvectors are $v_\chi = [\chi(g_1), \ldots, \chi(g_n)]^\top$, where $\chi \in X(G)$. Consider $M v_\chi$, the i-th item of the resulting vector is

$$(M v_\chi)_i = \sum_{\substack{g_j : g_i s = g_j \\ s \in S}} \chi(g_j) = \sum_{s \in S} \chi(g_i s) = \sum_{s \in S} \chi(g_i) \chi(s) = \chi(g_i) \sum_{s \in S} \chi(s)$$

Thus, we have

$$M v_\chi = \left(\sum_{s \in S} \chi(s) \right) v_\chi = \lambda_\chi v_\chi$$

where λ_χ denotes the eigenvalue associated with v_χ. $\qquad\qquad \square$

Let $\mathcal{G}$ be a directed graph and u, v be two vertices, the distance from u to v is the length (i.e. number of edges) of the shortest path from u to v. The diameter of $\mathcal{G}$, denoted by $\operatorname{diam}(\mathcal{G})$, is then the maximum distance among all pairs of vertices. It is not hard to see that S is a generating set of G if and only if $\mathcal{G}(G,S)$ has a finite diameter. In [3], Chung showed the diameter of a regular directed graph is related to the second largest eigenvalue (in absolute value) of its adjacency matrix.

Theorem 1 (Chung). *Suppose a directed graph $\mathcal{G}$ has n vertices and the out-degree of every vertex is k. If the eigenvectors of the adjacency matrix, M, of $\mathcal{G}$ form an orthogonal basis. Then we have*

$$\operatorname{diam}(\mathcal{G}) \leq \left\lceil \frac{\log(n-1)}{\log \frac{k}{\lambda}} \right\rceil$$

where λ is the second largest eigenvalue (in absolute value) of M.

A direct observation from this theorem is that as long as $\lambda < k$, the graph has a finite diameter. Combining Lemma 2 and Theorem 1, we obtain the following relationship between character sums and generating sets:

Theorem 2. *Let G be a finite abelian group, $S \subseteq G$ be a subset of elements of G. Then S is a generating set of G if*

$$\left| \sum_{s \in S} \chi(s) \right| < |S|$$

holds for every nontrivial $\chi \in X(G)$.

Proof. Consider the Cayley graph $\mathcal{G}(G, S)$. As we have seen in the proof for Lemma 2, the eigenvectors of the adjacency matrix of $\mathcal{G}$ are $\{[\chi(g_1), \ldots, \chi(g_n)]^\top : \chi \in X(G)\}$, which form an orthogonal basis by Lemma 1. By Lemma 2, the corresponding eigenvalues are $\sum_{s \in S} \chi(s)$ where $\chi \in X(G)$. Given the assumption that for any nontrivial eigenvalue λ, $\lambda < k = |S|$. Apply Theorem 1, we can see that $\text{diam}(\mathcal{G})$ is finite, which implies that $\langle S \rangle = G$. $\square$

We remark that when χ is trivial, we have $\left| \sum_{s \in S} \chi(s) \right| = |S|$. If S is a generating set, then one may derive the maximum number of elements in S needed to represent an element by Theorem 1.

3. Generalizing Chung's results to $\mathbb{F}_p[x]/f^e$

In this section, we study algebras of the form $A = \mathbb{F}_p[x]/f^e$, where p is a prime number, $e \geq 1$ is an integer, and $f \in \mathbb{F}_p[x]$ is an irreducible polynomial of degree $d \geq 2$. Notice that when $e = 1$, this reduces to Chung's situation, and therefore it is of more interest to study the case where $e > 1$. For convenience, in the rest of this paper, we will use A as the abbreviation for $\mathbb{F}_p[x]/f^e$.

3.1. *The structure of $(\mathbb{F}_p[x]/f^e)^\times$*

Lemma 3. *If $p \geq e$, then*

$$A^\times \simeq \mathbb{Z} \Big/ (p^d - 1)\mathbb{Z} \oplus \left(\bigoplus_{d(e-1)} \mathbb{Z}/p\mathbb{Z} \right)$$

Proof. Consider the map

$$\varphi : A^\times \to (\mathbb{F}_p[x]/f)^\times$$

where for each $a \in A^{\times}$,

$$\varphi(a) = a \pmod{f}$$

Clearly, φ is an onto function. We can see that the kernel of the map is precisely

$$\ker \varphi = \{1 + bf : b \in A \text{ where } 0 \leq \deg b \leq d(e-1) - 1\}$$

For every $1 + bf \in \ker \varphi$, since A as a ring has characteristic p, its p-th power is given by:

$$(1 + bf)^p = 1 + b^p f^p \pmod{f^e}$$

Given that $p \geq e$, we have $1 + b^p f^p = 1 \pmod{f^e}$. So, by the structure theorem of finitely generated abelian groups,

$$\ker \varphi \simeq \bigoplus_{d(e-1)} \mathbb{Z}/p\mathbb{Z}$$

Notice that $|\ker \varphi| = p^{d(e-1)}$, which is relatively prime to $|\mathrm{im}\varphi| = p^d - 1$, it follows that $A^{\times}$ is isomorphic to their direct product:

$$A^{\times} \simeq (\mathbb{F}_p[x]/f)^{\times} \times \ker \varphi \simeq \mathbb{Z}\Big/(p^d - 1)\mathbb{Z} \oplus \left(\bigoplus_{d(e-1)} \mathbb{Z}/p\mathbb{Z}\right)$$

$\square$

3.2. Regarding $\mathbb{F}_p[x]/f^e$ as an $\mathbb{F}_p$-algebra

From Lemma 3 we see that $A^{\times}$ is no longer cyclic when $p \geq e > 1$, which makes the structure different from the case when $e = 1$. Therefore, it is interesting to know whether (if yes, then under what situation does) the set $x + \mathbb{F}_p$ still generates $A^{\times}$. According to Theorem 2, in order for $x + \mathbb{F}_p$ to generate $A^{\times}$ one needs to show that for any nontrivial character on $A^{\times}$, the character sum over $x + \mathbb{F}_p$ (in absolute value) is smaller than p. It turns out that the following bound on character sums over the multiplicative groups of algebras over finite fields is closely related to our situation:

Theorem 3 (Weil). *Let B be an arbitrary finite n-dimensional commutative $\mathbb{F}_q$-algebra and x be an element of B. If χ is a character of the multiplicative group $B^{\times}$ (extended by zero to all of B) which is non-trivial on $F_q[x]$, then*

$$\left| \sum_{t \in \mathbb{F}_q} \chi(t - x) \right| \leq (n-1)\sqrt{q}$$

Theorem 3 was initially conjectured by Katz in [7]. It turns out that the theorem actually follows as a consequence of Weil's character sum estimate [12], as proven by Lenstra in his unpublished notes [8]. We refer readers to [11] and [10] for more details.

A can be naturally regarded as an $\mathbb{F}_p$-algebra of dimension de, and thus we obtain the following sufficient condition for the set $\mathbb{F}_p - x$ to generate $A^\times$.

Theorem 4. *If $\sqrt{p} > de - 1$, then $\mathbb{F}_p - x$ is a generating set for $A^\times$. Furthermore, every element $\alpha \in A^\times$ can be written as $\prod_{i=1}^m (a_i - x)$ where $a_i \in \mathbb{F}_p$ and*

$$m < 2de + 1 + \frac{4de \log(de - 1)}{\log p - 2\log(de - 1)}$$

Proof. Since A is an $\mathbb{F}_p$-algebra of dimension de, by Theorem 3, for any nontrivial $\chi \in X(A^\times)$, we have

$$\left| \sum_{t \in \mathbb{F}_p} \chi(t - x) \right| \leq (de - 1)\sqrt{p}$$

By Theorem 2, S is a generating set for $A^\times$ if $\sqrt{p} > de - 1$. In this case, according to Theorem 1, the number of elements to represent an element is at most

$$2de + 1 + \frac{4de \log(de - 1)}{\log p - 2\log(de - 1)}$$

$\square$

3.3. *Regarding $\mathbb{F}_p[x]/f^e$ as an $\mathbb{F}_q$-algebra*

Theorem 4 needs $\sqrt{p} > de - 1$, which could be too strict a requirement on p. In this situation, we would like to look for other types of generating sets. We notice that by changing the ground field from $\mathbb{F}_p$ to $\mathbb{F}_q = \mathbb{F}_p[x]/f$, A can actually be regarded as an $\mathbb{F}_q$-algebra.

Lemma 4. *Given arbitrary $a \in A$ which is written in the form $a = \sum_{i=0}^{e-1} a_i f^i$, where each $a_i \in A$ has degree less than d. Then $a \in A^\times$ if and only if $a_0 \neq 0 \pmod{f}$.*

Proof. For sufficiency, since $a \in A^\times$, there is $b = \sum_{i=0}^{e-1} b_i f^i \in A^\times$ such that $ab = 1$. That is,

$$\left(\sum_{i=0}^{e-1} a_i f^i \right) \left(\sum_{i=0}^{e-1} b_i f^i \right) = a_0 b_0 + (a_0 b_1 + a_1 b_0) f + \ldots = 1 \pmod{f^e}$$

and thus $a_0b_0 = 1$ (mod f). Therefore, $a_0 \neq 0$ (mod f).

For necessity, suppose $a_0 \neq 0$ (mod f), and we only need to show the existence of $b = a^{-1}$ in $A^\times$. Again, assume that $b = \sum_{i=0}^{e-1} b_i f^i$ (deg $b_i < d$) and their product is

$$\left(\sum_{i=0}^{e-1} a_i f^i\right)\left(\sum_{i=0}^{e-1} b_i f^i\right) = a_0b_0 + (a_0b_1 + a_1b_0)f + \ldots = c_0 + c_1f + \ldots$$

Since $a_0 \neq 0$ (mod f), there is b_0 such that $a_0b_0 = 1$ (mod f) and $b_0 \neq 0$ (mod f). Given b_0, b_1 is uniquely determined by the linear equation $a_0b_1 + a_1b_0 = 0$ (mod f) over $\mathbb{F}_q$. In general, each b_i ($1 \leq i \leq e-1$) is uniquely determined by the linear equation $c_i = 0$ (mod f) over $\mathbb{F}_q$ for fixed $b_0, \ldots, b_{i-1}$ values. Therefore, there is a unique b such that $ab = 1$ (mod f), and thus $a \in A^\times$. $\square$

Lemma 5. *For each $a_0 \in \mathbb{F}_q^\times$, there exists a unique $a \in A^\times$ which can be written as $a = \sum_{i=0}^{e-1} a_i f^i$, where each $a_i \in A$ has degree less than d, and and $a^{q-1} = 1$ (mod f^e).*

Proof. Since $a_0 \in \mathbb{F}_q^\times$, $a \in A^\times$ by Lemma 4. Suppose $a = \sum_{i=0}^{e-1} a_i f^i$, where each $a_i \in A$ has degree less than d. In order to make $a^{q-1} = 1$ (mod f^e), we need

$$a^{q-1} = \left(\sum_{i=0}^{e-1} f^i\right)^{q-1} = a_0^{q-1} + (q-1)a_0^{q-2}a_1 f + \ldots = 1 \quad (\text{mod } f^e)$$

From this equation we get

$$a_0^{q-1} + (q-1)a_0^{q-2}a_1 f = 1 \quad (\text{mod } f^2)$$

Because $a_0^{q-1} = 1$ (mod f), we have

$$a_0^{q-1} = 1 + A_0 f \quad (\text{mod } f^2)$$

where $A_0 \in A$ and deg $A_0 < d$. Plugin and simplify it, we see that a_1 is uniquely determined by the linear equation

$$A_0 + (q-1)a_0^{q-2}a_1 = 0 \quad (\text{mod } f)$$

over $\mathbb{F}_q$. We claim that this argument applies inductively to every a_i. Assume $a_0, a_1, \ldots, a_{k-1}$ are uniquely determined. In order to guarantee

$a^{q-1} = 1 \pmod{f^e}$, we need

$$\left(\sum_{i=0}^{e-1} a_i f^i\right)^{q-1} = \left(\sum_{i=0}^{k-1} a_i f^i + a_k f^k + \sum_{i=k+1}^{e-1} a_i f^i\right)^{q-1} \qquad (\text{mod } f^{k+1})$$

$$= \left(\sum_{i=0}^{k-1} a_i f^i + a_k f^k\right)^{q-1} \qquad (\text{mod } f^{k+1})$$

$$= \left(\sum_{i=0}^{k-1} a_i f^i\right)^{q-1} + (q-1)\left(\sum_{i=0}^{k-1} a_i f^i\right)^{q-2} a_k f^k \quad (\text{mod } f^{k+1})$$

$$= 1 \qquad (\text{mod } f^{k+1})$$

By induction, the first term can be written as

$$\left(\sum_{i=0}^{k-1} a_i f^i\right)^{q-1} = 1 + A_{k-1} f^k$$

for some $A_{k-1} \in A$ where $\deg A_{k-1} < d$. Then a_k is uniquely determined by the linear equation

$$A_{k-1} + (q-1)\left(\sum_{i=0}^{k-1} a_i f^i\right)^{q-2} a_k = 0 \pmod{f}$$

over $\mathbb{F}_q$. $\qquad \square$

Lemma 5 yields a well-defined function $\pi : \mathbb{F}_q^\times \to A^\times$, which can be extended to all of $\mathbb{F}_q$ by forcing $\pi(0) = 0$. We proved that π is essentially an embedding of $\mathbb{F}_q$ into A.

Lemma 6. *Let $\pi : \mathbb{F}_q^\times \to A$ be the function where for all $a_0 \in \mathbb{F}_q$,*

$$\pi(a_0) = \begin{cases} 0, & \text{if } a_0 = 0 \\ a \in A \text{ s.t. } a^{q-1} = 1, & \text{otherwise} \end{cases}$$

then $\pi(\mathbb{F}_q) \simeq \mathbb{F}_q$ as rings.

Proof. Assume $a_0, b_0 \in \mathbb{F}_q$, and that they can be written as $\pi(a_0) = \sum_{i=0}^{e-1} a_i f^i$, $\pi(b_0) = \sum_{i=0}^{e-1} b_i f^i$. We set off by showing $\pi(a_0 b_0) = \pi(a_0)\pi(b_0)$. When $a_0 = 0$ or $b_0 = 0$, this is obvious. Otherwise, notice that the first term of both sides are $a_0 b_0$, and we have $(\pi(a_0)\pi(b_0))^{q-1} = \pi(a_0)^{q-1}\pi(b_0)^{q-1} = 1$ By Lemma 5 , $\pi(a_0 b_0) = \pi(a_0)\pi(b_0)$. Next, we verify $\pi(a_0^{-1}) = \pi(a_0)^{-1}$ for all $a_0 \neq 0$. Since $a_0^{q-1} = 1$, $a_0^{-1} = a_0^{q-2}$. Therefore, $\pi(a_0^{-1}) = \pi(a_0^{q-2}) = \pi(a_0)^{q-2}$. Since $\pi(a_0)^{q-1} = 1$, $\pi(a_0)^{q-2} = \pi(a_0)^{-1}$. It remains to show

$\pi(a_0 + b_0) = \pi(a_0) + \pi(b_0)$. If $a_0 = 0$ or $b_0 = 0$, this is obvious. Otherwise, since the first term of both sides is $a_0 + b_0$, by Lemma 5, it suffices to show $(\pi(a_0) + \pi(b_0))^{q-1} = 1$. Denote the set $T = \{a \in A : a^{q-1} = 1\}$ and the set $T' = \{a \in A : a^q = a\} = S \cup \{0\}$. Since A is a ring of characteristic p, $(\pi(a_0)+\pi(b_0))^q = \pi(a_0)^q + \pi(b_0)^q = \pi(a_0)+\pi(b_0)$. That is, $\pi(a_0)+\pi(b_0) \in T'$, and hence either $\pi(a_0)+\pi(b_0) \in T$ or $\pi(a_0)+\pi(b_0) = 0$. In the first case, we are done; in the latter case, $a_0 = -b_0$, so we also have $\pi(a_0 + b_0) = \pi(0) = 0 = \pi(a_0) + \pi(b_0)$. $\qquad\square$

Theorem 5. *If $p \geq e$, then the set $\pi(\mathbb{F}_q) - x := \{\pi(t) - x : t \in \mathbb{F}_q\}$ is a generating set for $A^\times$. Every element $\alpha \in A^\times$ can be written as $\prod_{i=1}^m (\pi(a_i) - x)$ where $a_i \in \mathbb{F}_q$ and*

$$m < 2e + 1 + \frac{4e \log(e-1)}{d \log p - 2 \log(e-1)}$$

Proof. By Lemma 6, A can be regarded as a $\pi(\mathbb{F}_q)$-algebra of dimension e. By Theorem 3,

$$\left| \sum_{t \in \mathbb{F}_q} \chi\left(\pi(t) - x\right) \right| \leq (e-1)\sqrt{q}$$

The claim of theorem follows Theorem 2. The maximum length of each product, m, follows Theorem 1. $\qquad\square$

3.4. *Constructing small generating sets for* $(\mathbb{F}_p[x]/f^e)^\times$

Theorem 5 sets a relatively mild requirement, which only needs $p \geq e$, but this set is of size q, which might be more than is required in many cases. Only a small fraction of this set might already be sufficient to generate the group. Thus, in this section, we consider the problem of reducing the size of $\pi(F_q) - x$.

Let $K \subset \mathbb{F}_q$ be a subfield of size p^c, where $c|d$. From our previous discussions we see that A can be regarded as a $\pi(K)$-algebra of dimension de/c. Notice that Theorem 3 still applies, and we can similarly show that

Theorem 6. *If $p \geq e$ and $p^{c/2} > de/c - 1$, then $\pi(K) - x$ is a generating set for $A^\times$. Furthermore, every element $a \in A^\times$ can be written as $\prod_{i=1}^m (\pi(a_i) - x)$, where $a_i \in K$ and*

$$m < 2\frac{de}{c} + 1 + \frac{4\frac{de}{c} \log(\frac{de}{c} - 1)}{\frac{d}{c} \log p - 2 \log(\frac{de}{c} - 1)}$$

Algorithm 1 Genset(p, f, e)

Find $c = \min\{c \in \mathbb{Z} : c|d$ and $p^{c/2} > de/c - 1\}$, where $d = \deg f$;
Let K be the subfield of $\mathbb{F}_q$ of size p^c;
return $\pi(K) - x$

A small generating set for $A^\times$ can be constructed from the smallest ground field whose size satisfies the premises of Theorem 6.

One downside of this construction is that when d has few divisors, for example, d is a prime number, then $\mathbb{F}_p - x$ and $\pi(F_q) - x$ are the only two options. On the other extreme, when d has abundant divisors, we may be able to construct better generating sets. One special case would be such that the parameter d is a perfect power, say $d = b^n$. In this scenario, all divisors of d are of the form b^i for some $1 \leq i \leq n$.

Corollary 1. *Given fixed p and e such that $p \geq e$. If $d = b^n$ is a perfect power, where b is fixed. Then there is (constructively) a generating set of $A^\times$ of size $p^{O(\log d)}$.*

Proof. Notice that $c \geq 2\log_p d + 2$ would be sufficient for the condition of Theorem 6, $p^{c/2} > de/c - 1$, to hold. Let $m_0 \in \mathbb{R}$ be such that $b^{m_0} = 2\log_p d + 2$. Write $c = b^m$, we have $m \leq \lceil m_0 \rceil \leq m_0 + 1$. Thus, $c \leq 2b(\log_p d + 1) = O(\log d)$. □

4. Experimental Study

It is important to know whether the requirements on the parameters in Theorem 6 are too strict to be practical. Therefore, we run experiments to see whether the size of the generating set returned by Algorithm 1 can be substantially reduced in practice.

4.1. Algorithms for experiments

Suppose we were able to decompose every element in $A^\times$ according to Lemma 3, we would be able to verify whether a set generates the whole group by simply computing a matrix rank.

By Lemma 3, the group $A^\times$ can be written as $A^\times = H \times K$ where $H = \pi(\mathbb{F}_q) \simeq \mathbb{Z}/(q-1)\mathbb{Z}$ and $K = A^\times/H \simeq \bigoplus_{i=0}^{D} \mathbb{Z}/p\mathbb{Z}$, $D = d(e-1)$. Given any element $a \in A^\times$, a can be written as $a = \sum_{i=0}^{e-1} a_i f^i$ where

$\deg a_i < d$ for all i, and $a_0 \neq 0$. Observe that a can be written as.

$$a = \pi(a_0) \left(1 + \sum_{i=1}^{e-1} b_i f^i \right)$$

where each $b_i \in A$ is uniquely determined. Clearly, $\pi(a_0) \in H$ and $1 + \sum_{i=1}^{e-1} b_i f^i \in K$. Let α_0 denote the coordinate of a in H and let α_i ($1 \leq i \leq D$) denote the coordinate of a in the i-th component of K. In other words, $a \simeq \alpha_0 \oplus \alpha_1 \oplus \ldots \oplus \alpha_D$ according to Lemma 3. Then, finding α_0 is equivalent to finding the discrete-log of $\pi^{-1}(a_0)$ in $\mathbb{F}_q^\times$. There are many existing algorithms [1] and software (e.g. Sage) available. Thus, we will focus our discussion on the decomposition of elements in K.

Let K_i ($1 \leq i \leq e$) denote the subgroup of $A^\times$ of the form $\{1 + Pf^i \mod f^e : P \in \mathbb{F}_p[x]\}$. By definition, $K_1 = K$ and $K_e = \{1\}$. Then we obtain the following filtration of subgroups:

$$K_1 \supsetneq K_2 \supsetneq \ldots \supsetneq K_e$$

Proposition 1. *For each* $1 \leq i \leq e - 1$,

$$K_i / K_{i+1} = \prod_{j=0}^{d-1} \langle 1 + x^j f^i \rangle \simeq \bigoplus_{j=0}^{d-1} \mathbb{Z}/p\mathbb{Z}$$

Proof. Consider the map $K_i \to \mathbb{F}_p[x]/f$ sending $1 + Pf^i$ to $P \mod f$ for all $1 + Pf^i \in K_i$ with $\deg P < d$. It is easy to verify that this is a group homomorphism with K_{i+1} as the kernel. Thus we have

$$K_i / K_{i+1} \simeq \mathbb{F}_p[x]/f \simeq \bigoplus_{j=0}^{d-1} \mathbb{Z}/p\mathbb{Z}$$

whereby $1 + Pf^i$ is mapped to $\bigoplus_{j=0}^{d-1} p_j$ if $P = \sum_{j=0}^{d-1} p_j x^j$. Under the isomorphisms the basis $\{x^j \mod f : j = 0, \ldots, d-1\}$ for $\mathbb{F}_p[x]/f$ corresponds to the basis $\{1 + x^j f^i : j = 0, \ldots, d-1\}$ for K_i/K_{i+1}. $\qquad\square$

Given any element $k_i \in K_i$, we can write it as $k_i = 1 + \sum_{j=i}^{e-1} P_j f^j$. Under the isomorphism between K_i/K_{i+1} and $\mathbb{F}_p[x]/f$ in the proof of Proposition 1, we see that k_i, $1 + P_i f^i$, and $\prod_{j=0}^{d-1} (1 + x^j f^i)^{a_{i,j}}$ are all in the same class in K_i/K_{i+1}, where where $P_i = \sum_{j=0}^{d-1} a_{i,j} x^j$. By Proposition 1, the class of k_i modulo K_{i+1} is mapped to $\bigoplus_{j=0}^{d-1} a_{i,j}$, and

$$k_{i+1} := k_i \left(\prod_{j=0}^{d-1} (1 + x^j f^i)^{\bar{k}_{i,j}} \right)^{-1} \in K_{i+1}.$$

Given $a = 1 + \sum_{i=1}^{e-1} a_i f^i$. Starting with $k_1 = a$, our algorithm iteratively compute the decomposition of k_i in

$$K_i / K_{i+1} \simeq \bigoplus_{j=0}^{d-1} \mathbb{Z}/p\mathbb{Z}$$

and k_{i+1} for all $1 \le i \le e - 1$.

4.2. Analysis of experimental results

In our experiments, we consider the following three types of generating sets for $A^\times$:

- $\pi(\mathbb{F}_q) - x$, the size is equal to p^d;
- $\pi(K) - x$ obtained by Algorithm 1, whose size is p^c;
- The set constructed by adding random elements of $\pi(K) - x$ to $\emptyset$ one by one, until it generates $A^\times$. We write its size as p^b where $b \in \mathbb{R}$. By construction, we have $b \le c \le d$.

We first compare c and d. In this experiment, we set $p = 7$, $e = 5$ and $d = 2^1, 2^2, 2^3, \dots$. The result is shown in Figure 1.

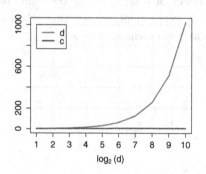
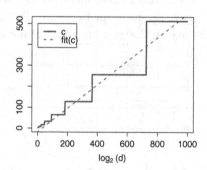

(a) Comparison between c and d (b) The growth of c with d

Fig. 1.   The growth of c and d

From Figure 1 we see that when d is a perfect power, c grows linearly with $\log d$, as stated in Corollary 1.

In the second experiment, we compare b and c with different parameters. We first set $e = 4$ and $d = 2^1, 2^2, 2^3, \dots$ and we increase p from 5 to 11.

From Figure 2, we observe a logarithmic growth from both b and c against d, and $c \approx 2b$. Also, we can see that when p increases, the growing speed of b and c decreases.

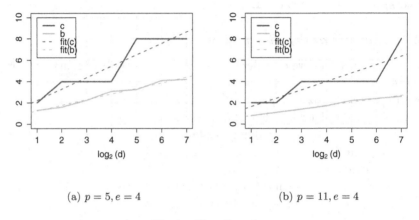

(a) $p = 5, e = 4$ (b) $p = 11, e = 4$

Fig. 2.   The effect of p

The third experiment studies the effect of e while fixing the value of $p = 7$, as shown in Figure 3. From this experiment, we see that when e increases, the growing speed of both b and c increases.

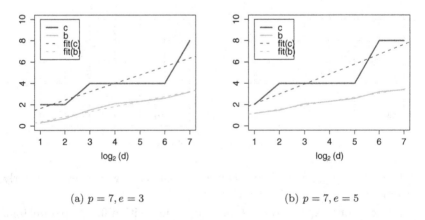

(a) $p = 7, e = 3$ (b) $p = 7, e = 5$

Fig. 3.   The effect of e

The observation $c \approx 2b$ implies that a square root number of elements from $\pi(K) - x$ might already be sufficient as a generating set. But the correctness of this observation remains an interesting open problem.

5. Explicit construction of expander graphs

As an application of our previous results, we present the explicit construction of a new family of expander graphs.

Consider the Cayley graph $\mathcal{G}(A^\times, S)$ where $S = \pi(K) - x$ and K is a subfield of $\mathbb{F}_q$ of size p^c. Let M denote the adjacency matrix of the graph. Clearly, the largest eigenvalue λ_1 of M (which corresponds to the eigenvector $\mathbb{1}$) equals p^c, the degree of each vertex. Let λ denote the second largest eigenvalue (in absolute value) of M, then we have

Theorem 7. *Let $0 < \gamma < 1$ be a constant. If $(1 - \gamma)p^{c/2} \geq de/c - 1$, then $\lambda \leq (1 - \gamma)\lambda_1$.*

Proof. A is a $\pi(K)$-algebra of dimension de/c. By Theorem 3,

$$\frac{\lambda}{\lambda_1} \leq \frac{(de/c - 1)\sqrt{p^c}}{p^c} \leq \frac{(1 - \gamma)(de/c - 1)}{de/c - 1} = 1 - \gamma$$

$\square$

Since the square matrix M has mutually orthogonal eigenvectors by Lemma 1, Caylay graphs that satisfy the premises of Theorem 7 has a spectral gap of at least γ. That is, $\mathcal{G}(A^\times, S)$ is an expander graph.

References

[1] Leonard M. Adleman and Ming-Deh A. Huang. Function field sieve method for discrete logarithms over finite fields. *Information and Computation*, 151(12):5 – 16, 1999.

[2] F. R. K. Chung. *Spectral Graph Theory*. American Mathematical Society, 1997.

[3] F.R.K.Chung. Diameters and eigenvalues. *American Mathematical Society*, 2(2):187–196, 1989.

[4] Shlomo Hoory, Nathan Linial, and Avi Wigderson. Expander graphs and their applications. *BULL. AMER. MATH. SOC.*, 43(4):439–561, 2006.

[5] Ming-Deh Huang and Anand Kumar Narayanan. Finding primitive elements in finite fields of small characteristic. *CoRR*, abs/1304.1206, 2013.

[6] Antoine Joux. A new index calculus algorithm with complexity $l(1/4 + o(1))$ in very small characteristic. Cryptology ePrint Archive, Report 2013/095, 2013.

[7] Nicholas M. Katz. An estimate for character sums. *Journal of the American Mathematical Society*, 2(2):pp. 197–200, 1989.

[8] H. W. Lenstra. Multiplicative groups generated by linear expressions.

[9] M. Lu, D. Wan, L.-P. Wang, and X.-D. Zhang. Algebraic cayley graphs over finite fields. *Finite Fields and Their Applications*, 28:43 – 56, 2014.

[10] Gary L. Mullen and Daniel Panario. *Handbook of Finite Fields*. Chapman & Hall/CRC, 1st edition, 2013.

[11] Daqing Wan. Generators and irreducible polynomials over finite fields. *Mathematics of Computation*, 66:1195–1212, 1997.

[12] A. Weil. *Basic number theory*. Grundlehren der mathematischen Wissenschaften. Springer-Verlag, 1974.

Nearly sparse linear algebra and application to discrete logarithms computations

Antoine Joux

Antoine Joux

Chaire de Cryptologie de la Fondation de l'UPMC
Sorbonne Universités, UPMC Univ Paris 06, CNRS, LIP6 UMR 7606
Paris, France
Antoine.Joux@m4x.org

Cécile Pierrot

Sorbonne Universités, UPMC Univ Paris 06, CNRS, LIP6 UMR 7606
Paris, France
Cecile.Pierrot@lip6.fr

In this article[1], we propose a method to perform linear algebra on a matrix with nearly sparse properties. More precisely, although we require the main part of the matrix to be sparse, we allow some dense columns with possibly large coefficients. This is achieved by modifying the Block Wiedemann algorithm. Under some precisely stated conditions on the choices of initial vectors in the algorithm, we show that our variation not only produces a random solution of a linear system but gives a full basis of the set of solutions. Moreover, when the number of heavy columns is small, the cost of dealing with them becomes negligible. In particular, this eases the computation of discrete logarithms in medium and high characteristic finite fields, where *nearly sparse matrices* naturally occur.

1. Introduction

Linear algebra is a widely used tool in both mathematics and computer science. At the boundary of these two disciplines, cryptography is no exception to this rule. Yet, one notable difference is that cryptographers mostly consider linear algebra over finite fields, bringing both drawbacks – the

---

[1]This work has been supported in part by the European Union's H2020 Programme under grant agreement number ERC-669891.

Cécile Pierrot has been funded by Direction Générale de l'Armement and CNRS.

notion of convergence is no longer available – and advantages – no sta
bility problems can occur. As in combinatory analysis or in the course of
solving partial differential equations, cryptography also presents the speci-
ficity of frequently dealing with sparse matrices. For instance, sparse linear
systems over finite fields appeared in cryptography in the late 70s when the
first sub-exponential algorithm to solve the discrete logarithm problem in
finite fields with prime order was designed [1]. Nowadays, every algorithm
belonging to the Index Calculus family deals with a sparse matrix [11, Section 3.4]. Hence, since both Frobenius Representation Algorithms (for small
characteristic finite fields) and discrete logarithm variants of the Number
Field Sieve (for medium and high characteristics) belong to this Index Calculus family, all recent discrete logarithm records on finite fields need to
find a solution of a sparse system of linear equations modulo a large integer.
Similarly, all recent record-breaking factorizations of composite numbers,
which are based on the Number Field Sieve, need to perform a sparse linear
algebra step modulo 2.

A sparse matrix is a matrix containing a relatively small number of coefficients that are not equal to zero. It often takes the form of a matrix
in which each row (or column) only has a small number of non-zero entries, compared to the dimension of the matrix. With sparse matrices, it is
possible to represent in computer memory much larger matrices, by giving
for each row (or column) the list of positions containing a non-zero coefficient, together with its value. When dealing with a sparse linear system
of equations, using plain Gaussian Elimination is often a bad idea, since
it does not consider nor preserve the sparsity of the input matrix. Indeed,
each pivoting step during Gaussian Elimination may increase the number
of entries in the matrix and, after a relatively small number of steps, it
overflows the available memory.

Thus, in order to deal with sparse systems, a different approach is required. Three main families of algorithms have been devised: the first one
adapts the ordinary Gaussian Elimination in order to choose pivots that
minimize the loss of sparsity and is generally used to reduce the initial
problem to a smaller and slightly less sparse problem. The two other algorithm families work in a totally different way. Namely, they do not try
to modify the input matrix but directly aim at finding a solution of the
sparse linear system by computing only matrix-by-vector multiplications.
One of these families consists of Krylov Subspace methods, adapted from
numerical analysis, and constructs sequences of mutually orthogonal vectors. For instance, this family contains the Lanczos and Conjugate Gradient

algorithms, adapted for the first time to finite fields in 1986 [7].

Throughout this article, we focus on the second family that contains Wiedemann algorithm and its generalizations. Instead of computing an orthogonal family of vectors, D. Wiedemann proposed in 1986 [20] to reconstruct the minimal polynomial of the considered matrix. This algorithm computes a sequence of scalars of the form $^twA^iv$ where v and w are two vectors and A is the sparse matrix of the linear algebra problem. It then tries to extract a linear recurrence relationship that holds for this sequence. In 1994, to achieve computations in realistic time, D. Coppersmith [6] adapted the Wiedemann algorithm over the finite field $\mathbb{F}_2$ for parallel and even distributed computations. One year later E. Kaltofen [12] not only generalized this algorithm to arbitrary finite fields but also gave a provable variant of Coppersmith's heuristic method. The main idea of Coppersmith's Block Wiedemann algorithm is to compute a sequence of matrices of the form $^tWA^iV$ where V and W are not vectors as previously but *blocks* of vectors. This step is parallelized by distributing the vectors of the block V to several processors or CPUs – let us say c of them. The asymptotic complexity of extracting the recursive relationships within the sequence of small matrices is in $\tilde{O}(cN^2)$ where N is the largest dimension of the input matrix. Another algorithm was presented by B. Beckerman and G. Labahn in 1994 [5] for performing the same task in subquadratic time and a further improvement was proposed by E. Thomé [19] in 2002: he reduced the complexity of finding the recursive relationships to $\tilde{O}(c^2N)$. The current fastest method is an application of the algorithm proposed by P. Giorgi, C-P. Jeannerod and G. Villard in 2003 [9] which runs in time $\tilde{O}(c^{\omega-1}N)$, where ω is the exponent of matrix multiplication. At the time of writing the best known[2] asymptotic value of this exponent is $\omega \approx 2.37286$. It comes from a slight improvement of Coppersmith-Winograd algorithm [8] due to F. Le Gall [14] and published in 2014.

Note that both Krylov Subspace methods and Wiedemann algorithms cost a number of matrix-by-vector multiplications equal to a small multiple of the matrix dimension: for a matrix containing λ entries per row in average, the cost of these matrix-by-vector multiplications is $O(\lambda N^2)$. With Block Wiedemann, it is possible to distribute the cost of these products up to c machines. In this case, the search for recursive relationships adds an extra cost of the form $\tilde{O}(c^{\omega-1}N)$. For a *nearly sparse matrix*, which

---

[2] Yet, for practical purposes, asymptotically fast matrix multiplication is unusable and working implementations of the algorithm of Giorgi, Jeannerod and Villard have complexity $\tilde{O}(c^2N)$.

Fig. 1.   A nearly sparse matrix

includes d dense columns in addition to its sparse part, the cost of matrix-by-vector multiplications increases. As a consequence, the total complexity becomes $O((\lambda + d)N^2)$ with an extra cost of $\tilde{O}(c^{\omega-1}N)$ for Block Wiedemann. Figure 1 provides a quick overview of the structure of such nearly sparse matrices.

In this article, we aim at adapting the Coppersmith's Block Wiedemann algorithm to improve the cost of linear algebra on matrices that have nearly sparse properties and reduce it to $O(\lambda N^2) + \tilde{O}(\max(c,d)^{\omega-1}N)$. In particular, when the number of dense columns is lower than the number of processors used for the matrix-by-vector steps, we show that the presence of these unwelcome columns does not affect the complexity of solving linear systems associated to these matrices. In practice, this result precisely applies to the discrete logarithm problem. Indeed, nearly sparse matrices appear in both medium and high characteristic finite fields discrete logarithm computations. To illustrate this claim, we recall the latest record [4] announced in June 2014 for the computation of discrete logarithms in a prime field $\mathbb{F}_p$, where p is a 180 digit prime number. It uses a matrix containing 7.28M rows and columns with an average weight of 150 non-zero coefficients per row and also presents 4 dense Schirokauer maps columns. These columns precisely give to the matrix the nearly sparse structure we study in the sequel.

Outline.   Section 2 makes a short recap on Coppersmith's Block Wiedemann algorithm, which is the currently best known algorithm to perform algebra on sparse linear systems while tolerating some amount of distributed computation. We propose in Section 4 the definition of a *nearly sparse matrix* and present then a rigorous algorithm to solve linear algebra problems associated to these matrices. In Section 4.7 we give a comparison of our method with preexisting linear algebra techniques and show that it is potentially competitive even with a tremendous number of dense columns. Section 5 ends by a practical application of this result: it explains how nearly sparse linear algebra eases discrete logarithm computations in medium and high characteristic finite fields.

2. A Reminder of Block Wiedemann Algorithm

This section first presents the classical problems of linear algebra that are encountered when dealing with sparse matrices. We then explain how the considered matrix is preconditioned into a square matrix. Section 2.2 draws the outline of the algorithm proposed by Wiedemann to solve linear systems given by a square matrix whereas Section 2.3 presents the parallelized variant due to Coppersmith. More precisely, the goal is to solve:

Problem 1. *Let* $\mathbb{K} = \mathbb{Z}/p\mathbb{Z}$ *be a prime finite field and* $S \in \mathcal{M}_{n \times N}(\mathbb{K})$ *be a (non necessarily square) sparse matrix with at most* λ *non-zero coefficients per row. Let* $\vec{v}$ *be a vector with* n *coefficients. The problem is to find a vector* $\vec{x}$ *with* N *coefficients such that* $S \cdot \vec{x} = \vec{v}$ *or, alternatively, a non-zero vector* $\vec{x}$ *such that* $S \cdot \vec{x} = 0$.

In practice, this problem is often generalized to rings $\mathbb{Z}/\mathcal{N}\mathbb{Z}$ for a modulus $\mathcal{N}$ of unknown factorization. However, for simplicity of exposition, and due to the fact that the algorithm of [9] is only proved over fields, we prefer to restrict ourselves to the prime field case.

2.1. *Preconditioning: making a sparse matrix square*

In order to be able to compute sequences of matrix-by-vector products of the form $(A^i \vec{y})_{i>0}$, both Wiedemann and Block Wiedemann algorithms need to work with a square matrix. Indeed, powers are only defined for square matrices. Consequently, if $N \neq n$, there is a necessary preliminary step to transform the given matrix into a square one. For example, it is possible to pad the matrix with zeroes and then apply the analysis of [13] for solving linear systems which do not have full rank using Wiedemann's method. This is done by multiplying on the left and right by random matrices and then by truncating the matrix to a smaller invertible square matrix with the same rank as the original one.

In practice, heuristic methods are used instead. Typically, one creates a random sparse matrix $R \in \mathcal{M}_{N \times n}(\mathbb{K})$ with at most λ non-zero coefficients per row, and transform afterwards the two problems into finding a vector $\vec{x}$

such that $(RS)\vec{x} = R\vec{v}$ or, alternatively, such that $(RS)\vec{x} = 0$. Setting $A = RS$ and $\vec{y} = R\vec{v}$, we can rewrite Problem 1 as finding a vector $\vec{x}$ such that:

$$A \cdot \vec{x} = \vec{y}$$

or such that:

$$A \cdot \vec{x} = 0$$

depending on the initial problem. In addition, in order to avoid the trivial solution when solving $A\vec{x} = 0$, one frequently computes $\vec{y} = A\vec{r}$ for a random vector $\vec{r}$, solves $A\vec{x} = \vec{y}$ and outputs $\vec{x} - \vec{r}$ as a kernel element.

We do not go further into the details of how preconditioning is usually performed. Indeed, we propose in Section 4.2 a simple alternative technique, that provably works with our algorithm for nearly sparse matrices, under some explicit technical conditions. Since sparse matrices are a special case of nearly sparse matrices, this alternative would also work for usual sparse matrices.

2.2. *Wiedemann algorithm*

Let us now consider a square matrix A of size $N \times N$ and denote m_A the number of operations required to compute the product of a vector of $\mathbb{K}^N$ by A. Wiedemann algorithm works by finding a non-trivial sequence of coefficients $(a_i)_{0 \leq i \leq N}$ such that:

$$\sum_{i=0}^{N} a_i A^i = 0. \tag{1}$$

Solving $A\vec{x} = \vec{y}$. If A is invertible, then we can assume $a_0 \neq 0$. Indeed, if $a_0 = 0$ we can rewrite $0 = \sum_{i=1}^{N} a_i A^i = A^{\delta}(\sum_{i=1}^{N} a_i A^{i-\delta})$ where a_{δ} is the first non zero coefficient. Multiplying by $(A^{-1})^{\delta}$ it yields the equality $\sum_{i=0}^{N-\delta} a_{i+\delta} A^i = 0$. So, shifting the coefficients until we find the first non-zero one allows to write $a_0 \neq 0$. Let us apply Equation (1) to the vector $\vec{x}$ we are seeking. It yields $-a_0\vec{x} = \sum_{i=1}^{N} a_i A^i \vec{x} = \sum_{i=1}^{N} a_i A^{i-1}(A\vec{x})$. Finally we recover $\vec{x} = -(1/a_0)\sum_{i=1}^{N} a_i A^{i-1}\vec{y}$. This last sum can be computed using N sequential multiplications of the initial vector $\vec{y}$ by the matrix A. The total cost to compute $\vec{x}$ as this sum is $O(N \cdot m_A)$ operations.

Algorithm 1 Wiedemann algorithm for $A\vec{x} = \vec{y}$

Input: A matrix A of size $N \times N$, $\vec{y} \neq 0$ a vector with N coefficients
Output: $\vec{x}$ such that $A \cdot \vec{x} = \vec{y}$.
   Computing a sequence of scalars
1: $\vec{v}_0 \leftarrow\in \mathbb{K}^N$, $\vec{w} \leftarrow\in \mathbb{K}^N$ two random vectors
2: **for** $i = 0, \cdots, 2N$ **do**
3:     $\lambda_i \leftarrow \vec{w} \cdot \vec{v}_i$
4:     $\vec{v}_{i+1} \leftarrow A\vec{v}_i$
5: **end for**
   Berlekamp-Massey algorithm
6: From $\lambda_0, \cdots, \lambda_{2N}$ recover coefficients $(a_i)_{0 \leq i \leq N}$ s.t. $\sum_{i=0}^{N} a_i A^i = 0$ and $a_0 \neq 0$.
   Resolution
7: **return** $-(1/a_0) \sum_{i=1}^{N-1} a_{i+1} A^i \vec{y}$.

Solving $A\vec{x} = 0$. Assuming that there exists a non-trivial element of the kernel of A, we deduce that $a_0 = 0$. Let again δ be the first index such that $a_\delta \neq 0$. Thus, for any vector $\vec{r}$ we have $0 = \sum_{i=\delta}^{N} a_i A^i \vec{r} = A^\delta(\sum_{i=\delta}^{N} a_i A^{i-\delta} \vec{r})$. We know that $\sum_{i=\delta}^{N} a_i A^{i-\delta} \neq 0$. Indeed, otherwise, $a_\delta \mathrm{Id} + \sum_{i=\delta+1}^{N} a_i A^{i-\delta} = a_\delta \mathrm{Id} + A(\sum_{i=\delta+1}^{N} a_i A^{i-\delta-1}) = 0$ would lead to $A(-(1/a_\delta) \sum_{i=\delta+1}^{N} a_i A^{i-\delta-1}) = \mathrm{Id}$, yet A is assumed non invertible. Thus, for a random vector $\vec{r}$, the sum $\sum_{i=\delta}^{N} a_i A^{i-\delta} \vec{r}$ is non zero with high probability: this vector is the zero vector if and only if $\vec{r}$ belongs to the kernel of the non null matrix $\sum_{i=\delta}^{N} a_i A^{i-\delta}$. Since the kernel of a non null matrix has at most dimension $N - 1$, the probability for a random vector to be in its kernel is upper bounded by $|\mathbb{K}|^{N-1}/|\mathbb{K}|^N = 1/|\mathbb{K}|$.

Now, computing iteratively $A(\sum_{i=\delta}^{N} a_i A^{i-\delta} \vec{r})$, $A^2(\sum_{i=\delta}^{N} a_i A^{i-\delta} \vec{r})$, $\cdots$, $A^\delta(\sum_{i=\delta}^{N} a_i A^{i-\delta} \vec{r})$ yields an element of the kernel of A in $O(N \cdot m_A)$ operations as well. Indeed, the first index j in $[\![1, \delta]\!]$ such that $A^j(\sum_{i=\delta}^{N} a_i A^{i-\delta} \vec{r}) = 0$ shows that $A^{j-1}(\sum_{i=\delta}^{N} a_i A^{i-\delta} \vec{r}) \neq 0$ belongs to the kernel of A. Thus, this method finds a non trivial element of $\mathrm{Ker}(A)$ with probability higher than $(|\mathbb{K}| - 1)/|\mathbb{K}|$, which quickly tends to 1 as the cardinality of the field grows.

How to find coefficients a_i verifying Equation (1). Cayley-Hamilton theorem testifies that the polynomial defined as $P = \det(A - X \cdot \mathrm{Id})$ annihilates the matrix A, *i.e.* $P(A) = 0$. So we know that there exists a polynomial of degree at most N whose coefficients satisfy Equation (1). Yet, directly computing such a polynomial would be too costly. The idea of Wiedemann algorithm is, in fact, to process by necessary conditions.

Let $(a_i)_{i \in [\![0, N]\!]}$ be such that $\sum_{i=0}^{N} a_i A^i = 0$. Then, for any arbitrary vector $\vec{v}$ we obtain $\sum_{i=0}^{N} a_i A^i \vec{v} = 0$. Again, for any arbitrary vector $\vec{w}$ and for any integer j we can write $\sum_{i=0}^{N} a_i {}^t\vec{w} A^{i+j} \vec{v} = 0$. Conversely, if $\sum_{i=0}^{N} a_i {}^t\vec{w} A^{i+j} \vec{v} = 0$ for any random vectors $\vec{v}$ and $\vec{w}$ and for any j in $[\![0, N]\!]$ then the probability to obtain coefficients verifying Equation (1) is high, assuming the cardinality of the field is sufficiently large [12]. Thus, Wiedemann algorithm seeks coefficients a_i that annihilate the sequence of scalars ${}^t\vec{w} A^i \vec{v}$. To do so, it can use the classical Berlekamp-Massey algorithm [2, 16] that finds the minimal polynomial of a recursive linear sequence in an arbitrary field. In a nutshell, the idea is to consider the generating function f of the sequence ${}^t\vec{w}\vec{v}, {}^t\vec{w} A\vec{v}, {}^t\vec{w} A^2\vec{v}, \cdots, {}^t\vec{w} A^{2N} \vec{v}$ and to find afterwards two polynomials g and h such that $f = g/h \mod X^{2N}$. Alternatively, the Berlekamp-Massey algorithm can be replaced by a half extended Euclidean algorithm, yielding a quasi-linear algorithm in the size of the matrix A.

2.3. *Coppersmith's Block Wiedemann algorithm*

The Block Wiedemann algorithm is a parallelization of the previous Wiedemann algorithm introduced by Don Coppersmith. It targets the context where sequences of matrix-vector products are computed on ℓ processors, instead of one. In this case, rather than solving Equation (1), it searches, given ℓ vectors $\vec{v}_1, \cdots, \vec{v}_\ell$, for coefficients a_{ij} such that:

$$\sum_{j=1}^{\ell} \sum_{i=0}^{\lceil N/\ell \rceil} a_{ij} A^i \vec{v}_j = 0 \qquad (2)$$

Note that the number of coefficients remains approximately the same as in the previous algorithm.

Solving $A\vec{x} = \vec{0}$. There, we choose ℓ random vectors $\vec{r}_1, \cdots, \vec{r}_\ell$ and set $\vec{v}_i = A\vec{r}_i$. Let δ denote the first index in $[\![1, \lceil N/\ell \rceil]\!]$ such that there exists j in $[\![1, \ell]\!]$ satisfying $a_{\delta j} \neq 0$. Equation (2) gives $\sum_{j=1}^{\ell} \sum_{i=\delta}^{\lceil N/\ell \rceil} a_{ij} A^{i+1} \vec{r}_j = \vec{0}$, i.e. $A^{\delta+1}(\sum_{j=1}^{\ell} \sum_{i=\delta}^{\lceil N/\ell \rceil} a_{ij} A^{i-\delta} \vec{r}_j) = \vec{0}$. Let $\vec{b}$ denote the vector $\sum_{j=1}^{\ell} \sum_{i=\delta}^{\lceil N/\ell \rceil} a_{ij} A^i \vec{r}_j$. According to [12], $\vec{b}$ is non zero with high probability. Hence, computing iteratively $A\vec{b}, A^2\vec{b}, \cdots, A^\delta\vec{b}$ yields an element of the kernel of A in $O(N \cdot m_A)$ operations again. Indeed, the first index k in $[\![1, \delta]\!]$ such that $A^k\vec{b} = 0$ shows that $A^{k-1}\vec{b}$ is a non trivial element of the kernel of A.

Algorithm 2 Block Wiedemann algorithm for $A\vec{x} = \vec{0}$

Input: A matrix A of size $N \times N$
Output: $\vec{x}$ such that $A \cdot \vec{x} = \vec{0}$.

    Computing a sequence of matrices
1:   $\vec{r}_1 \leftarrow\in \mathbb{K}^N, \cdots, \vec{r}_\ell \leftarrow\in \mathbb{K}^N$ and $\vec{w}_1 \leftarrow\in \mathbb{K}^N, \cdots, \vec{w}_\ell \leftarrow\in \mathbb{K}^N$
2:   $\vec{v}_1 \leftarrow A\vec{r}_1, \cdots, \vec{v}_\ell \leftarrow A\vec{r}_\ell$
3:   **for** any of the ℓ processors indexed by j **do**
4:      $u_0 \leftarrow v_j$
5:      **for** $i = 0, \cdots, 2\lceil N/\ell \rceil$ **do**
6:        **for** $k = 1, \cdots, \ell$ **do**
7:          $\lambda_{i,j,k} \leftarrow \vec{w}_k \cdot \vec{u}_i$
8:          $\vec{u}_{i+1} \leftarrow A\vec{u}_i$
9:        **end for**
10:     **end for**
11: **end for**
12: **for** $i = 0, \cdots, 2\lceil N/\ell \rceil$ **do**
13:    $M_i \leftarrow (\lambda_{i,j,k})$ the $\ell \times \ell$ matrix containing all the products of the form ${}^t\vec{w} A^i \vec{v}$
14: **end for**
    Thomé or Giorgi, Jeannerod, Villard's algorithm
15: From $M_0, \cdots, M_{2\lceil N/\ell \rceil}$ recover coefficients a_{ij} s.t. $\sum_{j=1}^{\ell} \sum_{i=0}^{\lceil N/\ell \rceil} a_{ij} A^i \vec{v}_j = \vec{0}$.
    Resolution
16: $\delta \leftarrow$ the first index in $[\![1, \lceil N/\ell \rceil]\!]$ such that there exists j in $[\![1, \ell]\!]$ satisfying $a_{\delta j} \neq 0$.
17: $\vec{b} \leftarrow \sum_{j=1}^{\ell} \sum_{i=\delta}^{\lceil N/\ell \rceil} a_{ij} A^i \vec{r}_j$.
18: $\vec{k} \leftarrow$ Error: trivial kernel element
19: **while** $\vec{b} \neq 0$ **do**
20:    $\vec{k} \leftarrow \vec{b}$
21:    $\vec{b} \leftarrow A\vec{k}$
22: **end while**
23: **return** $\vec{k}$

Solving $A\vec{x} = \vec{y}$. In order to solve $A\vec{x} = \vec{y}$, several different approaches are possible. For example, in [12] the size of A is increased by 1, adding $\vec{y}$ as an new column and adding a new zero row. It is then explained that a random kernel element, as produced by the above method, involves $\vec{y}$ and thus produces a solution of $A\vec{x} = \vec{y}$.

Another option is to set $\vec{v}_1 = \vec{y}$ and choose for $i \in [\![2, \ell]\!]$ the vectors $\vec{v}_i = A\vec{r}_i$, where each $\vec{r}_i$ is a random vector of the right size and to assume that $a_{01} \neq 0$. From Equation (2) we derive:

$$\sum_{i=0}^{\lceil N/\ell \rceil} a_{i1} A^i \vec{y} + \sum_{j=2}^{\ell} \sum_{i=0}^{\lceil N/\ell \rceil} a_{ij} A^{i+1} \vec{r}_j = 0.$$

Multiplying by the inverse of A, we obtain:

$$a_{01}\vec{x} + \sum_{i=1}^{\lceil N/\ell \rceil} a_{i1} A^i \vec{y} + \sum_{j=2}^{\ell} \sum_{i=0}^{\lceil N/\ell \rceil} a_{ij} A^i \vec{r}_j = 0.$$

Thus, we can recover $\vec{x}$ by computing:

$$(-1/a_{01}) \cdot \left(\sum_{i=1}^{\lceil N/\ell \rceil} a_{i1} A^{i-1} \vec{y} + \sum_{j=2}^{\ell} \sum_{i=0}^{\lceil N/\ell \rceil} a_{ij} A^{i} \vec{r}_j \right).$$

This can be done with a total cost of $O(N \cdot m_A)$ operations parallelized over the ℓ processors: each one is given one starting vector $\vec{v}$ and computes a sequence of matrix-by-vector products of the form $A^i \vec{v}$. The cost for each sequence is $O(N \cdot m_A/\ell)$ arithmetic operations. We do not deal here with the case where $a_{01} = 0$ since Section 4 covers all cases for nearly sparse matrices, thus for sparse matrices.

2.4. How to find coefficients a_i verifying Equation (2)

Let $\vec{v}_1, \cdots, \vec{v}_\ell$ be ℓ vectors and let consider the $\ell(\lceil N/\ell \rceil)$ elements obtained by the matrix-by-vector products of the form $A^i \vec{v}_j$ that appear in the sum of Equation (2). Since $\ell(\lceil N/\ell \rceil) > N$, all these vectors cannot be independent, so there exist coefficients satisfying (2). As for Wiedemann algorithm, we process by necessary conditions. More precisely, let $\vec{w}_1, \ldots,$ $\vec{w}_\ell$ be ℓ vectors. Assume that for any κ in $[\![0, \lceil N/\ell \rceil]\!]$ and k in $[\![1, \ell]\!]$ we have $\sum_{j=1}^{\ell} \sum_{i=0}^{\lceil N/\ell \rceil} a_{ij}\,^t\vec{w}_k A^{i+\kappa} \vec{v}_j = 0$, then the probability that the coefficients a_{ij} verify Equation (2) is close to 1 when $\mathbb{K}$ is large[3] (again see [12]). So Block Wiedemann algorithm looks for coefficients that annihilate the sequence of $2\lceil N/\ell \rceil$ small matrices of dimension $\ell \times \ell$ computed as $(^t\vec{w}_k A^\nu \vec{v}_j)$. Here, $\nu \in [\![0, 2\lceil N/\ell \rceil]\!]$ numbers the matrices, while k and j respectively denote the column and row numbers within each matrix. It is possible to compute the coefficients a_{ij} in subquadratic time (see Section 3 for details). For instance, Giorgi, Jeannerod, Villard give an efficient method with complexity $\tilde{O}(\ell^{\omega-1}N)$. This is the final component needed to write Block Wiedemann as Algorithm 2.

Moreover, putting together the matrix-by-vector products and the search for coefficients, the overall complexity can be expressed as $O(N \cdot m_A) + \tilde{O}(\ell^{\omega-1}N)$. Where the $O(N \cdot m_A)$ part can be distributed on up to ℓ processors and the $\tilde{O}(\ell^{\omega-1}N)$ part is computed sequentially.

Remark 1. In this section, we assumed that the number of sequences ℓ is equal to the number of processors c. This is the most natural choice in the classical application of Block Wiedemann, since increasing ℓ beyond

---

[3]When $\mathbb{K}$ is small, it is easy to make the probability close to 1 by increasing the number of vectors w beyond ℓ in the analysis as done in [6].

the number of processors can only degrades the overall performance. More precisely, the change leaves the $O(N \cdot m_A)$ contribution unaffected but increases $\tilde{O}(\ell^{\omega-1}N)$. However, since values of ℓ larger than c are considered in Section 4, it is useful to know that this can be achieved by sequentially computing several independent sequences on each processor. In this case, it is a good idea in practice to make the number of sequences a multiple of the number of processors, in order to minimize the wall clock running time of the matrix-by-vector multiplications.

3. Minimal Basis Computations

In this section, we recall an important result of Giorgi, Jeannerod and Villard [9], used in Section 2 for presenting Block Wiedemann. This result is a key ingredient for the algorithm we describe in Section 4. Let $\mathbb{K}$ be a finite field and G a matrix of power series over $\mathbb{K}$ of dimension $m \times n$ with $n < m$, i.e. an element of $K[[X]]^{m \times n}$. For an approximation order b, we consider m-dimensional row vectors $\vec{u}(X)$ of polynomials that satisfy the equation:

$$\vec{u} \cdot G \equiv \vec{0} \mod X^b. \tag{3}$$

For a vector of polynomial, we define its degree $\deg(\vec{u})$ as the maximum of the degree of the coordinates of $\vec{u}$.

Definition 1. A σ-basis of the set of solutions of Equation (3) is a square $m \times m$ matrix M of polynomials of $\mathbb{K}[X]$ such that:

- Every row vector $\vec{M}_i$ of M satisfies (3).
- For every solution $\vec{u}$ of (3), there exists a unique family of m polynomials $c_1, \ldots, c_m$ such that for each i:

$$\deg(c_i\vec{M}_i) \leq \deg(\vec{u}),$$

that, in addition, satisfies the relation:

$$\vec{u} = \sum_{i=1}^{m} c_i\vec{M}_i.$$

Giorgi, Jeannerod and Villard give an algorithm that computes a σ-basis for Equation (3) using $\tilde{O}(m^\omega b)$ algebraic operations in $\mathbb{K}$. Note that for practical implementations, especially with small values of m, ω should be replaced by 3, thus matching the complexity of the related algorithm given by Thomé [19].

4. Nearly Sparse Linear Algebra

In this section, our aim is twofold. We first aim at adapting the Block
Wiedemann algorithm to improve the resolution of some linear algebra
problems that are not exactly sparse but close enough to be treated similarly. We also give more precise conditions on the choices of the vectors
$\vec{v}_i$ and $\vec{w}_i$ that are made in Block Wiedemann algorithm. Rather than in-
sisting on random choices as in [12] we give explicit conditions on these
choices. When the conditions are satisfied, we show that our algorithm not
only recovers a random solution of the linear system of equations given as
input but, in fact, gives an explicit description of the full set of solutions.

The cornerstone of our method consists in working with the sparse part
of the matrix while forcing part of the initial vectors of the sequences computed by Block Wiedemann algorithm to be derived from the dense columns
of the matrix in addition to random initial vectors. In the rest of this section, we describe this idea in details.

4.1. *Nearly sparse matrices*

In the sequel we focus on linear algebra problems of the following form:

Problem 2. *Let M be a matrix of size $N \times (s + d)$ with coefficients in a
field $\mathbb{K}$. We assume that there exist two smaller matrices $M_s \in \mathcal{M}_{N \times s}(\mathbb{K})$
and $M_d \in \mathcal{M}_{N \times d}(\mathbb{K})$ such that :*

(1) $M = M_s | M_d$, where $|$ is the concatenation of matrices.[4]
(2) M_d is arbitrary.
*(3) M_s is sparse. Let us assume it has at most λ non-zero coefficients per
    row.*

*If $\vec{y}$ is a given vector with N coefficients, the problem is to find all vectors $\vec{x}$
with $s + d$ coefficients such that:*

$$M \cdot \vec{x} = \vec{y}$$

or, alternatively, such that:

$$M \cdot \vec{x} = \vec{0}.$$

*Such a matrix M is said to be d-nearly sparse, or as a shortcut, simply
nearly sparse when d is implied by context. Note that, in our definition,*

---

[4]Our method would also work for matrices with d dense columns located at any position.
It would suffice to reorder the columns of the linear algebra problem. However, for
simplicity of exposition, we assume the dense columns are the d final columns of M.

there is no a priori *restriction on the number of dense columns that appear in the matrix.*

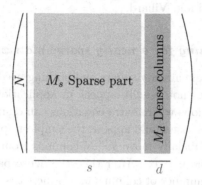

Fig. 2.   Parameters of the nearly spare linear algebra problem

An interesting consequence of the fact that we want to construct all the solutions of these linear algebra problems is that we only need to deal with the second (homogeneous) sub-problem of Problem 2. Indeed, it is easy to transform the resolution of $M \cdot \vec{x} = \vec{y}$ into the resolution of $M' \cdot \vec{x'} = \vec{0}$ for a nearly sparse matrix M' closely related to M. It suffices to set $M' = M|\vec{y}$ the matrix obtained by concatenating one additional dense column equal to $\vec{y}$ to the right of M. Now we see that $\vec{x}$ is a solution of $M \cdot \vec{x} = \vec{y}$ if and only if $\vec{x'} = {}^{t}({}^{t}\vec{x}| -1)$ is a solution of $M' \cdot \vec{x'} = \vec{0}$. Keeping this transformation in mind, in the sequel we only explain how to compute a basis of the kernel of a nearly sparse matrix. When solving the first (affine) sub-problem, we just need at the end to select in the kernel of M' the vectors with a -1 in the last position.

Thus, the two variants that appear in Problem 2 are more directly related in our context than their counterparts in Problem 1 are in the context of the traditional (Block) Wiedemann algorithm.

With such a nearly sparse matrix M, it would of course be possible to directly apply the usual Block Wiedemann algorithm and find a random element of its kernel. However, in this case, the cost of multiplying a vector by the matrix M becomes larger, of the order of $(\lambda + d)N$ operations. As a consequence, the complete cost of the usual Block Wiedemann algorithm becomes $O((\lambda + d) \cdot N^2) + \tilde{O}(\ell^{\omega-1} N)$ when using ℓ processors.

Figure 3 gives a roadmap of the various steps we go through in order to obtain an efficient bijection between the kernel of M and a subset of the solutions resulting from a minimal basis computation using the algorithm of Giorgi, Jeannerod and Villard.

4.2. *Preconditioning for a nearly sparse matrix*

If $N = s$, the matrix is already square and nothing needs to be done. Note the case $N < s$ does not usually appear in applications such as discrete logarithm computations where extra equations can easily be added. In the rare event where this case would appear, the simplest approach to deal with it is probably to artificially move $s - N$ columns from the sparse part to the dense part of the matrix. After this, the dense part becomes square ($N \times N$) while the number of columns in M_d increases to $d' = d + s - N$.

So in the sequel we focus on the case where the sparse part of M has more rows than columns, namely $N > s$. To turn the sparse part of M into a square matrix, a simple method consists in embedding the rectangular matrix M_s into a square one A by adding $N - s$ zero columns[5] at the right side of M_s.

Finding an element $^t(x_s, x_d)$ in the kernel of M is equivalent to finding a longer vector $^t(x_s, x_{tra}, x_d)$ in the kernel of $(A|M_d)$, where x_s, x_{tra} and x_d are respectively row vectors of $\mathbb{K}^s$, $\mathbb{K}^{N-s}$ and $\mathbb{K}^d$ (x_{tra} denotes the extraneous coordinates).

In the sequel, we focus on the matrix A regardless of how it has been constructed.

4.3. *Preliminary transformations with conditions*

We set B an integer to be determined later (see Section 4.5), let $\vec{\delta}_1, \ldots,$ $\vec{\delta}_d$ denote the column vectors of M_d and choose $\ell - d$ random vectors $\vec{r}_{d+1}, \cdots, \vec{r}_\ell$ in $\mathbb{K}^N$. From these vectors, we construct the family:

$$\mathcal{F} := \left\{ \begin{array}{l} \vec{\delta}_1, A\vec{\delta}_1, \cdots, A^{B-1}\vec{\delta}_1, \cdots, \vec{\delta}_d, A\vec{\delta}_d, \cdots, A^{B-1}\vec{\delta}_d, \\ \vec{r}_{d+1}, A\vec{r}_{d+1}, \cdots, A^{B-1}\vec{r}_{d+1}, \cdots, \vec{r}_\ell, A\vec{r}_\ell, \cdots, A^{B-1}\vec{r}_\ell \end{array} \right\}.$$

Our first condition is the assumption that $\mathcal{F}$ generates the full vector space $\mathbb{K}^N$. We discuss the validity of this assumption in Section 4.4.1.

---

[5] Although this zero-padding method fits the theoretical analysis well, other randomized preconditionning methods are also used in practice.

Condition on V.

To initialize the ℓ sequences the main idea is to force the first d ones to start from the d dense columns of M_d. In other words, by setting $\vec{v}_i = \vec{\delta}_i$ for i in $[\![1, d]\!]$ and $\vec{v}_i = \vec{r}_i$ for in $[\![d+1, \ell]\!]$. Then, we see that the assumption on $\mathcal{F}$ can be rewritten as:

$$\text{Vect}\left(\{A^i \vec{v}_j \mid {}^{i=0, \cdots, B-1}_{j=1, \cdots, \ell}\}\right) = \mathbb{K}^N. \tag{4}$$

Let ${}^t({}^t\vec{x} | x_1' | \cdots | x_d')$ be a vector in the kernel of $(A|M_d)$. If Equation (4) is satisfied, there exist, in particular, coefficients $\lambda_{ij} \in \mathbb{K}$ such that:

$$\vec{x} = \sum_{j=1}^{\ell} \sum_{i=0}^{B-1} \lambda_{ij} A^i \vec{v}_j$$

Thus we obtain:

$$(A|M_d) \, {}^t({}^t\vec{x} | x_1' | \cdots | x_d') \qquad \Leftrightarrow$$

$$A \sum_{j=1}^{\ell} \sum_{i=0}^{B-1} \lambda_{ij} A^i \vec{v}_j + M_d {}^t(x_1' | \cdots | x_d') = \vec{0} \qquad \Leftrightarrow$$

$$\sum_{j=1}^{d} \sum_{i=1}^{B} \lambda_{(i-1)j} A^i \vec{\delta}_j + \sum_{j=d+1}^{\ell} \sum_{i=1}^{B} \lambda_{(i-1)j} A^i \vec{r}_j + \sum_{j=1}^{d} x_j' \vec{\delta}_j = \vec{0} \Leftrightarrow$$

$$\sum_{j=1}^{\ell} \sum_{i=0}^{B} a_{ij} A^i \vec{v}_j = \vec{0}$$

where the coefficients a_{ij} are defined through:

$$a_{ij} = \begin{cases} \lambda_{(i-1)j} & \text{if } i > 0. \\ x_j' & \text{if } i = 0 \text{ and } j \leq d. \\ 0 & \text{if } i = 0 \text{ and } j > d. \end{cases}$$

To put it in a nutshell, as soon as the condition given by Equation (4) on the matrix $V = (\vec{v}_1 | \cdots | \vec{v}_\ell)$ is verified, every element of the kernel of $(A|M_d)$ gives a solution of:

$$\sum_{j=1}^{\ell} \sum_{i=0}^{B} a_{ij} A^i \vec{v}_j = \vec{0}, \tag{5}$$

where the coefficients a_{0j} are zeroes for $j > d$. Conversely (whether or not condition (4) is satisfied) any solution of Equation (5) with zeroes on these positions yields an element of the kernel of $(A|M_d)$. Thus, under condition (4), determining the kernel of $(A|M_d)$ is equivalent to finding a basis of the solutions of Equation (5) with the $\ell - d$ aforementioned zeroes.

Condition on W.

Of course, Equation (5) can be seen as a system of N linear equations over $\mathbb{K}$. However, solving it directly would not be more efficient than directly computing the kernel of M. Instead, we remark that for any matrix $W = (\vec{w}_1|\cdots|\vec{w}_\ell)$ consisting in ℓ columns of vectors in $\mathbb{K}^N$, a solution (a_{ij}) of Equation (5) leads to a solution of:

$$\sum_{j=1}^{\ell}\sum_{i=0}^{B} a_{ij}\,{}^t\vec{w}_k A^{i+\kappa}\vec{v}_j = 0 \tag{6}$$

for any $k \in [\![1, \ell]\!]$ and any $\kappa \in \mathbb{N}$.

In the reverse direction, assume that we are given a solution $(a_{ij})_{i\in[\![1,\ell]\!]}^{j\in[\![0,B]\!]}$ that satisfies Equation (6) for all $k \in [\![1, \ell]\!]$ and for all $\kappa \in [\![0, B-1]\!]$. Now assume that:

$$\text{Vect}\left(\{{}^t\vec{w}_j A^i|\,{}^{i\,=\,0,\,\cdots,\,B\,-\,1}_{j\,=\,1,\,\cdots,\,\ell}\}\right) = \mathbb{K}^N. \tag{7}$$

Under this condition, a_{ij} is also a solution of Equation (5). Indeed, by assumption, the vector $\sum_{j=1}^{\ell}\sum_{i=0}^{B} a_{ij} A^i \vec{v}_j$ is orthogonal to every vector in the basis of $\mathbb{K}^N$ listed in condition (7). Thus, it must be the zero vector.

Rewriting Equation (6) with matrix power series.

For a fixed value of κ, we can paste together the ℓ copies of Equation (6) for $k \in [\![1, \ell]\!]$. In order to do this, let $\vec{a}_i$ denote the vector ${}^t(a_{i1}, a_{i2}, \cdots, a_{il})$. With this notation, the ℓ equations can be grouped as:

$$\sum_{i=0}^{B}({}^tW A^{i+\kappa}V) \cdot \vec{a}_i = \vec{0}. \tag{8}$$

Let us now define the matrix power series $S(X)$ and the vector polynomial $P(X)$ as follows:

$$S(X) = \sum_{i\in\mathbb{N}}({}^tW A^i V)X^i \quad \text{and} \quad P(X) = \sum_{i=0}^{B}\vec{a}_i X^{B-i}.$$

Consider the product of $S(X)$ by $P(X)$. By definition of the multiplication for power series, we see that the coefficient corresponding to the monomial $X^{B+\kappa}$ in the product $S(X)P(X)$ is $\sum_{i=0}^{B}({}^tW A^{i+\kappa}V) \cdot \vec{a}_i$. According to Equation (8), this is $\vec{0}$ for all $\kappa \in \mathbb{N}$.

As a consequence, the vector power series $S(X)P(X)$ is in fact a vector polynomial $Q(X)$ of degree at most $B-1$. Thus, given $S(X)$ we search for vector polynomials $P(X)$ and $Q(X)$ of respective degrees at most B and at most $B-1$ such that $S(X)P(X) - Q(X) = \vec{0}$. To fit into the notations of Section 3, define $G(X) = (S(X)| - \mathrm{Id}(X))$ to be the $\ell \times 2\ell$ matrix power series formed by concatenating the opposite of the $\ell \times \ell$ identity matrix to $S(X)$. Denote $\vec{u}(X)$ the dimension 2ℓ row vector obtained by concatenating ${}^{t}P(X)$ and ${}^{t}Q(X)$. We now have $G(X){}^{t}\vec{u} = \vec{0}$, transpose and obtain:

$$\vec{u}(X) \cdot {}^{t}G(X) = \vec{0}. \tag{9}$$

Note that $\vec{u}(X)$ has degree at most B on its first ℓ coordinates and degree at most $B-1$ on the other coordinates. Furthermore, knowing that the coefficients a_{0j} are zeroes for $j > d$ leads to a zero constant coefficient for all polynomial coordinates from $d+1$ to ℓ.

In order to use the algorithm of Giorgi, Jeannerod and Villard, we prefer to work modulo a large monomial instead of dealing with power series. Indeed, we clearly have $\vec{u}(X) \cdot {}^{t}G(X) = \vec{0}$ modulo X^{b} for any integer b, with the same three constraints on $\vec{u}(X)$. We analyze the value of b permitting to claim that a solution of Equation (9) modulo X^{b}, and with the same constraints on the vector $\vec{u}$, can be transformed back into a solution of Equation (6) for any integer κ in $[\![0, B-1]\!]$.

Let us assume we have a vector $\vec{u} = {}^{t}(u^{(1)}, \cdots, u^{(2\ell)})$ solution of Equation (9) modulo X^{b} with:

- $\forall i \in [\![1, \ell]\!]$, $\deg u^{(i)}(X) \leq B$,
- $\forall i \in [\![\ell+1, 2\ell]\!]$, $\deg u^{(i)}(X) \leq B-1$,
- $\forall i \in [\![d+1, \ell]\!]$, $u^{(i)}(0) = 0$.

Since $\vec{u}$ consists of 2ℓ polynomials we can cut it into two separate parts and consider only its first ℓ polynomial terms, that are of degree at most B. There exists a canonical correspondence between this vector of polynomials and a polynomial $P(X)$ of degree at most B where the coefficients are vectors in $\mathbb{K}^{\ell}$. Writing $P(X) = \sum_{i=0}^{B} \vec{z}_{i} X^{i}$ with $\vec{z}_{i} \in \mathbb{K}^{\ell}$ we can define for all i in $[\![0, B]\!]$ and all j in $[\![1, \ell]\!]$:

$$a_{ij} = \text{the } j\text{-th coordinate of the vector } \vec{z}_{B-i}.$$

Since modulo X^{b} the product $P(X)S(X)$ is a vector polynomial of degree at most $B-1$ we deduce that for any integer κ such that $0 \leq \kappa \leq b - B - 1$ the coefficient in $P(X)S(X)$ related to the monomial $X^{B+\kappa}$ is the zero vector. Combining with $P(X) = \sum_{i=0}^{B} \vec{a}_{i} X^{B-i}$ and $S(X) = \sum_{i \in \mathbb{N}} ({}^{t}WA^{i}V)X^{i}$ it leads to $\sum_{i=0}^{B} ({}^{t}WA^{i+\kappa}V) \cdot \vec{a}_{i} = \vec{0}$. Hence multiplying V by the vectors $\vec{a}_{i}$ we get $\sum_{j=1}^{\ell} \sum_{i=0}^{B} {}^{t}WA^{i+\kappa}(a_{ij}\vec{v}_{j}) = \vec{0}$. Finally, if we consider each row ${}^{t}\vec{w}_{k}$

with k in $[\![1,\ell]\!]$ and any κ in $[\![0, b - B - 1]\!]$ we obtain coefficients a_{ij} that are solutions of Equation (6). Thus, to get Equation (6) for any κ in $[\![0, B - 1]\!]$ it suffices to set $b = 2B$.

Summary of the transformations.

To sum it up, we have transformed the problem of finding the kernel of M into the problem of finding all solutions of Equation (9) modulo X^{2B}, with degree at most B on the first ℓ coordinates, degree at most $B - 1$ on the other coordinates and a zero constant coefficient for coordinates $d + 1$ to ℓ. Under the two conditions (4) and (7), the above analysis directly gives a bijection between the set of solutions of the two problems, as illustrated in Figure 3.

4.4. Applying Giorgi, Jeannerod and Villard algorithm

Thanks to Giorgi, Jeannerod and Villard [9] we can compute a minimal σ-basis of the solution vectors of Equation (9) modulo X^{2B} in time $\tilde{O}(\ell^\omega B)$. However, we need to post-process this σ-basis to recover a basis of the kernel of M. More precisely, we need to derive an explicit description of all solution vectors of Equation (9) that have degree at most B on the first ℓ coordinates, degree at most $B - 1$ on the last ℓ coordinates and a zero constant coefficient for coordinates $d + 1$ to ℓ. We first show how to obtain all solution vectors that have degree at most B on the 2ℓ coordinates. A final filtering is then used to ensure that the stronger degree bound on the last ℓ coordinates holds and the $\ell - d$ constant coefficients are zeroes.

We first let $\vec{b_1}, \cdots, \vec{b_t}$ denote the t vectors[6] in the σ-base with degree at most B. Let $\vec{u}$ denote any solution vector of Equation (9) with degree at most B. From the minimality of the σ-base, we know that $\vec{u}$ can be written as linear combinations $\sum_{i=1}^{t} c_i \vec{b_i}$ where the c_i are polynomials in $\mathbb{K}[X]$ such that $\deg c_i + \deg b_i \le \deg \vec{u}$ for any $i = 1, \cdots, t$. Thus, the set of all solution vectors of Equation (9) with degree at most B is generated by the family:

$$\mathcal{E} := \bigcup_{i=1}^{t} \{\vec{b_i}, X\vec{b_i}, X^2\vec{b_i}, \cdots, X^{B - \deg \vec{b_i}} \vec{b_i}\}.$$

Note that this family is free and thus a basis of the subspace of solutions of Equation (9). Indeed, the t vectors $\vec{b_i}$ belong to a σ-basis and, thus,

---

[6]At most, there are 2ℓ such vectors. Note in practice, it is convenient to run the σ-basis algorithm on power series with precision slightly higher than $2B$ in order to have fewer vectors at this point (usually ℓ).

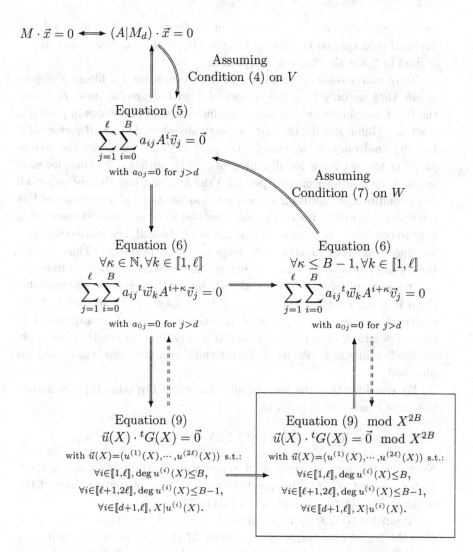

Fig. 3.   How the computation of the kernel of a nearly sparse matrix M reduces to the computation of the kernel of a power series matrix G. Equivalences and implications between various problems have to be read as follows: $A \Rightarrow B$ means that a solution of Equation A can be transformed into a solution of Equation B. The unknowns are denoted by x, a_{ij} or $\vec{u}$ whereas all the other variables $M, d, M_d, A, \ell, B, V = \{\vec{v}_1, \cdots, \vec{v}_\ell\}, W = \{\vec{w}_1, \cdots, \vec{w}_\ell\}$ and G are assumed to be known. Note that, even if it is true, we don't prove Equation (9) $\Rightarrow$ Equation (6) here since the others implications are already sufficient to conclude. Same remark for Equation (6) with all $\kappa \leq B-1 \Rightarrow$ Equation (9) mod X^{2B}.

are linearly independent. Moreover, multiplication by X induces a block diagonal structure on the matrix representing $\mathcal{E}$. Due to this structure, all vectors in $\mathcal{E}$ are also linearly independent.

To obtain a basis of the kernel of M, we now need a filtering step to ensure that we only keep the vectors of $\mathcal{E}$ with degree at most $B-1$ on the last ℓ coordinates and constant coefficients that are zeroes in positions $d+1$ to ℓ. Interestingly, the first property already holds for all vectors of $\mathcal{E}$ but the final multiple of each $\vec{b_i}$, i.e. $X^{B-\deg \vec{b_i}}\vec{b_i}$. Similarly, the second property already holds for all multiples of $X^j\vec{b_i}$ with $j \neq 0$. Thus, for each vector $\vec{b_i}$ all multiples except (possibly) the first and last already satisfy all extra condition. In addition, some linear combinations of these first and last multiples may satisfy the extra conditions and some may not. However, it is easy to construct the combinations that work. Indeed, the extra conditions are linear and only involve coefficients in $2\ell - d$ positions. Thus, to find these combinations, it suffices to extract the relevant coefficients from the polynomial multiples that do not already satisfy the condition and assemble them in a matrix of dimension $2\ell - d$ by at most[7] $2t$. The kernel of this matrix describes the desired combinations and it can be computed it in $O(\ell^\omega)$ operations asymptotically; in $O(\ell^3)$ operations in practice, especially for small values of ℓ. We let $\vec{b_i'}$ denote the t' combinations which are thus obtained.

We conclude that the basis of all solutions of Equation (9) that satisfy the three conditions is given by:

$$\mathcal{U} := \{\vec{b_1'}, \cdots, \vec{b_{t'}'}\} \cup \bigcup_{i=1}^{t} \{X\vec{b_i}, X^2\vec{b_i}, \cdots, X^{B-1-\deg \vec{b_i}}\vec{b_i}\}.$$

Note that this can be represented in a compact form, just by giving $t + t'$ vectors, with $t' \leq 2t$. This precisely gives a basis of the solutions of the equation highlighted by a frame in Figure 3.

Algorithm 3 sums up in pseudo-code the main steps that occur to compute the kernel of a nearly sparse matrix M that has been preconditioned into a matrix composed of a square matrix A concatenated with the dense part M_d of M.

4.4.1. *Checking condition* (4).

A benefit of this process is that it also checks the validity of (4). Indeed, looking back at the family $\mathcal{F}$ we see that is consists of ℓB vectors in $\mathbb{K}^N$.

---

[7]Indeed, vectors $\vec{b_i}$ with exact degree B appear once in the matrix while others appear twice.

Algorithm 3 Nearly sparse algorithm for $(A|M_d)\vec{x} = \vec{0}$

Input: A matrix A of size $N \times N$ and a matrix $M_d = (\vec{\delta}_1|\cdots|\vec{\delta}_d)$ of size $N \times d$

Output: A basis of $\mathrm{Ker}(A|M_d)$.

   Compute a sequence of matrices

1: $\vec{r}_{d+1} \leftarrow \in \mathbb{K}^N, \cdots, \vec{r}_\ell \leftarrow \in \mathbb{K}^N$ and $\vec{w}_1 \leftarrow \in \mathbb{K}^N, \cdots, \vec{w}_\ell \leftarrow \in \mathbb{K}^N$

2: $\vec{v}_1 \leftarrow \vec{\delta}_1, \cdots, \vec{v}_d \leftarrow \vec{\delta}_d$

3: $\vec{v}_{d+1} \leftarrow \vec{r}_{d+1}, \cdots, \vec{v}_\ell \leftarrow \vec{r}_\ell$

4: $B \leftarrow \lceil N/\ell \rceil$

5: **for** any of the ℓ processors indexed by j **do**

6:      $u_0 \leftarrow v_j$

7:      **for** $i = 0, \cdots, 2B$ **do**

8:          **for** $k = 1, \cdots, \ell$ **do**

9:              $\lambda_{i,j,k} \leftarrow \vec{w}_k \cdot \vec{u}_i$

10:              $\vec{u}_{i+1} \leftarrow A\vec{u}_i$

11:          **end for**

12:      **end for**

13: **end for**

14: **for** $i = 0, \cdots, 2B$ **do**

15:      $M_i \leftarrow (\lambda_{i,j,k})$ the $\ell \times \ell$ matrix containing all the products of the form ${}^t\vec{w}A^i\vec{v}$

16: **end for**

   Apply Giorgi, Jeannerod, Villard's algorithm

17: $S \leftarrow \sum_{i=0}^{2B-1} M_i X^i$.

18: Recover a σ−basis of the matrix ${}^t(S| - \mathrm{Id})$ modulo X^{2B}.

19: $\vec{b}_1, \cdots, \vec{b}_t \leftarrow$ the vectors in this σ−basis of degree lower than B.

20: $\vec{b}'_1, \cdots, \vec{b}'_{t'} \leftarrow$ a basis of the linear combinations of $\vec{b}_1, \cdots, \vec{b}_t, X^{B - \deg b_1}\vec{b}_1, \cdots,$
   $X^{B - \deg b_1}\vec{b}_t$ s.t. the ℓ last coordinates have degree lower than $B - 1$ and coordinates
   between $d + 1$ and ℓ are divisible by X.

21: $U \leftarrow [\vec{b}'_1, \cdots, \vec{b}'_{t'}, X\vec{b}_1, \cdots, X^{B-1-\deg b_1}\vec{b}_1, \cdots, X\vec{b}_t, \cdots, X^{B-1-\deg b_t}\vec{b}_t]$

   Resolution

22: Sol $\leftarrow []$

23: **for** $\vec{u} \in U$ **do**

24:      **for** $i = 0, \cdots, B$ **do**

25:          **for** $j = 1, \cdots, \ell$ **do**

26:              $a_{ij} \leftarrow$ the coefficient associated to the monomial X^{B-i} in the polynomial
   that is the j-th coefficient of $\vec{u}$.

27:          **end for**

28:      **end for**

29:      $\vec{x} \leftarrow {}^t({}^t(\sum_{j=1}^\ell \sum_{i=0}^{B-1} a_{(i+1)j} A^i \vec{v}_j)|a_{01}| \cdots |a_{0d})$

30:      Add $\vec{x}$ to Sol

31: **end for**

32: **return** Sol

The matrix corresponding to $\mathcal{F}$ has full rank if and only if the dimension of its kernel is $\ell B - N$. Yet, an element of this kernel is exactly a family of coefficients (a_{ij}) such that $\sum_{j=1}^\ell \sum_{i=0}^{B-1} a_{ij} A^i v_j = 0$. Note that it differs from Equation (5) from the fact that the sum ends at $B - 1$ and not B and nothing it said about the coefficients a_{0j}. Following the paths given in

Figure 3 we can derive a bijection between the kernel of this matrix and the set of solutions of Equation (9) with degree $B-1$ on the first ℓ coordinates and $B-2$ on the last ℓ coordinates. Since we already have computed a larger set of solutions of Equation (9), we can check if the dimension of the restricted set is $\ell B - N$. If not, the elements of the kernel of M that are obtained are still valid, but the basis of the kernel may be incomplete.

4.5. *Requirements on the parameters*

At this point, we need to choose the values of the parameters B and ℓ depending on the input parameters N, s and d. By construction, we already know that $\ell \geq d$. However, for conditions (4) and (7) to be satisfiable, there are additional restrictions. In particular, condition (7) requires a family of ℓB vectors to have rank N, thus we need:

$$ B \geq \left\lceil \frac{N}{\ell} \right\rceil . $$

There are other hidden implied requirements. Indeed, looking again at condition (7), we see that all vectors of $\left\{ {}^t\vec{w}_j A^i \mid {}^{i\,=\,1,\,\cdots,\,B\,-\,1}_{j\,=\,1,\,\cdots,\,\ell} \right\}$ belong to the image of A. Thus, the dimension of the vector space in condition (7) is upper bounded by $\mathrm{Rank}(A) + \ell$. Moreover, due to the preconditioning of Section 4.2, we know that the rank of A is at most s. This implies that the algorithm requires:

$$ \ell \geq \max(N - s, d). $$

Note that, over a large field $\mathbb{K}$, the dimension of the vector space in condition (7) for randomly chosen vectors $\vec{w}_i$ is $\mathrm{Rank}(A) + \ell$ with probability close to one.

The requirements associated to condition (4) do not give stronger arithmetic conditions on ℓ and B. However, the family of vectors in condition (4) also contains fixed vectors (derived from the dense columns of M), thus we cannot claim that the condition hold for random choices of V. However, since our algorithm also checks the validity of Condition (4), this is a minor drawback.

4.6. *Complexity analysis*

The total cost of our method contains two parts. One part is the complexity of the matrix-by-vector products whose sequential cost is $O(\lambda N^2)$ including the preparation of the sequence of $\ell \times \ell$ matrices and the final computation of the kernel basis. It can easily be distributed on several processors, especially when the number of sequences ℓ is equal to the number of processors c or a multiple of it. This minimizes the wall clock time of the matrix-by-vector phases at $O(\lambda N^2/c)$. Moreover, since $B \approx N/\ell$, the phase that recovers the coefficients a_{ij} has complexity $\tilde{O}(\ell^{\omega-1}N)$ using Giorgi, Jeannerod and Villard algorithm. The filtering step after this algorithm costs $O(\ell^w)$ and can thus be neglected (since obviously $\ell \leq N$).

To minimize the cost of Giorgi, Jeannerod and Villard algorithm, we let ℓ be the smallest multiple of c larger than d. In that case, the total sequencial cost of the algorithm becomes:

$$O(\lambda N^2) + \tilde{O}(\max(c,d)^{\omega-1}N).$$

This has to be compared with the previous $O((\lambda + d)N^2) + \tilde{O}(c^{\omega-1}N)$ obtained when combining Block Wiedemann algorithm with Giorgi, Jeannerod and Villard variant to solve the same problem. Note that the wall clock time also decreases from $O((\lambda+d)N^2/c) + \tilde{O}(c^{\omega-1}N)$ to $O(\lambda N^2/c) + \tilde{O}(\max(c,d)^{\omega-1}N)$.

If $d \leq c$ then the complexity of the variant we propose is clearly in $O(\lambda N^2) + \tilde{O}(c^{\omega-1}N)$, which is exactly the complexity obtained when combining Block Wiedeman algorithm with Giorgi, Jeannerod and Villard variant to solve a linear algebra problem on a (truly) sparse matrix of the same size. In a nutshell, when parallelizing on c processors, **it is possible to tackle up to c dense columns for free.**

4.7. *How dense can nearly sparse matrices be ?*

We already know that our nearly sparse algorithm behaves better than the direct adaptation of sparse methods. However, when the number d of dense columns becomes much larger, it makes more sense to compare to the complexity of dense methods, *i.e.* to compare our complexity with $O(N^\omega)$. In this case, we expect the number of processors to be smaller that the number of dense columns and thus replace $\max(c,d)$ by d in the complexity formulas.

Assume that $d \leq N^{1-\epsilon}$ with $\epsilon > 0$ then our complexity becomes $\tilde{O}(d^{\omega-1}N) = O(N^{\omega-\epsilon(\omega-1)}(\log N)^\alpha)$ for some $\alpha > 0$, which is asymptotically lower than $O(N^\omega)$. However, when the matrix is almost fully dense, i.e. for $d = \Omega(N)$, our technique becomes slower, by a logarithm factor, than the dense linear algebra methods.

5. Application to Discrete Logarithm Computations

In this section, we discuss the application of our adaptation of Block Wiedemann to discrete logarithm computations using the Number Field Sieve (NFS) algorithm [10, 15, 17], which applies to medium and high characteristic finite fields $\mathbb{F}_q$.

NFS contains several phases. First, a preparation phase constructs a commutative diagram of the form:

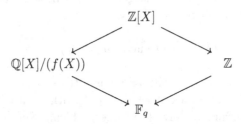

when using a rational side. Even if there also exists a generalization with a number field on each side of the diagram, for the sake of simplicity, we only sketch the description of the rational-side case.

The second phase builds multiplicative relations between the images in $\mathbb{F}q$ of products of ideals of small prime norms in the number field $\mathbb{Q}[X]/(f(X))$ and products of small primes. These relations are then transformed into linear relations between virtual logarithms of ideals and logarithms of primes modulo the multiplicative order of $\mathbb{F}_q^*$. Writing down these linear relations requires to get rid of a number of technical obstructions. In practice, this means that each relation is completed using a few extra unknowns in the linear system whose coefficients are computed from the so-called Schirokauer's maps [18]. Essentially, these maps represent the contribution of units from the number field in the equations. Due to the way they are computed, each of these maps introduces a dense column in the linear system of equations. The total number of such columns is upper-bounded by the degree of f (or the sum of the degrees when there are two number fields in the diagram).

The third phase is simply the resolution of the above linear system. In a final phase which we do not describe, NFS computes individual logarithms of field elements. An optimal candidate to apply our adaptation of Coppersmith's Block Wiedemann algorithm precisely lies in this third sparse linear algebra phase. Indeed, the number of dense columns is small enough to be smaller than the number of processors that one would expect to use in such a computation. Typically, in [4], the degree of the number field was 5, whereas the number of maps (so the number of dense columns) was 4, and the number of processors 12. Asymptotically, we know that in the range of application of NFS, the degree of the polynomials defining the number fields are at most $O((\log q/\log \log q)^{2/3})$. This is negligible compared to the size of the linear system, which is about $L_q(1/3) = \exp(O((\log q)^{1/3}(\log \log q)^{2/3}))$.

Thus, our new adaptation of Coppersmith's Block Wiedemann algorithm completely removes the difficulty of taking care of the dense columns that appear in this context. It is worth noting that these dense columns were a real practical worry and that other, less efficient, approaches have been tried to lighten the associated cost. For instance, in [3], the construction of the commutative diagram was replaced by a sophisticated method based on automorphisms to reduce the number of maps required in the computation. In this extend, this approach is no longer useful.

Moreover, since we generally have some extra processors in practice, it is even possible to consider the columns corresponding to very small primes or to ideals of very small norms as part of the dense part of the matrix and further reduce the cost of the linear algebra.

References

[1] Leonard Adleman, *A subexponential algorithm for the discrete logarithm problem with applications to cryptography*. 20th Annual Symposium on Foundations of Computer Science, 1979.

[2] Elwyn R. Berlekamp, *Nonbinary BCH decoding (Abstr.)*. IEEE Transactions on Information Theory, 1968.

[3] Razvan Barbulescu, Pierrick Gaudry, Aurore Guillevic and François Morain, *Improvements to the number field sieve for non-prime finite fields*. INRIA Hal Archive, Report 01052449, 2014.

[4] Cyril Bouvier, Pierrick Gaudry, Laurent Imbert, Hamza Jeljeli and Emmanuel Thomé, *Discrete logarithms in GF(p) – 180 digits*. Announcement to the NMBRTHRY list, item 003161, June 2014.

[5] Bernhard Beckermann and George Labahn *A Uniform Approach for the Fast Computation of Matrix-Type Padé Approximants*. SIAM Journal on Matrix Analysis and Applications, 1994.

[6] Don Coppersmith, *Solving homogeneous linear equations over GF(2) via block Wiedemann algorithm*. Mathematics of Computation, 1994.

[7] Don Coppersmith, Andrew Odlyzko and Richard Schroeppel, *Discrete Logarithms in GF(p)*. Algorithmica, 1986.

[8] Don Coppersmith and Shmuel Winograd, *Matrix Multiplication via Arithmetic Progressions*. Journal of Symbolic Computation, 1990.

[9] Pascal Giorgi and Claude-Pierre Jeannerod and Gilles Villard *On the*

complexity of polynomial matrix computations. Symbolic and Algebraic Computation, International Symposium ISSAC, 2003.

[10] Antoine Joux, Reynald Lercier, Nigel Smart and Frederik Vercauteren, *The number field sieve in the medium prime case.* Advances in Cryptology-CRYPTO 2006.

[11] Antoine Joux and Andrew Odlyzko and Cécile Pierrot, *The past, evolving present and future of discrete logarithm.* Open Problems in Mathematics and Computational Sciences, C. K. Koc, ed. Springer, 2014.

[12] Erich Kaltofen, *Analysis of Coppersmith's Block Wiedemann Algorithm for the Parallel Solution of Sparse Linear Systems.* Mathematics of Computation, 1995.

[13] Erich Kaltofen and David Saunders, *On Wiedemann's Method of Solving Sparse Linear Systems.* Applied Algebra, Algebraic Algorithms and Error-Correcting Codes, New Orleans, USA, October 7-11, 1991.

[14] François Le Gall, *Powers of tensors and fast matrix multiplication.* International Symposium on Symbolic and Algebraic Computation, ISSAC '14, Kobe, Japan, July 23-25, 2014.

[15] Arjen K. Lenstra and Hendrik W. Lenstra, Jr., *The development of the number field sieve.* Springer-Verlag, Lecture Notes in Mathematics, 1993.

[16] James L. Massey, *Shift-register synthesis and BCH decoding.* IEEE Transactions on Information Theory. 1969.

[17] Cécile Pierrot, *The Multiple Number Field Sieve with Conjugation and Generalized Joux-Lercier Methods.* Advances in Cryptology - EUROCRYPT 2015 - 34th Annual International Conference on the Theory and Applications of Cryptographic Techniques, Sofia, Bulgaria, April 26-30, 2015.

[18] Oliver Schirokauer, *Discrete logarithm and local units.* Philosophical Transactions of the Royal Society of London, 1993.

[19] Emmanuel Thomé, *Subquadratic Computation of Vector Generating Polynomials and Improvement of the Block Wiedemann Algorithm.* J. Symb. Comput., 2002.

[20] Douglas H. Wiedemann *Solving sparse linear equations over finite fields.* IEEE Transactions on Information Theory, 1986.

Full degree two del Pezzo surfaces over small finite fields

A. Knecht

Villanova University, Villanova, PA, USA
amanda.knecht@villanova.edu

K. Reyes

University of Michigan, Ann Arbor, MI, USA
kgre@umich.edu

1. Introduction

A smooth two dimensional projective variety X defined over a field k is called a *del Pezzo* surface if its anticanonical divisor $-\omega_X$ is ample. The *degree* d of a del Pezzo surface is the self-intersection number of its canonical class and is always between 1 and 9. Over algebraically closed fields, del Pezzo surfaces admit a description that is simple to understand and is adequate for the purposes of this paper.[Man86, IV.24.4] [Dem80, Thm 1].

Theorem 1 (Manin, Demazure). *Let X_d be a del Pezzo surface of degree d defined over an algebraically closed field.*

(i) X_9 is isomorphic to $\mathbb{P}^2$.

(ii) X_8 is isomorphic to $\mathbb{P}^1 \times \mathbb{P}^1$ or to the blow-up of $\mathbb{P}^2$ at one point.

(iii) If $1 \leq d \leq 7$, then X_d is the blow-up of $\mathbb{P}^2$ at $9 - d$ points, no three of which lie on a line, no six of which lie on a conic, and no eight lie on a singular cubic having one of the points as the singularity.

Because any four points in $\mathbb{P}^2$ can be moved to any four other points via an automorphism of $\mathbb{P}^2$, for $5 \leq d \leq 7$, there is only one del Pezzo surface of that degree up to isomorphism [Man86, IV.24.4.1]. Thus, from a moduli theoretic point of view, the most interesting del Pezzo surfaces are those of degree less than five. Over arbitrary fields these del Pezzo surfaces may not be the blow-up of points in $\mathbb{P}^2$, so we need a simple description that does not require the field k to be algebraically closed. Such a description is provided by Kollár [Kol96, III.3.5].

Theorem 2 (Kollár). *Let X_d be a degree d del Pezzo surface defined over an arbitrary field k.*

(i) *X_4 is isomorphic to the intersection of two quadrics in $\mathbb{P}^4$.*

(ii) *X_3 is isomorphic to a cubic surface in $\mathbb{P}^3$.*

(iii) *X_2 is isomorphic to a hypersurface of degree four in weighted projective space $\mathbb{P}(2,1,1,1)$.*

(iv) *X_1 is isomorphic to a hypersurface of degree six in weighted projective space $\mathbb{P}(3,2,1,1)$.*

Del Pezzo surfaces contain a finite number of surprisingly rigid curves that encode much of their geometry. They are called *exceptional curves* and are defined as irreducible genus zero curves with self-intersection -1, and they are the only irreducible curves on X_d with negative self-intersection. A surface X_d is called *split* over the field k if it can be realized as the blow-up of $\mathbb{P}^2$ at $9 - d$ points. Split surfaces are of interest to us because when a surface X_d is split, all of its exceptional curves are defined over k. The exceptional curves on X_d can easily be described when $2 \leq d \leq 7$.

Theorem 3 (Manin). *Let $f : X_d \to \mathbb{P}^2$ be the blow-up of the plane at the points $x_1, \ldots, x_{9-d}$ with $2 \leq d \leq 7$ and let $L \subset X_d$ be an exceptional curve. The image $f(L)$ in $\mathbb{P}^2$ is one of the following:*

(a) *one of the points x_i;*

(b) *a line passing through two of the points x_i;*

(c) *a conic passing through five of the points x_i;*

(d) *a cubic passing through seven of the points x_i such that one of them is a double point.*

The number of exceptional curves on X_d is given in the following table:

degree d	7	6	5	4	3	2
number of exceptional curves	3	6	10	16	27	56

For example, the 56 exceptional curves on degree 2 del Pezzo surfaces are mapped by $f : X_2 \to \mathbb{P}^2$ to:

(i)   7 points $x_1, x_2, \ldots, x_7$ that are blown up
(ii) 21 lines passing through 2 of the points x_i and x_j
(iii) 21 conics passing through 5 of the points
(iv) 7 singular cubics passing through all 7 points and having a double point
at one of them.

For the rest of the paper, X_d denotes a split degree d del Pezzo surface defined over a finite field $\mathbb{F}_q$, and $L_{d,q}$ will denote the elements of $X_d(\mathbb{F}_q)$ contained on the exceptional curves of X_d. Because the intersections of the exceptional curves on high degree del Pezzo surfaces are simple, an easy counting argument lets us fill in the following table when $4 \le d \le 7$:

degree d	7	6	5	4	3	2
number of exceptional curves	3	6	10	16	27	56
$L_{d,q}$	$3q+1$	$6q$	$10q-5$	$16q-24$	*	*

There do not exist formulas depending only on d and q for degree 3 and 2 del Pezzo surfaces, because the intersections of the exceptional curves are more complicated. For cubic surfaces the exceptional curves can intersect three at a time in places called *Eckardt points*. If we let e denote the number of Eckardt points on a cubic surface X_3 then Hirschfeld found that $L_{3,q} = 27(q-4) + e$ [Hir81]. A classical result by André Weil gives us the number of $\mathbb{F}_q$-rational points on split del Pezzo surfaces [Wei56]:

$$|X_d(\mathbb{F}_q)| = q^2 + (10-d)q + 1.$$

Hirschfeld defined *full* del Pezzo surfaces as split surfaces whose $\mathbb{F}_q$-rational points are all contained on the exceptional curves, i.e. $|X_d(\mathbb{F}_q)| = L_{d,q}$. Combining the table above with Weil's formula leads to the following table:

degree d	$|X_d(\mathbb{F}_q)|$	$L_{d,q}$	$|X_d(\mathbb{F}_q)| - L_{d,q}.$
7	$q^2 + 3q + 1$	$3q + 1$	q^2
6	$q^2 + 4q + 1$	$6q$	$(q-1)^2$
5	$q^2 + 5q + 1$	$10q - 5$	$(q-2)(q-3)$
4	$q^2 + 6q + 1$	$16q - 24$	$(q-5)^2$
3	$q^2 + 7q + 1$	$27(q-4) + e$	$(q-10)^2 + 9 - e$

Using the table above and an in depth analysis of cubic surfaces, Hirschfeld classified the full del Pezzo surfaces of degree nine through three [Hir82].

Theorem 4 (Hirschfeld). *Let X_d be a split del Pezzo surface of degree d defined over a finite field $\mathbb{F}_q$.*

(i) X_d is never full when $d \in \{6, 7, 8, 9\}$.
(ii) X_5 is full if and only if $q = 2$ or 3.
(iii) X_4 is full if and only if $q = 5$.
(iv) Full cubic surfaces exist only over the fields $\mathbb{F}_4, \mathbb{F}_7, \mathbb{F}_8, \mathbb{F}_9, \mathbb{F}_{11}, \mathbb{F}_{13}$, and $\mathbb{F}_{16}$.

This paper gives a partial classification of full degree two del Pezzo surfaces over fields of odd characteristic.

Theorem 5. *Let X_2 be a full degree two del Pezzo surface over $\mathbb{F}_q$ where q is odd. Then $9 \leq q \leq 37$ and up to isomorphism:*

(i) there is a unique X_2 defined over $\mathbb{F}_9$ and $\mathbb{F}_{11}$;
(ii) there are two X_2's defined over $\mathbb{F}_{13}$;
(iii) there are six X_2's defined over $\mathbb{F}_{17}$;
(iv) there five X_2's defined over $\mathbb{F}_{19}$;
(v) there are at least two X_2's defined over $\mathbb{F}_{23}$.

Theoretically there can exist full degree two del Pezzo surfaces over $\mathbb{F}_{25}, \mathbb{F}_{27}, \mathbb{F}_{29}, \mathbb{F}_{31}$, and $\mathbb{F}_{37}$, but we have not found them yet.

2. Degree Two del Pezzo Surfaces

From now on we will work over a finite field of odd characteristic, $\mathbb{F}_q$. As stated above, a split degree two del Pezzo surface X_2 can be thought of as the blow-up of the plane at seven points in general position or a degree four hypersurface in weighted projective space $\mathbb{P}(2, 1, 1, 1)$. The weighted projective space we are considering has variables w, x, y, z where the w is given weight two and the other variables are weight one. This means for our surface is that we can write X_2 as the zero set of an equation of the form

$$w^2 = f_4(x, y, z)$$

where f_4 is a homogeneous degree four polynomial in three variables with coefficients in $\mathbb{F}_q$.

   This representation of X_2 yields a third way of thinking of degree two del Pezzo surfaces. The polynomial $f_4(x, y, z)$ defines a quartic curve Q in the plane $\mathbb{P}^2$, and X_2 is a double cover of the plane branched over Q. Thus, there is a surjective map $\varphi : X_2 \to \mathbb{P}^2$ such that the preimage of a point in the plane is two points in X_2 unless the point is contained in the quartic Q, in which case the preimage is one point. The map φ is the rational map given by the anticanonical linear system, $|-K_{X_d}|$, and happens to be a morphism in this case [Kol96, III.3.5]. If L is a line in the plane, then by Bezout's Theorem L intersects Q at four points counting multiplicity and we write $L \cap Q = P_1 + P_2 + P_3 + P_4$. The line L is *tangent* to Q at a point P_1 if we can say $L \cap Q = 2P_1 + P_2 + P_3$. A *bitangent* is a line L such that $L \cap Q = 2P_1 + 2P_2$ or $L \cap Q = 4P_1$. When L and Q intersect only at one point P_1 we call that point a *hyperflex*. Over a field of characteristic greater than three, the number of hyperflexes is bounded by 12 [SV86]. There are only two quartic curves with exactly 12 hyperflexes, the Fermat quartic defined by $x^4 + y^4 + z^4 = 0$ and the curve defined by $x^4 + y^4 + z^4 + 3(x^2y^2 + x^2z^2 + y^2z^2) = 0$ [KK79]. Over $\mathbb{F}_9$, the Fermat quartic has 28 hyperflexes.

The fifty-six exceptional curves on X_2 come from the twenty-eight bitangents to the quartic curve Q. Each bitangent to Q lifts to a pair of exceptional curves in X_2 that intersect transversely at two points of Q or, in the case of a hyperflex, are tangent to each other as pictured below.

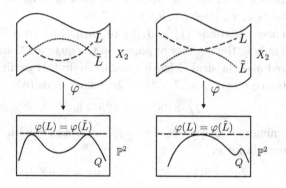

Since the bitangents to Q are lying in the plane $\mathbb{P}^2$, each bitangent is guaranteed to intersect each of the other 27 bitangents once. Similarly to the case of cubic surfaces, now up to four bitangents can intersect at once [TVAV09, Proof of Lemma 4.1]. This corresponds to four of the exceptional curves in X_2 intersecting at one point. We call such a point in X_2 a *generalized Eckardt point*. As with cubic surfaces, a point on X_2 where three exceptional curves meet is called an Eckardt point. In order to find a formula for $L_{2,q}$ similar to Hirschfeld's formula for $L_{3,q}$ we need to establish a notation for the points on the bitangent lines to Q.

Let L_i be a bitangent to Q for $1 \leq i \leq 28$ and let:

h_i = number of points on L_i where exactly 4 bitangents meet,
e_i = number of points on L_i where exactly 3 bitangents meet,
f_i = number of points on L_i where exactly 2 bitangents meet,
g_i = number of points on L_i that are not on any other bitangents,
c_i = number of times L_i intersects Q over $\mathbb{F}_q$. Thus $c_i \in \{0, 1, 2\}$.
Then,

$$|L_i(\mathbb{F}_q)| = q + 1 = h_i + e_i + f_i + g_i + c_i \tag{1}$$

and

$$27 = 3h_i + 2e_i + f_i. \tag{2}$$

Now let $h = \frac{1}{4}\sum h_i$, $e = \frac{1}{3}\sum e_i$, $f = \frac{1}{2}\sum f_i$, $g = \sum g_i$, $c = \sum c_i$.
Since each L_i will lift to two exceptional curves in X_2 that intersect only
over $\mathbb{F}_q$ points of Q, we get the formula

$$L_{2,q} = 2h + 2e + 2f + 2g + c. \tag{3}$$

Notice that we do not double count the c points because those are places
where the map φ is one-to-one. Also notice that in order for X_2 to have a
chance at fullness, we need $c = |Q(\mathbb{F}_q)| \le 56$.

We can use equations (1) and (2) to turn (3) into a formula which
depends on only q, the number of points on the quartic Q, and the number
of generalized and classical Eckardt points. To do this we first manipulate
(2) to get the equality $f_i = 27 - 3h_i - 2e_i$ which leads to

$$2f = 28(27) - 12h - 6e. \tag{4}$$

Then combining $g_i = q + 1 - h_i - e_i - f_i - c_i$ from (1) with (4) we get the
following formula for g,

$$2g = 28(2q - 52) + 16h + 6e - 2|Q(\mathbb{F}_q)|. \tag{5}$$

Using equations (4) and (5) to eliminate the f and g in (6) yields the desired
formula

$$L_{2,q} = 6h + 2e + 28(2q - 25) - |Q(\mathbb{F}_q)|.$$

In order for X_2 to be full, the equation

$$(q^2 + 8q + 1) - (6h + 2e + 28(2q - 25) - |Q(\mathbb{F}_q)|) = 0$$

must be satisfied. Or we can solve

$$(q - 24)^2 + |Q(\mathbb{F}_q)| = 6h + 2e - 125. \tag{6}$$

Lemma 1. *Full degree two del Pezzo surfaces can only exist over $\mathbb{F}_q$ when*
$9 \le q \le 37$.

Proof. In order to find a lower bound for the size of the field, we must find the smallest field in which a degree two del Pezzo can split. We discover this bound by looking at how the bitangents of a quartic plane curve intersect. Since each of the bitangents must intersect the 27 others and at most four can intersect simultaneously, each bitangent must contain at least 9 points. The number of points on a line defined over $\mathbb{F}_q$ is $q + 1$, hence $q \geq 9$.

The upper bound is found by analyzing equation (6). There are at most 126 generalized Eckardt points on a degree two del Pezzo surface since each bitangent line contains at most 9 of them. Hence, $2h \leq \frac{28*9}{2} = 126, h \leq 63$. There are also at most 242 Eckardt points on a degree 2 del Pezzo surface since each bitangent can contain at most 13 Eckardt points and not all 28 of them can. Even so, assuming all 28 lines contain 13 Eckardt points yields $e \leq 121$. Thus, we can see that the maximum value for the right hand side of (6) is 253 when $h = 63, e = 0$. In order to maximize the q on the left hand side of (6), we can assume that the quartic Q does not have any $\mathbb{F}_q$-rational points. Thus we see that in order for a degree two del Pezzo surface defined over $\mathbb{F}_q$ to have the possibility of being full, we need $(q - 24)^2 \leq 253$ or $q \leq 37$. When $q = 37$, the Weil bounds tell us that Q will contain at least 2 and at most 74 rational points, so the case $h = 63, e = 0, |Q(\mathbb{F}_q)| = 0$ does not apply. In this case (6) becomes $296 \leq 6h + 2e \leq 350$. One solution to this equality is $h = 21, e = 85$, which we have not yet ruled out as a possibility.

$\square$

The degree two del Pezzo surfaces we experiment with all contain generalized Eckardt points. Over small finite fields, split surfaces always have generalized Eckardt points. But Kaplan [Kap13, Prop 76] has found two surfaces over $\mathbb{F}_{23}$ that do not contain these points.

Lemma 2. *Over* $\mathbb{F}_9, \mathbb{F}_{11}$ *and* $\mathbb{F}_{13}$, *split degree two del Pezzo surfaces contain generalized Eckardt points.*

Proof. Over a field with q elements, every bitangent to the quartic Q contains exactly $q + 1$ $\mathbb{F}_q$-rational points. When $q = 9$, $h_i = 9$ for each $1 \leq i \leq 28$, so X_2 contains $2 \cdot (\frac{28 \cdot 9}{4}) = 126$ generalized Eckardt points. When $q = 11$, there are 12 points on each bitangent so $h_i \geq 1$ for each $1 \leq i \leq 28$. Thus X_2 contains at least $2 \cdot (\frac{28 \cdot 1}{4}) = 14$ generalized Eckardt points.

Now suppose by way of contradiction that the size of the ground field is 13 and the del Pezzo does not contain any generalized Eckardt points. Then, every bitangent to Q must intersect 2 lines simultaneously at 13 points, $e_i = 13, f_i = 1$ for all $1 \leq i \leq 28$. This means that the number of Eckardt points on the corresponding del Pezzo surface is $2 \cdot (\frac{28 \cdot 13}{3}) = \frac{364}{3} \notin \mathbb{Z}$. Thus, the surface must contain a generalized Eckardt point. $\square$

3. Kuwata Curves

One reason we restrict ourselves to surfaces with generalized Eckardt points is that over an algebraically closed field, a quartic curve Q admitting a generalized Eckardt point is equivalent to Q admitting an involution [Kuw05, Theorem 5.1]. Thus there is nice symmetry between the lines. Kuwata gives explicit equations for the quartic curves which admit a pair of commuting involutions, and these are the surfaces we study [Kuw05].

Theorem 6 (Kuwata). *Let Q be a quartic plane curve with a pair of commuting involutions. By a change of coordinates, possibly defined over a field extension, Q can be defined by an equation of the form*

$$a_1 x^4 + a_2 y^4 + a_3 z^4 + a_4 x^2 y^2 + a_5 y^2 z^2 + a_6 z^2 x^2 = 0.$$

Kuwata uses the equation above to find a family of quartic curves whose twenty-eight bitangent lines are all defined over the base field which does not need to be algebraically closed. We call the curves $C_{\lambda\mu\nu}$ defined below *Kuwata curves* [Kuw05].

Theorem 7 (Kuwata). *Let λ, μ, ν be three elements in $\mathbb{F}_q$ satisfying*

$$(1 - \lambda^2)(1 - \mu^2)(1 - \nu^2)(1 - \mu^2\lambda^2)(1 - \nu^2\lambda^2)(1 - \mu^2\nu^2)(1 - \lambda^2\mu^2\nu^2) \neq 0.$$

Let $C_{\lambda\mu\nu}$ be the quartic plane curve given by

$$((1 - \mu^2\nu^2)x^2 + (1 - \nu^2\lambda^2)y^2 + (1 - \mu^2\lambda^2)z^2)^2$$
$$- 4(1 - \lambda^2\mu^2\nu^2)\left((1 - \nu^2)x^2y^2 + (1 - \lambda^2)y^2z^2 + (1 - \mu^2)z^2x^2\right) = 0.$$

The twenty-eight double tangent lines of $C_{\lambda\mu\nu}$ are given by

$$x \pm \lambda y = 0, \qquad y \pm \mu x = 0, \qquad z \pm \nu x = 0,$$
$$x \pm \lambda z = 0, \qquad y \pm \mu z = 0, \qquad z \pm \nu y = 0,$$
$$(1 + \mu\nu)x \pm (1 + \nu\lambda)y \pm (1 - \lambda\mu)z = 0,$$
$$(1 + \mu\nu)x \pm (1 - \nu\lambda)y \pm (1 + \lambda\mu)z = 0,$$
$$(1 - \mu\nu)x \pm (1 + \nu\lambda)y \pm (1 + \lambda\mu)z = 0,$$
$$(1 - \mu\nu)x \pm (1 - \nu\lambda)y \pm (1 - \lambda\mu)z = 0.$$

We use the computer algebra system MAGMA to find all Kuwata curves over our desired fields. Then we use the equations of the lines to determine the intersection form (h, e, f, g, c) of each line. With this information, we are able to find full degree two del Pezzo surfaces. The following section lists the results we were able to find while experimenting with Kuwata curves in MAGMA.

Another recent result involving degree two del Pezzo surfaces with generalized Eckardt points is due to Salgado, Testa, and Várilly-Alvarado [STVA14]. Let q_2, q_4 be homogeneous polynomials of degrees two and four respectively such that the polynomial $q_2^2 - 4q_4$ has distinct roots. When the characteristic of the ground field is not two, the surface X_2 in the weighted projective space $\mathbb{P}(1, 1, 1, 2)$ with equation $w^2 = x^4 + q_2(y, z)x^2 + q_4(y, z)$ is a smooth del Pezzo surface of degree two with an involution given by $x \mapsto -x$ and contains at least two generalized Eckardt points $[1, 0, 0, 1]$ and $[1, 0, 0, -1]$. In fact, every del Pezzo surface of degree two with a point contained in four exceptional curves has an involution and is of the form described above.

4. Full Surfaces over Small Finite Fields

In this section we present the results we found using MAGMA and Kuwata curves. For each of the finite fields we list the full degree two del Pezzo surfaces and how each of the bitangents to the quartic Q intersects the other bitangents in the form (h, e, f, g, c) defined above.

Our first result on full degree 2 del Pezzo surfaces was first found by Hirschfeld when he was studying full cubic surfaces[Hir67]. He found that there is only one cubic surface over $\mathbb{F}_9$ with exactly one rational point not on the lines. This point can be blown-up to create a full degree two del Pezzo.

Theorem 8. *Over the field* $\mathbb{F}_9$ *there exists a unique full degree two del Pezzo surface up to isomorphism. It is given by the equation*

$$w^2 = x^4 + y^4 + z^4$$

and all of the lines on X contain nine generalized Eckardt points.

All bitangents are all of type $(9,0,0,0,1)$ and the quartic Q has exactly 28 $\mathbb{F}_9$-rational points, all of which are hyperflexes.

Theorem 9. *Over the field $\mathbb{F}_{11}$ there exists a unique full degree two del Pezzo surface up to isomorphism. It is given by the equation*

$$w^2 = x^4 + y^4 + z^4 + (x^2y^2 + x^2z^2 + y^2z^2).$$

All of the bitangents to Q are of type $(3,9,0,0,0)$. The quartic equation $x^4 + y^4 + z^4 + (x^2y^2 + x^2z^2 + y^2z^2)$ does not have any solutions over $\mathbb{F}_{11}$.

Theorem 10. *Over the field $\mathbb{F}_{13}$ there exist exactly two full degree two del Pezzo surfaces up to isomorphism.*

$$w^2 = x^4 + y^4 + z^4 + 8(x^2y^2 + x^2z^2 + y^2z^2) \tag{7}$$

and

$$w^2 = x^4 + y^4 + z^4 - x^2y^2. \tag{8}$$

The bitangents corresponding to equation (7) come in two intersection types. There are 24 bitangents of type $(1,11,2,0,0)$ and 4 of type $(3,9,0,0,2)$. The quartic curve $x^4 + y^4 + z^4 + 8(x^2y^2 + x^2z^2 + y^2z^2)$ contains eight $\mathbb{F}_{13}$-rational points.

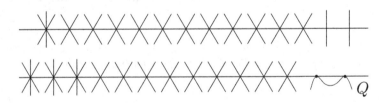

The bitangents corresponding to equation (8) also come in two intersection types. There are 24 bitangents of type $(1,11,2,0,0)$ and 4 of type $(1,12,0,0,1)$. The quartic curve $x^4+y^4+z^4-x^2y^2$ contains four $\mathbb{F}_{13}$-rational points, all hyperflexes.

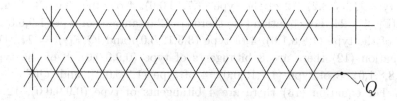

We know that these are the only full degree two del Pezzos over the fields $\mathbb{F}_9, \mathbb{F}_{11}, \mathbb{F}_{13}$ because Nathan Kaplan has found all the split degree two del Pezzo surfaces over these fields [Kap13, Prop 70,71,72], and our results agree.

Corollary 1. *Over fields of size* $9, 11,$ *and* $13,$ *a degree two del Pezzo surface is full if and only if it is split.*

Theorem 11. *Over the field* $\mathbb{F}_{17}$ *there are exactly six full degree two del Pezzo surfaces up to isomorphism.*

$$w^2 = x^4 + y^4 + z^4, \tag{9}$$

$$w^2 = x^4 + y^4 + z^4 - x^2y^2 + 7x^2z^2 + 7y^2z^2, \tag{10}$$

$$w^2 = x^4 + y^4 + 2z^4 + 13x^2y^2 + 6x^2z^2 + 6y^2z^2, \tag{11}$$

$$w^2 = x^4 + y^4 + 2z^4 + 13x^2y^2 + x^2z^2 + y^2z^2, \tag{12}$$

$$w^2 = x^4 + y^4 + 2z^4 + 5x^3y + 15x^3z + 13x^2y^2 + 5x^2yz + 13x^2z^2$$
$$+ 13xy^3 + 15xy^2z + 5xyz^2 + 5xz^3 + 2y^3z + 13y^2z^2 + 12yz^3, \tag{13}$$

$$w^2 = x^4 + 2y^4 + z^4 + 2x^3y + 6x^2y^2 + 16x^2yz + 11x^2z^2 + 5xy^3$$
$$+ 7xy^2z + 3xyz^2 + 6xz^3 + 6y^3z + 10y^2z^2 + 8yz^3. \tag{14}$$

In [Kap13, Prop 73], Kaplan gives seven isomorphism types for split degree two del Pezzo surfaces defined over $\mathbb{F}_{17}$. His first four equations correspond to the first four full X_2's we list above that come from Kuwata's equations. His sixth equation is isomorphic to the Kuwata curve C_{234} which is not full. Thus, once the field has at least 17 elements, there is enough room for a degree 2 del Pezzo surface to be split but not full. His fifth and seventh equations do not correspond to Kuwata curves, but are indeed full and are isomorphic to the last two equations in the theorem.

The bitangents corresponding to equation (9) come in two intersection types. There are 12 bitangents of type $(1,8,8,0,1)$, 16 of the type $(3,3,12,0,0)$, and $|Q(\mathbb{F}_{17})| = 12$. For equation (10) there are 12 bitangents

of type (1,9,6,0,2), 12 of the type (1,7,10,0,0), 4 of type (3,6,6,3,0), and
$|Q(\mathbb{F}_{17})| = 24$. For equation (11) there are 12 bitangents of type (1,9,6,0,2),
12 of the type (1,8,8,1,0), 4 of type (3,3,12,0,0), and $|Q(\mathbb{F}_{17})| = 24$. For
equation (12) there are 8 bitangents of type (0,9,9,0,0), 8 of the type
(1,8,8,1,0), 8 of type (1,7,10,0,0), 4 of type (1,9,6,0,2), and $|Q(\mathbb{F}_{17})| =$
8. For equation (13) there are 4 bitangents of type (0,9,9,0,0), 12 of
the type (0,10,7,1,0), 6 of type (1,7,10,0,0), 6 of type (1,9,6,0,2), and
$|Q(\mathbb{F}_{17})| = 12$. For equation (14) there is 1 bitangent of type (1,9,6,0,2),
2 of type (1,9,6,2,0), 8 of type (0,9,9,0,0), 10 of the type (0,10,7,1,0), 1
of type (1,7,10,0,0), 2 of type (0,10,7,0,1), 4 of type (0, 11, 5, 0, 2), and
$|Q(\mathbb{F}_{17})| = 12$.

Theorem 12. *Over the field $\mathbb{F}_{19}$ there exist five full degree two del Pezzo
surfaces up to isomorphism.*

$$w^2 = x^4 + y^4 + z^4 + 4x^2y^2 + 4x^2z^2 + 5y^2z^2, \tag{15}$$

$$w^2 = x^4 + y^4 + z^4 - x^2y^2 + 7x^2z^2 + 7y^2z^2, \tag{16}$$

$$w^2 = x^4 + y^4 + z^4 + 12x^3y + 2x^3z + 15x^2y^2 + 14x^2yz + 11x^2z^2$$
$$+ 7xy^3 + 10xy^2z + 5xyz^2 + 10xz^3 + 2y^3z + 13y^2z^2 + 8yz^3, \tag{17}$$

$$w^2 = x^4 + y^4 + z^4 + 11x^3y + 14x^3z + 18x^2y^2 + 10x^2yz + 9x^2z^2$$
$$+ 11xy^3 + 16xy^2z + 18xyz^2 + 5xz^3 + 3y^3z + 10y^2z^2 + 2yz^3, \tag{18}$$

$$w^2 = x^4 + y^4 - z^4 + x^3y + 9x^3z + 7x^2y^2 + 16x^2yz + x^2z^2$$
$$+ xy^3 + 7xy^2z + 5xyz^2 + 6y^3z + 11y^2z^2. \tag{19}$$

In [Kap13, Prop 75], Kaplan gives fourteen isomorphism types for split
degree two del Pezzo surfaces defined over $\mathbb{F}_{19}$. His 8th and 12th equations correspond to the equations (15) and (16) in the list above. His
$5^{th}, 6^{th}, 10^{th}, 13^{th}$, and 14th equations are isomorphic to the five other
Kuwata curves and are not full. Kaplan's $1^{st}, 2^{nd}, 9^{th}$ and 11th equations are not Kuwata curves and are not full. But his $3^{rd}, 4^{th}$, and 7th
curves define full del Pezzos that are not Kuwata and are isomorphic to the
last three equations in the theorem above.

The bitangents corresponding to the quartic in equation (15) come
in five different types. There are 8 bitangents of type (1,7,10,2,0), 4 of
type (1,7,10,0,2), 8 of the type (0,7,13,0,0), 4 of type (1,8,8,1,2), 4 of type
(1,8,8,3,0), and $|Q(\mathbb{F}_{19})| = 16$. For equation (16) there are 12 bitangents
of type (1,6,12,1,0), 12 of type (1,5,14,0,0), 4 of the type (3,6,6,3,2), and
$|Q(\mathbb{F}_{19})| = 8$. For equation (17) there are 2 bitangents of type (0,10,7,3,0),

2 of type (0,10,7,1,2), 8 of type (0,9,9,2,0), 4 of type (0,9,9,0,2), 2 of type (0,9,9,1,1), 6 of type (0,8,11,1,0), 2 of type (1,6,12,1,0), 1 of type (1,9,6,2,2), 1 of type (1,5,14,0,0), and $|Q(\mathbb{F}_{19})| = 16$. For equation (18) there are 2 bitangents of type (0,10,7,2,1), 4 of type (0,10,7,1,2), 5 of type (0,7,13,0,0), 5 of type (0,9,9,2,0), 8 of type (0,8,11,1,0), 1 of type (1,7,10,0,2), 1 of type (1,9,6,4,0), 2 of type (1,7,10,2,0), and $|Q(\mathbb{F}_{19})| = 12$. For equation (19) there are 6 bitangents of type (0,7,13,0,0), 6 of type (0,8,11,1,0), 6 of type (0,9,9,2,0), 6 of type (1,7,10,2,0), 3 of type (1,7,10,0,2), 1 of type (3,3,12,0,2), and $|Q(\mathbb{F}_{19})| = 8$.

Theorem 13. *Over the field* $\mathbb{F}_{23}$ *there exist at least two full degree two del Pezzo surfaces up to isomorphism.*

$$w^2 = x^4 + y^4 + z^4 + 6(x^2y^2 + x^2z^2 + y^2z^2) \qquad (20)$$

and

$$w^2 = x^4 + y^4 + z^4 - (x^2y^2 + x^2z^2 + y^2z^2). \qquad (21)$$

The bitangents corresponding to the quartic in equation (20) come in three different types. There are 12 bitangents of type (1,16,12,3,2), 12 of type (1,5,14,4,0), 4 of type (3,3,12,6,0), and $|Q(\mathbb{F}_{23})| = 24$. For equation (21) all 28 bitangents are of type (3,0,18,3,0) and $Q(\mathbb{F}_{23}) = \emptyset$.

There are twelve other isomorphism classes of Kuwata curves that correspond to split but not full degree two del Pezzo surfaces over $\mathbb{F}_{23}$. In [Kap13, Prop 76], Kaplan says that there are at least 19 isomorphism classes of split degree two del Pezzo surfaces over $\mathbb{F}_{23}$. He gives two classes that are not isomorphic to Kuwata curves because they have no non-trivial automorphisms. Thus, Kaplan has found two split surfaces over $\mathbb{F}_{23}$ that contain no generalized Eckardt points. His surfaces are not full. Since Kaplan only gives the equations for two classes, and we know the equations for 14 Kuwata classes, there are still at least three isomorphism classes of split X_2's to be found that could possibly be full.

Once the field size is larger than 23, none of the del Pezzo surfaces corresponding to Kuwata curves are full. Thus, a full X_2 defined over $\mathbb{F}_q$ with $25 \le q \le 37$ cannot have two commuting involutions in its automorphism group. So we are left with the open question:

Question 1. *Are there full degree two del Pezzo surfaces defined over* $\mathbb{F}_{25}, \mathbb{F}_{27}, \mathbb{F}_{29}, \mathbb{F}_{31}$ *and* $\mathbb{F}_{37}$?

Another question we have yet to answer involves finding full degree two del Pezzos that do not come from Kuwata curves:

Question 2. *Are there other full degree two del Pezzo surfaces defined over* $\mathbb{F}_{23}$?

Once these questions are answered there remains the harder open problem involving smooth sextic hypersurfaces in the weighted projective space $\mathbb{P}(3,2,1,1)$:

Question 3. *What are the full degree one del Pezzo surfaces?*

This question is extremely computationally heavy because there are 240 exceptional curves to consider and $q^2 + 9q + 1$ points on the surface.

Acknowledgment The authors are greatly indebted to the referee for writing a MAGMA program that helped us analyze the bitangents to the curves given by Kaplan. We would also like to thank Robert Lazarsfeld for introducing them at the University of Michigan and Zachary Scherr for discussing this problem with us.

References

[Dem80] M. Demazure. *Surfaces de Del Pezzo II*, Séminaire sur les Singularités des Surfaces, Lecture Notes in Mathematics, vol. 777, pp. 23–69, Springer, Berlin, 1980.

[Hir67] J. W. P. Hirschfeld. *Classical configurations over finite fields. I. The double-six and the cubic surface with* 27 *lines*. Rend. Mat. e Appl. (5), 26:115–152, 1967.

[Hir81] J. W. P. Hirschfeld. *Cubic surfaces whose points all lie on their* 27 *lines*. I Finite geometries and designs (Proc. Conf., Chelwood Gate, 1980), vol. 49, London Math. Soc. Lecture Note Ser., pp. 169–171, Cambridge Univ. Press, Cambridge-New York, 1981.

[Hir82] J. W. P. Hirschfeld. *del Pezzo surfaces over finite fields*. Tensor (N.S.), 37(1):79–84, 1982.

[Kap13] Nathan Kaplan. *Rational Point Counts for del Pezzo Surfaces over Finite Fields and Coding Theory*. ProQuest LLC, Ann Arbor, MI, 2013. Thesis (Ph.D.)–Harvard University.

[KK79] Akikazu Kuribayashi and Kaname Komiya. On Weierstrass points and automorphisms of curves of genus three. In *Algebraic geometry (Proc. Summer Meeting, Univ. Copenhagen, Copenhagen, 1978)*, volume 732 of *Lecture Notes in Math.*, pp. 253–299. Springer, Berlin, 1979.

[Kol96] J. Kollár. *Rational curves on algebraic varieties*, volume 32 of *Ergebnisse der Mathematik und ihrer Grenzgebiete. 3. Folge. A Series of*

Modern Surveys in Mathematics [Results in Mathematics and Related Areas. 3rd Series. A Series of Modern Surveys in Mathematics]. Springer-Verlag, Berlin, 1996.

[Kuw05]   Masato Kuwata. Twenty-eight double tangent lines of a plane quartic curve with an involution and the Mordell-Weil lattices. *Comment. Math. Univ. St. Pauli*, 54(1):17–32, 2005.

[Man86]   Yu. I. Manin. *Cubic forms*, volume 4 of *North-Holland Mathematical Library*. North-Holland Publishing Co., Amsterdam, second edition, 1986.

[STVA14]   Cecília Salgado, Damiano Testa, and Anthony Várilly-Alvarado. *On the unirationality of del Pezzo surfaces of degree 2*, J. Lond. Math. Soc. (2), 90(1):121–139, 2014.

[SV86]   Karl-Otto Stöhr and José Felipe Voloch. *Weierstrass points and curves over finite fields*. Proc. London Math. Soc. (3), 52(1):1–19, 1986.

[TVAV09]   Damiano Testa, Anthony Várilly-Alvarado, and Mauricio Velasco. *Cox rings of degree one del Pezzo surfaces*. Algebra Number Theory, 3(7), pp. 729–761, 2009.

[Wei56]   André Weil. *Abstract versus classical algebraic geometry*. In *Proceedings of the International Congress of Mathematicians, 1954, Amsterdam, vol. III*, pp. 550–558. Erven P. Noordhoff N.V., Groningen; North-Holland Publishing Co., Amsterdam, 1956.

Diameter of some monomial digraphs

A. Kodess

University of Rhode Island, Kingston, RI, USA kodess@uri.edu

F. Lazebnik, S. Smith and J. Sporre

University of Delaware, Newark, DE, USA
{fellaz,smithsj,jsporre}@udel.edu

1. Introduction

For all terms related to digraphs which are not defined below, see Bang-Jensen and Gutin [1]. In this paper, by a *directed graph* (or simply *digraph*) D we mean a pair (V, A), where $V = V(D)$ is the set of vertices and $A = A(D) \subseteq V \times V$ is the set of arcs. For an arc (u, v), the first vertex u is called its *tail* and the second vertex v is called its *head*; we also denote such an arc by $u \to v$. If (u, v) is an arc, we call v an *out-neighbor* of u, and u an *in-neighbor* of v. The number of out-neighbors of u is called the *out-degree* of u, and the number of in-neighbors of u — the *in-degree* of u. For an integer $k \geq 2$, a *walk* W from x_1 to x_k in D is an alternating sequence $W = x_1 a_1 x_2 a_2 x_3 \ldots x_{k-1} a_{k-1} x_k$ of vertices $x_i \in V$ and arcs $a_j \in A$ such that the tail of a_i is x_i and the head of a_i is x_{i+1} for every i, $1 \leq i \leq k - 1$. Whenever the labels of the arcs of a walk are not important, we use the notation $x_1 \to x_2 \to \cdots \to x_k$ for the walk, and say that we have an $x_1 x_k$-walk. In a digraph D, a vertex y is *reachable* from a vertex x if D has a walk from x to y. In particular, a vertex is reachable from itself. A digraph D is *strongly connected* (or, just *strong*) if, for every pair x, y of distinct vertices in D, y is reachable from x and x is reachable from y. A *strong component* of a digraph D is a maximal induced subdigraph of D that is strong. If x and y are vertices of a digraph D, then the *distance from x to y* in D, denoted dist(x, y), is the minimum length of an xy-walk, if y is reachable from x, and otherwise $\text{dist}(x, y) = \infty$. The *distance from a set* X *to a set* Y of vertices in D is

$$\text{dist}(X, Y) = \max\{\text{dist}(x, y) \colon x \in X, y \in Y\}.$$

The *diameter* of D is $\text{diam}(D) = \text{dist}(V, V)$.

Let p be a prime, e a positive integer, and $q = p^e$. Let $\mathbb{F}_q$ denote the finite field of q elements, and $\mathbb{F}_q^* = \mathbb{F}_q \setminus \{0\}$.

Let $\mathbb{F}_q^2$ denote the Cartesian product $\mathbb{F}_q \times \mathbb{F}_q$, and let $f \colon \mathbb{F}_q^2 \to \mathbb{F}_q$ be an arbitrary function. We define a digraph $D = D(q; f)$ as follows: $V(D) = \mathbb{F}_q^2$, and there is an arc from a vertex $\mathbf{x} = (x_1, x_2)$ to a vertex $\mathbf{y} = (y_1, y_2)$ if and only if

$$x_2 + y_2 = f(x_1, y_1).$$

If (x, y) is an arc in D, then $\mathbf{y}$ is uniquely determined by $\mathbf{x}$ and y_1, and $\mathbf{x}$ is uniquely determined by $\mathbf{y}$ and x_1. Hence, each vertex of D has both its in-degree and out-degree equal to q.

By Lagrange's interpolation, f can be uniquely represented by a bivariate polynomial of degree at most $q - 1$ in each of the variables. If $f(x, y) = x^m y^n$, $1 \le m, n \le q - 1$, we call D a *monomial* digraph, and denote it also by $D(q; m, n)$. Digraph $D(3; 1, 2)$ is depicted in Fig. 1.1. It is clear, that $\mathbf{x} \to \mathbf{y}$ in $D(q; m, n)$ if and only if $\mathbf{y} \to \mathbf{x}$ in $D(q; n, m)$. Hence, one digraph is obtained from the other by reversing the direction of every arc. In general, these digraphs are not isomorphic, but if one of them is strong then so is the other and their diameters are equal. As this paper is concerned only with the diameter of $D(q; m, n)$, it is sufficient to assume that $1 \le m \le n \le q - 1$.

The digraphs $D(q; f)$ and $D(q; m, n)$ are directed analogues of some algebraically defined graphs, which have been studied extensively and have many applications. See Lazebnik and Woldar [18] and references therein; for some subsequent work see Viglione [24], Lazebnik and Mubayi [14], Lazebnik and Viglione [17], Lazebnik and Verstraëte [16], Lazebnik and Thomason [15], Dmytrenko, Lazebnik and Viglione [7], Dmytrenko, Lazebnik and Williford [8], Ustimenko [23], Viglione [25], Terlep and Williford [22], Kronenthal [13], Cioabă, Lazebnik and Li [3], Kodess [11], and Kodess and Lazebnik [12].

The questions of strong connectivity of digraphs $D(q; f)$ and $D(q; m, n)$ and descriptions of their components were completely answered in [12]. Determining the diameter of a component of $D(q; f)$ for an arbitrary prime power q and an arbitrary f seems to be out of reach, and most of our

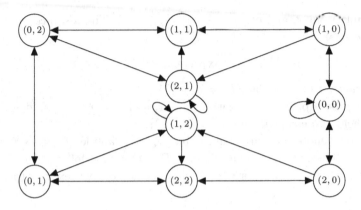

Fig. 1.   The digraph $D(3;1,2)$: $x_2 + y_2 = x_1 y_1^2$.

results below are concerned with some instances of this problem for strong monomial digraphs. The following theorems are the main results of this paper.

Theorem 1.1. *Let p be a prime, e, m, n be positive integers, $q = p^e$, $1 \le m \le n \le q - 1$, and $D_q = D(q; m, n)$. Then the following statements hold.*

(1) If D_q is strong, then $\mathrm{diam}(D_q) \ge 3$.
(2) If D_q is strong, then
  - *for $e = 2$, $\mathrm{diam}(D_q) \le 96\sqrt{n+1} + 1$;*
  - *for $e \ge 3$, $\mathrm{diam}(D_q) \le 60\sqrt{n+1} + 1$.*

(3) If $\gcd(m, q-1) = 1$ or $\gcd(n, q-1) = 1$, then $\mathrm{diam}(D_q) \le 4$. If $\gcd(m, q-1) = \gcd(n, q-1) = 1$, then $\mathrm{diam}(D_q) = 3$.
(4) If p does not divide n, and $q > (n^2 - n + 1)^2$, then $\mathrm{diam}(D(q; 1, n)) = 3$.
(5) If D_q is strong, then:

  (a) If $q > n^2$, then $\mathrm{diam}(D_q) \le 49$.
  (b) If $q > (m-1)^4$, then $\mathrm{diam}(D_q) \le 13$.
  (c) If $q > (n-1)^4$, then $\mathrm{diam}(D(q; n, n)) \le 9$.

Remark 1. The converse to either of the statements in part (3) of Theorem 1.1 is not true. Consider, for instance, $D(9; 2, 2)$ of diameter 4, or $D(29; 7, 12)$ of diameter 3.

Remark 2. The result of part 5a can hold for some $q \le m^2$.

For prime q, some of the results of Theorem 1.1 can be strengthened.

Theorem 1.2. *Let p be a prime, $1 \leq m \leq n \leq p-1$, and $D_p = D(p; m, n)$. Then D_p is strong and the following statements hold.*

(1) $\text{diam}(D_p) \leq 2p - 1$ *with equality if and only if $m = n = p - 1$.*
(2) If $(m, n) \notin \{((p-1)/2, (p-1)/2), ((p-1)/2, p-1), (p-1, p-1)\}$, then
$\text{diam}(D_p) \leq 120\sqrt{m} + 1.$
(3) If $p > (m-1)^3$, then $\text{diam}(D_p) \leq 19$.

The paper is organized as follows. In section 2 we present all results which are needed for our proofs of Theorems 1.1 and 1.2 in sections 3 and 4, respectively. Section 5 contains concluding remarks and open problems.

2. Preliminary results

We begin with a general result that gives necessary and sufficient conditions for a digraph $D(q; m, n)$ to be strong.

Theorem 2.1. *[[12], Theorem 2] $D(q; m, n)$ is strong if and only if $\gcd(q - 1, m, n)$ is not divisible by any $q_d = (q-1)/(p^d - 1)$ for any positive divisor d of e, $d < e$. In particular, $D(p; m, n)$ is strong for any m, n.*

Every walk of length k in $D = D(q; m, n)$ originating at (a, b) is of the form

$$
\begin{aligned}
(a, b) &\rightarrow (x_1, -b + a^m x_1^n) \\
&\rightarrow (x_2, b - a^m x_1^n + x_1^m x_2^n) \\
&\rightarrow \cdots \\
&\rightarrow (x_k, x_{k-1}^m x_k^n - x_{k-2}^m x_{k-1}^n + \cdots + (-1)^{k-1} a^m x_1^n + (-1)^k b).
\end{aligned}
$$

Therefore, in order to prove that $\text{diam}(D) \leq k$, one can show that for any choice of $a, b, u, v \in \mathbb{F}_q$, there exists $(x_1, \ldots, x_k) \in \mathbb{F}_q^k$ so that

$$(u, v) = (x_k, x_{k-1}^m x_k^n - \cdots + (-1)^{k-1} a^m x_1^n + (-1)^k b). \tag{1}$$

In order to show that $\text{diam}(D) \geq l$, one can show that there exist $a, b, u, v \in \mathbb{F}_q$ such that (1) has no solution in $\mathbb{F}_q^k$ for any $k < l$.

2.1. *Waring's problem*

In order to obtain an upper bound on $\text{diam}(D(q; m, n))$ we will use some results concerning Waring's problem over finite fields.

Waring's number $\gamma(r,q)$ over $\mathbb{F}_q$ is defined as the smallest positive integer s (should it exist) such that the equation

$$x_1^r + x_2^r + \cdots + x_s^r = a$$

has a solution $(x_1, \ldots, x_s) \in \mathbb{F}_q^s$ for any $a \in \mathbb{F}_q$. Similarly, $\delta(r,q)$ is defined as the smallest positive integer s (should it exist) such that for any $a \in \mathbb{F}_q$, there exists $(\epsilon_1, \ldots, \epsilon_s)$, each $\epsilon_i \in \{-1, 1\} \subseteq \mathbb{F}_q$, for which the equation

$$\epsilon_1 x_1^r + \epsilon_2 x_2^r + \cdots + \epsilon_s x_s^r = a$$

has a solution $(x_1, \ldots, x_s) \in \mathbb{F}_q^s$. It is easy to argue that $\delta(r,q)$ exists if and only if $\gamma(r,q)$ exists, and in this case $\delta(r,q) \le \gamma(r,q)$.

A criterion on the existence of $\gamma(r,q)$ is the following theorem by Bhashkaran [2].

Theorem 2.2. [[2], Theorem G] *Waring's number $\gamma(r,q)$ exists if and only if r is not divisible by any $q_d = (q-1)/(p^d - 1)$ for any positive divisor d of e, $d < e$.*

The study of various bounds on $\gamma(r,q)$ has drawn considerable attention. We will use the following two upper bounds on Waring's number due to J. Cipra [5].

Theorem 2.3. [[5], Theorem 4] *If $e = 2$ and $\gamma(r,q)$ exists, then $\gamma(r,q) \le 16\sqrt{r+1}$. Also, if $e \ge 3$ and $\gamma(r,q)$ exists, then $\gamma(r,q) \le 10\sqrt{r+1}$.*

Corollary 2.1. [[5], Corollary 7] *If $\gamma(r,q)$ exists and $r < \sqrt{q}$, then $\gamma(r,q) \le 8$.*

For the case $q = p$, the following bound will be of interest.

Theorem 2.4. [Cochrane, Pinner [6], Corollary 10.3] *If $|\{x^k : x \in \mathbb{F}_p^*\}| > 2$, then $\delta(k,p) \le 20\sqrt{k}$.*

The next two statements concerning very strong bounds on Waring's number in large fields follow from the work of Weil [26], and Hua and Vandiver [10].

Theorem 2.5. [Small [20]] *If $q > (k-1)^4$, then $\gamma(k,q) \le 2$.*

Theorem 2.6. [Cipra [4], p. 4] *If $p > (k-1)^3$, then $\gamma(k,p) \le 3$.*

For a survey on Waring's number over finite fields, see Castro and Rubio (Section 7.3.4, p. 211), and Ostafe and Winterhof (Section 6.3.2.3, p. 175) in Mullen and Panario [19]. See also Cipra [4].

We will need the following technical lemma.

Lemma 2.1. *Let $\delta = \delta(r, q)$ exist, and $k \geq 2\delta$. Then for every $a \in \mathbb{F}_q$ the equation*

$$x_1^r - x_2^r + x_3^r - \cdots + (-1)^{k+1}x_k^r = a \qquad (2)$$

has a solution $(x_1, \ldots, x_k) \in \mathbb{F}_q^k$.

Proof. Let $a \in \mathbb{F}_q$ be arbitrary. There exist $\varepsilon_1, \ldots, \varepsilon_\delta$, each $\varepsilon_i \in \{-1, 1\} \subseteq \mathbb{F}_q$, such that the equation $\sum_{i=1}^{\delta} \varepsilon_i y_i^r = a$ has a solution $(y_1, \ldots, y_\delta) \in \mathbb{F}_q^\delta$. As $k \geq 2\delta$, the alternating sequence $1, -1, 1, \ldots, (-1)^k$ with k terms contains the sequence $\varepsilon_1, \ldots, \varepsilon_\delta$ as a subsequence. Let the indices of this subsequence be $j_1, j_2, \ldots, j_\delta$. For each l, $1 \leq l \leq k$, let $x_l = 0$ if $l \neq j_i$ for any i, and $x_l = y_i$ for $l = j_i$. Then $(x_1, \ldots, x_k)$ is a solution of (2). $\square$

2.2. *The Hasse-Weil bound*

In the next section we will use the Hasse-Weil bound, which provides a bound on the number of $\mathbb{F}_q$-points on a plane non-singular absolutely irreducible projective curve over a finite field $\mathbb{F}_q$. If the number of points on the curve C of genus g over the finite field $\mathbb{F}_q$ is $|C(\mathbb{F}_q)|$, then

$$\left| |C(\mathbb{F}_q)| - q - 1 \right| \leq 2g\sqrt{q}. \qquad (3)$$

It is also known that for a non-singular curve defined by a homogeneous polynomial of degree k, $g = (k-1)(k-2)/2$. Discussion of all related notions and a proof of this result can be found in Hirschfeld, Korchmáros, Torres [9] (Theorem 9.18, p. 343) or in Szőnyi [21] (p. 197).

3. Proof of Theorem 1.1

(1). As there is a loop at $(0, 0)$, and there are arcs between $(0, 0)$ and $(x, 0)$ in either direction, for every $x \in \mathbb{F}_q^*$, the number of vertices in D_q which are at distance at most 2 from $(0, 0)$ is at most $1 + (q-1) + (q-1)^2 < q^2$. Thus, there are vertices in D_q which are at distance at least 3 from $(0, 0)$, and so $\operatorname{diam}(D_q) \geq 3$.

(2). As D_q is strong, by Theorem 2.1, for any positive divisor d of e, $d < e$, $q_d \nmid \gcd(p^e - 1, m, n)$. As, clearly, $q_d \mid (p^e - 1)$, either $q_d \nmid m$ or $q_d \nmid n$. This implies by Theorem 2.2 that either $\gamma(m, q)$ or $\gamma(n, q)$ exists.

Let (a, b) and (u, v) be arbitrary vertices of D_q. By (1), there exists a walk of length at most k from (a, b) to (u, v) if the equation

$$v = x_{k-1}^m u^n - x_{k-2}^m x_{k-1}^n + \cdots + (-1)^{k-1}a^m x_1^n + (-1)^k b \qquad (4)$$

has a solution $(x_1, \ldots, x_k) \in \mathbb{F}_q^k$.

Assume first that $\gamma_m = \gamma(m, q)$ exists. Taking $k = 6\gamma_m + 1$, and $x_i = 0$ for $i \equiv 1 \mod 3$, and $x_i = 1$ for $i \equiv 0 \mod 3$, we have that (4) is equivalent to

$$-x_{k-2}^m + x_{k-5}^m - \cdots + (-1)^k x_5^m + (-1)^{k-1} x_2^m = v - (-1)^k b - u^n.$$

As the number of terms on the left is $(k-1)/3 = 2\gamma_m$, this equation has a solution in $\mathbb{F}_q^{2\gamma_m}$ by Lemma 2.1. Hence, (4) has a solution in $\mathbb{F}_q^k$.

If $\gamma_n = \gamma(n, q)$ exists, then the argument is similar: take $k = 6\gamma_n + 1$, $x_i = 0$ for $i \equiv 0 \mod 3$, and $x_i = 1$ for $i \equiv 1 \mod 3$.

The result now follows from the bounds on $\gamma(r, q)$ in Theorem 2.3.

Remark 3. As $m \le n$, if $\gamma(m, q)$ exists, the upper bounds in Theorem 1.1, part **(2)**, can be improved by replacing n by m. Also, if a better upper bound on $\delta(m, q)$ than $\gamma(m, q)$ (respectively, on $\delta(n, q)$ than $\gamma(n, q)$) is known, the upper bounds in Theorem 1.1, **(2)**, can be further improved: use $k = 6\delta(m, q) + 1$ (respectively, $k = 6\delta(n, q) + 1$) in the proof. Similar comments apply to other parts of Theorem 1.1 as well as Theorem 1.2.

(3). Recall the basic fact $\gcd(r, q-1) = 1 \Leftrightarrow \{x^r : x \in \mathbb{F}_q\} = \mathbb{F}_q$.

Let $k = 4$. If $\gcd(m, q-1) = 1$, a solution to (1) of the form $(0, x_2, 1, u)$ is seen to exist for any choice of $a, b, u, v \in \mathbb{F}_q$. If $\gcd(n, q-1) = 1$, there exists a solution of the form $(1, x_2, 0, u)$. Hence, $\operatorname{diam}(D_q) \le 4$.

Let $k = 3$, and $\gcd(m, q-1) = \gcd(n, q-1) = 1$. If $a = 0$, then a solution to (1) of the form $(x_1, 1, u)$ exists. If $a \ne 0$, a solution of the form $(x_1, 0, u)$ exists. Hence, D_q is strong and $\operatorname{diam}(D_q) \le 3$. Using the lower bound from part **(1)**, we conclude that $\operatorname{diam}(D_q) = 3$.

(4). As was shown in part 3, for any n, $\operatorname{diam}(D(q; 1, n)) \le 4$. If, additionally, $\gcd(n, q-1) = 1$, then $\operatorname{diam}(D(q; 1, n)) = 3$. It turns out that if p does not divide n, then only for finitely many q is the diameter of $D(q; 1, n)$ actually 4.

For $k = 3$, (1) is equivalent to

$$(u, v) = (x_3, x_2 x_3^n - x_1 x_2^n + a x_1^n - b), \tag{5}$$

which has solution $(x_1, x_2, x_3) = (0, u^{-n}(b + v), u)$, provided $u \ne 0$.

Suppose now that $u = 0$. Aside from the trivial case $a = 0$, the question of the existence of a solution to (5) shall be resolved if we prove that the equation

$$a x^n - x y^n + c = 0 \tag{6}$$

has a solution for any $a, c \in \mathbb{F}_q^*$ (for $c = 0$, (6) has solutions). The projective curve corresponding to this equation is the zero locus of the homogeneous polynomial

$$F(X, Y, Z) = aX^n Z - XY^n + cZ^{n+1}.$$

It is easy to see that, provided p does not divide n,

$$F = F_X = F_Y = F_Z = 0 \quad \Leftrightarrow \quad X = Y = Z = 0,$$

and thus the curve has no singularities and is absolutely irreducible.

Counting the two points $[1 : 0 : 0]$ and $[0 : 1 : 0]$ on the line at infinity $Z = 0$, we obtain from (3), the inequality $N \geq q - 1 - 2g\sqrt{q}$, where $N = N(c)$ is the number of solutions of (6). As $g = n(n-1)/2$, solving the inequality $q - 1 - n(n-1)\sqrt{q} > 0$ for q, we obtain a lower bound on q for which $N \geq 1$.

(5a). The result follows from Corollary 2.1 by an argument similar to that of the proof of part **(2)**.

(5b). For $k = 13$, (1) is equivalent to

$$(u, v) = (x_{13}, -b + a^m x_1^n - x_1^m x_2^n + x_2^m x_3^n - \cdots - x_{11}^m x_{12}^n + x_{12}^m x_{13}^n).$$

If $q > (m-1)^4$, set $x_1 = x_4 = x_7 = x_{10} = 0$, $x_3 = x_6 = x_9 = x_{12} = 1$. Then $v - u^n + b = -x_{11}^m + x_8^m - x_5^m + x_2^m$, which has a solution $(x_2, x_5, x_8, x_{11}) \in \mathbb{F}_q^4$ by Theorem 2.5 and Lemma 2.1.

(5c). For $k = 9$, (1) is equivalent to

$$(u, v) = (x_9, -b + a^n x_1^n - x_1^n x_2^n + x_2^n x_3^n - \cdots - x_7^m x_8^n + x_8^n x_9^n).$$

If $q > (n-1)^4$, set $x_1 = x_4 = x_5 = x_8 = 0$, $x_3 = x_7 = 1$. Then $v + b = x_2^n + x_6^n$, which has a solution $(x_2, x_6) \in \mathbb{F}_q^2$ by Theorem 2.5.

4. Proofs of Theorem 1.2

Lemma 4.1. *Let $D = D(q; m, n)$. Then, for any $\lambda \in \mathbb{F}_q^*$, the function $\phi : V(D) \to V(D)$ given by $\phi((a, b)) = (\lambda a, \lambda^{m+n} b)$ is a digraph automorphism of D.*

The proof of the lemma is straightforward. It amounts to showing that ϕ is a bijection and that it preserves adjacency: $\mathbf{x} \to \mathbf{y}$ if and only if $\phi(\mathbf{x}) \to \phi(\mathbf{y})$. We omit the details. Due to Lemma 4.1, any walk in D initiated at a vertex (a, b) corresponds to a walk initiated at a vertex $(0, b)$

if $a = 0$, or at a vertex $(1, b')$, where $b' = a^{-m-n}b$, if $a \neq 0$. This implies
that if we wish to show that diam$(D_p) \leq 2p-1$, it is sufficient to show that
the distance from any vertex $(0, b)$ to any other vertex is at most $2p - 1$,
and that the distance from any vertex $(1, b)$ to any other vertex is at most
$2p - 1$.

First we note that by Theorem 2.1, $D_p = D(p; m, n)$ is strong for any
choice of m, n.

For $a \in \mathbb{F}_p$, let integer $\overline{a}$, $0 \leq \overline{a} \leq p - 1$, be the representative of the
residue class a.

It is easy to check that diam$(D(2; 1, 1)) = 3$. Therefore, for the remainder of the proof, we may assume that p is odd.

(1). In order to show that diam$(D_p) \leq 2p - 1$, we use (1) with $k = 2p - 1$,
and prove that for any two vertices (a, b) and (u, v) of D_p there is always a
solution $(x_1, \ldots, x_{2p-1}) \in \mathbb{F}_q^{2p-1}$ of

$$(u, v) = (x_{2p-1}, -b + a^m x_1^n - x_1^m x_2^n + x_2^m x_3^n - \cdots - x_{2p-3}^m x_{2p-2}^n + x_{2p-2}^m x_{2p-1}^n),$$

or, equivalently, a solution $\mathbf{x} = (x_1, \ldots, x_{2p-2}) \in \mathbb{F}_q^{2p-2}$ of

$$a^m x_1^n - x_1^m x_2^n + x_2^m x_3^n - \cdots - x_{2p-3}^m x_{2p-2}^n + x_{2p-2}^m u^n = b + v. \quad (7)$$

As the upper bound $2p - 1$ on the diameter is exact and holds for all p, we
need a more subtle argument compared to the ones we used before. The only
way we can make it is (unfortunately) by performing a case analysis on $\overline{b + v}$
with a nested case structure. In most of the cases we just exhibit a solution
$\mathbf{x}$ of (7) by describing its components x_i. It is always a straightforward
verification that $\mathbf{x}$ satisfies (7), and we will suppress our comments as cases
proceed.

Our first observation is that if $\overline{b + v} = 0$, then $\mathbf{x} = (0, \ldots, 0)$ is a solution to (7). We may assume now that $\overline{b + v} \neq 0$.

Case 1.1: $\overline{b + v} \geq \frac{p-1}{2} + 2$
We define the components of $\mathbf{x}$ as follows:
    if $1 \leq i \leq 4(p - (\overline{b + v}))$, then $x_i = 0$ for $i \equiv 1, 2 \mod 4$, and $x_i = 1$ for
$i \equiv 0, 3 \mod 4$;
    if $4(p - (\overline{b + v})) < i \leq 2p - 2$, then $x_i = 0$.
    Note that $x_i^m x_{i+1}^n = 0$ unless $i \equiv 3 \mod 4$, in which case $x_i^m x_{i+1}^n = 1$.
If we group the terms in groups of four so that each group is of the form

$$-x_i^m x_{i+1}^n + x_{i+1}^m x_{i+2}^n - x_{i+2}^m x_{i+3}^n + x_{i+3}^m x_{i+4}^n,$$

where $i \equiv 1 \mod 4$, then assuming i, $i+1$, $i+2$, $i+3$, and $i+4$ are within the range of $1 \leq i < i+4 \leq 4(\overline{b}+v)$, it is easily seen that one group contributes -1 to

$$a^m x_1^n - x_1^m x_2^n + x_2^m x_3^n - \cdots - x_{2p-3}^m x_{2p-2}^n + x_{2p-2}^m x_{2p-1}^n.$$

There are $\frac{4(p-(\overline{b}+v))}{4} = p-(\overline{b}+v)$ such groups, and so the solution provided adds -1 exactly $p - (\overline{b}+v)$ times. Hence, $\mathbf{x}$ is a solution to (7).

For the remainder of the proof, solutions to (7) will be given without justification as the justification is similar to what's been done above.

<u>Case 1.2:</u> $\overline{b}+v \leq \frac{p-1}{2}$
We define the components of $\mathbf{x}$ as follows:
    if $1 \leq i \leq 4(\overline{b}+v) - 1$, then $x_i = 0$ for $i \equiv 0,1 \mod 4$, and $x_i = 1$ for $i \equiv 2,3 \mod 4$;
    if $4(\overline{b}+v) - 1 < i \leq 2p - 2$, then $x_i = 0$.

<u>Case 1.3:</u> $\overline{b}+v = \frac{p-1}{2} + 1$
    This case requires several nested subcases.

<u>Case 1.3.1:</u> $u = x_{2p-1} = 0$
    Here, there is no need to restrict x_{2p-2} to be 0. The components of a solution $\mathbf{x}$ of (7) are defined as:
    if $1 \leq i \leq 2p - 2$, then $x_i = 0$ for $i \equiv 1,2 \mod 4$, and $x_i = 1$ for $i \equiv 0,3 \mod 4$.

<u>Case 1.3.2:</u> $a = 0$
    Here, there is no need to restrict x_1 to be 0. Therefore, the components of a solution $\mathbf{x}$ of (7) are defined as:
    if $1 \leq i \leq 2p - 2$, then $x_i = 0$ for $i \equiv 0,3 \mod 4$, and $x_i = 1$ for $i \equiv 1,2 \mod 4$.

<u>Case 1.3.3:</u> $u \neq 0$ and $a \neq 0$
    Because of Lemma 4.1, we may assume without loss of generality that $a = 1$. Let $x_{2p-2} = 1$, so that $x_{2p-2}^m u^n = u^n \neq 0$ and let $t = \overline{b}+v - u^n$. Note that $t \neq \frac{p-1}{2} + 1$.

<u>Case 1.3.3.1:</u> $t = 0$
    The components of a solution $\mathbf{x}$ of (7) are defined as: $x_{2p-2} = 1$, and if $1 \leq i < 2p - 2$, then $x_i = 0$.

Case 1.3.3.2: $0 < t \le \frac{p-1}{2}$

The components of a solution $\mathbf{x}$ of (7) are defined as: $x_{2p-2} = 1$, and if $1 \le i \le 4(t-1) + 1$, then $x_i = 0$ for $i \equiv 2, 3 \mod 4$, and $x_i = 1$ for $i \equiv 0, 1 \mod 4$;

if $4(t-1) + 1 < i < 2p - 2$, then $x_i = 0$.

Case 1.3.3.3: $t \ge \frac{p-1}{2} + 2$

The components of a solution $\mathbf{x}$ of (7) are defined as: $x_{2p-2} = 1$, and if $1 \le i \le 4(p-t)$, then $x_i = 0$ for $i \equiv 1, 2 \mod 4$, and $x_i = 1$ for $i \equiv 0, 3 \mod 4$;

if $4(p-t) < i < 2p - 2$, then $x_i = 0$.

The whole range of possible values $\overline{b + v}$ has been checked. Hence, $\mathrm{diam}(D) \le 2p - 1$.

We now show that if $\mathrm{diam}(D) = 2p - 1$, then $m = n = p - 1$. To do so, we assume that $m \ne p - 1$ or $n \ne p - 1$ and prove the contrapositive. Specifically, we show that $\mathrm{diam}(D) \le 2p - 2 < 2p - 1$ by again using (1) but with $k = 2p - 2$. We prove that for any two vertices (a, b) and (u, v) of D_p there is always a solution $(x_1, \dots, x_{2p-2}) \in \mathbb{F}_q^{2p-2}$ of

$$(u, v) = (x_{2p-2}, b - a^m x_1^n + x_1^m x_2^n - \dots - x_{2p-4}^m x_{2p-3}^n + x_{2p-3}^m x_{2p-2}^n),$$

or, equivalently, a solution $\mathbf{x} = (x_1, \dots, x_{2p-3}) \in \mathbb{F}_q^{2p-3}$ of

$$- a^m x_1^n + x_1^m x_2^n - x_2^m x_3^n + \dots - x_{2p-4}^m x_{2p-3}^n + x_{2p-3}^m u^n = -b + v. \quad (8)$$

We perform a case analysis on $\overline{-b + v}$.

Our first observation is that if $\overline{-b + v} = 0$, then $\mathbf{x} = (0, \dots, 0)$ is a solution to (8). We may assume for the remainder of the proof that $\overline{-b + v} \ne 0$.

Case 2.1: $\overline{-b + v} \le \frac{p-1}{2} - 1$

We define the components of $\mathbf{x}$ as follows:

if $1 \le i \le 4(\overline{-b + v})$, then $x_i = 0$ for $i \equiv 1, 2 \mod 4$, and $x_i = 1$ for $i \equiv 0, 3 \mod 4$;

if $4(\overline{-b + v}) < i \le 2p - 3$, then $x_i = 0$.

Case 2.2: $\overline{-b + v} \ge \frac{p-1}{2} + 2$

We define the components of $\mathbf{x}$ as follows:

if $1 \le i \le 4(p - (\overline{-b + v})) - 1$, then $x_i = 0$ for $i \equiv 0, 1 \mod 4$, and $x_i = 1$ for $i \equiv 2, 3 \mod 4$;

if $4(p - (\overline{-b+v})) - 1 < i \leq 2p - 3$, then $x_i = 0$.

Case 2.3: $\overline{-b+v} = \frac{p-1}{2}$

Case 2.3.1: $a = 0$

We define the components of $\mathbf{x}$ as:

if $1 \leq i \leq 2p - 3$, then $x_i = 0$ for $i \equiv 0, 3 \mod 4$, and $x_i = 1$ for $i \equiv 1, 2 \mod 4$.

Case 2.3.2: $a \neq 0$

Here, we may assume without loss of generality that $a = 1$ by Lemma (4.1).

Case 2.3.2.1: $n \neq p - 1$

If $n \neq p - 1$, then there exists $\beta \in \mathbb{F}_p^*$ such that $\beta^n \notin \{0, 1\}$. For such a β, let $x_1 = \beta$ and consider $t = \overline{-b+v+a^m x_1^n} = \overline{-b+v+\beta^n} \notin \{\frac{p-1}{2}, \frac{p-1}{2}+1\}$.

Case 2.3.2.1.1: $t = 0$

We define the components of $\mathbf{x}$ as: $x_1 = \beta$ and

if $2 \leq i \leq 2p - 3$, then $x_i = 0$.

Case 2.3.2.1.2: $t \leq \frac{p-1}{2} - 1$

We define the components of $\mathbf{x}$ as: $x_1 = \beta$ and

if $2 \leq i \leq 4t$, then $x_i = 0$ for $i \equiv 1, 2 \mod 4$, and $x_i = 1$ for $i \equiv 0, 3 \mod 4$;

if $4t < i \leq 2p - 3$, then $x_i = 0$.

Case 2.3.2.1.3: $t \geq \frac{p-1}{2} + 2$

We define the components of $\mathbf{x}$ as: $x_1 = \beta$ and

if $2 \leq i \leq 4(p - t) + 1$, then $x_i = 0$ for $i \equiv 2, 3 \mod 4$, and $x_i = 1$ for $i \equiv 0, 1 \mod 4$;

if $4(p - t) + 1 < i \leq 2p - 3$, then $x_i = 0$.

Case 2.3.2.2: $n = p - 1$

Case 2.3.2.2.1: $u \in \mathbb{F}_p^*$

Here, we have that $u^n = 1$, so that the components of a solution $\mathbf{x}$ of (8) are defined as:

if $1 \leq i \leq 2p - 3$, then $x_i = 0$ for $i \equiv 1, 2 \mod 4$, and $x_i = 1$ for $i \equiv 0, 3 \mod 4$.

Case 2.3.2.2.2: $u = 0$

Since $n = p - 1$, it must be the case that $m \neq p - 1$ so that there exists $\alpha \in \mathbb{F}_p^*$ such that $\alpha^m \notin \{0.1\}$. For such an α, let $x_2 = \alpha, x_3 = 1$ and consider $t = \overline{-b + v + x_2^m x_3^n} = \overline{-b + v + \alpha^m} \notin \{\frac{p-1}{2}, \frac{p-1}{2} + 1\}$.

<u>Case 2.3.2.2.2.1</u>: $t = 0$
We define the components of $\mathbf{x}$ as: $x_1 = 0, x_2 = \alpha, x_3 = 1$ and
if $4 \leq i \leq 2p - 3$, then $x_i = 0$.

<u>Case 2.3.2.2.2.2</u>: $t \leq \frac{p-1}{2} - 1$
We define the components of $\mathbf{x}$ as: $x_1 = 0, x_2 = \alpha, x_3 = 1$ and
if $4 \leq i \leq 4t$, then $x_i = 0$ for $i \equiv 1, 2 \mod 4$, and $x_i = 1$ for $i \equiv 0, 3$ mod 4;
if $4t < i \leq 2p - 3$, then $x_i = 0$.

<u>Case 2.3.2.2.2.3</u>: $t \geq \frac{p-1}{2} + 2$
We define the components of $\mathbf{x}$ as: $x_1 = 0, x_2 = \alpha, x_3 = 1$ and
if $4 \leq i \leq 4(p - t) + 3$, then $x_i = 0$ for $i \equiv 0, 1 \mod 4$, and $x_i = 1$ for $i \equiv 2, 3 \mod 4$;
if $4(p - t) + 3 < i \leq 2p - 3$, then $x_i = 0$.

<u>Case 2.4</u>: $\overline{-b + v} = \frac{p-1}{2} + 1$

<u>Case 2.4.1</u>: $u = 0$
We define the components of $\mathbf{x}$ as:
if $1 \leq i \leq 2p - 3$, then $x_i = 0$ for $i \equiv 0, 1 \mod 4$, and $x_i = 1$ for $i \equiv 2, 3$ mod 4.

<u>Case 2.4.2</u>: $u \neq 0$
Here, we may assume without loss of generality that $u = 1$ by Lemma (4.1).

<u>Case 2.4.2.1</u>: $m \neq p - 1$
If $m \neq p - 1$, then there exists $\alpha \in \mathbb{F}_p^*$ such that $\alpha^m \notin \{0, 1\}$. For such an α, let $x_{2p-3} = \alpha$ and consider $t = \overline{-b + v - x_{2p-3}^m u^n} = \overline{-b + v - \alpha^m} \notin \{\frac{p-1}{2}, \frac{p-1}{2} + 1\}$.

<u>Case 2.4.2.1.1</u>: $t = 0$
We define the components of $\mathbf{x}$ as: $x_{2p-3} = \alpha$ and
if $1 \leq i \leq 2p - 4$, then $x_i = 0$.

<u>Case 2.4.2.1.2</u>: $t \leq \frac{p-1}{2} - 1$
We define the components of $\mathbf{x}$ as: $x_{2p-3} = \alpha$ and

if $1 \le i \le 4t$, then $x_i = 0$ for $i \equiv 1, 2 \mod 4$, and $x_i = 1$ for $i \equiv 0, 3 \mod 4$;

if $4t < i \le 2p - 4$, then $x_i = 0$.

Case 2.4.2.1.3: $t \ge \frac{p-1}{2} + 2$

We define the components of $\mathbf{x}$ as: $x_{2p-3} = \alpha$ and

if $1 \le i \le 4(p - t) - 1$, then $x_i = 0$ for $i \equiv 0, 1 \mod 4$, and $x_i = 1$ for $i \equiv 2, 3 \mod 4$;

if $4(p - t) - 1 < i \le 2p - 4$, then $x_i = 0$.

Case 2.4.2.2: $m = p - 1$

Case 2.4.2.2.1: $a \in \mathbb{F}_p^*$

Here, we have that $a^m = 1$, so that the components of a solution $\mathbf{x}$ of (8) are defined as:

if $1 \le i \le 2p - 5$, then $x_i = 0$ for $i \equiv 2, 3 \mod 4$, and $x_i = 1$ for $i \equiv 0, 1 \mod 4$.

Case 2.4.2.2.2: $a = 0$

Since $m = p - 1$, it must be the case that $n \ne p - 1$ so that there exists $\beta \in \mathbb{F}_p^*$ such that $\beta^n \notin \{0.1\}$. For such a β, let $x_{2p-5} = 1, x_{2p-4} = \beta$ and consider $t = \overline{-b + v - x_{2p-5}^m x_{2p-4}^n} = \overline{-b + v - \beta^n} \notin \{\frac{p-1}{2}, \frac{p-1}{2} + 1\}$.

Case 2.4.2.2.2.1: $t = 0$

We define the components of $\mathbf{x}$ as: $x_{2p-5} = 1, x_{2p-4} = \beta, x_{2p-3} = 0$ and

if $1 \le i \le 2p - 6$, then $x_i = 0$.

Case 2.4.2.2.2.2: $t \le \frac{p-1}{2} - 1$

We define the components of $\mathbf{x}$ as: $x_{2p-5} = 1, x_{2p-4} = \beta, x_{2p-3} = 0$ and

if $1 \le i \le 4t - 2$, then $x_i = 0$ for $i \equiv 0, 3 \mod 4$, and $x_i = 1$ for $i \equiv 1, 2 \mod 4$;

if $4t - 2 < i \le 2p - 6$, then $x_i = 0$.

Case 2.4.2.2.2.3: $t \ge \frac{p-1}{2} + 2$

We define the components of $\mathbf{x}$ as: $x_{2p-5} = 1, x_{2p-4} = \beta, x_{2p-3} = 0$ and

if $1 \le i \le 4(p - t) - 1$, then $x_i = 0$ for $i \equiv 0, 1 \mod 4$, and $x_i = 1$ for $i \equiv 2, 3 \mod 4$;

if $4(p - t) - 1 < i \le 2p - 6$, then $x_i = 0$.

All cases have been checked, so if $m \ne p-1$ or $n \ne p-1$, then $\mathrm{diam}(D) < 2p - 1$.

We now prove that if $m = n = p - 1$, then $d := \operatorname{diam}(D(p; m, n)) = 2p - 1$. In order to do this, we explicitly describe the structure of the digraph $D(p; p - 1, p - 1)$, from which the diameter becomes clear. In this description, we look at sets of vertices of a given distance from the vertex $(0, 0)$, and show that some of them are at distance $2p - 1$. We recall the following important general properties of our digraphs that will be used in the proof.

- Every out-neighbor (u, v) of a vertex (a, b) of $D(q; m, n)$ is completely determined by its first component u.
- Every vertex of $D(q; m, n)$ has its out-degree and in-degree equal q.
- In $D(q; m, m)$, $\mathbf{x} \to \mathbf{y}$ if and only if $\mathbf{y} \to \mathbf{x}$

In $D(p; p - 1, p - 1)$, we have that $(x_1, y_1) \to (x_2, y_2)$ if and only if

$$y_1 + y_2 = x_1^{p-1} x_2^{p-1} = \begin{cases} 0 & \text{if } x_1 = 0 \text{ or } x_2 = 0, \\ 1 & \text{if } x_1 \text{ and } x_2 \text{ are non-zero.} \end{cases}$$

For notational convenience, we set

$$(*, a) = \{(x, a) : x \in \mathbb{F}_p^*\}$$

and, for $1 \le k \le d$, let

$$N_k = \{v \in V(D(p; m, n)) : \operatorname{dist}((0, 0), v) = k\}.$$

We assume that $N_0 = \{(0, 0)\}$. It is clear from this definition that these $d + 1$ sets N_k partition the vertex set of $D(p; p-1, p-1)$; for every k, $1 \le k \le d-1$, every out-neighbor of a vertex from N_k belongs to $N_{k-1} \cup N_k \cup N_{k+1}$, and N_{k+1} is the set of all out-neighbors of all vertices from N_k which are not in $N_{k-1} \cup N_k$.

Thus we have $N_0 = \{(0, 0)\}$, $N_1 = (*, 0)$, $N_2 = (*, 1)$, $N_3 = \{(0, -1)\}$. If $p > 2$, $N_4 = \{(0, 1)\}$, $N_5 = (*, -1)$. As there exist two (opposite) arcs between each vertex of $(*, x)$ and each vertex $(*, -x + 1)$, these subsets of vertices induce the complete bipartite subdigraph $\overrightarrow{K}_{p-1, p-1}$ if $x \ne -x + 1$, and the complete subdigraph $\overrightarrow{K}_{p-1}$ if $x = -x + 1$. Note that our $\overrightarrow{K}_{p-1, p-1}$ has no loops, but $\overrightarrow{K}_{p-1}$ has a loop on every vertex. Digraph $D(5; 4, 4)$ is depicted in Fig. 1.2.

The structure of $D(p; p - 1, p - 1)$ for any other prime p is similar. We can describe it as follows: for each $t \in \{0, 1, \ldots, (p - 1)/2\}$, let

$$N_{4\bar{t}} = \{(0, t)\}, \quad N_{4\bar{t}+1} = (*, -t),$$

and for each $t \in \{0, 1, \ldots, (p - 3)/2\}$, let

$$N_{4\bar{t}+2} = (*, t + 1), \quad N_{4\bar{t}+3} = \{(0, -t - 1)\}.$$

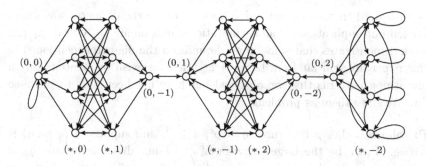

Fig. 2.   The digraph $D(5; 4, 4)$: $x_2 + y_2 = x_1^4 y_1^4$.

Note that for $0 \leq \bar{t} < (p-1)/2$, $N_{4\bar{t}+1} \neq N_{4\bar{t}+2}$, and for $\bar{t} = (p-1)/2$, $N_{2p-1} = (*, (p+1)/2)$. Therefore, for $p \geq 3$, $D(p; p-1, p-1)$ contains $(p-1)/2$ induced copies of $\overrightarrow{K}_{p-1,p-1}$ with partitions $N_{4\bar{t}+1}$ and $N_{4\bar{t}+2}$, and a copy of $\overrightarrow{K}_{p-1}$ induced by N_{2p-1}. The proof is a trivial induction on $\bar{t}$. Hence, $\mathrm{diam}(D(p; p-1, p-1)) = 2p - 1$. This ends the proof of Theorem 1.2 (1).

(2). We follow the argument of the proof of Theorem 1.1, part **(2)** and use Lemma 2.1, with $k = 6\delta(m, p) + 1$. We note, additionally, that if $m \notin \{p, (p-1)/2\}$, then $\gcd(m, p-1) < (p-1)/2$, which implies $|\{x^m : x \in \mathbb{F}_p^*\}| > 2$. The result then follows from Theorem 2.4.

(3). We follow the argument of the proof of Theorem 1.1, part **(5b)** and use Lemma 2.1 and Theorem 2.6.

This ends the proof of Theorem 1.2.

5. Concluding remarks

Many results in this paper follow the same pattern: if Waring's number $\delta(r, q)$ exists and is bounded above by δ, then one can show that $\mathrm{diam}(D(q; m, n)) \leq 6\delta + 1$. Determining the exact value of $\delta(r, q)$ is an open problem, and it is likely to be very hard. Also, the upper bound $6\delta + 1$ is not exact in general. Out of all partial results concerning $\delta(r, q)$, we used only those ones which helped us deal with the cases of the diameter of $D(q; m, n)$ that we considered, especially where the diameter was small. We left out applications of all asymptotic bounds on $\delta(r, q)$. Our computer work demonstrates that some upper bounds on the diameter mentioned in this paper are still far from being tight. Here we wish to mention only a few strong patterns that we observed but have not been able to prove so far. We state them as problems.

Problem 1. Let p be prime, $q = p^e$, $e \geq 2$, and suppose $D(q; m, n)$ is strong. Let r be the largest divisor of $q - 1$ not divisible by any $q_d = (p^e - 1)/(q^d - 1)$ where d is a positive divisor of e smaller than e. Is it true that

$$\max_{1 \leq m \leq n \leq q-1} \{\operatorname{diam}(D(q; m, n))\} = \operatorname{diam}(D(q; r, r))?$$

Find an upper bound on $\operatorname{diam}(D(q; r, r))$ better than the one of Theorem 1.1, part **(5c)**.

Problem 2. Is it true that for every prime p and $1 \leq m \leq n$, $(m, n) \neq (p-1, p-1)$, $\operatorname{diam}(D(p; m, n)) \leq (p+3)/2$ with the equality if and only if $(m, n) = ((p-1)/2, (p-1)/2)$ or $(m, n) = ((p-1)/2, p-1)$?

Problem 3. Is it true that for every prime p, $\operatorname{diam}(D(p; m, n))$ takes only one of two consecutive values which are completely determined by $\gcd((p-1, m, n)$?

Acknowledgement

The authors are thankful to the anonymous referee whose careful reading and thoughtful comments led to a number of significant improvements in the paper.

References

[1] J. Bang-Jensen, G. Gutin, Digraphs: Theory, Algorithms and Applications, Springer 2009.

[2] M. Bhaskaran, Sums of m-th powers in algebraic and abelian number fields, Arch. Math. (Basel) 17 (1966), 497-504; Correction, ibid. 22 (1972), 370-371.

[3] S.M. Cioabă, F. Lazebnik and W. Li, On the Spectrum of Wenger Graphs, J. Combin. Theory Ser. B 107: (2014), 132–139.

[4] J. Cipra, Waring's number in finite fields, Doctoral Thesis, Kansas State University, 2010.

[5] J. Cipra. Waring's number in a finite field, Integers 8 2009.

[6] T. Cochrane, C. Pinner, Sum-product estimates applied to Waring's problem mod p, Integers 8 (2008), A46.

[7] V. Dmytrenko, F. Lazebnik and R. Viglione, An Isomorphism Criterion for Monomial Graphs, J. Graph Theory 48 (2005), 322–328.

[8] V. Dmytrenko, F. Lazebnik and J. Williford, On monomial graphs of girth eight, Finite Fields Appl. 13 (2007), 828–842.

[9] J.W.P. Hirschfeld, G. Korchmáros, F. Torres, Algebraic Curves over a Finite Field, Princeton Series in Applied Mathematics, 2008.

[10] L.K. Hua, H.S. Vandiver, Characters over certain types of rings with applications to the theory of equations in a finite field, Proc. Natl. Acad. Sci. U.S.A. 35 (1949), 94–99.

[11] A. Kodess, Properties of some algebraically defined digraphs, Doctoral Thesis, University of Delaware, 2014.

[12] A. Kodess, F. Lazebnik, Connectivity of some algebraically defined digraphs, Electron. J. Combin., 22(3) (2015), #P3.27, 1–11.

[13] B.G. Kronenthal, Monomial graphs and generalized quadrangles, Finite Fields Appl. 18 (2012), 674–684.

[14] F. Lazebnik, D. Mubayi, New lower bounds for Ramsey numbers of graphs and hypergraphs, Adv. Appl. Math. 8 (3/4) (2002), 544–559.

[15] F. Lazebnik, A. Thomason, Orthomorphisms and the construction of projective planes, Math. Comp. 73 (247) (2004), 1547–1557.

[16] F. Lazebnik, J. Verstraëte, On hypergraphs of girth five, Electron. J. Combin. 10 (R25) (2003), 1–15.

[17] F. Lazebnik, R. Viglione, An infinite series of regular edge- but not vertex-transitive graphs, J. Graph Theory 41 (2002), 249–258.

[18] F. Lazebnik, A.J. Woldar, General properties of some families of graphs defined by systems of equations, J. Graph Theory 38 (2) (2001), 65–86.

[19] G.L. Mullen and D. Panario, Handbook of Finite Fields, CRC Press, Taylor & Francis Group, 2013.

[20] C. Small, Sums of powers in large fields, Proc. Amer. Math. Soc. 65 (1977), p. 3535.

[21] T. Szőnyi, Some applications of algebraic curves in finite geometry and combinatorics. In Surveys in Combinatorics, Edited by R.A. Bailey, pp. 197–236, 1997 (London), vol. 241 of London Math. Soc. Lecture Note Ser., Cambridge Univ. Press, Cambridge, 1997.

[22] T.A. Terlep, J. Williford, Graphs from generalized Kac-Moody algebras, SIAM J. Discrete Math. 26 no. 3 (2012), 1112–1120.

[23] V.A. Ustimenko, On the extremal regular directed graphs without commutative diagrams and their applications in coding theory and cryptography, Albanian J. Math. 1 (01/2007), 283–295.

[24] R. Viglione, Properties of some algebraically defined graphs, Doctoral Thesis, University of Delaware, 2002.

[25] R. Viglione, On Diameter of Wenger Graphs, Acta Appl. Math. 104 (2) (11/2008), 173–176.

[26] A. Weil, Number of solutions of equations in finite fields, Bull. Amer. Math. Soc. 55 (1949), 497–508.

Permutation polynomials of the form $X + \gamma \operatorname{Tr}(X^k)$

Gohar Kyureghyan

Otto-von-Guericke University, Magdeburg, Germany
gohar.kyureghyan@ovgu.de

Michael Zieve

University of Michigan, Ann Arbor, MI, USA
zieve@umich.edu

1. Introduction

Let $\mathbb{F}_q$ be the finite field of size q, and let p be the characteristic of $\mathbb{F}_q$. A *permutation polynomial* over $\mathbb{F}_q$ is a polynomial $F(X) \in \mathbb{F}_q[X]$ for which the induced function $x \mapsto f(x)$ is a permutation of $\mathbb{F}_q$. Permutation polynomials are used in various applications of finite fields, where special importance is attached to permutation polynomials which have few terms. Due in part to this, and also due to the intrinsic interest of the problem, the study of permutation polynomials with few terms has thrived for well over a century. In this paper we make a new contribution to this topic by analyzing permutation polynomials of the form $X + \gamma \operatorname{Tr}_{q^n/q}(X^k)$ where n and k are positive integers, $\gamma \in \mathbb{F}_{q^n}^*$, and $\operatorname{Tr}_{q^n/q}(X) := X^{q^{n-1}} + X^{q^{n-2}} + \cdots + X^q + X$ (so that $\operatorname{Tr}_{q^n/q}$ induces the trace map from $\mathbb{F}_{q^n}$ to $\mathbb{F}_q$). We produce the following families of permutation polynomials:

Theorem 1. *The polynomial $X + \gamma \operatorname{Tr}_{q^n/q}(X^k)$ is a permutation polynomial over $\mathbb{F}_{q^n}$ in each of the following cases:*

(a) $n = 2$, $q \equiv \pm 1 \pmod 6$, $\gamma = -1/3$, $k = 2q - 1$
(b) $n = 2$, $q \equiv 5 \pmod 6$, $\gamma^3 = -1/27$, $k = 2q - 1$
(c) $n = 2$, $q \equiv 1 \pmod 3$, $\gamma = 1$, $k = (q^2 + q + 1)/3$
(d) $n = 2$, $q \equiv 1 \pmod 4$, $(2\gamma)^{(q+1)/2} = 1$, $k = (q+1)^2/4$

178

(e) $n = 2$, $q = Q^2$, $Q > 0$ odd, $\gamma = -1$, $k = Q^3 - Q + 1$
(f) $n = 2$, $q = Q^2$, $Q > 0$ odd, $\gamma = -1$, $k = Q^3 + Q^2 - Q$
(g) $n = 3$, q odd, $\gamma = 1$, $k = (q^2 + 1)/2$
(h) $n = 3$, q odd, $\gamma = -1/2$, $k = q^2 - q + 1$
(i) $n = 2\ell r$, $\gamma^{q^{2\ell}-1} = -1$, $k = q^\ell + 1$ *for some positive integers ℓ and r.*

We remark that "at random" one would not expect there to be permutation polynomials of the form $X + \gamma \operatorname{Tr}_{q^n/q}(X^k)$ over $\mathbb{F}_{q^n}$ when $Q := q^n$ is sufficiently large and $\gamma \in \mathbb{F}_Q^*$, since the polynomials of this form induce fewer than Q^2 distinct functions on $\mathbb{F}_Q$ while the proportion of permutations amongst all functions $\mathbb{F}_Q \to \mathbb{F}_Q$ is $Q!/Q^Q$. Hence it seems that any infinite sequence of examples such as those in Theorem 1 should exist for some "good reason", and it is an interesting challenge to understand the different sorts of reasons which can account for such unexpected examples to occur.

An intriguing feature of the present paper is that we need several different methods in order to prove the various cases of Theorem 1. In cases (a)–(d) and (g) we use a well-known result (Proposition 2) asserting that $Xh(X^{q-1})$ permutes $\mathbb{F}_{q^n}$ if and only if $Xh(X)^{q-1}$ permutes the set $\mu_{(q^n-1)/(q-1)}$ of $(q^n - 1)/(q - 1)$-th roots of unity. We then reformulate the latter condition in a simpler way (see Proposition 4), so that cases (a) and (b) boil down to showing that a certain degree-3 rational function permutes μ_{q+1}, and case (c) boils down to showing that a certain power map X^N permutes μ_{q+1}. Case (d) is more interesting, since we wind up showing that $Xh(X)^{q-1}$ permutes μ_{q+1} by showing that it acts as $c_1^2 X$ on the non-squares in μ_{q+1} and $c_2^2 X^N$ on the squares, for certain elements $c_1, c_2 \in \mu_{q+1}$; in this case it seems complicated to analyze $Xh(X)^{q-1}$ at once on all elements of μ_{q+1}, rather than treating the squares and non-squares separately. In case (g) we use a double application of Proposition 2, by first saying that the given polynomial permutes $\mathbb{F}_{q^3}$ if and only if a certain associated polynomial permutes μ_{q^2+q+1}, then composing the associated polynomial with certain power maps X^N which permute μ_{q^2+q+1}, and then reducing the composition mod $X^{q^2+q+1} - 1$ to obtain a polynomial which (again by Proposition 2) permutes μ_{q^2+q+1} if and only if a certain additive polynomial permutes $\mathbb{F}_{q^3}$; we then finish the proof by directly verifying this last bijectivity. This kind of approach does not seem to work in cases (e) and (f), so for these cases we use a variant of Dobbertin's method,[3] which involves computing Gröbner bases of several ideals in a multivariate polynomial ring. This variant involves new features compared to previous applications of Dobbertin's approach, and may be of independent interest. We prove case (h) by composing with certain bijective power maps and additive polynomials to obtain a function which permutes both the squares and the non-squares in $\mathbb{F}_{q^3}$; this method was inspired by the notion of affine equivalence of planar functions. Finally, in case (i) we use an additive analogue of Proposition 2 (see Proposition 1) in order to show that the stated polynomial permutes $\mathbb{F}_{q^n}$ if and only if certain associated polynomials permute $\mathbb{F}_q$, and we calculate that each of these associated polynomials induces the same function on $\mathbb{F}_q$ as does $X + c$ for some $c \in \mathbb{F}_q$.

We computed all permutation polynomials over fields $\mathbb{F}_{q^n}$ of the form $X + \gamma \operatorname{Tr}_{q^n/q}(X^k)$ with $\gamma \in \mathbb{F}_{q^n}^*$, q odd, $n > 1$, and $q^n < 5000$. Here the hypothesis $n > 1$ is needed to distinguish our study from that of permutation binomials. It is easy to see that, for fixed q, n, γ, we can replace k by qk or by $k + q^n - 1$ without affecting whether $X + \gamma \operatorname{Tr}_{q^n/q}(X^k)$ permutes $\mathbb{F}_{q^n}$. Modulo this equivalence, the only permutation polynomials arising in our computation which are not listed in Theorem 1 are:

- Cases with $k = p^i$, where the polynomial is automatically a member of the much-studied class of *additive* permutation polynomials.
- Cases with $p \mid n$ and $(q^n - 1) \mid k(q^{n/p} - 1)$, where the polynomial induces the identity map on $\mathbb{F}_{q^n}$.
- $q = 7$, $n = 2$, $k = 10$, $\gamma^4 = 1$
- $q = 9$, $n = 2$, $k = 33$, $\gamma^2 - \gamma = 1$
- $q = 27$, $n = 2$, $k = 261$, $(\gamma - 1)^{13} = \gamma^{13}$
- $q = 9$, $n = 3$, $k \in \{11, 19, 33, 57\}$, $\gamma^4 = -1$
- $q = 49$, $n = 2$, $k = 385$, $\gamma^5 = -1$.

Thus, it seems that Theorem 1 may explain the bulk of all permutation polynomials of this form, so long as we exclude the simple cases where the polynomial is additive or induces the identity map.

Permutation polynomials of the form $X + \gamma \operatorname{Tr}_{q^n/q}(X^k)$ were studied in.[1,2] All examples in those papers had the special feature that γ is a linear translator of $f(X) := \operatorname{Tr}_{q^n/q}(X^k)$, in the sense that there is some $\delta \in \mathbb{F}_q$ for which

$$f(x + u\gamma) - f(x) = u\delta$$

for all $x \in \mathbb{F}_{q^n}$ and $u \in \mathbb{F}_q$. This property is satisfied (with $\delta = 0$) for the polynomials in case (i) of Theorem 1, as we show in Section 10. However, the permutation polynomials in cases (a)–(h) of Theorem 1 are of a different nature, since for these polynomials γ is *not* a linear translator of $\operatorname{Tr}_{q^n/q}(X^k)$.

This paper is organized as follows. In Section 2 we prove some general properties about maps of the form $X + \gamma \operatorname{Tr}_{q^n/q}(X^k)$. Then in Sections 3–10 we prove Theorem 1 in each of cases (a)–(i). In addition, we note that Theorems 2, 6, 7 and 9 list further families of sparse rational functions which permute either a finite field or a group of roots of unity in a finite field.

2. General remarks and equivalent statements

We begin with two propositions which reformulate the condition that $X + \gamma \operatorname{Tr}_{q^n/q}(X^k)$ should permute $\mathbb{F}_{q^n}$ in terms of properties of some associated functions on $\mathbb{F}_q$. These reformulations play a crucial role in our proof of Theorem 1. In fact the reformulations apply to a more general class of functions.

Proposition 1. *For any function $f \colon \mathbb{F}_{q^n} \to \mathbb{F}_q$ with $n \geq 2$, and any $\gamma \in \mathbb{F}_{q^n}^*$, the following three statements are equivalent:*

(a) *The map $F \colon x \mapsto x + \gamma \cdot f(x)$ is bijective on $\mathbb{F}_{q^n}$.*
(b) *For each $\alpha \in \mathbb{F}_{q^n}$ the map $x \mapsto x + f(\alpha + \gamma \cdot x)$ is bijective on $\mathbb{F}_q$.*
(c) *For each $\alpha \in \mathbb{F}_{q^n}$ there is a unique $x \in \mathbb{F}_q$ for which $x + f(\alpha + \gamma \cdot x) = 0$.*

Proof. For each $\alpha \in \mathbb{F}_{q^n}$, the function F maps the line $\alpha + \gamma \mathbb{F}_q$ into itself, so that F permutes $\mathbb{F}_{q^n}$ if and only if F induces a permutation on each such line. Explicitly, for $u \in \mathbb{F}_q$ we have

$$F(\alpha + \gamma u) = \alpha + \gamma(u + f(\alpha + \gamma u)),$$

so that F permutes the line $\alpha + \gamma \mathbb{F}_q$ if and only if the function $u \mapsto u + f(\alpha + \gamma u)$ permutes $\mathbb{F}_q$. Thus (a) and (b) are equivalent. Since (b) immediately implies (c), it remains only to show that (c) implies (b), or equivalently that if (b) did not hold then (c) would not hold. So suppose there is some $\alpha' \in \mathbb{F}_{q^n}$ for which the function $x \mapsto x + f(\alpha' + \gamma \cdot x)$ is not bijective on $\mathbb{F}_q$. Then there are distinct elements $u_1, u_2 \in \mathbb{F}_q$ which have the same image y under this function. Hence for $i = 1, 2$ we have

$$y = u_i + f(\alpha' + \gamma \cdot u_i),$$

or equivalently

$$(u_i - y) + f(\alpha' + \gamma \cdot (u_i - y) + \gamma y) = 0.$$

Thus $x_i := u_i - y$ satisfies $x_i + f(\alpha' + \gamma y + \gamma x_i) = 0$; since x_1 and x_2 are distinct elements of $\mathbb{F}_q$, this contradicts (c) for the value $\alpha := \alpha' + y\gamma$. $\square$

The next result is a special case of [6, Lemma 2.1].

Proposition 2. *For any* $h(X) \in \mathbb{F}_{q^n}[X]$*, the polynomial* $Xh(X^{q-1})$ *permutes* $\mathbb{F}_{q^n}$ *if and only if* $Xh(X)^{q-1}$ *permutes the set of* $(q^n - 1)/(q-1)$*-th roots of unity in* $\mathbb{F}_{q^n}^*$*.*

We now give further reformulations in case $n = 2$.

Proposition 3. *Let* $\gamma, \omega \in \mathbb{F}_{q^2}$ *be linearly independent over* $\mathbb{F}_q$*, and let* $f \colon \mathbb{F}_{q^2} \to \mathbb{F}_q$ *be a function satisfying* $f(u \cdot x) = u \cdot f(x)$ *for each* $u \in \mathbb{F}_q$ *and* $x \in \mathbb{F}_{q^2}$*. Then* $x \mapsto x + \gamma \cdot f(x)$ *permutes* $\mathbb{F}_{q^2}$ *if and only if* $f(\gamma) \neq -1$ *and* $x \mapsto x + f(\omega + \gamma x)$ *permutes* $\mathbb{F}_q$*.*

Proof. By Proposition 1, $x \mapsto x + \gamma \cdot f(x)$ permutes $\mathbb{F}_{q^2}$ if and only if $F_\alpha \colon x \mapsto x + f(\alpha + \gamma \cdot x)$ permutes $\mathbb{F}_q$ for each $\alpha \in \mathbb{F}_{q^2}$. The elements $\alpha \in \mathbb{F}_{q^2}$ are precisely the elements $\alpha := u \cdot \gamma + v \cdot \omega$ with $u, v \in \mathbb{F}_q$, so we treat each choice of u, v in turn. If $v = 0$ then for $x \in \mathbb{F}_q$ we have

$$F_\alpha(x) = x + f((u + x)\gamma) = x + (u + x)f(\gamma) = (1 + f(\gamma))x + uf(\gamma),$$

so that F_α permutes $\mathbb{F}_q$ if and only if $f(\gamma) \neq -1$. If $v \neq 0$ then for $x \in \mathbb{F}_q$ we have

$$F_\alpha(x) = x + f(v\omega + (u + x)\gamma) = v\left(\frac{x}{v} + f\left(\omega + \frac{u + x}{v}\gamma\right)\right),$$

so that F_α permutes $\mathbb{F}_q$ if and only if $x \mapsto x + f(\omega + (x + \frac{u}{v})\gamma)$ permutes $\mathbb{F}_q$, or equivalently $x \mapsto x + f(\omega + x\gamma)$ permutes $\mathbb{F}_q$. $\square$

Proposition 4. *Let* $k := (q - 1)N + 1$ *for some integer* $N \geq 0$*, and pick any* $\gamma \in \mathbb{F}_{q^2}^*$*. Then the polynomial* $F(X) := X + \gamma \operatorname{Tr}_{q^2/q}(X^k)$ *permutes* $\mathbb{F}_{q^2}$ *if and only if*

$$H(X) := \frac{X^N + \gamma^q(1 + X^{2N-1})}{X^{N-1} + \gamma(X^{2N-1} + 1)}$$

permutes the set μ_{q+1} *of* $(q + 1)$*-th roots of unity in* $\mathbb{F}_{q^2}^*$*, or equivalently this rational function is injective on* μ_{q+1} *and its denominator has no roots in* μ_{q+1}*.*

Proof. By Proposition 2, $F(X)$ permutes $\mathbb{F}_{q^2}$ if and only if

$$G(X) := X\bigl(1 + \gamma(X^N + X^{qN+1})\bigr)^{q-1}$$

permutes μ_{q+1}. For any $x \in \mu_{q+1}$ such that $G(x) \neq 0$, we have

$$
\begin{aligned}
\frac{G(x)}{x} &= \left(1 + \gamma(x^N + x^{qN+1})\right)^{q-1} \\
&= \frac{1 + \gamma^q(x^{Nq} + x^{q^2N+q})}{1 + \gamma(x^N + x^{qN+1})} \\
&= \frac{1 + \gamma^q(x^{-N} + x^{N-1})}{1 + \gamma(x^N + x^{-N+1})} \\
&= \frac{H(x)}{x}.
\end{aligned}
$$

Therefore $G(X)$ permutes μ_{q+1} if and only if $H(X)$ permutes μ_{q+1}. Finally, if the denominator of $H(X)$ has no roots in μ_{q+1} then the above computation shows that G and H agree on μ_{q+1}, and thus $H(\mu_{q+1}) = G(\mu_{q+1}) \subseteq \mu_{q+1}$, so that bijectivity of H on μ_{q+1} follows from injectivity. □

3. The case that $n = 2$ and $k = 2q - 1$

In this section we prove cases (a) and (b) of Theorem 1. We begin with a proof of (a), which also produces some degree-3 rational functions which permute $\mathbb{F}_q$, as well as some degree-3 rational functions which permute the set μ_{q+1} of $(q+1)$-th roots of unity in $\mathbb{F}_{q^2}^*$.

Theorem 2. *If* $\gcd(q,6) = 1$ *then the following are true:*

(a) $F_1(X) := X - \dfrac{1}{3}\mathrm{Tr}_{q^2/q}(X^{2q-1})$ *permutes* $\mathbb{F}_{q^2}$.

(b) *For any non-square* $\nu \in \mathbb{F}_q$, *the function* $\dfrac{X(X^2 - 9\nu)}{X^2 - \nu}$ *permutes* $\mathbb{F}_q$.

(c) $g(X) := \dfrac{X^3 - 3X^2 + 1}{X^3 - 3X + 1}$ *permutes* μ_{q+1}.

Proof. First we show that (a) and (b) are equivalent. Pick $\omega \in \mathbb{F}_{q^2}$ with $\omega^q = -\omega \neq 0$. Then ω and $-1/3$ are linearly independent over $\mathbb{F}_q$. We apply Proposition 3 with $\gamma = -\frac{1}{3}$ and $f(X) := \mathrm{Tr}_{q^2/q}(X^{2q-1})$, noting that $f(a \cdot x) = a \cdot f(x)$ for $a \in \mathbb{F}_q$ and $x \in \mathbb{F}_{q^2}$ since $2q - 1 \equiv 1 \pmod{q-1}$. Since $f(-\frac{1}{3}) = -\frac{2}{3} \neq -1$, it follows that (a) holds if and only if

$$
h(X) := X + \mathrm{Tr}_{q^2/q}\!\left(\left(\omega - \frac{1}{3}X\right)^{2q-1}\right)
$$

permutes $\mathbb{F}_q$. For $x \in \mathbb{F}_q$ we have

$$h(3x) = 3x + \frac{(\omega - x)^{2q}}{\omega - x} + \frac{(\omega - x)^{2q^2}}{(\omega - x)^q}$$

$$= 3x + \frac{(-\omega - x)^2}{\omega - x} + \frac{(\omega - x)^2}{(-\omega - x)}$$

$$= \frac{x^3 - 9\omega^2 x}{x^2 - \omega^2}.$$

Since the above equivalence holds for each ω with $\omega^q = -\omega \neq 0$, and the set of all corresponding values ω^2 coincides with the set of non-squares in $\mathbb{F}_q$, this shows that (a) and (b) are equivalent. The equivalence of (a) and (c) follows from Proposition 4 with $N = 2$ and $\gamma = -\frac{1}{3}$. So it is enough to verify (c). By Proposition 4, it suffices to show that g is both well-defined and injective on μ_{q+1}. If some $\alpha \in \mu_{q+1}$ is a root of $r(X) := X^3 - 3X + 1$, then also α^q is a root of $r(X)$; since neither 1 nor -1 is a root of $r(X)$, it follows that $\alpha \neq \alpha^q$. Since the product of the roots of $r(X)$ is -1, the third root of $r(X)$ must be $-1/\alpha^{q+1} = -1$, which is not the case since $r(-1) \neq 0$. Thus g is well-defined on μ_{q+1}. The numerator of $g(X) - g(Y)$ is

$$(X^3 - 3X^2 + 1)(Y^3 - 3Y + 1) - (Y^3 - 3Y^2 + 1)(X^3 - 3X + 1)$$

$$= 3(X - Y)(XY - X + 1)(XY - Y + 1).$$

Hence if $g(\alpha) = g(\beta)$ for some distinct $\alpha, \beta \in \mu_{q+1}$ then $\alpha\beta \in \{\alpha - 1, \beta - 1\}$. Assume without loss that $\alpha\beta = \alpha - 1$, so that $(\alpha - 1)^{q+1} = 1$. But

$$(\alpha - 1)^{q+1} = \alpha^{q+1} - \alpha^q - \alpha + 1 = 2 - \frac{1}{\alpha} - \alpha,$$

so that $\alpha + \frac{1}{\alpha} = 1$ and thus $\beta = 1 - \frac{1}{\alpha} = \alpha$, a contradiction. $\square$

We now show that the permutation property of the polynomials in case (b) of Theorem 1 follows at once from the analogous property in case (a).

Theorem 3. *If $q \equiv 5 \pmod 6$ and $\gamma \in \mathbb{F}_{q^2}$ satisfies $\gamma^3 = -\frac{1}{27}$, then*

$$F_2(X) := X + \gamma \operatorname{Tr}_{q^2/q}(X^{2q-1})$$

permutes $\mathbb{F}_{q^2}$.

Proof. Since $\omega := -3\gamma$ satisfies $\omega^3 = 1$, we have $\omega^{2q-1} = 1$ and thus

$$F_2(\omega X) = \omega X - \frac{1}{3}\omega \operatorname{Tr}_{q^2/q}(\omega^{2q-1} X^{2q-1})$$

$$= \omega\left(X - \frac{1}{3} \operatorname{Tr}_{q^2/q}(X^{2q-1})\right),$$

so the result follows from Theorem 2. $\square$

Remark 1. A different proof of bijectivity of F_1 was given in the recent paper.[4]

4. The case that $n = 2$ and $k = (q^2 + q + 1)/3$

We now prove case (c) of Theorem 1.

Theorem 4. *If* $q \equiv 1 \pmod 3$, *then*

$$F_3(X) := X + \operatorname{Tr}_{q^2/q}(X^{(q^2+q+1)/3})$$

permutes $\mathbb{F}_{q^2}$.

Proof. By Proposition 4 with $N = \frac{q+2}{3}$, it suffices to show that

$$g(X) := \frac{X^N + 1 + X^{2N-1}}{X^{N-1} + X^{2N-1} + 1}$$

permutes μ_{q+1}. Here $3N \equiv 1 \pmod{q+1}$, so by putting $Y = X^N$ our condition becomes that

$$\frac{Y + 1 + \frac{1}{Y}}{\frac{1}{Y^2} + \frac{1}{Y} + 1}$$

should permute μ_{q+1}. This function is the identity function $y \mapsto y$ so long as μ_{q+1} contains no roots of $Y^2 + Y + 1$, which is the case since those roots have order 3. Hence $g(X)$ induces the same function on μ_{q+1} as does X^N, so that g permutes μ_{q+1}. $\qquad\square$

5. The case that $n = 2$ and $k = (q+1)^2/4$

We now prove case (d) of Theorem 1.

Theorem 5. *Let* $q \equiv 1 \pmod 4$, *and let* $\gamma \in \mathbb{F}_{q^2}$ *satisfy* $(2\gamma)^{(q+1)/2} = 1$. *Then*

$$F_4(X) := X + \gamma \operatorname{Tr}_{q^2/q}(X^{(q+1)^2/4})$$

permutes $\mathbb{F}_{q^2}$.

Proof. By Proposition 2, it suffices to show that

$$g(X) := X(\gamma^{-1} + X^N + X^{qN+1})^{q-1}$$

induces a bijection on μ_{q+1} in case $N = \frac{q+3}{4}$. Note that every element of μ_{q+1} can be written in exactly one way as $\pm y^2$ with $y \in \mu_{(q+1)/2}$. By hypothesis 2γ is a square in μ_{q+1}, so that -2γ is a non-square in μ_{q+1}, and thus $\gamma^{-1} + 2y$ is nonzero for each $y \in \mu_{(q+1)/2}$. For $y \in \mu_{(q+1)/2}$ we compute

$$(y^2)^N = y^{(q+3)/2} = y \quad \text{and} \quad (y^2)^{qN+1} = y^{q+2} = y$$

so that

$$g(-y^2) = -y^2(\gamma^{-1} + y(-1)^N + y(-1)^{qN+1})^{q-1}$$
$$= -y^2\gamma^{1-q} = -(2y\gamma)^2,$$

and

$$g(y^2) = y^2(\gamma^{-1} + y + y)^{q-1} = y^2\frac{(\gamma^{-1} + 2y)^q}{\gamma^{-1} + 2y}$$

$$= y^2\frac{4\gamma + 2y^{-1}}{\gamma^{-1} + 2y} = 2\gamma y.$$

Since 2γ is in $\mu_{(q+1)/2}$, and squaring is a bijective map on $\mu_{(q+1)/2}$, it follows that g induces a bijection on $\mu_{(q+1)/2}$ and also a bijection on $-\mu_{(q+1)/2}$, so that g induces a bijection on μ_{q+1} as desired. ☐

Remark 2. In fact, the polynomial F_4 fixes each non-square in $\mathbb{F}_{q^2}$. Moreover, we have

$$F_4(x) = \begin{cases} x + 2\gamma \cdot x^{\frac{(q+1)^2}{4}} & \text{if } x \text{ is a square in } \mathbb{F}_{q^2} \\ x & \text{if } x \text{ is a non-square in } \mathbb{F}_{q^2}. \end{cases}$$

To see this, note that for any $x \in \mathbb{F}_{q^2}$ the element $x^{(q^2-1)/4}$ is either zero or a fourth root of unity, and since $q \equiv 1 \pmod{4}$ it follows that $x^{(q^2-1)/4}$ lies in the subfield $\mathbb{F}_q$. Thus for $x \in \mathbb{F}_{q^2}$ we have

$$F_4(x) = x + \gamma \operatorname{Tr}_{q^2/q}(x^{\frac{q^2-1}{4}} x^{\frac{q+1}{2}}) = x + \gamma x^{\frac{q^2-1}{4}} \operatorname{Tr}_{q^2/q}(x^{\frac{q+1}{2}}).$$

Our claimed expression for $F_4(x)$ now follows from the fact that

$$\operatorname{Tr}_{q^2/q}(x^{\frac{q+1}{2}}) = x^{\frac{q+1}{2}} + x^{\frac{q^2+q}{2}}$$

$$= x^{\frac{q+1}{2}}(1 + x^{\frac{q^2-1}{2}})$$

$$= \begin{cases} 2x^{\frac{q+1}{2}} & \text{if } x \text{ is a square in } \mathbb{F}_{q^2} \\ 0 & \text{if } x \text{ is a non-square in } \mathbb{F}_{q^2}. \end{cases}$$

6. The case that $n = 2$ and $q = Q^2$ where Q is an odd prime power and $k = Q^3 - Q + 1$

In this section we prove case (e) of Theorem 1, by showing that

$$F_5(X) := X - \operatorname{Tr}_{Q^4/Q^2}(X^{Q^3-Q+1})$$

is a permutation on $\mathbb{F}_{Q^4}$, when Q is an odd prime power. Our proof relies on a variant of Dobbertin's method.[3] We also exhibit certain sparse rational functions which permute either $\mathbb{F}_{Q^2}$ or the set of (Q^2+1)-th roots of unity in $\mathbb{F}_{Q^4}^*$.

Theorem 6. *For any odd prime power Q, we have:*

(a) $F_5(X) = X - \mathrm{Tr}_{Q^4/Q^2}(X^{Q^3-Q+1})$ *permutes* $\mathbb{F}_{Q^4}$.

(b) $\dfrac{X^{Q+2} + 3\nu X^Q + 4\nu^{(Q+1)/2}X}{X^2 - \nu}$ *permutes* $\mathbb{F}_{Q^2}$, *where* ν *is any non-square in* $\mathbb{F}_{Q^2}^*$.

(c) $\dfrac{X^{2Q-1} - X^Q + 1}{X^{2Q-1} - X^{Q-1} + 1}$ *permutes the set of* $(Q^2 + 1)$-*th roots of unity in* $\mathbb{F}_{Q^4}^*$.

We begin with some simple lemmas about the preimages under F_5 of some special values.

Lemma 1. *The only root of* $F_5(X)$ *in* $\mathbb{F}_{Q^4}$ *is* 0.

Proof. Suppose to the contrary that $x \in \mathbb{F}_{Q^4}^*$ is a root of F_5. Then

$$x^{Q^3-Q} + x^{-Q^3+Q^2+Q-1} = 1,$$

and $y := x^{Q^2-1}$ is an element of μ_{Q^2+1} satisfying

$$y^Q + y^{1-Q} = 1. \tag{1}$$

In particular, y cannot be ± 1. Since $y^{Q^2} = 1/y$, we obtain

$$y^{-1} + y^{Q+1} = (y^Q + y^{1-Q})^Q = 1^Q = 1^{Q^3} = (y^Q + y^{1-Q})^{Q^3} = y + y^{-Q-1},$$

or equivalently

$$(y^{Q+2} + 1) \cdot (y^Q - 1) = 0.$$

Since $y \neq 1$, we obtain $y^{Q+2} = -1$, so that $y^{2Q+4} = 1$. Hence the order of y divides

$$\gcd(2Q + 4, Q^2 + 1) = 2\gcd(Q + 2, Q^2 + 1) = 2\gcd(Q + 2, 5),$$

so that $y^{10} = 1$. Since $y^{Q+2} = -1$, it follows that $y^5 = -1$. Now (1) simplifies to

$$-y^{-2} - y^3 = 1,$$

and since $y^5 = -1$ this yields the contradiction $0 = 1$. □

Lemma 2. *The set* S *of elements* $x \in \mathbb{F}_{Q^4}^*$ *for which* $x^{2Q^2} - x^{Q^2+1} + x^2 = 0$ *is nonempty only when* $3 \mid Q$, *in which case* S *consists of the* $(Q^2 - 1)$-*th roots of* -1 *and* F_5 *fixes each element of* S.

Proof. Write $u := x^{Q^2-1}$ with $x \in S$, so that $u^{Q^2+1} = 1$ (since plainly $x \neq 0$). If $3 \nmid Q$ then the equation $u^2 - u + 1 = 0$ implies that u is a primitive sixth root of unity, which is impossible since $6 \nmid (Q^2+1)$. Hence $3 \mid Q$, so that $X^{2Q^2} - X^{Q^2+1} + X^2 = (X^{Q^2} + X)^2$ and thus S consists of the (Q^2-1)-th roots of -1. Therefore for $x \in S$ and $k := Q^3 - Q + 1$ we have $F_5(x) = x - x^k - x^{Q^2 k} = x - x^k - (-x)^k = x$. $\qquad\square$

Proof of Theorem 6. Write $q := Q^2$. The equivalence of (a), (b) and (c) follows from Propositions 3 and 4 in the same manner as in our previous results. In the remainder of the proof we show that F_5 is bijective on $\mathbb{F}_{Q^4}$. In light of Lemma 1, it suffices to show that each $d \in \mathbb{F}_{Q^4}^*$ has at most one preimage x in $\mathbb{F}_{Q^4}^*$.

Pick any d in $\mathbb{F}_{Q^4}^*$, and write

$$e := d^Q, \ f := d^{Q^2}, \ g := d^{Q^3}.$$

Let x be an element of $\mathbb{F}_{Q^4}^*$ for which $F_5(x) = d$, and write

$$y := x^Q, \ z := x^{Q^2}, \ w := x^{Q^3}.$$

The equations $F_5(x^{Q^i}) = d^{Q^i}$ for $i = 0, 1, 2, 3$ may be written as

$$x - \frac{xw}{y} - \frac{yz}{w} = d \tag{2}$$

$$y - \frac{yx}{z} - \frac{zw}{x} = e \tag{3}$$

$$z - \frac{zy}{w} - \frac{wx}{y} = f \tag{4}$$

$$w - \frac{wz}{x} - \frac{xy}{z} = g. \tag{5}$$

By Lemma 2, if the set $S := \{u \in \mathbb{F}_{Q^4} : u^{2Q^2} - u^{Q^2+1} + u^2 = 0\}$ is nonempty then $3 \mid Q$ and S consists of the (Q^2-1)-th roots of -1, each of which is fixed by F_5. We claim that if $d \in S$ then $x \in S$. For, if $d \in S$ then $3 \mid Q$ and $f = -d$, so that the sum of the left sides of (2) and (4) is zero. The numerator of this sum factors as $(w+y)(wx+zy)$, so we must have either $w = -y$ or $wx = -zy$. If $w = -y$ then $y^{Q^2-1} = -1$, so that $y \in S$ and hence $x = y^{Q^3} \in S$. If $wx = -zy$ then $(yz)^{Q^2-1} = -1$, so that $y^{(Q+1)(Q^2-1)} = -1$; but since $y \in \mathbb{F}_{Q^4}^*$ it follows that the order of y divides $\gcd(2(Q+1)(Q^2-1), Q^4-1) = 2(Q^2-1)$, so that $y^{(Q+1)(Q^2-1)} = 1 \neq -1$, a contradiction. Thus if $d \in S$ then $x \in S$.

Since F_5 fixes every element of S, and maps $T := \mathbb{F}_{Q^4}^* \setminus S$ into itself, it remains to show that each $d \in T$ has at most one preimage in T under F_5.

Thus we assume henceforth that $d, x \in T$, so that $f^2 - df - d^2 \neq 0$ and $z^2 - xz + x^2 \neq 0$, and raising to the Q-th power yields $g^2 - eg - e^2 \neq 0$ and $w^2 - yw + y^2 \neq 0$. Since $z \neq 0$ we can use (3) to solve for w, obtaining

$$w = (y - \frac{yx}{z} - e)\frac{x}{z}. \tag{6}$$

Substituting (6) into (5) and simplifying yields

$$y = -z\frac{e(x-z) + gz}{x^2 + z^2 - xz}. \tag{7}$$

Substituting (6) and (7) into (2) and (4) yields $A, B \in \mathbb{F}_{Q^4}(X, Z)$ such that $A(x, z) = B(x, z) = 0$. Now compute a Gröbner basis for the ideal of $\mathbb{F}_{Q^4}[Z', X', V', Z, X]$ generated by the numerators of $A(X, Z)$ and $B(X, Z)$ as well as the elements $ZZ' - 1$, $XX' - 1$ and $(X^2 - XZ + Z^2)V' - 1$. Since each element of this ideal vanishes when we substitute $\frac{1}{z}, \frac{1}{x}, \frac{1}{x^2 - xz + z^2}, z, x$ for Z', X', V', Z, X, we may make this substitution into each element of the Gröbner basis to conclude that

$$z = x - d + f \tag{8}$$

and $C(x) = 0$ for a certain degree-5 polynomial $C(X) \in \mathbb{F}_{Q^4}[X]$.

Suppose that $x' \in T \setminus \{x\}$ satisfies $F_5(x') = F_5(x)$; then we must have $C(x') = 0$. The coefficients of $C(X)$ are rational functions in d, e, f, g; by using (2)–(5) we may replace these by rational functions in x, y, z, w, and after multiplying by a suitable polynomial in x, y, z, w we can rewrite the resulting polynomial as $(X - x)D(X)E(X)$ where

$$D(X) := yw(x^2 - xz + z^2)X^2 + xyz(y - w)(x - z)^2$$
$$+ (z - x)(x^2yw + xy^2z - xyzw - xzw^2 + yz^2w)X$$
$$E(X) := (x^2y^2 - xyzw + z^2w^2)X^2 + wz(y - w)(x - z)^3$$
$$+ (z - x)(x^2yw + xy^2z - xyzw - xzw^2 - yz^2w + 2z^2w^2)X.$$

Hence x' is a root of at least one of $D(X)$ and $E(X)$. We must have $z \neq x$ (and equivalently $w \neq y$), since otherwise $D(X) = x^2y^2X^2 = E(X)$, which is impossible since x' is a nonzero root of one of these polynomials. Also $x^2y^2 - xyzw + z^2w^2 \neq 0$, since otherwise $(xy)^2 - (xy)^{Q^2+1} + (xy)^{2Q^2} = 0$ so that $(xy)^{Q^2-1} = -1$, but then the order of x divides both $2(Q+1)(Q^2-1)$ and $Q^4 - 1$ and hence divides $2(Q^2 - 1)$, which yields the contradiction $(xy)^{Q^2-1} = 1 \neq -1$. Thus both D and E have degree 2.

If $D(x') = 0$ then $D(x')^Q = 0$, so that $(x')^Q$ is a root of the polynomial $D_2(X)$ obtained from $D(X)$ by replacing each coefficient by its Q-th power. Equations (7) and (8) imply that $y = G(x)$ where

$$G(X) := -(X - d + f)\frac{e(d - f) + g(X - d + f)}{X^2 + (X - d + f)^2 - X(X - d + f)}.$$

Since these equations were deduced from the identity $\Gamma_5(x) = d$, it follows that $(x')^Q = G(x')$, so that x' is a root of the numerator of $D_2(G(X))$. This numerator factors as $E(X)H(X)$ where

$$H(X) := (x^2y^2 - xy^2z + xyzw - xzw^2 + z^2w^2)X^2 + x^2y(x - z)^2(y - w)$$
$$+ (z - x)(2x^2y^2 - x^2yw - xy^2z + xyzw - xzw^2 + yz^2w)X.$$

Thus x' is a root of at least one of $E(X)$ and $H(X)$. If $E(x') = 0$ then x' is a common root of $D(X)$ and $E(X)$, so the resultant of $D(X)$ and $E(X)$ must vanish. This resultant equals $-xz(x-z)^4(y-w)^2(x^2y^2-xyzw+z^2w^2)u^{Q+1}$ where

$$u := x^2y^2 - x^2yw + xyzw - yz^2w + z^2w^2,$$

so we must have $u = 0$. But then a routine computation shows that $D(X) = yw(x^2 - xz + z^2)(X - x)^2$, which yields the contradiction $x' = x$. Thus if $D(x') = 0$ then we must have $H(x') = 0$. If H has degree 2 then as above the resultant of $D(X)$ and $H(X)$ must vanish, so that

$$0 = xyzw(x - z)^6(y - w)^2(x^2y^2 - xyzw + z^2w^2)^{Q+1},$$

which we know is false. Thus $\deg H < 2$. Since the constant term of $H(X)$ is nonzero, the condition $H(x') = 0$ forces $\deg H = 1$, so the resultant of $D(X)$ and $H(X)$ is $xy(x - z)^4(w - y)v$ where

$$v := x^5y^3w - 2x^4y^4z + x^4y^3zw - 2x^4y^2zw^2 + 3x^3y^4z^2 - 3x^3y^3z^2w$$
$$+ 6x^3y^2z^2w^2 - 2x^3yz^2w^3 + x^3z^2w^4 - x^2y^4z^3 - 2x^2y^3z^3w - x^2y^2z^3w^2$$
$$- x^2z^3w^4 + 2xy^3z^4w - xy^2z^4w^2 + 2xyz^4w^3 - y^2z^5w^2.$$

Thus we must have $v = 0$, and also the coefficient of X^2 in $H(X)$ must vanish. But we can express $yz^6w^2(y-w)^3(y^2+w^2)$ as a sum of the products of these two quantities with certain polynomials in x, y, z, w, so that this expression vanishes and thus $y^2 = -w^2$, whence $y^{2Q^2-2} = -1$. However, there are no $(2Q^2 - 2)$-th roots of -1 in $\mathbb{F}_{Q^4}^*$, since $2Q^2 - 2$ is divisible by the largest power of 2 which divides $Q^4 - 1$.

This completes the proof when $D(x') = 0$, and the proof when $E(x') = 0$ is similar. □

7. The case that $n = 2$ and $q = Q^2$ where Q is an odd prime power and $k = Q^3 + Q^2 - Q$

Case (f) of Theorem 1 is contained in the following result.

Theorem 7. *For any odd prime power Q, we have:*

(a) $F_6(X) := X - \operatorname{Tr}_{Q^4/Q^2}(X^{Q^3+Q^2-Q})$ *permutes* $\mathbb{F}_{Q^4}$.

(b) $\dfrac{X^{Q+2} + 3\nu X^Q - 4\nu^{(Q+1)/2}X}{X^2 - \nu}$ *permutes* $\mathbb{F}_{Q^2}$, *where* ν *is any non-square in* $\mathbb{F}_{Q^2}$.

(c) $\dfrac{X^{2Q+1} - X^{Q+1} + 1}{X^{2Q+1} - X^Q + 1}$ *permutes the set of* $(Q^2 + 1)$-*th roots of unity in* $\mathbb{F}_{Q^4}^*$.

We note that assertions (b) and (c) of this result are extremely similar to the corresponding assertions in Theorem 6. However, we have not been able to find a direct proof that these similar assertions are logically equivalent to one another. If we could find such a proof, then Theorem 7 would follow from Theorem 6. At present, the best we can do is to prove Theorem 7 via a similar argument to our proof of Theorem 6.

Proof. The equivalence of (a), (b) and (c) follows from Propositions 3 and 4 in the same manner as in our previous results. The proof of assertion (a) is nearly identical to the proof of assertion (a) in Theorem 6, so the details can safely be left to the reader.   □

8. The case that $n = 3$ and $k = (q^2 + 1)/2$

We now prove case (g) of Theorem 1.

Theorem 8. *If q is odd, then*

$$F_7(X) := X + \operatorname{Tr}_{q^3/q}\left(X^{\frac{q^2+1}{2}}\right)$$

permutes $\mathbb{F}_{q^3}$.

Proof. By Proposition 2, it suffices to show that

$$g(X) := X(1 + X^{(q+1)/2} + X^{(q^2+q+2)/2} + X^{(q^2+2)(q+1)/2})^{q-1}$$

permutes the set μ_{q^2+q+1} of $(q^2 + q + 1)$-th roots of unity in $\mathbb{F}_{q^3}^*$. For $x \in \mu_{q^2+q+1}$ we compute

$$\begin{aligned}
g(x^{2q}) &= x^{2q}\left(1 + x^{-1} + x^q + x^{q-1}\right)^{q-1} \\
&= x^{2q} \cdot (1 + x^{-1})^{q-1} \cdot (1 + x^q)^{q-1} \\
&= x^{q^2+q} \cdot (1 + x^{-1})^{q^2-1} = x^{-1}(1 + x^{-1})^{q^2-1},
\end{aligned}$$

so that (since $1 + x^{-1} \neq 0$)

$$g(x^{2q})^{-q} = x^q(1 + x^{-1})^{-1+q} = x(x + 1)^{q-1}.$$

Again by Proposition 2, $X(X+1)^{q-1}$ permutes μ_{q^2+q+1} if and only if $X^q + X$ permutes $\mathbb{F}_{q^3}$, which is indeed the case since $X^q + X$ is an additive polynomial having no nonzero roots in $\mathbb{F}_{q^3}$. Therefore $g(X^{2q})^{-q}$ permutes μ_{q^2+q+1}, so that also $g(X)$ permutes μ_{q^2+q+1}, whence F_7 permutes $\mathbb{F}_{q^3}$. □

9. The case that $n = 3$ and $k = q^2 - q + 1$

In this section we prove case (h) of Theorem 1.

Theorem 9. *If q is odd then*

$$F_8(X) := X - \frac{1}{2}\operatorname{Tr}_{q^3/q}\left(X^{q^2-q+1}\right)$$

permutes $\mathbb{F}_{q^3}$.

Proof. Define

$$H(X) := 2X^{(q^4+q^2)/2} - \operatorname{Tr}_{q^3/q}(X)$$
$$L(X) := 2X^q - \operatorname{Tr}_{q^3/q}(X)$$
$$G(X) := L(H(X)).$$

First note that $L(X)$ has no nonzero roots in $\mathbb{F}_{q^3}$, since any such root x would satisfy $2x^q = \operatorname{Tr}_{q^3/q}(x) \in \mathbb{F}_q$ so that $x \in \mathbb{F}_q$, whence $L(x) = -x$. Since $L(X)$ is an additive polynomial in $\mathbb{F}_{q^3}[X]$, it follows that $L(X)$ permutes $\mathbb{F}_{q^3}$. Next, for every $x \in \mathbb{F}_{q^3}$ we have

$$G(x^2) = L(x)^2.$$

Writing S for the set of squares in $\mathbb{F}_{q^3}$, it follows that

$$G(S) = \left(L(\mathbb{F}_{q^3})\right)^2 = S,$$

since L permutes $\mathbb{F}_{q^3}$. Let w be any non-square in $\mathbb{F}_q^*$, so that also w is a non-square in $\mathbb{F}_{q^3}^*$ and thus $\mathbb{F}_{q^3} = S \cup wS$. Since all terms of both L and H have degree congruent to 1 (mod $q - 1$), the same is true of G, so the hypothesis $w^{q-1} = 1$ implies that

$$G(wS) = wG(S) = wS.$$

Therefore G permutes both S and wS, so G permutes $S \cup wS = \mathbb{F}_{q^3}$. Hence also H permutes $\mathbb{F}_{q^3}$, and since $\gcd(q^2 - q + 1, q^3 - 1) = 1$ it follows that $H(X^{q^2-q+1})$ permutes $\mathbb{F}_{q^3}$. It follows that $F_8(X^q)$ (and hence $F_8(X)$) permutes $\mathbb{F}_{q^3}$ since $H(0) = 0 = F_8(0)$ and for $x \in \mathbb{F}_{q^3}^*$ we have

$$H(x^{q^2-q+1}) = 2x^q - \operatorname{Tr}_{q^3/q}(x^{q^2-q+1}) = 2F_8(x^q). □$$

Our proof of bijectivity of H is inspired by the proof of [5, Thm. 3.7], which also distinguished the behavior of a function on squares and non-squares. The key identity $L(H(x^2)) = L(x)^2$ in our proof says that $x \mapsto H(x^2)$ is a planar function which is affine equivalent to $x \mapsto x^2$. Since the nucleus of any semifield induced by $x \mapsto H(x^2)$ is $\mathbb{F}_{q^3}$, the hypotheses of [5, Thm. 3.7] do not apply in our situation; however, it appears that those hypotheses can be relaxed to cover both our situation and several others. Our proof of Theorem 9 uses our key identity to avoid the bulk of the work involved in the proof of [5, Thm. 3.7].

10. The case that $k = q^\ell + 1$ with $\ell \mid n$

We now prove case (i) of Theorem 1.

Theorem 10. *For any prime power q and any positive integers ℓ, n with $2\ell \mid n$, if $\gamma \in \mathbb{F}_{q^n}$ satisfies $\gamma^{q^{2\ell}-1} = -1$ then $F_9(x) := X + \gamma \operatorname{Tr}_{q^n/q}(X^{q^\ell+1})$ permutes $\mathbb{F}_{q^n}$.*

Note that if q is odd then $\mathbb{F}_{q^n}$ contains elements γ as in Theorem 10 if and only if n is divisible by 4ℓ.

Proof. By Proposition 1, F_9 permutes $\mathbb{F}_{q^n}$ if and only if for each $\alpha \in \mathbb{F}_{q^n}$ the polynomial $h_\alpha(X) := X + \operatorname{Tr}_{q^n/q}\big((\alpha + \gamma X)^{q^\ell+1}\big)$ permutes $\mathbb{F}_q$. For $x \in \mathbb{F}_q$ and $Q := q^\ell$ we have
$$(\alpha + \gamma x)^{Q+1} = \gamma^{Q+1} x^{Q+1} + \alpha \gamma^Q x^Q + \alpha^Q \gamma x + \alpha^{Q+1}$$
$$= \gamma^{Q+1} x^2 + \alpha \gamma^Q x + \alpha^Q \gamma x + \alpha^{Q+1},$$
so that
$$\operatorname{Tr}_{q^n/q}((\alpha+\gamma x)^{Q+1}) = \operatorname{Tr}_{q^n/q}(\gamma^{Q+1})x^2 + \operatorname{Tr}_{q^n/q}(\alpha\gamma^Q + \alpha^Q\gamma)x + \operatorname{Tr}_{q^n/q}(\alpha^{Q+1}).$$
Since $\gamma^{Q^2-1} = -1$, we have $\gamma^{Q+Q^2} = -\gamma^{Q+1}$ and thus
$$\operatorname{Tr}_{q^n/q}(\gamma^{Q+1}) = \operatorname{Tr}_{q^n/Q^2}(\operatorname{Tr}_{Q/q}(\operatorname{Tr}_{Q^2/Q}(\gamma^{Q+1}))) = 0.$$
Likewise,
$$\operatorname{Tr}_{q^n/q}(\alpha\gamma^Q + \alpha^Q\gamma) = \operatorname{Tr}_{q^n/q}(\alpha^Q\gamma^{Q^2} + \alpha^Q\gamma) = \operatorname{Tr}_{q^n/q}(0) = 0.$$
Thus h_α induces the same function on $\mathbb{F}_q$ as does the polynomial $X + \operatorname{Tr}_{q^n/q}(\alpha^{Q+1})$, so that indeed h_α permutes $\mathbb{F}_q$ as required. $\qquad\square$

Remark 3. The proof of Theorem 10 shows that for every $\alpha \in \mathbb{F}_{q^n}$ we have
$$\operatorname{Tr}_{q^n/q}((\alpha + \gamma x)^{Q+1}) - \operatorname{Tr}_{q^n/q}(\alpha^{Q+1}) = 0,$$
so that γ is a linear translator of $\operatorname{Tr}_{q^n/q}(x^{Q+1})$. Theorem 10 is a generalization of [2, Thm. 6], which addressed the case of prime q.

Acknowledgments   The authors thank the referee for providing several corrections and helpful suggestions. The second author was partially supported by NSF grant DMS-1162181.

References

1. P. Charpin and G. Kyureghyan (2008). On a class of permutation polynomials over $\mathbb{F}_{2^n}$, *Sequences and their Applications—SETA 2008*, Lecture Notes in Comput. Sci. 5203, Springer, pp. 368–376.
2. P. Charpin and G. Kyureghyan (2010). Monomial functions with linear structure and permutation polynomials, *Finite Fields: Theory and Applications*, Contemp. Math. 518, Amer. Math. Soc., pp. 99–111.
3. H. Dobbertin (2002). Uniformly representable permutation polynomials, *Sequences and their Applications (Bergen, 2001)*, Springer, pp. 1–22.
4. X.-D. Hou (2014). Determination of a type of permutation trinomials over finite fields, *Acta Arith.* 166, pp. 253–278.
5. G. Weng and X. Zeng (2012). Further results on planar DO functions and commutative semifields, *Des. Codes Cryptogr.* 63, pp. 413–423.
6. M. E. Zieve (2008). Some families of permutation polynomials over finite fields, *Int. J. Number Theory* 4, pp. 851–857.

Scattered spaces in Galois geometry

Michel Lavrauw

Università degli Studi di Padova   `michel.lavrauw@unipd.it`

1. Introduction and motivation

Given a set Ω and a set S of subsets of Ω, a subset $U \subset \Omega$ is called *scattered* with respect to S if U intersects each element of S in at most one element of Ω. In the context of Galois Geometry this concept was first studied in 2000 [4], where the set Ω was the set of points of the projective space $\mathrm{PG}(11, q)$ and S was a 3-spread of $\mathrm{PG}(11, q)$. The terminology of a *scattered space*[1] was introduced later in [8]. The paper [4] was motivated by the theory of blocking sets, and it was shown that there exists a 5-dimensional subspace, whose set of points $U \subset \Omega$ is scattered with respect to S, which then led to an interesting construction of a $(q+1)$-fold blocking set in $\mathrm{PG}(2, q^4)$. The notion of "being scattered" has turned out to be a useful concept in Galois Geometry. This paper is motivated by the recent developments in Galois Geometry involving scattered spaces. The first part aims to give an overview of the known results on scattered spaces (mainly from [4],[8], and [31]) and the second part gives a survey of the applications.

2. Notation and terminology

A *t-spread* of a vector space V is a partition of $V \setminus \{0\}$ by subspaces of constant dimension t. Equivalently, a $(t-1)$-spread of $\mathrm{PG}(V)$ (the projective space associated to V) is a set of $(t-1)$-dimensional subspaces partitioning the set of points of $\mathrm{PG}(V)$. Sometimes the shorter term *spread* is used,

---

[1] This notion of scattered spaces is not to be confused with scattered spaces in Topology, where a space A is called *scattered* if every non-empty subset $T \subset A$ contains a point isolated in T.

when the dimension is irrelevant or clear from the context. Two spreads S_1, S_2 of PG(V) (respectively V) are called *equivalent* if there exists a collineation α of PG(V) (respectively, an element α of ΓL(V)) such that $S_1^\alpha = S_2$.

A standard construction of a spread (going back to Segre [48]) is the following. We sketch the construction in the context of the larger framework of *field reduction* techniques on which we will elaborate in Section 4.3. Consider any $\mathbb{F}_q$-vector space isomorphism from $\mathbb{F}_{q^t} \to \mathbb{F}_q^t$, and extend this to an isomorphism φ between the $\mathbb{F}_q$-vector spaces $\mathbb{F}_{q^t}^r$ and $\mathbb{F}_q^{rt}$. For each nonzero vector $v \in V(r, q^t)$ consider the vector space $S_v = \{\varphi(\lambda v) : \lambda \in \mathbb{F}_{q^t}\}$. One easily verifies that the set $\mathcal{D}_{r,t,q} := \{S_v : v \in V(r, q^t)\}$ defines a t-spread in $V(rt, q)$. A spread S is called *Desarguesian* if S is equivalent to $\mathcal{D}_{r,t,q}$ for some r, t and q.

Let D be any set of subspaces in $V(n, q)$. A subspace W of V is called *scattered with respect to D* if W intersects each element of D in at most a 1-dimensional subspace. Equivalently, a subspace of a projective space PG(V) is called scattered with respect to a set of subspaces D of PG(V) if it intersects each element of D in at most a point. In this paper D will typically be a spread. We note that we will often switch between vector spaces and projective spaces, assuming that the reader is familiar with both terminologies. To avoid overcomplicating the notation, we will use the same symbol (for instance D) for subset of the subspaces of a projective space and its associated object in the underlying vector space. We will make sure that there is no ambiguity concerning vector space dimension and projective dimension.

If D is any set of subspaces of a vector space or a projective space, and W is a subspace of the same space, then by $\mathcal{B}_D(W)$ we denote the set of elements of D which have a nontrivial intersection with W. If there is no confusion possible, then we also use the simplified notation $\mathcal{B}(W)$.

3. Scattered spaces

We start this section with a number of examples illustrating some of the difficulties that arise in the study of scattered spaces with respect to spreads.

Example 1.

(1) If D is a spread of lines in PG($3, q$) then every line not contained in the spread is scattered w.r.t. D. Also, since $|D| = q^2 + 1$ and a plane

contains $q^2 + q + 1$ points, no plane of $PG(3, q)$ is scattered w.r.t. D.

(2) If D is a spread of planes in $PG(5, q)$ then every line not contained in an element of the spread is scattered w.r.t. D. Also since $|D| = q^3 + 1$ and a solid (a 3-dimensional projective space) contains $q^3 + q^2 + q + 1$ points, no solid of $PG(3, q)$ is scattered w.r.t. D. The existence of a scattered plane is not immediately clear, but this will follow from one of the results in the next sections (Theorem 1).

(3) If D is a spread of lines in $PG(5, q)$ then it is easy to see that the dimension of a scattered subspace of $PG(5, q)$ w.r.t. D cannot exceed 3 (a solid). Also any such spread allows a scattered line (trivial), and a scattered plane (Theorem 1). On the other hand, the existence of a scattered solid depends on the spread. We will see that a Desarguesian spread does not allow scattered solids (by Theorem 2), but there are spreads which do (by Theorem 3).

3.1. *Maximally vs maximum scattered*

It is important to observe the distinction between the following two definitions. A subspace U is called *maximally scattered* w.r.t. a spread D if U is not contained in a larger scattered space. A subspace U is called *maximum scattered* w.r.t. a spread D if any scattered space T w.r.t. D satisfies $\dim T \leq \dim U$. As we will see in the following example, there exist maximally scattered spaces which are not maximum scattered.

Example 2. Consider the irreducible polynomial $f(x) = x^6 + x^4 + x^3 + x + 1 \in \mathbb{F}_2[x]$, put $\mathbb{F}_{2^6} = \mathbb{F}_2[x]/(f(x))$ and consider the set of subspaces $D = \{S_{u,v} : u, v \in \mathbb{F}_2^6\}$, where $S_{u,v} = \langle (M^k u, M^k v) : k \in \{0, 1, \ldots, 5\} \rangle$ of $V(12, 2)$, where M is the companion matrix of $f(x)$. This is the standard construction (by field reduction see Section 4.3) of a Desarguesian spread, in this case a Desarguesian 6-spread in $V(12, 2)$. The subspace U_5 spanned by the rows of the matrix

$$\begin{bmatrix} 1 & 0 & 0 & 0 & 0 & 0 & 0 & 0 & 0 & 1 & 1 & 0 \\ 0 & 1 & 0 & 0 & 0 & 0 & 0 & 0 & 1 & 0 & 0 & 0 \\ 0 & 0 & 1 & 0 & 0 & 0 & 0 & 0 & 0 & 0 & 1 & 1 \\ 0 & 0 & 0 & 1 & 0 & 1 & 0 & 0 & 0 & 0 & 0 & 0 \\ 0 & 0 & 0 & 0 & 1 & 1 & 1 & 1 & 1 & 1 & 1 & 0 \end{bmatrix}$$

is a maximally scattered 5-dimensional subspace of $V(12, 2)$. If $\varphi : \mathbb{F}_{2^6} \to \mathbb{F}_2^6$ is any vector space isomorphism, then the subspace $U_6 := \{(\varphi(\alpha), \varphi(\alpha^2)) : \alpha \in \mathbb{F}_2^6\}$ is maximum scattered. It follows that U_5 is maximally scattered but not maximum scattered. This example was constructed using the GAP-package FinInG (see [24],[23]).

3.2. *A lower bound on the dimension of maximally scattered spaces*

It is obvious that every line of a projective space, which is not contained in an element of a spread D, is scattered with respect to D, so a maximally scattered space has vector dimension at least 2. The following theorem gives a lower bound on the dimension of a maximally scattered space in terms of the dimension of the space and the dimension of the spread elements. Its proof (see [8]) is purely combinatorial and gives a method to extend a scattered space in case the bound is not attained.

Theorem 1. [8, Theorem 2.1]
If U is a maximally scattered subspace w.r.t. a t-spread in $V(rt, q)$, then $\dim U \geq \lceil (rt - t)/2 \rceil + 1$.

Theorem 1 implies the existence of a scattered plane w.r.t. a plane spread in $PG(5, q)$, answering one of the questions from Example 1.

3.3. *An upper bound on the dimension of scattered spaces*

Let S be a $(t-1)$-spread in $PG(rt-1, q)$. The number of spread elements is $(q^{rt} - 1)/(q^t - 1) = q^{(r-1)t} + q^{(r-2)t} + \ldots + q^t + 1$. Since a scattered subspace can contain at most one point of every spread element, the number of points in a scattered space must be less than or equal to the number of spread elements. This gives the following trivial upper bound.

Theorem 2. [8, Theorem 3.1]
If U is scattered w.r.t. a t-spread in $V(rt, q)$, then $\dim U \leq rt - t$.

Note that for a line spread in $PG(3, q)$ the upper and lower bound coincide. But this is quite exceptional. In fact, excluding trivial cases we may assume that t and r are both at least 2, and it follows that the exact dimension of a maximum scattered space is determined by the lower and upper bounds only for $(r, t) \in \{(2, 2), (2, 3)\}$, i.e., a line spread in $PG(3, q)$ and a plane spread in $PG(5, q)$. The projective dimension of a maximum scattered space in these cases is respectively 1, the dimension of a line, and 2, the dimension of a plane. There is a large variety of spreads and there is not much one can say about the possible dimension of a scattered subspace with respect to an arbitrary spread (see Section 3.4). This is one of the reasons to consider scattered spaces with respect to a Desarguesian spread (see Section 3.5). Another reason is the correspondence between the elements of a Desarguesian spread and the points of a projective space over an extension field (so-called *field reduction*, see Section 4.3). However before we proceed, in the following section we show that the upper bound from Theorem 2 cannot be improved without restrictions on the spread.

3.4. *Scattering spreads with respect to a subspace*

A spread D is called *a scattering spread with respect to a subspace U*, if this subspace U is scattered with respect to the spread D.

Theorem 3. [8, Theorem 3.2]
If W is an $(rt - t - 1)$-dimensional subspace of $\mathrm{PG}(rt - 1, q)$, $r \geq 2$, then there exists a scattering $(t - 1)$-spread S with respect to W.

This theorem shows that there is no room for improvement of Theorem 2 without assuming some extra properties on the spread. Up to now, the only spreads that have been investigated in detail are the Desarguesian spreads (see Section 3.5).

3.5. *Scattered spaces w.r.t. Desarguesian spreads*

Let S be a $(t-1)$-spread of $\mathrm{PG}(rt-1, q)$ and consider the following incidence structure. First embed $\mathrm{PG}(rt - 1, q)$ as a hyperplane H in $\Sigma = \mathrm{PG}(rt, q)$. Denote the set of points of $\Sigma \backslash H$ by $\mathcal{P}$ and the set of t-dimensional subspaces of Σ which intersect H in an element of S by $\mathcal{L}$. Define an incidence relation $\mathcal{I}$ on $(\mathcal{P} \times \mathcal{L}) \cup (\mathcal{L} \times \mathcal{P})$ by symmetric containment. Then the incidence structure $\mathcal{D}(S) = (\mathcal{P}, \mathcal{L}, \mathcal{I})$ is a design with parallelism, also called *Sperner space* or *S-space* (see e.g.[5]). More precisely, $\mathcal{D}(S)$ is a $2 - (q^{rt}, q^t, 1)$ design such that for each anti-flag (x, U), there exists exactly one element of $\mathcal{L}$ which is incident with x and parallel to U (two elements of $\mathcal{L}$ are called *parallel* if their intersection is an element of S). Moreover, the design $\mathcal{D}(S)$ is a Desarguesian affine space if and only if S is a Desarguesian spread (see [5, Theorem 2]). This correspondence is crucial in the proof of the following theorem and is of central importance for many of the applications of scattered spaces. We remind the reader that $\mathcal{D}_{r,t,q}$ denotes the Desarguesian t-spread in $V(rt, q)$.

Theorem 4. [8, Theorem 4.3]
If U is a scattered subspace w.r.t. $\mathcal{D}_{r,t,q}$, then $\dim U \leq rt/2$.

It was also shown in [8] that this upper bound is tight whenever r is even.

Theorem 5. [8] *If r is even, then there exists a scattered subspace w.r.t. $\mathcal{D}_{r,t,q}$ in $V(rt, q)$ of dimension $rt/2$.*

For r odd, the exact dimension of a maximum scattered space is in general not known. The following theorem gives a lower bound on the dimension of a maximum scattered space.

Theorem 6. [8] *The dimension of a maximum scattered subspace w.r.t. $\mathcal{D}_{r,t,q}$ in $V(rt, q)$ is at least $r'k$ where $r'|r$, $(r', t) = 1$, and $r'k$ is maximal such that*

$$k < (rt - t + 3)/2 \text{ for } q = 2 \text{ and } r' = 1$$

and

$$r'k < (rt - t + r' + 3)/2 \text{ otherwise.}$$

We conclude this section with an overview of the values of r, t, q, with r odd and t even, for which maximum scattered spaces w.r.t. $\mathcal{D}_{r,t,q}$ have been constructed. Note that for $t = 2$, the existence of a scattered r-space w.r.t. $\mathcal{D}_{r,2,q}$ easily follows by considering an appropriate maximal subspace lying on the Segre variety $S_{r,2}(q)$, or equivalently a subspace $\mathbb{F}_q^r \otimes v$ of $\mathbb{F}_q^r \otimes \mathbb{F}_q^2$, for some nonzero vector $v \in \mathbb{F}_q^2$ (see e.g. Section 1.6 and in particular Theorem 1.6.4 of [31] for more details). For $t = 4$ and $r = 3$, a 6-dimensional scattered space w.r.t. $\mathcal{D}_{3,4,q}$ was constructed in [4] (see more on this in Section 4.6). A much more general result (including the construction from [4]) was recently obtained by Bartoli et al. in [6]. They constructed scattered linear sets of rank $rt/2$ (see Section 4.3 below for definitions) in $\mathrm{PG}(r - 1, q^t)$ for many parameters r, t and q. As a corollary one obtains the existence of maximum scattered spaces w.r.t. $\mathcal{D}_{r,t,q}$ in the following cases.

Corollary 1. (From [6]) *There exist scattered spaces w.r.t. $\mathcal{D}_{r,t,q}$, t even, of dimension $rt/2$ in the following cases: (i) $q = 2$ and $t \geq 4$; (ii) $q \geq 2$ and $t \not\equiv 0 \mod 3$; (iii) $q \equiv 1 \mod 3$ and $t \equiv 0 \mod 3$.*

Apart from the computational examples from [6], for $t = 6$ and $q \in \{3, 4, 5\}$, the existence of scattered spaces of dimension $rt/2$ w.r.t. $\mathcal{D}_{r,t,q}$, with r odd and t even, remains open for $t \equiv 0 \mod 3$, $q \not\equiv 1 \mod 3$, and $q > 2$.

4. Applications

4.1. *Translation hyperovals*

We start the section with one of the earliest applications of scattered spaces: hyperovals of translation planes. We assume the reader is familiar with the notion of a projective plane of order q. All necessary background (and much more) can be found in for instance [17] or [26]. A *hyperoval* in a projective plane of order q is a set $\mathcal{H}$ of $q + 2$ points, no three of which are collinear, i.e. no line of the plane contains three points of $\mathcal{H}$. The existence of a hyperoval in a projective plane π implies the order q of π to be even.

Let π be a projective plane. A *perspectivity* of π is a collineation α for which there exists a point-line pair (x, ℓ) such that α fixes each line on x and each point on ℓ, and α is then also called an (x, ℓ)-*perspectivity*. If x is on ℓ then α is called an (x, ℓ)-*elation*, otherwise α is called an (x, ℓ)-*homology*. The point x is the *center* and the line ℓ is the *axis* of α. A *translation plane* is a projective plane which contains a line ℓ_∞, such that for each point x on ℓ_∞ the group of elations with center x and axis ℓ_∞ acts transitively on the points of $m \setminus \{x\}$ for each line m on x. Equivalently, the automorphism group G of the plane is called (x, ℓ_∞)-*transitive* for each point x on ℓ_∞.

A hyperoval $\mathcal{H}$ is called a *translation hyperoval* if there exists a group G of q elations with common axis ℓ, fixing $\mathcal{H}$. The line ℓ is a 2-secant, called the *translation line* of $\mathcal{H}$, and the group G acts transitively on the points of $\mathcal{H} \setminus \ell$. Translation hyperovals in the Desarguesian projective plane $\mathrm{PG}(2, q)$ were classified by Payne in 1971 [45]. For non-Desarguesian projective planes, and in particular for translation planes, the classification remains open. Denniston [19] and Korchmáros [30] constructed translation ovals in non-Desarguesian translation planes, while Jha and Johnson [27] proved that for each non-prime integer $N > 3$, there exists a non-Desarguesian translation plane of order 2^N admitting a translation hyperoval. Note that not every projective plane admits translation hyperovals since some planes (e.g. Figueroa planes) don't have enough translations.

The study of translation planes is equivalent to the study of spreads of a vector space whose dimension is twice the dimension of the elements of the spread. This correspondence is due to the André-Bruck-Bose (ABB) construction of a translation plane from a spread and vice versa. In fact, the ABB-construction is a special case of the construction of the design with parallelism $\mathcal{D}(S)$ mentioned above (with $r = 2$). In this case, $\mathcal{D}(S)$ is a $2-$ $(q^{2t}, q^t, 1)$ design with parallelism, i.e. an affine plane. The corresponding projective plane is denoted by $\pi(S)$. The equivalence between spreads and translation planes was obtained in 1954 by André [1], using a group-theoretic point of view. The geometric construction as presented above was published by Bruck and Bose [11] in 1964. The following theorem shows the equivalence between translation hyperovals in translation planes (sharing the same axis) and scattered spaces, and is comparable to [27, Theorem 5], although the terminology in [27] is quite different.

Theorem 7. *A translation plane $\pi(S)$ of order 2^t with translation line ℓ_∞ contains a translation hyperoval with translation line ℓ_∞ if and only if the spread S in $V(2t, 2)$ admits a scattered space of dimension t.*

Proof. We prove the first part of the theorem in a projective setting. Let S be a $(t-1)$-spread in $\Sigma = \mathrm{PG}(2t-1, 2)$, and U a scattered $(t-1)$-space w.r.t S. Embed Σ as a hyperplane in Σ^* and choose a t-dimensional subspace K_U of Σ^* with $K_U \cap \Sigma = U$. Let $\mathcal{H}$ denote the set of points in $K_U \setminus U$ together with the two points (call them x and y) at infinity corresponding to the spread elements which are disjoint from U. We claim that $\mathcal{H}$ is a hyperoval of $\pi(S)$. To prove this claim, consider a line ℓ in $\pi(S)$. If ℓ contains x (respectively y) then the t-space of Σ^* corresponding to ℓ intersects K_U in exactly one point z. In this case, the line ℓ contains exactly two points of $\mathcal{H}$, namely z and x (respectively y). If ℓ does not contain x or y, then the t-space $\bar{\ell}$ of Σ^* corresponding to ℓ intersects Σ in an element $R \in S$ which intersects U in a point u. There are two possibilities: either $\bar{\ell}$ intersects K_U only in the point u, or $\bar{\ell}$ intersects K_U in a line $\{u, u', u''\}$. In the first case, ℓ is external to $\mathcal{H}$. In the latter case, the line ℓ intersects $\mathcal{H}$ in exactly two points u' and u''. This shows that no three points of $\mathcal{H}$ are collinear in $\pi(S)$ (note that the line ℓ_∞ meets $\mathcal{H}$ in the two points x and y). It follows that $\mathcal{H}$ is a set of $2^t + 2$ points, no three of which are collinear, i.e. $\mathcal{H}$ is a hyperoval. The existence of the group H of q elations, with the line at infinity as the common axis, immediately follows from the construction, as H corresponds to the translation group stabilising $K_U \setminus U$.

Conversely, suppose that $\pi(S)$ contains a translation hyperoval $\mathcal{H}$. Let T denote the translation group of $\pi(S)$ and $H \leq T$ the translation group of $\mathcal{H}$. If $T_z \leq T$ denotes the group of (z, ℓ_∞)-elations, then the spread S coincides with the set $\{T_z \ : \ z \in \ell_\infty\}$. All groups T, H and $T_z \in S$ are elementary abelian 2-groups, and we will consider them as $\mathbb{F}_2$-vector spaces. Note that $|T| = q^{2t}$, $|H| = 2^t$, and $|T_z| = 2^t$. Let x and y be the two points of $\mathcal{H}$ on the line ℓ_∞. Then, for $z \in \ell_\infty \setminus \{x, y\}$, it follows that $|T_z \cap H| = 2$, since a non-trivial (z, ℓ_∞)-elation fixing $\mathcal{H}$ is necessarily an involution. This suffices to conclude that H intersects each element of $S \setminus \{T_x, T_y\}$ in a one-dimensional subspace, and therefore H is a scattered space with respect to S of dimension t in $V(2t, 2)$. $\qquad\square$

4.2. Translation caps in affine spaces

The next application is a generalisation to higher dimensional spaces of the correspondence between translation hyperovals and scattered spaces. This comes from recent work [6], to which we refer for the details. Here we only give a sketch of this correspondence.

As mentioned above, the ABB construction is a special case, with $r = 2$, of the more general construction of the $2 - (q^{rt}, q^t, 1)$ design $\mathcal{D}(S)$, and when the spread S is Desarguesian, the design $\mathcal{D}(S)$ is an affine space AG$(r - 1, q^t)$. Generalising the previous construction, of a translation hyperoval from a scattered space in PG$(2t - 1, 2)$, now starting from a scattered space U w.r.t $\mathcal{D}_{r,t,2}$ in PG$(rt - 1, 2)$, one obtains a set of points $\mathcal{K} = K_U \setminus U$ in the affine space AG$(r - 1, 2^t)$. Again this set of points satisfies the property that no three of them are collinear. Such a set is called a *cap*, and this particular construction gives a *translation cap*. Translating this correspondence from [6] into our terminology gives the following.

Theorem 8. *A scattered subspace w.r.t. $\mathcal{D}_{r,t,2}$, $t > 1$, corresponds to a translation cap in* AG$(r - 1, 2^t)$ *and viceversa.*

This correspondence leads to the existence of complete caps whose cardinality is close to the theoretical lower bound for complete caps. See [6] for further details.

4.3. Linear sets

Linear sets have many interesting aspects, and as it would take us too much time to elaborate on all of these. We refer to [41] and [47] for surveys on the topic. Before we explain some of the applications of scattered spaces to the theory of linear sets, we briefly introduce the notion of a linear set using the notation and terminology of field reduction which was formalised in [41]. The technique called "field reduction" is based on the well understood concept of subfields in a finite field, and, maybe surprisingly, has proved to be a very powerful tool in Galois Geometry. Consider the *field reduction map* $\mathcal{F}_{r,t,q}$ as in [41] from PG$(r - 1, q^t)$ to PG$(rt - 1, q)$. Points of PG$(r - 1, q^t)$ are mapped onto $(t - 1)$-spaces of PG$(rt - 1, q)$, and in particular the image of the set $\mathcal{P}$ of points of $\mathrm{PG}(r-1, q^t)$ forms a Desarguesian $(t-1)$-spread $\mathcal{D}_{r,t,q}$ of $\mathrm{PG}(rt-1, q)$. If U is a subspace of $\mathrm{PG}(rt-1, q)$ then by $\mathcal{B}(U)$ we denote the set of points corresponding to the spread elements which have non-trivial intersection with U, i.e.

$$\mathcal{B}(U) = \{x \in \mathcal{P} \ : \ \mathcal{F}_{r,t,q}(x) \cap U \neq \emptyset\}. \tag{1}$$

Here the set $\mathcal{B}(U)$ is considered as a set of points in $\mathrm{PG}(r-1, q^t)$, but using the one-to-one correspondence between $\mathcal{P}$ and $\mathcal{D}_{r,t,q}$ given by the field reduction map $\mathcal{F}_{r,t,q}$, sometimes $\mathcal{B}(U)$ is also considered as a subset of $\mathcal{D}_{r,t,q}$, consistent with the notation we used in the previous sections. This just means that the sets $\mathcal{B}(U)$ and $\mathcal{F}_{r,t,q}(\mathcal{B}(U))$ are sometimes identified. The context (i.e. the ambient space) should always clarify if $\mathcal{B}(U)$ is considered as a subset of $\mathcal{P}$ or as a subset of $\mathcal{D}_{r,t,q}$.

A set of points L in $\mathrm{PG}(r-1, q^t)$ is called an $\mathbb{F}_q$-*linear set* if there exists a subspace U in $\mathrm{PG}(rt-1, q)$ such that $L = \mathcal{B}(U)$. An $\mathbb{F}_q$-linear set $\mathcal{B}(U)$ is said to have *rank* m if U has projective dimension $m-1$. These definitions immediately lead to the following proposition.

Proposition 1. *An $\mathbb{F}_q$-linear set $\mathcal{B}(U)$ in $\mathrm{PG}(r-1, q^t)$ has maximal size (w.r.t. its rank) if and only if U is scattered w.r.t. $\mathcal{D}_{r,t,q}$.*

An $\mathbb{F}_q$-linear set L of rank m has at most $(q^m - 1)/(q - 1)$ points and if this bound is reached then L is called a *scattered linear set*. If a $\mathcal{B}(U)$ is an $\mathbb{F}_q$-linear set in $\mathrm{PG}(r-1, q^t)$ and U is maximum (respectively maximally) scattered w.r.t. $\mathcal{D}_{r,t,q}$, then $\mathcal{B}(U)$ is called a *maximum* (respectively *maximally) scattered $\mathbb{F}_q$-linear set.*

In [47] Polverino introduced the notion of the dual linear set. If β is a non-degenerate sesquilinear form on $\mathbb{F}_{q^t}^r$ and Tr denotes the trace map from $\mathbb{F}_{q^t}$ to $\mathbb{F}_q$, then $Tr \circ \beta$ defines a non-degenerate form from $\mathbb{F}_{q^t}^r$ to $\mathbb{F}_q$. If $\perp$ denotes the corresponding polarity in $\mathrm{PG}(rt-1, q)$, and $\mathcal{B}(U)$ is an $\mathbb{F}_q$-linear set in $\mathrm{PG}(r-1, q^t)$, then $\mathcal{B}(U^\perp)$ is called the *dual linear set with respect to β*. If $\mathcal{B}(U)$ has rank m then $\mathcal{B}(U^\perp)$ has rank $rt - m$. For maximum scattered linear sets we have the following theorem.

Theorem 9. *[47, Theorem 3.5.] If rt is even and $\mathcal{B}(U)$ is a maximum scattered $\mathbb{F}_q$-linear set of $\mathrm{PG}(r-1, q^t)$, then the dual linear set with respect to any polarity of $\mathrm{PG}(r-1, q^t)$ is a maximum scattered $\mathbb{F}_q$-linear set as well.*

One of the important questions regarding linear sets is the *equivalence problem*. Contrary to linear subspaces (which are equivalent if and only if they have the same dimension), two linear sets of the same rank are not necessarily equivalent under the action of the projective group or the collineation group of the ambient projective space. This is of course not surprising since two linear sets of the same rank might even have different cardinalities. In few cases the equivalence problem has been solved.

Theorem 10. *[39]* [2] *All scattered $\mathbb{F}_q$-linear sets of rank 3 in $\mathrm{PG}(1, q^3)$ and $\mathrm{PG}(1, q^4)$ are equivalent under $\mathrm{PGL}(2, q^4)$. All scattered $\mathbb{F}_q$-linear sets of rank 3 in $\mathrm{PG}(1, 2^5)$ are equivalent under $\mathrm{P\Gamma L}(2, 2^5)$.*

The equivalence of maximum scattered $\mathbb{F}_q$-linear sets in $\mathrm{PG}(1, q^3)$ can be generalised to all projective spaces of odd dimension over $\mathbb{F}_{q^3}$. The following was shown for $n = 2$ in [44, Proposition 2.7] and for general n in [40, Theorem 4].

Theorem 11. *All maximum scattered $\mathbb{F}_q$-linear sets in $\mathrm{PG}(2n - 1, q^3)$ are $\mathrm{P\Gamma L}$-equivalent.*

This equivalence does however not generalise to maximum scattered $\mathbb{F}_q$-linear sets of odd dimensional projective spaces over extension fields of degree > 3. For instance, in [43] it was shown that in $\mathrm{PG}(2n - 1, q^t)$, $q > 3$, $t \geq 4$, there exist inequivalent maximum scattered linear sets.

Another interesting question concerning linear sets is the *intersection problem*. Given two linear sets L_1 and L_2 of given rank in a given projective space, what are the possibilities for the intersection $L_1 \cap L_2$? Again the answer is trivial for subspaces, and the question has also been answered for subgeometries (see [21, Theorem 1.3]), but for linear sets the problem is much more complicated and only few results are known. The following theorem gives an answer to the intersection problem for an $\mathbb{F}_q$-linear set of rank k and a scattered $\mathbb{F}_q$-linear set of rank two (i.e. an $\mathbb{F}_q$-subline).

Theorem 12. *[39, Theorem 8 and 9]* *An $\mathbb{F}_q$-subline intersects an $\mathbb{F}_q$-linear set of rank k of $\mathrm{PG}(1, q^h)$ in $0, 1, \ldots, \min\{q + 1, k\}$ or $q + 1$ points and for every subline $L \cong \mathrm{PG}(1, q)$ of $\mathrm{PG}(1, q^h)$, there is a linear set S of rank k, $k \leq h$ and $k \leq q + 1$, intersecting L in exactly j points, for all $0 \leq j \leq k$.*

In [22, Proposition 5.2] the authors determined the intersection of two scattered $\mathbb{F}_q$-linear sets of rank $t + 1$ in $\mathrm{PG}(2, q^t)$. Further results on the

---

[2]It was pointed out in [14] that one of the conditions of Theorem 3 in [39] is not necessary for the equivalence of two linear sets. The condition is however sufficient, and hence does not affect the equivalences stated here.

intersection of (not necessarily scattered) linear sets can be found in [40] and [46].

4.4. *Two-intersection sets*

A two-intersection set w.r.t. k-dimensional spaces in $\mathrm{PG}(V)$ is a set Ω of points such that the size of the intersection of the set Ω with a k-space only takes two different values, say m_1 and m_2. The numbers m_1 and m_2 are called the *intersection numbers* of the set Ω. A fundamental result which makes scattered spaces particularly interesting is the following.

Theorem 13.[8] *If U is a scattered space w.r.t. $\mathcal{D}_{r,t,q}$ with $\dim U = rt/2$, then $B(U)$ is a two-intersection set w.r.t. hyperplanes in $\mathrm{PG}(r-1,q^t)$, with intersection numbers*

$$m_1 = \frac{q^{\frac{rt}{2}-t} - 1}{q-1} \ \text{and}\ m_2 = \frac{q^{\frac{rt}{2}-t+1} - 1}{q-1}.$$

If t is even, then this set has the same parameters as the union of $(q^{t/2} - 1)/(q-1)$ pairwise disjoint Baer subgeometries isomorphic to $\mathrm{PG}(r-1,q^{t/2})$ (call such a set of *type I*). If t is odd, then this set has the same parameters as the union of $(q^t - 1)/(q-1)$ elements of an $(r/2-1)$-spread in $\mathrm{PG}(r-1,q^t)$. We call these two-intersection sets of *type II*. It was proved for $r = 3$ and $t = 4$ in [4] and for general rt even in [9] that the two-intersection sets obtained from scattered spaces are not of these types.

Theorem 14.[9] *A scattered $\mathbb{F}_q$-linear set of rank $rt/2$ in $\mathrm{PG}(r-1,q^t)$ is inequivalent to the two-intersection sets of type I or type II.*

4.5. *Two-weight codes*

An $\mathbb{F}_q$-linear $[n,k]$-code C is a k-dimensional subspace of $\mathbb{F}_q^n$. Vectors belonging to C are called *codewords* and the *weight* $\mathrm{wt}(c)$ of a codeword is the number of nonzero coordinates of c with respect to some fixed basis of $\mathbb{F}_q^n$. The *distance* $d(c_1, c_2)$ between two codewords is the number of positions in which they have different coordinates and is thus equal to $\mathrm{wt}(c_1 - c_2)$. If the nonzero minimum distance of C is d, then the code is called an $\mathbb{F}_q$-*linear* $[n,k,d]$-*code*. In this case the code C is an e-error correcting code with $e = \lfloor \frac{1}{2}(d-1) \rfloor$.

Given a two-intersection set $B(U)$ from a maximum scattered space U in $\mathrm{PG}(rt - 1, q)$, we can obtain a two-weight code as follows. We briefly sketch the construction and refer to Calderbank et al.[12] for further details.

Put $n = |\mathcal{B}(U)|$ and define the code C_U as the subspace generated by the columns of the $(n \times r)$-matrix M_U whose rows are the coordinates of the points of $\mathcal{B}(U)$ with respect to some fixed frame of $\mathrm{PG}(r-1, q^t)$. Then C_U has length n and dimension r (for this we use that U is maximum scattered). Since $\mathcal{B}(U)$ has intersection numbers m_1 and m_2, the code C_U is a two-weight code with weights $n - m_1$ and $n - m_2$. Hence we have the following theorem.

Theorem 15.[8] *If U is a scattered space of dimension $m = rt/2$ w.r.t. $\mathcal{D}_{r,t,q}$, then C_U is an $\mathbb{F}_q$-linear $[(q^m - 1)(q - 1), r]$-code with weights*

$$q^{m-t}\left(\frac{q^t - 1}{q - 1}\right) \quad and \quad q^{m-t+1}\left(\frac{q^{t-1} - 1}{q - 1}\right).$$

4.6. *Blocking sets*

Blocking sets have received a tremendous amount of attention in the past decades, and it is through research in this area that scattered spaces came into the spotlight. In particular, the results by Blokhuis et al. from [10] where it was shown that an s-fold blocking set in $\mathrm{PG}(2, q^4)$ of size $s(q^4+1)+c$, with s and c small enough, contains the union of s disjoint Baer subplanes, motivated the paper by Ball et al.[4] from 2000. In the latter paper the authors constructed a scattered linear set of rank 6, thus obtaining a $(q+1)$-fold blocking set of size $(q+1)(q^4 + q^2 + 1)$ in $\mathrm{PG}(2, q^4)$, and they proved that it is not the union of Baer subplanes (see also Theorem 14).

A more general result on scattered spaces and blocking sets is the following theorem from [8]. This shows that also scattered spaces which are not maximum generate blocking sets.

Theorem 16.[8]
A scattered subspace U of dimension m, with respect to a Desarguesian t-spread, in $\mathrm{V}(rt, q)$ induces a $(\theta_{k-1}(q))$-fold blocking set $\mathcal{B}(U)$, with respect to $(\frac{rt-m+k}{t} - 1)$-dimensional subspaces in $\mathrm{PG}(r - 1, q^t)$, of size $\theta_{m-1}(q)$, where $1 \leq k \leq m$ such that $t \mid (m - k)$.

4.7. *Embeddings of Segre varieties*

The *Segre variety* $S_{t,t}(q)$ is an algebraic variety in $\mathrm{PG}(t^2 - 1, q) \cong \mathrm{PG}(\mathbb{F}_q^t \otimes \mathbb{F}_q^t)$, whose points correspond to the fundamental tensors in $\mathbb{F}_q^t \otimes \mathbb{F}_q^t$. The geometry of points and lines lying on the Segre variety $S_{t,t}(q)$ is a *semilinear space* (also called a *product space* representing the product

$\mathrm{PG}(t-1,q) \times \mathrm{PG}(t-1,q))$. A *projective embedding* of a semilinear space is
an injective map into a projective space mapping lines into lines. So, $S_{t,t}(q)$
is a projective embedding of the product space $\mathrm{PG}(t-1,q) \times \mathrm{PG}(t-1,q)$
in $\mathrm{PG}(t^2-1,q)$. By [50], any embedded product space is an injective projection of a Segre variety. Since the image of any embedding of
$\mathrm{PG}(t-1,q) \times \mathrm{PG}(t-1,q)$ into a projective space $\mathrm{PG}(m,F)$ contains two
disjoint $(t-1)$-subspaces, it holds that $m \geq 2t-1$. Therefore, a projec-
tive embedding of $\mathrm{PG}(t-1,q) \times \mathrm{PG}(t-1,q)$ into $\mathrm{PG}(2t-1,F)$ is called a
minimum embedding. In [38] a construction is given of such a minimum em-
bedding using a maximum scattered subspace w.r.t. a Desarguesian spread.

Theorem 17.[38] *If U is a maximum scattered subspace w.r.t. $\mathcal{D}_{2,t,q}$, then
$\mathcal{B}(U) \subset \mathrm{PG}(2t-1,q)$ is a minimum embedding of the Segre variety $S_{t,t}(q)$.*

The smallest non-trivial example of a Segre variety $S_{t,t}(q)$ is the hyperbolic
quadric $\mathrm{PG}(3,q)$ for $t=2$. In [38], it is shown that there exists an embedding $\mathcal{B}(U)$ of $S_{t,t}(q)$ which is also a hypersurface of degree t in $\mathrm{PG}(2t-1,q)$,
extending the properties of the hyperbolic quadric in $\mathrm{PG}(3,q)$. By construction this embedding is covered by two systems of maximum subspaces (in
this case $(t-1)$-dimensional). However, unlike the Segre variety, it turns
out that the embedding $\mathcal{B}(U)$ contains t systems of maximum subspaces,
and hence for $t>2$, contrary to what one might expect, there exist systems
of maximum subspaces which are not the image of maximum subspaces of
the Segre variety, see [38, Theorem 6].

4.8. Pseudoreguli

The concept of a pseudoregulus is a generalisation of the concept of a regulus. If A, B, C are three distinct $(n-1)$-dimensional subspaces contained
in a common $(2n-1)$-space, then through each point of any of these three
subspaces there is exactly one line intersecting each of the spaces A, B
and C. Such a line is called a *transversal line* w.r.t. A, B, and C. If
each transversal line ℓ is given coordinates with respect to the frame $A \cap \ell$,
$B \cap \ell$, and $C \cap \ell$, then the points on all the transversals with the same
coordinates form an $(n-1)$-dimensional subspace. The set of these subspaces is called the *regulus $R(A,B,C)$ determined by A, B and C*, and the
transversal lines w.r.t. A, B and C are also called *transversal lines of the
regulus $R(A,B,C)$*. A regulus consisting of $(n-1)$-dimensional subspaces
is also called an $(n-1)$-*regulus*. Equivalently, the regulus $R(A,B,C)$ is

the family of maximal subspaces containing A, B, and C of the unique Segre variety $S_{2,n}(q)$ containing A, B and C. The transversal lines form the other family of maximal subspaces of $S_{2,n}(q)$. If $n = 2$ these become the two families of $q + 1$ lines lying on a hyperbolic quadric in $\mathrm{PG}(3, q)$. In 1980, Freeman constructed a set of $q^2 + 1$ lines in $\mathrm{PG}(3, q^2)$ which have exactly 2 transversal lines, called a *pseudoregulus*. The $q^2 + 1$ lines are the extended lines of a Desarguesian spread in a Baer subgeometry $\mathrm{PG}(3, q)$, and the transversal lines are the two conjugate lines defining the spread. This idea was extended by Marino et al. in [44], to a set of $q^3 + 1$ lines in $\mathrm{PG}(3, q^3)$ using a maximum scattered $\mathbb{F}_q$-linear set (of rank 6), obtaining the following.

Proposition 2.[44] *To any scattered $\mathbb{F}_q$-linear set L of rank 6 of $\mathrm{PG}(3, q^3)$ is associated an $\mathbb{F}_q$-pseudoregulus L consisting of all $(q^2 + q + 1)$-secant lines of L.*

Instead of a Baer subgeometry in $\mathrm{PG}(3, q^2)$ the authors of [44] considered a subgeometry $\mathrm{PG}(5, q)$ in $\mathrm{PG}(5, q^3)$ and a Desarguesian plane spread $\mathcal{D}_{4,3,q}$ of $\mathrm{PG}(5, q)$ together with the three conjugate lines ℓ, ℓ^ω and ℓ^{ω^2} defining the spread $\mathcal{D}_{4,3,q}$. Projecting the subgeometry from ℓ onto the 3-dimensional space $\langle \ell^\omega, \ell^{\omega^2} \rangle$ gives a scattered linear set of rank 6. The transversals to the associated $\mathbb{F}_q$-pseudoregulus are the lines ℓ^ω and ℓ^{ω^2}.

Remark 1. Note that in $\mathrm{PG}(3, q^3)$ the pseudoregulus can thus be reconstructed from the scattered linear set of rank 6. Simply take all the $(q^2 + q + 1)$-secants. This was not the case in the previous situation in $\mathrm{PG}(3, q^2)$, where the pseudoregulus cannot be reconstructed from the Baer subgeometry $\mathrm{PG}(3, q)$ (or equivalently a maximum scattered linear set of rank 4). In this case any Desarguesian spread of the Baer subgeometry gives a pseudoregulus.

These ideas were further developed in [40] in higher dimensional projective spaces, and the following theorem was proved.

Theorem 18.[40] *If L is a scattered $\mathbb{F}_q$-linear set of rank $3n$ in $\mathrm{PG}(2n - 1, q^3)$, $n \geq 2$, then a line of $\mathrm{PG}(2n - 1, q^3)$ meets L in $0, 1, q+1$ or $q^2 + q + 1$ points and every point of L lies on exactly one $(q^2 + q + 1)$-secant to L. Two different $(q^2 + q + 1)$-secants to L are disjoint and there exist exactly two $(n - 1)$-spaces, meeting each of the $(q^2 + q + 1)$-secants in a point.*

So in this case, the transversals are no longer lines, but $(n - 1)$-dimensional subspaces, and again the pseudoregulus is uniquely determined by the maximum scattered linear set. This leads to the following questions. Does every pseudoregulus (uniquely) determine a scattered subspace? And how do we recognise a pseudoregulus, given a set of mutually disjoint lines? The following theorem gives a geometric characterisation of a regulus and pseudoregulus in $\mathrm{PG}(3, q^3)$, giving a partial answer to the second question. In the following results, $\tilde{\mathcal{L}}$ denotes the set of points contained in the lines of the set $\mathcal{L}$.

Theorem 19.[40, Theorem 24] *Let $\mathcal{L}$ be a set of $q^3 + 1$ mutually disjoint lines in $\mathrm{PG}(3, q^3)$, $q > 2$. If each subline defined by three collinear points of $\tilde{\mathcal{L}}$ is contained in $\tilde{\mathcal{L}}$, then $\mathcal{L}$ is a regulus or a pseudoregulus.*

Also the first question was answered in [40]: it is possible to reconstruct the maximum scattered $\mathbb{F}_q$-linear set from an $\mathbb{F}_q$-pseudoregulus in $\mathrm{PG}(2n - 1, q^3)$, but this scattered linear set is not unique.

Theorem 20.[40] *Let $q > 2$, $n \geq 2$. Let $\mathcal{L}$ be a pseudoregulus in $\mathrm{PG}(2n - 1, q^3)$, let P be a point of $\tilde{\mathcal{L}}$, on the line ℓ of $\mathcal{L}$, not lying on one of the transversal spaces to $\mathcal{L}$. Let $T = \{\ell_1, \ell_2, \ldots\}$ be the set of $(q + 1)$-secants through P to $\tilde{\mathcal{L}}$, let $P(T)$ be the set of points on the lines of T in $\tilde{\mathcal{L}}$. Let π_i be the plane $\langle \ell, \ell_i \rangle$, and let D_i be the set of directions on ℓ, determined by the intersection of π_i with $\tilde{\mathcal{L}}$. Then $D_i = D_1$, for all i, and $P(T)$, together with the points of D_1, form a scattered $\mathbb{F}_q$-linear set of rank $3n$ determining the pseudoregulus $\mathcal{L}$.*

It follows from the proof of the above theorem in [40], that the scattered linear set is not uniquely determined by the pseudoregulus. In fact we have the following.

Corollary 2. *Let $q > 2$. If $\mathcal{L}$ is a pseudoregulus in $\mathrm{PG}(2n - 1, q^3)$, then there are $q - 1$ scattered $\mathbb{F}_q$-linear sets having $\mathcal{L}$ as associated pseudoregulus.*

The property that any maximum scattered $\mathbb{F}_q$-linear set in $\mathrm{PG}(2n - 1, q^t)$ gives rise to a pseudoregulus no longer holds for $t > 3$. This was further investigated in [43], introducing *maximum scattered linear sets of pseudoregulus type*. Also, all maximum scattered $\mathbb{F}_q$-linear sets in $\mathrm{PG}(2n - 1, q^t)$ are no longer (projectively) equivalent for $t > 3$. Indeed the following was proved in [43].

Theorem 21.[43] *For $n \geq 2$ and $t > 3$, (i) there exist maximum scattered $\mathbb{F}_q$-linear sets which are not of pseudoregulus type, and (ii) there are $\varphi(t)/2$*

orbits of maximum scattered $\mathbb{F}_q$*-linear sets of* $\mathrm{PG}(2n-1, q^t)$ *of pseudoregulus type under the action of the collineation group* $\mathrm{P\Gamma L}(2n, q^t)$.

In the statement of this theorem $\varphi(t)$ denotes Euler's totient function, i.e. the number of integers smaller than t and relatively prime to t. Also the maximum scattered $\mathbb{F}_q$-linear sets in the theorem are always of rank nt, and all maximum scattered $\mathbb{F}_q$-linear sets of pseudoregulus type are of the form $L_{\rho,f}$, see [43].

The authors of [43] also introduced maximum scattered $\mathbb{F}_q$-linear sets of pseudoregulus type on a projective line, and these were further investigated in [15]. Of course there is no longer a pseudoregulus associated to such a linear set, but the definition is stimulated from the algebraic formula for maximum scattered $\mathbb{F}_q$-linear sets $L_{\rho,f}$ of pseudoregulus type in higher dimensions.

4.9. Semifield theory

The term "semifield" is a short term introduced by Knuth in 1965 [29] for a non-associative division algebra. The algebraic structures themselves were first studied by Dickson in 1906 [20]. We restrict ourselves to the finite case. A *finite semifield* $\mathbb{S}$ is an algebra with at least two elements, and two binary operations $+$ and $\circ$, satisfying the following axioms.

(S1) $(\mathbb{S}, +)$ is a group with identity element 0.
(S2) $x \circ (y + z) = x \circ y + x \circ z$ and $(x + y) \circ z = x \circ z + y \circ z$, for all $x, y, z \in \mathbb{S}$.
(S3) $x \circ y = 0$ implies $x = 0$ or $y = 0$.
(S4) $\exists 1 \in \mathbb{S}$ such that $1 \circ x = x \circ 1 = x$, for all $x \in \mathbb{S}$.

An algebra satisfying all of the axioms of a semifield except (S4) is called a *pre-semifield*. An important example of a finite pre-semifield is the Generalized Twisted Field (GTF) due to Albert [3], obtained by defining the multiplication $x \circ y = xy - cx^\alpha y^\beta$ on the finite field $\mathbb{F}_{q^n}$, where $\alpha, \beta \in \mathrm{Aut}(\mathbb{F}_{q^n})$ with $Fix(\alpha) = Fix(\beta) = \mathbb{F}_q$, and $c \in \mathbb{F}_{q^n}$ with $N_{\mathbb{F}_{q^n}/\mathbb{F}_q}(c) \neq 1$.

Finite semifields have been studied intensively by a variety of mathematicians and they have many connections with structures in Galois Geometry. We refer to [29],[28],[33] and to the chapter [36] and its references for the necessary background, an overview of the results, connections, and further reading. A semifield $\mathbb{S}$ defines a projective plane $\pi(\mathbb{S})$, called a *semifield plane*, and isomorphism classes of semifield planes correspond to *isotopism classes* of semifields, a result from Albert [2]. We restrict ourselves here to what we call scattered semifields, which we define as follows.

Suppose $\mathbb{S}$ is a semifield which is l-dimensional over its left nucleus and ls-dimensional over its center $\mathbb{F}_q$. Consider the set $R(\mathbb{S})$ of endomorphisms of $\mathbb{F}_{q^s}^l$ corresponding to right multiplication in the semifield $\mathbb{S}$. This set is called a *semifield spread set*, and $R(\mathbb{S})$ is an $\mathbb{F}_q$-linear set of rank ls in the projective space $\mathrm{PG}(l^2 - 1, q^s)$. Now if $R(\mathbb{S})$ is a scattered $\mathbb{F}_q$-linear set, then $\mathbb{S}$ is called a *scattered semifield*. It is known that $R(\mathbb{S})$ is disjoint from the $(l-2)$–th secant variety $\Omega(S_{l,l}(q^s))$ of a Segre variety $S_{l,l}(q^s)$ and this connection also gives a nice interpretation of the isotopism classes. [3]

Theorem 22 (from [32]). *The isotopism class of a semifield $\mathbb{S}$ corresponds to the orbit of $R(\mathbb{S})$ under the action of the group $\mathcal{H} \leq \mathrm{PGL}(l^2, q^s)$ preserving the two systems of maximal subspaces contained in the Segre variety $S_{l,l}(q^s)$ in $\mathrm{PG}(l^2 - 1, q^s)$.*

It follows from Theorem 22 that being "scattered" is an isotopism invariant, and this makes it a useful tool to investigate the isotopism problem, which is usually a very hard problem: given two semifields $\mathbb{S}_1$ and $\mathbb{S}_2$, decide whether they are isotopic or not.

The power of this geometric approach is illustrated in [13] ($l = s = 2$), and [44] ($l = 2$, $s = 3$), and by the recent results from [34] and [35], where the structure of $R(\mathbb{S})$ is used to solve the isotopism problem regarding the semifields constructed by Dempwolff in [18].

4.10. *Splashes of subgeometries*

Given a subgeometry π_0 and a line l_∞ in a projective space π, by extending the hyperplanes of π_0 to hyperplanes of π and intersecting these with the line l_∞, one obtains a set of points on the projective line l_∞. Precisely, if we denote the set of hyperplanes of a projective space π by $\mathcal{H}(\pi)$, and $\overline{U}$ denotes the extension of a subspace U of the subgeometry π_0 to a subspace of π, then we obtain the set of points $\{l_\infty \cap \overline{H} \ : \ H \in \mathcal{H}(\pi_0)\}$. These sets have been studied in [42] generalising the initial studies in [7] where the *splash of π_0 on l_∞* was introduced for Desarguesian planes and cubic extensions, i.e. for a subplane $\pi_0 \cong \mathrm{PG}(2,q)$ in $\pi \cong \mathrm{PG}(2,q^3)$. If l_∞ is tangent to (respectively, disjoint from) π_0, then a splash is called the *tangent splash* (respectively, *external* or *exterior splash*) of π_0 on l_∞. Note that when l_∞ is secant to π_0, the splash of π_0 on l_∞ is just a subline.

---

[3] A relation between the isotopism classes of semifields and equivalence classes of *embeddings* of Segre varieties can be found in [37].

One of the main results of [42] shows the equivalence between splashes and linear sets on a projective line.

Theorem 23. *Let $r, n > 1$. If $S = S(\pi_0, l_\infty)$ is the splash of the q-subgeometry π_0 of $\mathrm{PG}(r - 1, q^n)$ on the line l_∞, then S is an $\mathbb{F}_q$-linear set of rank r. Conversely, if S is an $\mathbb{F}_q$-linear set of rank r on the line $l_\infty \cong \mathrm{PG}(1, q^n)$, then there exists an embedding of l_∞ in $\mathrm{PG}(r - 1, q^n)$ and a q-subgeometry π_0 of $\mathrm{PG}(r - 1, q^n)$ such that $S = S(\pi_0, l_\infty)$.*

Also, it is shown that the number of hyperplanes through a point determines the weight of that point in the linear set. This leads to the following characterisation of scattered linear sets.

Theorem 24. *Let S be the splash of a subgeometry $\pi_0 \cong \mathrm{PG}(r - 1, q)$ of $\pi \cong \mathrm{PG}(r - 1, q^n)$ on $l_\infty \cong \mathrm{PG}(1, q^n)$. Then S is a scattered linear set if and only if S is an external splash, where every point of S is on exactly one hyperplane of π_0.*

4.11. *MRD-codes*

A *rank metric code* is a set of $(m \times n)$-matrices over some field where the distance between two codewords is defined as the matrix rank of their difference. A *maximum rank distance code* (*MRD* code) is a rank metric code of maximum size with respect to its distance. Precisely, if $C \subset M_{m \times n}(\mathbb{F}_q)$, with minimum distance d, then C is an MRD code if and only if $|C| = q^{nk}$ with $k = m - d + 1$. Such codes were constructed by Delsarte [16] and Gabidulin [25] for every m, n, and d.

Recently, Sheekey constructed new families of MRD codes in [49]. They generalise the previously known Gabidulin codes. These codes are called *twisted Gabidulin codes* because of their analogy with the generalized twisted fields constructed by Albert (see Section 4.9).

In the same paper [49], the author gives an interesting connection between maximum scattered linear sets on a projective line $\mathrm{PG}(1, q^t)$, and MRD-codes of dimension $2t$ over $\mathbb{F}_q$ and minimum distance $t - 1$. If U is a t-dimensional subspace of $\mathbb{F}_q^{2t}$, then U can be represented by $\{(x, f(x)) : x \in \mathbb{F}_{q^t}\}$ for some $\mathbb{F}_q$-linear polynomial $f(Y) \in \mathbb{F}_q[Y]$. Now U is scattered w.r.t. the Desarguesian spread $D_{2,t,q}$ if for each $a \in \mathbb{F}_{q^t}^*$, the dimension of the intersection of U with $D_a = \{(x, ax) : x \in \mathbb{F}_{q^t}\}$ has dimension at most 1. This is equivalent with the condition that the rank of the $\mathbb{F}_q$-linear map $x \mapsto ax + bf(x)$ is at least $t - 1$, for all $a, b \in \mathbb{F}_{q^t}$,

$(a,b) \neq (0,0)$. Then $C_U = \{\varphi(ax + bf(x)) \; : \; a,b \in \mathbb{F}_{q^t}, (a,b) \neq (0,0)\}$ is an $\mathbb{F}_q$-subspace of the vector space of $(t \times t)$-matrices, where φ is an isomorphism associating a $(t \times t)$-matrix M_g to each linearized polynomial $g(x)$ w.r.t. some fixed basis.

Theorem 25.[49] *If U is a maximum scattered space w.r.t. $\mathcal{D}_{2,t,q}$, then C_U is an $\mathbb{F}_q$-linear MRD-code of dimension $2t$ and minimum distance $t-1$, and conversely.*

This one-to-one correspondence is particularly nice since it preserves equivalence.

Theorem 26.[49] *Two scattered $\mathbb{F}_q$-linear sets $\mathcal{B}(U)$ and $\mathcal{B}(U')$ of rank t are equivalent in $\mathrm{PG}(1,q^t)$ if and only if C_U and $C_{U'}$ are equivalent MRD-codes.*

We refer to [49] for examples and further details.

Acknowledgment The author thanks the members of the Algebra group of Sabancı University for their hospitality and the anonymous referees for their comments and suggestions.

References

[1] André, J. Über nicht-Desarguessche Ebenen mit transitiver Translationsgruppe. (German) Math. Z. 60 (1954). 156–186.

[2] Albert, A. A. Finite division algebras and finite planes. 1960 Proc. Sympos. Appl. Math., Vol. 10 pp. 53–70 American Mathematical Society, Providence, R.I.

[3] Albert, A. A. Generalized twisted fields. Pacific J. Math. 11 (1961) 1–8.

[4] Ball, S.; Blokhuis, A.; Lavrauw, M. Linear $(q+1)$-fold blocking sets in $\mathrm{PG}(2,q^4)$. Finite Fields Appl. 6 (2000), no. 4, 294–301.

[5] Barlotti, A.; Cofman, J. Finite Sperner spaces constructed from projective and affine spaces. Abh. Math. Sem. Univ. Hamburg 40 (1974), 231–241.

[6] Bartoli, D.; Giulietti, M.; Marino, G.; Polverino, O. Maximum scattered linear sets and complete caps in Galois spaces. Preprint arXiv:1512.07467v1.

[7] Barwick, S. G.; Jackson, W. An investigation of the tangent splash of a subplane of $\mathrm{PG}(2,q^3)$. Des. Codes Cryptogr. 76 (2015), no. 3, 451–468.

[8] Blokhuis, A.; Lavrauw, M. Scattered spaces with respect to a spread in $\mathrm{PG}(n,q)$. Geom. Dedicata 81 (2000), no. 1-3, 231–243.

[9] Blokhuis, A.; Lavrauw, M. On two-intersection sets with respect to hyperplanes in projective spaces. J. Combin. Theory Ser. A 99 (2002), no. 2, 377–382.

[10] Blokhuis, A.; Storme, L.; Szőnyi, T. Lacunary polynomials, multiple blocking sets and Baer subplanes. J. London Math. Soc. (2) 60 (1999), no. 2, 321–332.

[11] Bruck R.H.; Bose R.C. The construction of translation planes from projective spaces. J. Algebra 1 (1964) 85–102.

[12] Calderbank, R.; Kantor, W. M. The geometry of two-weight codes. Bull. London Math. Soc. 18 (1986), no. 2, 97–122.

[13] Cardinali, I.; Polverino, O.; Trombetti, R. Semifield planes of order q^4 with kernel $\mathbb{F}_{q^2}$ and center $\mathbb{F}_q$. European J. Combin. 27 (2006), no. 6, 940–961.

[14] Csajbók, B.; Zanella C. On the equivalence of linear sets. arXiv:1501.03441

[15] Csajbók, B.; Zanella C. On scattered linear sets of pseudoregulus type in $PG(1, q^t)$. arXiv:1506.08875

[16] Delsarte, Ph. Bilinear forms over a finite field, with applications to coding theory. J. Combin. Theory Ser. A 25 (1978), no. 3, 226–241.

[17] Dembowski, P. Finite geometries. Ergebnisse der Mathematik und ihrer Grenzgebiete, Band 44 Springer-Verlag, Berlin-New York 1968 xi+375 pp.

[18] Dempwolff, U. More translation planes and semifields from Dembowski-Ostrom polynomials, Des. Codes Cryptogr. 68(1–3) (2013), 81–103.

[19] Denniston, R. H. F. Some non-Desarguesian translation ovals. Ars Combin. 7 (1979), 221–222.

[20] Dickson, L. E. Linear algebras in which division is always uniquely possible. Trans. Amer. Math. Soc. 7 (1906), no. 3, 370–390.

[21] Donati, G.; Durante, N. On the intersection of two subgeometries of $PG(n, q)$, Des. Codes Cryptogr., vol. 46, no. 3 (2008), 261–267.

[22] Donati, G.; Durante, N. Scattered linear sets generated by collineations between pencils of lines, J. Algebr. Comb., vol. 40, no. 4 (2014), 1121–1134.

[23] FinInG – a GAP package - *Finite Incidence Geometry*, version 1.0, 2014. Bamberg, J.; Betten, A.; Cara, Ph.; De Beule, J.; Lavrauw, M. and Neunhöffer, M.

[24] The GAP Group, GAP - Groups, Algorithms, and Programming, Version 4.7.8; 2015 (http://www.gap-system.org)

[25] Gabidulin, É. M. Theory of codes with maximum rank distance. (Russian) Problemy Peredachi Informatsii 21 (1985), no. 1, 3–16.

[26] Hughes, D. R.; Piper, F. C. Projective planes. Graduate Texts in Mathematics, Vol. 6. Springer-Verlag, New York-Berlin, 1973. x+291 pp.

[27] Jha, V.; Johnson, N. L. On the ubiquity of Denniston-type translation ovals in generalized André planes. Combinatorics '90 (Gaeta, 1990), 279–296, Ann. Discrete Math., 52, North-Holland, Amsterdam, 1992.

[28] Kantor W.M. Finite semifields. In: Hulpke, A., Liebler, R., Penttila, T., Seress, À, (eds.) Finite Geometries, Groups, and Computation, pp. 103–114. Walter de Gruyter GmbH & Co. KG, Berlin (2006).

[29] Knuth D.E. Finite semifields and projective planes. J. Algebra 2 (1965), 182–217.

[30] Korchmáros, G. Inherited arcs in finite affine planes. J. Combin. Theory Ser. A 42 (1986), no. 1, 140–143.

[31] Lavrauw, M. Scattered spaces with respect to spreads, and eggs in

216 *Lavrauw*

finite projective spaces. Dissertation, Technische Universiteit Eindhoven, Eindhoven, 2001. Eindhoven University of Technology, Eindhoven, 2001. viii+115 pp.

[32] Lavrauw, M. Finite semifields with a large nucleus and higher secant varieties to Segre varieties. Adv. Geom. 11 (2011), no. 3, 399–410.

[33] Lavrauw, M. Finite semifields and nonsingular tensors. Des. Codes Cryptogr. 68 (2013), no. 1-3, 205–227.

[34] Lavrauw, M.; Marino, G.; Polverino, O.; Trombetti, R. Solution to an isotopism question concerning rank 2 semifields. J. Combin. Des. 23 (2015), no. 2, 60–77.

[35] Lavrauw, M.; Marino, G.; Polverino, O.; Trombetti, R. The isotopism problem of a class of 6-dimensional rank 2 semifields and its solution. Finite Fields Appl. 34 (2015), 250–264.

[36] Lavrauw, M.; Polverino, O. Finite semifields. Chapter in *Current Research Topics in Galois Geometry (Editors J. De Beule and L. Storme)* Nova Academic Publishers (2011), pp. 131-160.

[37] Lavrauw, M.; Zanella, C. Segre embeddings and finite semifields. Finite Fields Appl. 25 (2014), 8–18.

[38] Lavrauw, M.; Sheekey, J.; Zanella, C. On embeddings of minimum dimension of $PG(n,q) \times PG(n,q)$. Des. Codes Cryptogr. 74 (2015), no. 2, 427–440.

[39] Lavrauw, M.; Van de Voorde, G. On linear sets on a projective line. *Des. Codes Cryptogr.* 56 (2-3) (2010), 89–104.

[40] Lavrauw, M.; Van de Voorde, G. Scattered linear sets and pseudoreguli. Electron. J. Combin. 20 (2013), no. 1, Paper 15, 14 pp.

[41] Lavrauw, M.; Van de Voorde, G. Field reduction and linear sets in finite geometry. Topics in finite fields, 271–293, Contemp. Math., 632, Amer. Math. Soc., Providence, RI, 2015.

[42] Lavrauw, M.; Zanella, C. Subgeometries and linear sets on a projective line. Finite Fields Appl. 34 (2015), 95–106.

[43] Lunardon, G.; Marino, G.; Polverino, O.; Trombetti, R. Maximum scattered linear sets of pseudoregulus type and the Segre variety $S_{n,n}$. J. Algebraic Combin. 39 (2014), no. 4, 807–831.

[44] Marino, G.; Polverino, O.; Trombetti, R. On F_q-linear sets of $PG(3,q^3)$ and semifields. J. Combin. Theory Ser. A 114 (2007), no. 5, 769–788.

[45] Payne, S. E. A complete determination of translation ovoids in finite Desarguesian planes, Atti Acad. Naz. Lincei Rend., 51 (1971), 328–331.

[46] Pepe, V. On the algebraic variety $V_{r,t}$. Finite Fields Appl. 17 (2011), no. 4, 343–349.

[47] Polverino, O. Linear sets in finite projective spaces. Discrete Math. 310 (2010), no. 22, 3096–3107.

[48] Segre, B. Teoria di Galois, fibrazioni proiettive e geometrie non desarguesiane. (Italian) Ann. Mat. Pura Appl. (4) 64 (1964), 1–76.

[49] Sheekey, J. A new family of linear maximum rank distance codes, arXiv:1504.01581.

[50] Zanella, C. Universal properties of the Corrado Segre embedding. Bull. Belg. Math. Soc. Simon Stevin 3 (1996), no. 1, 65–79.

On the value set of small families of polynomials over a finite field, III

Guillermo Matera

Universidad Nacional de General Sarmiento and CONICET, Buenos Aires, Argentina gmatera@ungs.edu.ar

Mariana Pérez and Melina Privitelli

Universidad Nacional de General Sarmiento, Buenos Aires, Argentina {vperez,mprivite}@ungs.edu.ar

1. Introduction

Let $\mathbb{F}_q$ be the finite field of $q := p^s$ elements, where p is a prime number, let $\overline{\mathbb{F}}_q$ denote its algebraic closure, and let T be an indeterminate over $\overline{\mathbb{F}}_q$. For $f \in \mathbb{F}_q[T]$, its value set is the image of the mapping from $\mathbb{F}_q$ to $\mathbb{F}_q$ defined by f (cf.[20]). We shall denote its cardinality by $\mathcal{V}(f)$, namely $\mathcal{V}(f) := |\{f(c) : c \in \mathbb{F}_q\}|$.

In a seminal paper, Birch and Swinnerton–Dyer [1] showed that, for fixed $d \geq 1$, if $f \in \mathbb{F}_q[T]$ is a "general" polynomial of degree d, then
$$\mathcal{V}(f) = \mu_d\, q + \mathcal{O}(q^{\frac{1}{2}}), \tag{1}$$
where $\mu_d := \sum_{r=1}^{d} (-1)^{r-1}/r!$ and the $\mathcal{O}$–constant depends only on d.

Uchiyama [25] and Cohen ([10],[9]) were concerned with estimates for the average cardinality of the value set when f ranges over all monic polynomials of degree d in $\mathbb{F}_q[T]$. In particular, in [9] the problem of estimating the average cardinality of the value set on linear families of monic polynomials of $\mathbb{F}_q[T]$ of degree d is addressed. More precisely, it is shown that, for a linear family $\mathcal{A}$ of codimension $m \leq d - 2$ satisfying certain conditions,
$$\mathcal{V}(\mathcal{A}) = \mu_d\, q + \mathcal{O}(q^{\frac{1}{2}}), \tag{2}$$
where $\mathcal{V}(\mathcal{A})$ denotes the average cardinality of the value set of the elements in $\mathcal{A}$. As a particular case we have the classical case of polynomials with prescribed coefficients, where simpler conditions are obtained.

A difficulty with (2) is that the hypotheses on the linear family $\mathcal{A}$ seem complicated and not easy to verify. A second concern is that (2) imposes the restriction $p > d$, which inhibits its application to fields of small characteristic. For these reasons, in [8] and [21] we obtained explicit estimates for any family of monic polynomials of $\mathbb{F}_q[T]$ of degree d with certain consecutive coefficients prescribed, which are valid for $p > 2$. In this paper we develop a framework which allows us to significantly generalize these results to rather general (eventually nonlinear) families of monic polynomials of $\mathbb{F}_q[T]$ of degree d.

More precisely, let d, m be positive integers with $q > d \geq m + 2$, and let $A_{d-1}, \ldots, A_0$ be indeterminates over $\overline{\mathbb{F}}_q$. Let $G_1, \ldots, G_m \in \mathbb{F}_q[A_{d-1}, \ldots, A_0]$ be polynomials of degree $d_1, \ldots, d_m$ and $\mathcal{A} := \mathcal{A}(G_1, \ldots, G_m)$ the family

$$\mathcal{A} := \left\{ T^d + \sum_{j=0}^{d-1} a_j T^j \in \mathbb{F}_q[T] : G_i(a_{d-1}, \ldots, a_0) = 0 \ (1 \leq i \leq m) \right\}. \quad (3)$$

Denote by $\mathcal{V}(\mathcal{A})$ the average value of $\mathcal{V}(f)$ for f ranging in $\mathcal{A}$, that is,

$$\mathcal{V}(\mathcal{A}) := \frac{1}{|\mathcal{A}|} \sum_{f \in \mathcal{A}} \mathcal{V}(f). \quad (4)$$

Our main result establishes rather general conditions on $G_1, \ldots, G_m$ under which the asymptotic behavior of $\mathcal{V}(\mathcal{A})$ agrees with that of the general case, as predicted in (1) and (2). More precisely, we prove that

$$|\mathcal{V}(\mathcal{A}) - \mu_d\, q| \leq 2^d \delta (3D + d^2) q^{\frac{1}{2}} + 67 \delta^2 (D + 2)^2\, d^{d+5} e^{2\sqrt{d}-d},$$

where $\delta := \prod_{i=1}^{m} d_i$ and $D := \sum_{i=1}^{m} (d_i - 1)$.

Our approach relies on tools of algebraic geometry in the same vein as [8] and [21]. In Section 2 we recall the basic notions and notations of algebraic geometry we use. In Section 3 we provide a combinatorial expression for $\mathcal{V}(\mathcal{A})$ in terms of certain numbers $\mathcal{S}_r^{\mathcal{A}}$ of "interpolating sets" with $1 \leq r \leq d$ and we relate each $\mathcal{S}_r^{\mathcal{A}}$ to the number of $\mathbb{F}_q$–rational points of an incidence variety Γ_r^* of $\overline{\mathbb{F}}_q^{d+r}$. In Section 4 we show that Γ_r^* is an $\mathbb{F}_q$–definable normal complete intersection, and establish a number of geometric properties of Γ_r^*. To estimate the number of $\mathbb{F}_q$–rational points of Γ_r^* is necessary to discuss the behavior of Γ_r^* at "infinity", which is done in Section 5. Finally, the results of Sections 4 and 5 allow us to estimate, in Section 6, the number of $\mathbb{F}_q$–rational points of Γ_r^*, and therefore determine the asymptotic behavior of $\mathcal{V}(\mathcal{A})$. Linear and nonlinear families of polynomials satisfying our requirements are exhibited.

2. Basic notions of algebraic geometry

In this section we collect the basic definitions and facts of algebraic geometry that we need in the sequel. We use standard notions and notations which can be found in, e.g.,[18],[24].

Let $\mathbb{K}$ be any of the fields $\mathbb{F}_q$ or $\overline{\mathbb{F}}_q$. We denote by $\mathbb{A}^n$ the affine n–dimensional space $\overline{\mathbb{F}}_q^n$ and by $\mathbb{P}^n$ the projective n–dimensional space over $\overline{\mathbb{F}}_q^{n+1}$. Both spaces are endowed with their respective Zariski topologies over $\mathbb{K}$, for which a closed set is the zero locus of a set of polynomials of $\mathbb{K}[X_1, \ldots, X_n]$, or of a set of homogeneous polynomials of $\mathbb{K}[X_0, \ldots, X_n]$.

A subset $V \subset \mathbb{P}^n$ is a *projective variety defined over* $\mathbb{K}$ (or a projective $\mathbb{K}$–variety for short) if it is the set of common zeros in $\mathbb{P}^n$ of homogeneous polynomials $F_1, \ldots, F_m \in \mathbb{K}[X_0, \ldots, X_n]$. Correspondingly, an *affine variety of* $\mathbb{A}^n$ *defined over* $\mathbb{K}$ (or an affine $\mathbb{K}$–variety) is the set of common zeros in $\mathbb{A}^n$ of polynomials $F_1, \ldots, F_m \in \mathbb{K}[X_1, \ldots, X_n]$. We think a projective or affine $\mathbb{K}$–variety to be equipped with the induced Zariski topology. We shall denote by $\{F_1 = 0, \ldots, F_m = 0\}$ or $V(F_1, \ldots, F_m)$ the affine or projective $\mathbb{K}$–variety consisting of the common zeros of $F_1, \ldots, F_m$.

In the remaining part of this section, unless otherwise stated, all results referring to varieties in general should be understood as valid for both projective and affine varieties.

A $\mathbb{K}$–variety V is *irreducible* if it cannot be expressed as a finite union of proper $\mathbb{K}$–subvarieties of V. Further, V is *absolutely irreducible* if it is $\overline{\mathbb{F}}_q$–irreducible as a $\overline{\mathbb{F}}_q$–variety. Any $\mathbb{K}$–variety V can be expressed as an irredundant union $V = \mathcal{C}_1 \cup \cdots \cup \mathcal{C}_s$ of irreducible (absolutely irreducible) $\mathbb{K}$–varieties, unique up to reordering, which are called the *irreducible* (*absolutely irreducible*) $\mathbb{K}$–*components* of V.

For a $\mathbb{K}$–variety V contained in $\mathbb{P}^n$ or $\mathbb{A}^n$, we denote by $I(V)$ its *defining ideal*, namely the set of polynomials of $\mathbb{K}[X_0, \ldots, X_n]$, or of $\mathbb{K}[X_1, \ldots, X_n]$, vanishing on V. The *coordinate ring* $\mathbb{K}[V]$ of V is defined as the quotient ring $\mathbb{K}[X_0, \ldots, X_n]/I(V)$ or $\mathbb{K}[X_1, \ldots, X_n]/I(V)$. The *dimension* $\dim V$ of V is the length r of a longest chain $V_0 \subsetneq V_1 \subsetneq \cdots \subsetneq V_r$ of nonempty irreducible $\mathbb{K}$–varieties contained in V. We say that V has *pure dimension* r if every irreducible $\mathbb{K}$–component of V has dimension r. A $\mathbb{K}$–variety of $\mathbb{P}^n$ or $\mathbb{A}^n$ of pure dimension $n - 1$ is called a $\mathbb{K}$–*hypersurface*. It turns out that a $\mathbb{K}$–hypersurface of $\mathbb{P}^n$ (or $\mathbb{A}^n$) is the set of zeros of a single nonzero polynomial of $\mathbb{K}[X_0, \ldots, X_n]$ (or of $\mathbb{K}[X_1, \ldots, X_n]$).

The *degree* $\deg V$ of an irreducible $\mathbb{K}$–variety V is the maximum number of points lying in the intersection of V with a linear space L of codimension

$\dim V$, for which $V \cap L$ is a finite set. More generally, following [17] (see also [14]), if $V = \mathcal{C}_1 \cup \cdots \cup \mathcal{C}_s$ is the decomposition of V into irreducible $\mathbb{K}$–components, we define the degree of V as

$$\deg V := \sum_{i=1}^{s} \deg \mathcal{C}_i.$$

The degree of a $\mathbb{K}$–hypersurface V is the degree of a polynomial of minimal degree defining V. For our estimates we shall use the following *Bézout inequality* (see [17],[14],[26]): if V and W are $\mathbb{K}$–varieties of the same ambient space, then the following inequality holds:

$$\deg(V \cap W) \leq \deg V \cdot \deg W. \tag{5}$$

Let $V \subset \mathbb{A}^n$ be a $\mathbb{K}$–variety and $I(V) \subset \mathbb{K}[X_1, \ldots, X_n]$ its defining ideal. Let x be a point of V. The *dimension* $\dim_x V$ *of* V *at* x is the maximum of the dimensions of the irreducible $\mathbb{K}$–components of V that contain x. If $I(V) = (F_1, \ldots, F_m)$, the *tangent space* $\mathcal{T}_x V$ to V at x is the kernel of the Jacobian matrix $(\partial F_i / \partial X_j)_{1 \leq i \leq m, 1 \leq j \leq n}(x)$ of $F_1, \ldots, F_m$ with respect to $X_1, \ldots, X_n$ at x. We have the inequality $\dim \mathcal{T}_x V \geq \dim_x V$ (see, e.g.,[24, page 94]). The point x is *regular* if $\dim \mathcal{T}_x V = \dim_x V$. Otherwise, the point x is called *singular*. The set of singular points of V is the *singular locus* $\mathrm{Sing}(V)$ of V; it is a closed $\mathbb{K}$–subvariety of V. A variety is called *nonsingular* if its singular locus is empty. For a projective variety, the concepts of tangent space, regular and singular point can be defined by considering an affine neighborhood of the point under consideration.

Let V and W be irreducible affine $\mathbb{K}$–varieties of the same dimension and let $f : V \to W$ be a regular map for which $\overline{f(V)} = W$ holds, where $\overline{f(V)}$ denotes the closure of $f(V)$ with respect to the Zariski topology of W. Such a map is called *dominant*. Then f induces a ring extension $\mathbb{K}[W] \hookrightarrow \mathbb{K}[V]$ by composition with f. We say that the dominant map f is *finite* if this extension is integral, namely each element $\eta \in \mathbb{K}[V]$ satisfies a monic equation with coefficients in $\mathbb{K}[W]$. A basic fact is that a dominant finite morphism is necessarily closed. Another fact we shall use is that the preimage $f^{-1}(S)$ of an irreducible closed subset $S \subset W$ under a dominant finite morphism f is of pure dimension $\dim S$ (see, e.g.,[11, §4.2, Proposition]).

2.1. Rational points

Let $\mathbb{P}^n(\mathbb{F}_q)$ be the n–dimensional projective space over $\mathbb{F}_q$ and let $\mathbb{A}^n(\mathbb{F}_q)$ be the n–dimensional $\mathbb{F}_q$–vector space $\mathbb{F}_q^n$. For a projective variety $V \subset \mathbb{P}^n$ or an affine variety $V \subset \mathbb{A}^n$, we denote by $V(\mathbb{F}_q)$ the set of $\mathbb{F}_q$–rational points of V, namely $V(\mathbb{F}_q) := V \cap \mathbb{P}^n(\mathbb{F}_q)$ in the projective case and $V(\mathbb{F}_q) := V \cap \mathbb{A}^n(\mathbb{F}_q)$ in the affine case. For an affine variety V of dimension r and degree δ, we have the following bound (see, e.g.,[2, Lemma 2.1]):

$$|V(\mathbb{F}_q)| \le \delta q^r. \tag{6}$$

On the other hand, if V is a projective variety of dimension r and degree δ, we have the following bound (see [15, Proposition 12.1] or [3, Proposition 3.1]; see [19] for more precise upper bounds):

$$|V(\mathbb{F}_q)| \le \delta\, p_r, \tag{7}$$

where $p_r := q^r + q^{r-1} + \cdots + q + 1 = |\mathbb{P}^r(\mathbb{F}_q)|$.

2.2. *Complete intersections*

Elements $F_1, \ldots, F_{n-r}$ in $\mathbb{K}[X_1, \ldots, X_n]$ or $\mathbb{K}[X_0, \ldots, X_n]$ form a *regular sequence* if F_1 is nonzero and no F_i is zero or a zero divisor in the quotient ring $\mathbb{K}[X_1, \ldots, X_n]/(F_1, \ldots, F_{i-1})$ or $\mathbb{K}[X_0, \ldots, X_n]/(F_1, \ldots, F_{i-1})$ for $2 \le i \le n - r$. In that case, the (affine or projective) $\mathbb{K}$–variety $V := V(F_1, \ldots, F_{n-r})$ is called a *set–theoretic complete intersection*. We remark that V is necessarily of pure dimension r. Furthermore, V is called an *ideal–theoretic complete intersection* if its ideal $I(V)$ over $\mathbb{K}$ can be generated by $n - r$ polynomials.

If $V \subset \mathbb{P}^n$ is an ideal–theoretic complete intersection defined over $\mathbb{K}$ of dimension r, and $F_1, \ldots, F_{n-r}$ is a system of homogeneous generators of $I(V)$, the degrees $d_1, \ldots, d_{n-r}$ depend only on V and not on the system of generators. Arranging the d_i in such a way that $d_1 \ge d_2 \ge \cdots \ge d_{n-r}$, we call $(d_1, \ldots, d_{n-r})$ the *multidegree* of V. In this case, a stronger version of (5) holds, called the *Bézout theorem* (see, e.g.,[16, Theorem 18.3]):

$$\deg V = d_1 \cdots d_{n-r}.$$

In what follows we shall deal with a particular class of complete intersections, which we now define. A complete intersection V is called *normal* if it is *regular in codimension 1*, that is, the singular locus $\mathrm{Sing}(V)$ of V has codimension at least 2 in V, namely $\dim V - \dim \mathrm{Sing}(V) \ge 2$ (actually, normality is a general notion that agrees on complete intersections with the one we define here). A fundamental result for projective complete intersections is the Hartshorne connectedness theorem (see, e.g.,[18, Theorem VI.4.2]): if $V \subset \mathbb{P}^n$ is a complete intersection defined over $\mathbb{K}$ and $W \subset V$ is any $\mathbb{K}$–subvariety of codimension at least 2, then $V \setminus W$ is connected in the Zariski topology of $\mathbb{P}^n$ over $\mathbb{K}$. Applying the Hartshorne connectedness theorem with $W := \mathrm{Sing}(V)$, one deduces the following result.

Theorem 1. *If $V \subset \mathbb{P}^n$ is a normal complete intersection, then V is absolutely irreducible.*

3. A geometric approach to estimate value sets

Let m and d be positive integers with $q > d \geq m + 2$, and let $\mathcal{A}$ be the family of (3). We may assume without loss of generality that $G_1, \ldots, G_m$ are elements of $\mathbb{F}_q[A_{d-1}, \ldots, A_1]$. Indeed, let $\Pi : \mathcal{A} \to \mathbb{F}_q$ be the mapping $\Pi(T^d + a_{d-1}T^{d-1} + \cdots + a_0) := a_0$. Denote $\mathcal{A}_{a_0} := \Pi^{-1}(a_0)$. We have

$$\frac{1}{|\mathcal{A}|} \sum_{f \in \mathcal{A}} \mathcal{V}(f) - \mu_d q = \frac{1}{\sum_{a_0 \in \mathbb{F}_q} |\mathcal{A}_{a_0}|} \sum_{a_0 \in \mathbb{F}_q} |\mathcal{A}_{a_0}| \left(\frac{1}{|\mathcal{A}_{a_0}|} \sum_{f \in \mathcal{A}_{a_0}} \mathcal{V}(f) - \mu_d q \right).$$

As a consequence, if there exists a constant $E(d_1, \ldots, d_m, d)$ such that

$$\left| \frac{1}{|\mathcal{A}_{a_0}|} \sum_{f \in \mathcal{A}_{a_0}} \mathcal{V}(f) - \mu_d q \right| \leq E(d_1, \ldots, d_m, d) q^{\frac{1}{2}}$$

holds for any $a_0 \in \mathbb{F}_q$, then we conclude that

$$\left| \frac{1}{|\mathcal{A}|} \sum_{f \in \mathcal{A}} \mathcal{V}(f) - \mu_d q \right| \leq \frac{1}{\sum_{a_0 \in \mathbb{F}_q} |\mathcal{A}_{a_0}|} \sum_{a_0 \in \mathbb{F}_q} |\mathcal{A}_{a_0}| E(d_1, \ldots, d_m, d) q^{\frac{1}{2}}$$

$$\leq E(d_1, \ldots, d_m, d) q^{\frac{1}{2}}.$$

Further, as $\mathcal{V}(f) = \mathcal{V}(f + a_0)$ for any $f \in \mathcal{A}$, we shall also assume that $f(0) = 0$ for any $f \in \mathcal{A}$.

For any $f \in \mathcal{A}$, we have that $\mathcal{V}(f)$ equals the number of $a_0 \in \mathbb{F}_q$ for which $f + a_0$ has at least one root in $\mathbb{F}_q$. If $\mathbb{K}$ is any of the fields $\mathbb{F}_q$ or $\overline{\mathbb{F}}_q$, by $\mathbb{K}[T]_d$ we denote the set of monic polynomials of $\mathbb{K}[T]$ of degree d. Let $\mathcal{N} : \mathbb{F}_q[T]_d \to \mathbb{Z}_{\geq 0}$ be the counting function of the number of roots in $\mathbb{F}_q$ and $\mathbf{1}_{\{\mathcal{N}>0\}} : \mathbb{F}_q[T]_d \to \{0,1\}$ the characteristic function of the set of polynomials having at least one root in $\mathbb{F}_q$. We deduce that

$$\sum_{f \in \mathcal{A}} \mathcal{V}(f) = \sum_{a_0 \in \mathbb{F}_q} \sum_{f \in \mathcal{A}} \mathbf{1}_{\{\mathcal{N}>0\}}(f + a_0)$$

$$= \left| \{ f + a_0 \in \mathcal{A} + \mathbb{F}_q : \mathcal{N}(f + a_0) > 0 \} \right|.$$

For a set $\mathcal{X} \subset \mathbb{F}_q$, we define $\mathcal{S}_{\mathcal{X}}^{\mathcal{A}} \subset \mathbb{F}_q[T]_d$ as the set of polynomials $f + a_0 \in \mathcal{A} + \mathbb{F}_q$ vanishing on $\mathcal{X}$, namely

$$\mathcal{S}_{\mathcal{X}}^{\mathcal{A}} := \{ f + a_0 \in \mathcal{A} + \mathbb{F}_q : (f + a_0)(x) = 0 \text{ for any } x \in \mathcal{X} \}.$$

For $r \in \mathbb{N}$ we shall use the symbol $\mathcal{X}_r$ to denote a subset of $\mathbb{F}_q$ of r elements. Our approach to determine the asymptotic behavior of $\mathcal{V}(\mathcal{A})$ relies on the following combinatorial result.

Lemma 1. *Given $d, m \in \mathbb{N}$ with $q > d \geq m + 2$, we have*

$$\mathcal{V}(\mathcal{A}) = \frac{1}{|\mathcal{A}|} \sum_{r=1}^{d} (-1)^{r-1} \sum_{\mathcal{X}_r \subset \mathbb{F}_q} |\mathcal{S}_{\mathcal{X}_r}^{\mathcal{A}}|. \tag{8}$$

Proof. Given a subset $\mathcal{X}_r := \{\alpha_1, \ldots, \alpha_r\} \subset \mathbb{F}_q$, consider the set $\mathcal{S}_{\mathcal{X}_r}^{\mathcal{A}} \subset \mathbb{F}_q[T]_d$ defined as above. It is easy to see that $\mathcal{S}_{\mathcal{X}_r}^{\mathcal{A}} = \bigcap_{i=1}^{r} \mathcal{S}_{\{\alpha_i\}}^{\mathcal{A}}$ and

$$\left| \{ f + a_0 \in \mathcal{A} + \mathbb{F}_q : \mathcal{N}(f + a_0) > 0 \} \right| = \left| \bigcup_{x \in \mathbb{F}_q} \mathcal{S}_{\{x\}}^{\mathcal{A}} \right|.$$

Therefore the inclusion–exclusion principle implies

$$\mathcal{V}(\mathcal{A}) = \frac{1}{|\mathcal{A}|} \left| \bigcup_{x \in \mathbb{F}_q} \mathcal{S}_{\{x\}}^{\mathcal{A}} \right| = \frac{1}{|\mathcal{A}|} \sum_{r=1}^{q} (-1)^{r-1} \sum_{\mathcal{X}_r \subset \mathbb{F}_q} |\mathcal{S}_{\mathcal{X}_r}^{\mathcal{A}}|.$$

Now $|\mathcal{S}_{\mathcal{X}_r}^{\mathcal{A}}| = 0$ for $r > d$, because a polynomial of degree d cannot vanish on more than d elements of $\mathbb{F}_q$. This readily implies the lemma. □

Lemma 1 shows that the behavior of $\mathcal{V}(\mathcal{A})$ is determined by that of

$$\mathcal{S}_r^{\mathcal{A}} := \sum_{\mathcal{X}_r \subset \mathbb{F}_q} |\mathcal{S}_{\mathcal{X}_r}^{\mathcal{A}}|, \tag{9}$$

for $1 \leq r \leq d$, which are the subject of the next sections.

3.1. A geometric approach to estimate $\mathcal{S}_r^{\mathcal{A}}$

Fix r with $1 \leq r \leq d$. Let $A_{d-1}, \ldots, A_0$ be indeterminates over $\overline{\mathbb{F}}_q$ and let $G_1, \ldots, G_m \in \mathbb{F}_q[A_{d-1}, \ldots, A_1]$ be the polynomials defining the family $\mathcal{A}$ of (3). Set $\boldsymbol{A} := (A_{d-1}, \ldots, A_1)$ and $\boldsymbol{A}_0 := (\boldsymbol{A}, A_0)$. To estimate $\mathcal{S}_r^{\mathcal{A}}$, we introduce the following definitions and notations. Let $T, T_1, \ldots, T_r$ be new indeterminates over $\overline{\mathbb{F}}_q$ and denote $\boldsymbol{T} := (T_1, \ldots, T_r)$. Consider the polynomial $F \in \mathbb{F}_q[\boldsymbol{A}_0, T]$ defined as

$$F(\boldsymbol{A}_0, T) := T^d + A_{d-1} T^{d-1} + \cdots + A_1 T + A_0. \tag{10}$$

For $\boldsymbol{a}_0 \in \mathbb{F}_q^d$ we may write $F(\boldsymbol{a}_0, T) = f + a_0$, with $f \in \mathbb{F}_q[T]_d$ and $f(0) = 0$.

Consider the affine quasi-$\mathbb{F}_q$-variety $\Gamma_r \subset \mathbb{A}^{d+r}$ defined as follows:

$$\Gamma_r := \{(\boldsymbol{a}, a_0, \boldsymbol{\alpha}) \in \mathbb{A}^d \times \mathbb{A}^r : \alpha_i \neq \alpha_j \ (1 \leq i < j \leq r),$$
$$F(\boldsymbol{a}_0, \alpha_i) = 0 \ (1 \leq i \leq r), \ G_k(\boldsymbol{a}) = 0 \ (1 \leq k \leq m)\}.$$

Our next result relates $|\Gamma_r(\mathbb{F}_q)|$ to the numbers $\mathcal{S}_r^{\mathcal{A}}$ $(1 \le i \le d)$.

Lemma 2. *Let r be an integer with $1 \le r \le d$. Then*

$$\frac{|\Gamma_r(\mathbb{F}_q)|}{r!} = \mathcal{S}_r^{\mathcal{A}}.$$

Proof. Let $(\boldsymbol{a}_0, \boldsymbol{\alpha})$ be a point of $\Gamma_r(\mathbb{F}_q)$ and $\sigma : \{1, \dots, r\} \to \{1, \dots, r\}$ an arbitrary permutation. Let $\sigma(\boldsymbol{\alpha})$ be the image of $\boldsymbol{\alpha}$ by the linear mapping induced by σ. Then it is clear that $(\boldsymbol{a}_0, \sigma(\boldsymbol{\alpha}))$ belong to $\Gamma_r(\mathbb{F}_q)$. Furthermore, $\sigma(\boldsymbol{\alpha}) = \boldsymbol{\alpha}$ if and only if σ is the identity permutation. This shows that $\mathbb{S}_r$, the symmetric group of r elements, acts over the set $\Gamma_r(\mathbb{F}_q)$ and each orbit under this action has $r!$ elements.

The orbit of an arbitrary point $(\boldsymbol{a}_0, \boldsymbol{\alpha}) \in \Gamma_r(\mathbb{F}_q)$ uniquely determines a polynomial $F(\boldsymbol{a}_0, T) = f + a_0$ with $f \in \mathcal{A}$ and a set $\mathcal{X}_r := \{\alpha_1, \dots, \alpha_r\} \subset \mathbb{F}_q$ with $|\mathcal{X}_r| = r$ such that $(f + a_0)|_{\mathcal{X}_r} \equiv 0$. Therefore, each orbit uniquely determines a set $\mathcal{X}_r \subset \mathbb{F}_q$ with $|\mathcal{X}_r| = r$ and an element of $\mathcal{S}_{\mathcal{X}_r}^{\mathcal{A}}$. Reciprocally, to each element of $\mathcal{S}_{\mathcal{X}_r}^{\mathcal{A}}$ corresponds a unique orbit of $\Gamma_r(\mathbb{F}_q)$. This implies

$$\text{number of orbits of } \Gamma_r(\mathbb{F}_q) = \sum_{\mathcal{X}_r \subseteq \mathbb{F}_q} |\mathcal{S}_{\mathcal{X}_r}^{\mathcal{A}}|,$$

and finishes the proof of the lemma. $\qquad\qquad\qquad\qquad\qquad\qquad\square$

To estimate the quantity $|\Gamma_r(\mathbb{F}_q)|$ we shall consider the Zariski closure $\mathrm{cl}(\Gamma_r)$ of $\Gamma_r \subset \mathbb{A}^{d+r}$. Our aim is to provide explicit equations defining $\mathrm{cl}(\Gamma_r)$. For this purpose, we shall use the following notation. Let $X_1, \dots, X_{l+1}$ be indeterminates over $\overline{\mathbb{F}}_q$ and $f \in \overline{\mathbb{F}}_q[T]$ a polynomial of degree at most l. For notational convenience, we define the 0th divided difference $\Delta^0 f \in \overline{\mathbb{F}}_q[X_1]$ of f as $\Delta^0 f := f(X_1)$. Further, for $1 \le i \le l$ we define the ith divided difference $\Delta^i f \in \overline{\mathbb{F}}_q[X_1, \dots, X_{i+1}]$ of f as

$$\Delta^i f(X_1, \dots, X_{i+1}) = \frac{\Delta^{i-1} f(X_1, \dots, X_i) - \Delta^{i-1} f(X_1, \dots, X_{i-1}, X_{i+1})}{X_i - X_{i+1}}.$$

With these notations, let $\Gamma_r^* \subset \mathbb{A}^{d+r}$ be the $\mathbb{F}_q$-variety defined as

$$\Gamma_r^* := \{(\boldsymbol{a}_0, \boldsymbol{\alpha}) \in \mathbb{A}^d \times \mathbb{A}^r : \Delta^{i-1} F(\boldsymbol{a}_0, \alpha_1, \dots, \alpha_i) = 0 \ (1 \le i \le r),$$

$$G_k(\boldsymbol{a}_0) = 0 \ (1 \le k \le m)\},$$

where $\Delta^{i-1} F(\boldsymbol{a}_0, T_1, \dots, T_i)$ denotes the $(i-1)$th divided difference of $F(\boldsymbol{a}_0, T) \in \overline{\mathbb{F}}_q[T]$. Next we relate the varieties Γ_r and Γ_r^*.

Lemma 3. *With notations and assumptions as above, we have*

$$\Gamma_r = \Gamma_r^* \cap \{(\boldsymbol{a}_0, \boldsymbol{\alpha}) : \alpha_i \ne \alpha_j \ (1 \le i < j \le r)\}. \qquad (11)$$

Proof. Let (a_0, α) be a point of Γ_r. By the definition of the divided differences of $F(a_0, T)$ we easily conclude that $(a_0, \alpha) \in \Gamma_r^*$. On the other hand, let (a_0, α) be a point of the set in the right–hand side of (11). We claim that $F(a_0, \alpha_i) = 0$ for $1 \leq i \leq r$. Observe that $F(a_0, \alpha_1) = \Delta^0 F(a_0, \alpha_1) = 0$. Arguing inductively, suppose that we have $F(a_0, \alpha_1) = \cdots = F(a_0, \alpha_{i-1}) = 0$. By definition $\Delta^{i-1} F(a_0, \alpha_1, \dots, \alpha_i)$ can be expressed as a linear combination with nonzero coefficients of the differences $F(a_0, \alpha_{j+1}) - F(a_0, \alpha_j)$ with $1 \leq j \leq i - 1$. Combining the inductive hypothesis with the fact that $\Delta^{i-1} F(a_0, \alpha_1, \dots, \alpha_i) = 0$, we easily conclude that $F(a_0, \alpha_i) = 0$, finishing thus the proof of the claim. $\square$

4. On the geometry of the variety Γ_r^*

In this section we establish several properties of the geometry of the affine $\mathbb{F}_q$–variety Γ_r^*, assuming that the polynomials $G_1, \dots, G_m$ and the affine variety $V \subset \mathbb{A}^d$ they define satisfy certain the conditions that we now state. The first conditions allow us to estimate the cardinality of $\mathcal{A}$:

- (H_1) $G_1, \dots, G_m$ form a regular sequence and generate a radical ideal of $\mathbb{F}_q[A_{d-1}, \dots, A_0]$.
- (H_2) The variety $V \subset \mathbb{A}^d$ defined by $G_1, \dots, G_m$ is normal.
- (H_3) Let $G_1^{d_1}, \dots, G_m^{d_m}$ denote the homogeneous parts of higher degree of $G_1, \dots, G_m$. Then $G_1^{d_1}, \dots, G_m^{d_m}$ satisfy (H_1) and (H_2).

As stated in the introduction, we are interested in families $\mathcal{A}$ for which $\mathcal{V}(\mathcal{A}) = \mu_d q + \mathcal{O}(q^{\frac{1}{2}})$. If many of the polynomials in $\mathcal{A} + \mathbb{F}_q$ are not square–free, then $\mathcal{V}(\mathcal{A})$ might not behave as in the general case. For $\mathcal{B} \subset \overline{\mathbb{F}}_q[T]_d$, the set of elements of $\mathcal{B}$ which are not square–free is called the *discriminant locus* $\mathcal{D}(\mathcal{B})$ of $\mathcal{B}$. With a slight abuse of notation, in what follows we identify each $f_{a_0} = T^d + a_{d-1} T^{d-1} + \cdots + a_0 \in \mathcal{B}$ with the d–tuple $a_0 := (a_{d-1}, \dots, a_0)$, and consider $\mathcal{B}$ as a subset of $\mathbb{A}^d$. For $f_{a_0} \in \mathcal{B}$, let $\mathrm{Disc}(f_{a_0}) := \mathrm{Res}(f_{a_0}, f'_{a_0})$ denote the discriminant of f_{a_0}, that is, the resultant of f_{a_0} and its derivative f'_{a_0}. Since f_{a_0} has degree d, by basic properties of resultants it follows that

$$\mathrm{Disc}(f_{a_0}) = \mathrm{Disc}(F(A_0, T))|_{A_0 = a_0}$$
$$:= \mathrm{Res}(F(A_0, T), \Delta^1 F(A_0, T, T), T)|_{A_0 = a_0},$$

where the expression Res in the right–hand side denotes resultant with respect to T. Observe that $\mathcal{D}(\mathcal{B}) = \{a_0 \in \mathcal{B} : \mathrm{Disc}(F(A_0, T))|_{A_0 = a_0} = 0\}$.

We shall need further to consider first subdiscriminant loci. The *first subdiscriminant locus* $\mathcal{S}_1(\mathcal{B})$ of $\mathcal{B} \subset \overline{\mathbb{F}}_q[T]_d$ is the set of $a_0 \in \mathcal{B}$ for which the first subdiscriminant $\mathrm{Subdisc}(f_{a_0}) := \mathrm{Subres}(f_{a_0}, f'_{a_0})$ vanishes, where $\mathrm{Subres}(f_{a_0}, f'_{a_0})$ denotes the first subresultant of f_{a_0} and f'_{a_0}. Since f_{a_0} has degree d, basic properties of subresultants imply

$$\mathrm{Subdisc}(f_{a_0}) = \mathrm{Subdisc}(F(A_0, T))|_{A_0 = a_0}$$
$$:= \mathrm{Subres}(F(A_0, T), \Delta^1 F(A_0, T, T), T))|_{A_0 = a_0},$$

where Subres in the right–hand side denotes first subresultant with respect to T. We have $\mathcal{S}_1(\mathcal{B}) = \{a_0 \in \mathcal{B} : \mathrm{Subdisc}(F(A_0, T))|_{A_0 = a_0} = 0\}$.

Our next condition requires that the discriminant and the first subdiscriminant locus intersect well V. More precisely, we require the condition:

(H$_4$) $V \cap \mathcal{D}(V)$ has codimension one in V, and $V \cap \mathcal{D}(V) \cap \mathcal{S}_1(V)$ has codimension two in V.

We shall prove that Γ_r^* is a set–theoretic complete intersection, whose singular locus has codimension at least 2. This will allow us to conclude that Γ_r^* is an ideal–theoretic complete intersection.

Lemma 4. *Γ_r^* is a set–theoretic complete intersection of dimension $d - m$.*

Proof. By hypothesis (H$_1$), $G_1, \ldots, G_m$ form a regular sequence. To show that $G_1, \ldots, G_m, \Delta^{i-1} F(A_0, T_1, \ldots, T_i)$ $(1 \le i \le r)$ form a regular sequence, we argue by induction on i.

For $i = 1$, we observe that the set of common zeros of $G_1, \ldots, G_m$ in $\mathbb{A}^d \times \mathbb{A}^r$ is $V \times \mathbb{A}^r$, and each irreducible component of $V \times \mathbb{A}^r$ is of the form $\mathcal{C} \times \mathbb{A}^r$, where $\mathcal{C}$ is an irreducible component of V. As $\Delta^0 F(A_0, T_1) = F(A_0, T_1)$ is of degree d in T_1, it cannot vanish identically on any component $\mathcal{C} \times \mathbb{A}^r$, which implies that it cannot be a zero divisor modulo $G_1, \ldots, G_m$.

Now suppose that the assertion is proved for $1 \le j \le r - 1$, that is, $G_1, \ldots, G_m, \Delta^{i-1} F(A_0, T_1, \ldots, T_i)$ $(1 \le i \le j)$ form a regular sequence. These are all elements of $\mathbb{F}_q[A_0, T_1, \ldots, T_j]$. On the other hand, the monomial T_{j+1}^{d-j} occurs in the dense representation of $\Delta^j F(A_0, T_1, \ldots, T_{j+1})$ with a nonzero coefficient. We deduce that $\Delta^j F(A_0, T_1, \ldots, T_{j+1})$ cannot be a zero divisor modulo $G_1, \ldots, G_m, \Delta^{i-1} F(A_0, T_1, \ldots, T_i)$ $(1 \le i \le j)$, which proves our assertion. This implies the statement of the lemma. $\square$

4.1. The dimension of the singular locus of Γ_r^*

Next we show that the singular locus of Γ_r^* has codimension at least 2 in Γ_r^*. For this purpose, we carry out an analysis of the singular locus of Γ_r^* which generalizes that of [21, Section 4.1] to this (eventually nonlinear) setting. We start with the following criterion of nonsingularity.

Lemma 5. *Let $J_{G,F} \in \overline{\mathbb{F}}_q[A_0, T]^{(m+r) \times (d+r)}$ be the Jacobian matrix of $G := (G_1, \ldots, G_m)$ and $F(A_0, T_i)$ $(1 \leq i \leq r)$ with respect to A_0, T, and let $(a_0, \alpha) \in \Gamma_r^*$. If $J_{G,F}(a_0, \alpha)$ has full rank, then (a_0, α) is a nonsingular point of Γ_r^*.*

Proof. Considering the Newton form of the polynomial interpolating $F(a_0, T)$ at $\alpha_1, \ldots, \alpha_r$ we easily deduce that $F(a_0, \alpha_i) = 0$ for $1 \leq i \leq r$. This shows that $F(A_0, T_i)$ vanishes on Γ_r^* for $1 \leq i \leq r$. As a consequence, any element of the tangent space $\mathcal{T}_{(a_0, \alpha)}\Gamma_r^*$ of Γ_r^* at (a_0, α) belongs to the kernel of the Jacobian matrix $J_{G,F}(a_0, \alpha)$.

By hypothesis, the $(m + r) \times (d + r)$ matrix $J_{G,F}(a_0, \alpha)$ has full rank $m + r$, and thus its kernel has dimension $d - m$. We conclude that the tangent space $\mathcal{T}_{(a_0, \alpha)}\Gamma_r^*$ has dimension at most $d - m$. Since Γ_r^* is of pure dimension $d - m$, it follows that (a_0, α) is a nonsingular point of Γ_r^*. $\square$

Let $(a_0, \alpha) := (a_0, \alpha_1, \ldots, \alpha_r)$ be an arbitrary point of Γ_r^* and let $f_{a_0} := F(a_0, T)$. Then the Jacobian matrix $J_{G,F}(a_0, \alpha)$ has the following form:

$$
J_{G,F}(a_0, \alpha) := \begin{pmatrix} \dfrac{\partial G}{\partial A_0}(a_0, \alpha) & 0 \\[2ex] \dfrac{\partial F}{\partial A_0}(a_0, \alpha) & \dfrac{\partial F}{\partial T}(a_0, \alpha) \end{pmatrix}.
$$

Observe that $(\partial F/\partial T)(a_0, \alpha)$ is a diagonal matrix whose ith diagonal entry is $f'_{a_0}(\alpha_i)$. If $(\partial G/\partial A_0)(a_0, \alpha)$ has full rank and all the roots in $\overline{\mathbb{F}}_q$ of f_{a_0} are simple, then $J_{G,F}(a_0, \alpha)$ has full rank and (a_0, α) is a regular point of Γ_r^*. Therefore, to prove that the singular locus of Γ_r^* is a subvariety of codimension at least 2 in Γ_r^*, it suffices to consider the set of points $(a_0, \alpha) \in \Gamma_r^*$ for which either a_0 is a singular point of V, or at least one coordinate of α is a multiple root of f_{a_0}. To deal with the first of these cases, consider the morphism of $\mathbb{F}_q$–varieties defined as follows:

$$
\Psi_r : \quad \begin{array}{c} \Gamma_r^* \to V \\ (a_0, \alpha) \mapsto a_0, \end{array} \tag{12}
$$

We have the following result.

Lemma 6. Ψ_r *is a finite dominant morphism.*

Proof. It is easy to see that Ψ_r is a surjective mapping. Therefore, it suffices to show that the coordinate function t_i of $\overline{\mathbb{F}}_q[\Gamma_r^*]$ defined by T_i satisfies a monic equation with coefficients in $\overline{\mathbb{F}}_q[V]$ for $1 \leq i \leq r$. Denote by ξ_j the coordinate function of V defined by A_j for $0 \leq j \leq d-1$, and set $\boldsymbol{\xi}_0 := (\xi_{d-1}, \ldots, \xi_0)$. Since the polynomial $F(A_0, T_i)$ vanishes on Γ_r^* for $1 \leq i \leq r$, and is a monic element of $\overline{\mathbb{F}}_q[A_0][T_i]$, we deduce that the monic element $F(\boldsymbol{\xi}_0, T_i)$ of $\overline{\mathbb{F}}_q[V][T_i]$ annihilates t_i in Γ_r^* for $1 \leq i \leq r$. This shows that the ring extension $\overline{\mathbb{F}}_q[V] \hookrightarrow \overline{\mathbb{F}}_q[\Gamma_r^*]$ is integral, namely Ψ_r is a finite dominant mapping. $\qquad\square$

A first consequence of Lemma 6 is that the set of points $(\boldsymbol{a}_0, \boldsymbol{\alpha}) \in \Gamma_r^*$ with $\boldsymbol{a}_0$ singular are under control.

Corollary 1. *The set $\mathcal{W}_0$ of points $(\boldsymbol{a}_0, \boldsymbol{\alpha}) \in \Gamma_r^*$ with $\boldsymbol{a}_0 \in \mathrm{Sing}(V)$ is contained in a subvariety of codimension 2 of Γ_r^*.*

Proof. Hypotheses (H_1) and (H_2) imply that V is a normal complete intersection. It follows that $\mathrm{Sing}(V)$ has codimension at least two in V. Then $\mathcal{W}_0 = \Psi_r^{-1}(\mathrm{Sing}(V))$ has codimension at least two in Γ_r^*. $\qquad\square$

By Corollary 1 it suffices to consider the set of singular points $(\boldsymbol{a}_0, \boldsymbol{\alpha})$ of Γ_r^* with $\boldsymbol{a}_0 \in V \setminus \mathrm{Sing}(V)$. By the remarks before Lemma 6, if $(\boldsymbol{a}_0, \boldsymbol{\alpha})$ is such a singular point, then $f_{\boldsymbol{a}_0}$ must have multiple roots. We start considering the "extreme" case where $f'_{\boldsymbol{a}_0}$ is the zero polynomial.

Lemma 7. *The set $\mathcal{W}_1$ of points $(\boldsymbol{a}_0, \boldsymbol{\alpha}) \in \Gamma_r^*$ with $f'_{\boldsymbol{a}_0} = 0$ is contained in a subvariety of codimension 2 of Γ_r^*.*

Proof. The condition $f'_{\boldsymbol{a}_0} = 0$ implies $\boldsymbol{a}_0$ belongs to both the discriminant locus $\mathcal{D}(V)$ and the first subdiscriminant locus $\mathcal{S}_1(V)$, namely $\mathcal{W}_1 \subset \Psi_r^{-1}(\mathcal{D}(V) \cap \mathcal{S}_1(V))$. Hypothesis (H_4) asserts that $\mathcal{D}(V) \cap \mathcal{S}_1(V)$ has codimension two in V. Therefore, taking into account that Ψ_r is a finite morphism we deduce that $\mathcal{W}_1$ has codimension at least two in Γ_r^*. $\qquad\square$

In what follows we shall assume that $\boldsymbol{a}_0$ is a regular point of V, $f'_{\boldsymbol{a}_0}$ is nonzero and $f_{\boldsymbol{a}_0}$ has multiple roots. We analyze the case where exactly one of the coordinates of $\boldsymbol{\alpha}$ is a multiple root of $f_{\boldsymbol{a}_0}$.

Lemma 8. *Suppose that there exists a unique coordinate α_i of $\boldsymbol{\alpha}$ which is a multiple root of $f_{\boldsymbol{a}_0}$. Then $(\boldsymbol{a}_0, \boldsymbol{\alpha})$ is a regular point of Γ_r^*.*

Proof. Assume without loss of generality that α_1 is the only multiple root of f_{a_0} among the coordinates of α. According to Lemma 5, it suffices to show that the Jacobian matrix $J_{G,F}(a_0, \alpha)$ has full rank. For this purpose, consider the $r \times (r+1)$–submatrix $\partial F / \partial(A_0, T)(a_0, \alpha)$ of $J_{G,F}(a_0, \alpha)$ consisting of the entries of its last r rows and its last $r + 1$ columns:

$$\frac{\partial F}{\partial(A_0, T)}(a_0, \alpha) := \begin{pmatrix} 1 & 0 & 0 & 0 & \cdots & 0 \\ 1 & 0 & f'_{a_0}(\alpha_2) & 0 & \cdots & 0 \\ \vdots & \vdots & & 0 & \ddots & \vdots \\ \vdots & \vdots & \vdots & & \ddots & 0 \\ 1 & 0 & 0 & \cdots & 0 & f'_{a_0}(\alpha_r) \end{pmatrix}.$$

Since by hypothesis α_i is a simple root of f'_{a_0} for $i \geq 2$, we have $f'_{a_0}(\alpha_i) \neq 0$ for $i \geq 2$, and thus $\partial F / \partial(A_0, T)(a_0, \alpha)$ is of full rank r.

On the other hand, since the matrix $(\partial G / \partial A_0)(a_0, \alpha)$ has rank m and its last column is zero, denoting by $(\partial G / \partial A)(a_0, \alpha)$ the submatrix of $(\partial G / \partial A_0)(a_0, \alpha)$ obtained by deleting its last column, we see that $J_{G,F}(a_0, \alpha)$ can be expressed as the following block matrix:

$$J_{G,F}(a_0, \alpha) = \begin{pmatrix} \dfrac{\partial G}{\partial A}(a_0, \alpha) & 0 \\ * & \dfrac{\partial F}{\partial(A_0, T)}(a_0, \alpha) \end{pmatrix}.$$

Since both $(\partial G / \partial A)(a_0, \alpha)$ and $(\partial F / \partial(A_0, T))(a_0, \alpha)$ have full rank, we conclude that $J_{G,F}(a_0, \alpha)$ has rank $m + r$. $\qquad\square$

Now we analyze the case where two distinct multiple roots of f_{a_0} occur among the coordinates of α.

Lemma 9. *Let $\mathcal{W}_2$ be the set of points $(a_0, \alpha) \in \Gamma_r^*$ for which there exist $1 \leq i < j \leq r$ such that $\alpha_i \neq \alpha_j$ and α_i, α_j are multiple roots of f_{a_0}. Then $\mathcal{W}_2$ is contained in a subvariety of codimension 2 of Γ_r^*.*

Proof. Let (a_0, α) be an arbitrary point of $\mathcal{W}_2$. Since f_{a_0} has at least two distinct multiple roots, the greatest common divisor of f_{a_0} and f'_{a_0} has degree at least 2. This implies that

$$\mathrm{Disc}(f_{a_0}) = \mathrm{Subdisc}(f_{a_0}) = 0.$$

As a consequence, $\mathcal{W}_2 \subset \Psi_r^{-1}(\mathcal{Z})$, where Ψ_r is the morphism of (12), $\mathcal{Z} := \mathcal{D}(V) \cap \mathcal{S}_1(V)$ and $\mathcal{D}(V)$ and $\mathcal{S}_1(V)$ are the discriminant locus and the first subdiscriminant locus of V. Hypothesis (H_4) proves that $\mathcal{Z}$ has codimension two in V. It follows that $\dim \mathcal{Z} = d - m - 2$, and hence $\dim \Psi_r^{-1}(\mathcal{Z}) = d - m - 2$. The statement of the lemma follows. $\square$

Next we consider the case where only one multiple root of f_{a_0} occurs among the coordinates of $\boldsymbol{\alpha}$, and at least two distinct coordinates of $\boldsymbol{\alpha}$ take this value. Then we have either that all the remaining coordinates of $\boldsymbol{\alpha}$ are simple roots of f_{a_0}, or there exists at least a third coordinate whose value is the same multiple root. We now deal with the first of these cases.

Lemma 10. *Let $(\boldsymbol{a}_0, \boldsymbol{\alpha}) \in \Gamma_r^*$ be a point satisfying the conditions:*

- *there exist $1 \le i < j \le r$ such that $\alpha_i = \alpha_j$ and α_i is a multiple root of f_{a_0};*
- *for any $k \notin \{i, j\}$, α_k is a simple root of f_{a_0}.*

Then either $(\boldsymbol{a}_0, \boldsymbol{\alpha})$ is regular point of Γ_r^* or it is contained in a subvariety $\mathcal{W}_3$ of codimension two of Γ_r^*.

Proof. We may assume without loss of generality that $i = 1$ and $j = 2$. Observe that $\Delta^1 F(\boldsymbol{A}_0, T_1, T_2)$ and $F(\boldsymbol{A}_0, T_i)$ $(2 \le i \le r)$ vanish on Γ_r^*. Therefore, the tangent space $\mathcal{T}_{(\boldsymbol{a}_0, \boldsymbol{\alpha})} \Gamma_r^*$ of Γ_r^* at $(\boldsymbol{a}_0, \boldsymbol{\alpha})$ is included in the kernel of the Jacobian matrix $J_{\boldsymbol{G}, \Delta^1, \boldsymbol{F}^*}(\boldsymbol{a}_0, \boldsymbol{\alpha})$ of $\boldsymbol{G}$, $\Delta^1 F(\boldsymbol{A}_0, T_1, T_2)$ and $F(\boldsymbol{A}_0, T_i)$ $(2 \le i \le r)$ with respect to $\boldsymbol{A}_0, \boldsymbol{T}$. We claim that either $J_{\boldsymbol{G}, \Delta^1, \boldsymbol{F}^*}(\boldsymbol{a}_0, \boldsymbol{\alpha})$ has rank $r + m$, or $(\boldsymbol{a}_0, \boldsymbol{\alpha})$ is contained in a subvariety of codimension two of Γ_r^*.

Now we prove the claim. We may express $J_{\boldsymbol{G}, \Delta^1, \boldsymbol{F}^*}(\boldsymbol{a}_0, \boldsymbol{\alpha})$ as

$$
J_{\boldsymbol{G}, \Delta^1, \boldsymbol{F}^*}(\boldsymbol{a}_0, \boldsymbol{\alpha}) = \begin{pmatrix} \dfrac{\partial \boldsymbol{G}}{\partial \boldsymbol{A}}(\boldsymbol{a}_0, \boldsymbol{\alpha}) & \boldsymbol{0} \\[2mm] * & \dfrac{\partial(\Delta^1, \boldsymbol{F}^*)}{\partial(A_0, \boldsymbol{T})}(\boldsymbol{a}_0, \boldsymbol{\alpha}) \end{pmatrix},
$$

where $(\partial \boldsymbol{G} / \partial \boldsymbol{A})(\boldsymbol{a}_0, \boldsymbol{\alpha}) \in \mathbb{F}_q^{m \times (d-1)}$ is the Jacobian matrix of $\boldsymbol{G}$ with respect to $\boldsymbol{A}$ and $(\partial(\Delta^1, \boldsymbol{F}^*) / \partial(A_0, \boldsymbol{T}))(\boldsymbol{a}_0, \boldsymbol{\alpha}) \in \mathbb{F}_q^{r \times (r+1)}$ is the Jacobian matrix of $\Delta^1 F(\boldsymbol{A}_0, T_1, T_2)$ and $F(\boldsymbol{A}_0, T_i)$ $(2 \le i \le r)$ with respect to $A_0, \boldsymbol{T}$:

$$
\frac{\partial(\Delta^1, \boldsymbol{F}^*)}{\partial(A_0, \boldsymbol{T})}(\boldsymbol{a}_0, \boldsymbol{\alpha}) := \begin{pmatrix} 0 & \lambda_1 & \lambda_2 & 0 & \cdots & 0 \\ 1 & 0 & 0 & 0 & \cdots & 0 \\ 1 & 0 & 0 & f_{a_0}'(\alpha_3) & \ddots & \vdots \\ \vdots & \vdots & \vdots & \vdots & \ddots & 0 \\ 1 & 0 & 0 & 0 & \cdots & f_{a_0}'(\alpha_r) \end{pmatrix}.
$$

Now we determine $\lambda_i := (\partial \Delta^1 F(\boldsymbol{A}_0, T_1, T_2)/\partial T_i)(\boldsymbol{a}_0, \boldsymbol{\alpha})$ for $i = 1, 2$. Observe that

$$\frac{\partial}{\partial T_2}\left(\frac{T_2^j - T_1^j}{T_2 - T_1}\right) = \frac{jT_2^{j-1}(T_2 - T_1) - (T_2^j - T_1^j)}{(T_2 - T_1)^2} = \sum_{k=2}^{j}\binom{j}{k}T_2^{j-k}(T_1 - T_2)^{k-2}.$$

It follows that

$$\lambda_2 := \frac{\partial \Delta^1 F(\boldsymbol{A}_0, T_1, T_2)}{\partial T_2}(\boldsymbol{a}_0, \boldsymbol{\alpha}) = \sum_{j=2}^{d} a_j \frac{j(j-1)}{2}\alpha_1^{j-2} = \Delta^2 f_{\boldsymbol{a}_0}(\alpha_1, \alpha_1).$$

Furthermore, it is easy to see that $\lambda_1 = -\lambda_2$.

Recall that α_i is a simple root of $f_{\boldsymbol{a}_0}$ for $i \geq 3$, which implies $f'_{\boldsymbol{a}_0}(\alpha_i) \neq 0$ for $i \geq 3$. If $\Delta^2 f_{\boldsymbol{a}_0}(\alpha_1, \alpha_1) \neq 0$, then there exists an $(r \times r)$–submatrix of $(\partial(\Delta^1, \boldsymbol{F}^*)/\partial(\boldsymbol{A}_0, \boldsymbol{T}))(\boldsymbol{a}_0, \boldsymbol{\alpha})$ with rank r. Thus, $J_{G, \Delta^1, \boldsymbol{F}^*}(\boldsymbol{a}_0, \boldsymbol{\alpha})$ has rank $r + m$, namely the first assertion of the claim holds. It follows that the kernel of $J_{G, \Delta^1, \boldsymbol{F}^*}(\boldsymbol{a}_0, \boldsymbol{\alpha})$ has dimension $d - m$. This implies that $\dim \mathcal{T}_{(\boldsymbol{a}_0, \boldsymbol{\alpha})}\Gamma_r^* \leq d - m$, which proves that $(\boldsymbol{a}_0, \boldsymbol{\alpha})$ is regular point of Γ_r^*.

On the other hand, for $(\boldsymbol{a}_0, \boldsymbol{\alpha}) \in \Gamma_r^*$ as in the statement of the lemma with $\Delta^2 f_{\boldsymbol{a}_0}(\alpha_1, \alpha_1) = 0$, we have that α_1 is root of multiplicity at least three of $f_{\boldsymbol{a}_0}$. Then the set $\mathcal{W}_3$ of such points is contained in $\Psi_r^{-1}(\mathcal{D}(V) \cap \mathcal{S}_1(V))$. The lemma follows arguing as in the proof of Lemma 9. $\qquad\square$

Finally, we analyze the set of points of Γ_r^* such that the value of at least three distinct coordinates of $\boldsymbol{\alpha}$ is the same multiple root of $f_{\boldsymbol{a}_0}$.

Lemma 11. *Let $\mathcal{W}_4 \subset \Gamma_r^*$ be the set of points $(\boldsymbol{a}_0, \boldsymbol{\alpha})$ for which there exist $1 \leq i < j < k \leq r$ such that $\alpha_i = \alpha_j = \alpha_k$ and α_i is a multiple root of $f_{\boldsymbol{a}_0}$. Then $\mathcal{W}_4$ is contained in a subvariety of codimension 2 of Γ_r^*.*

Proof. Let $(\boldsymbol{a}_0, \boldsymbol{\alpha})$ be an arbitrary point of $\mathcal{W}_4$. Without loss of generality we may assume that $\alpha_1 = \alpha_2 = \alpha_3$ is the multiple root of $f_{\boldsymbol{a}_0}$ of the statement of the lemma. Since $(\boldsymbol{a}_0, \boldsymbol{\alpha})$ satisfies the equations

$$F(\boldsymbol{A}_0, T_1) = \Delta F(\boldsymbol{A}_0, T_1, T_2) = \Delta^2 F(\boldsymbol{A}_0, T_1, T_2, T_3) = 0,$$

we see that α_1 is a common root of $f_{\boldsymbol{a}_0}$, $\Delta F(\boldsymbol{a}_0, T, T)$ and $\Delta^2 F(\boldsymbol{a}_0, T, T, T)$. It follows that α_1 is a root of multiplicity at least 3 of $f_{\boldsymbol{a}_0}$, and thus the greatest common divisor of $f_{\boldsymbol{a}_0}$ and $f'_{\boldsymbol{a}_0}$ has degree at least 2. As a consequence, the proof follows by the arguments of the proof of Lemma 9. $\qquad\square$

Now we can prove the main result of this section. According to Corollary 1 and Lemmas 7, 8, 9, 10 and 11, the set of singular points of Γ_r^* is contained in the set $\mathcal{W}_0 \cup \mathcal{W}_1 \cup \mathcal{W}_2 \cup \mathcal{W}_3 \cup \mathcal{W}_4$, where $\mathcal{W}_0, \mathcal{W}_1, \mathcal{W}_2, \mathcal{W}_3$ and $\mathcal{W}_4$ are defined in the statement of Corollary 1 and Lemmas 7, 9, 10 and 11. Since each set $\mathcal{W}_i$ is contained in a subvariety of codimension 2 of Γ_r^*, we obtain the following result.

Theorem 2. *Let $q > d \geq m + 2$. The singular locus of Γ_r^* has codimension at least 2 in Γ_r^*.*

In the next result we deduce important consequences of Theorem 2.

Corollary 2. *With assumptions as in Theorem 2, the ideal $J \subset \mathbb{F}_q[\boldsymbol{A}_0, \boldsymbol{T}]$ generated by $G_1, \ldots, G_m$ and $\Delta^{i-1} F(\boldsymbol{A}_0, T_1, \ldots, T_i)$ $(1 \leq i \leq r)$ is radical. Moreover, Γ_r^* is an ideal–theoretic complete intersection of dimension $d - m$.*

Proof. We prove that J is a radical ideal. Denote by $J_{\boldsymbol{G}, \Delta}(\boldsymbol{A}_0, \boldsymbol{T})$ the Jacobian matrix of $G_1, \ldots, G_m$ and $\Delta^{i-1} F(\boldsymbol{A}_0, T_1, \ldots, T_i)$ $(1 \leq i \leq r)$ with respect to $\boldsymbol{A}_0, \boldsymbol{T}$. By Lemma 4, these polynomials form a regular sequence. Hence, according to [12, Theorem 18.15], it is sufficient to prove that the set of points $(\boldsymbol{a}_0, \boldsymbol{\alpha}) \in \Gamma_r^*$ for which $J_{\boldsymbol{G}, \Delta}(\boldsymbol{a}_0, \boldsymbol{\alpha})$ does not have full rank is contained in a subvariety of Γ_r^* of codimension at least 1.

In the proof of Lemma 5 we show that $F(\boldsymbol{A}_0, T_i) \in J$ for $1 \leq i \leq r$. This implies that each gradient $\nabla F(\boldsymbol{a}_0, \alpha_i)$ is a linear combination of the gradients of $G_1, \ldots, G_m$ and $\Delta^{i-1} F(\boldsymbol{A}_0, T_1, \ldots, T_i)$ $(1 \leq i \leq r)$. We conclude that rank $J_{\boldsymbol{G}, \boldsymbol{F}}(\boldsymbol{a}_0, \boldsymbol{\alpha}) \leq$ rank $J_{\boldsymbol{G}, \Delta}(\boldsymbol{a}_0, \boldsymbol{\alpha})$.

Let $(\boldsymbol{a}_0, \boldsymbol{\alpha})$ be an arbitrary point of Γ_r^* such that $J_{\boldsymbol{G}, \Delta}(\boldsymbol{a}_0, \boldsymbol{\alpha})$ does not have full rank. By Corollary 1 we may assume without loss of generality that $\boldsymbol{a}_0 \in V \setminus \mathrm{Sing}(V)$. Then $J_{\boldsymbol{G}, \boldsymbol{F}}(\boldsymbol{a}_0, \boldsymbol{\alpha})$ does not have full rank and thus $f_{\boldsymbol{a}_0}$ has multiple roots. Observe that the set of points $(\boldsymbol{a}_0, \boldsymbol{\alpha}) \in \Gamma_r^*$ for which $f_{\boldsymbol{a}_0}$ has multiple roots is equal to $\Psi_r^{-1}(\mathcal{D}(V))$, where $\mathcal{D}(V)$ is discriminant locus of V. According to hypothesis (H_4), $\mathcal{D}(V)$ has codimension one in V, which implies that $\Psi_r^{-1}(\mathcal{D}(V))$ has codimension 1 of Γ_r^*. It follows that the set of points $(\boldsymbol{a}_0, \boldsymbol{\alpha}) \in \Gamma_r^*$ for which $J_{\boldsymbol{G}, \Delta}(\boldsymbol{a}_0, \boldsymbol{\alpha})$ does not have full rank is contained in a subvariety of Γ_r^* of codimension at least 1. Hence, J is a radical ideal, which in turn implies that Γ_r^* is an ideal–theoretic complete intersection of dimension $d - m$. $\square$

5. The geometry of the projective closure of Γ_r^*

To estimate the number of $\mathbb{F}_q$–rational points of Γ_r^* we need information on the behavior of Γ_r^* at infinity. For this purpose, we shall analyze the projective closure of Γ_r^*, whose definition we now recall. Consider the embedding of $\mathbb{A}^{d+r}$ into the projective space $\mathbb{P}^{d+r}$ which assigns to any point $(\boldsymbol{a_0}, \boldsymbol{\alpha}) \in \mathbb{A}^{d+r}$ the point $(a_{d-1} : \cdots : a_0 : 1 : \alpha_1 : \cdots : \alpha_r) \in \mathbb{P}^{d+r}$. The closure in the Zariski topology of $\mathbb{P}^{d+r}$ of the image of Γ_r^* under this embedding is called the *projective closure* $\mathrm{pcl}(\Gamma_r^*) \subset \mathbb{P}^{d+r}$ of Γ_r^*. The points of $\mathrm{pcl}(\Gamma_r^*)$ lying in $\{T_0 = 0\}$ are called the points of $\mathrm{pcl}(\Gamma_r^*)$ *at infinity*.

It is well–known that $\mathrm{pcl}(\Gamma_r^*)$ is the $\mathbb{F}_q$–variety of $\mathbb{P}^{d+r}$ defined by the homogenization $F^h \in \mathbb{F}_q[\boldsymbol{A_0}, T_0, \boldsymbol{T}]$ of each polynomial F in the ideal $J \subset \mathbb{F}_q[\boldsymbol{A_0}, \boldsymbol{T}]$ generated by $G_1, \ldots, G_m$ and $\Delta^{i-1}F(\boldsymbol{A_0}, T_1, \ldots, T_i)$ $(1 \leq i \leq r)$. We denote by J^h the ideal generated by all the polynomials F^h with $F \in J$. Since J is radical it turns out that J^h is also radical (see, e.g.,[18, §I.5, Exercise 6]). Furthermore, $\mathrm{pcl}(\Gamma_r^*)$ is of pure dimension $d - m$ (see, e.g.,[18, Propositions I.5.17 and II.4.1]) and degree equal to $\deg \Gamma_r^*$ (see, e.g.,[6, Proposition 1.11]).

Lemma 12. *The polynomials* $G_1^h, \ldots, G_m^h$ *and* $\Delta^{i-1}F(\boldsymbol{A_0}, T_1, \ldots, T_i)^h$ $(1 \leq i \leq r)$ *form a regular sequence of* $\mathbb{F}_q[\boldsymbol{A_0}, T_0, \boldsymbol{T}]$.

Proof. We claim that $G_1^h, \ldots, G_m^h$ form a regular sequence of $\mathbb{F}_q[\boldsymbol{A_0}, T_0, \boldsymbol{T}]$. Indeed, the projective subvariety of $\mathbb{P}^{d+r}$ defined by T_0 and $G_1^h, \ldots, G_m^h$ is isomorphic to the subvariety of $\mathbb{P}^{d+r-1}$ defined by $G_1^{d_1}, \ldots, G_m^{d_m}$. Since hypothesis (H_3) implies that the latter is of pure dimension $d + r - m - 1$, we conclude that the subvariety of $\mathbb{P}^{d+r}$ defined by $G_1^h, \ldots, G_m^h$ is of pure dimension $d + r - m$. It follows that $G_1^h, \ldots, G_m^h$ form a regular sequence. Then the lemma follows arguing as in the proof of Lemma 4. $\qquad\square$

Proposition 1. *The projective variety defined by* $G_1^h, \ldots, G_m^h$ *and* $\Delta^{i-1}F(\boldsymbol{A_0}, T_1, \ldots, T_i)^h$ $(1 \leq i \leq r)$ *is* $\mathrm{pcl}(\Gamma_r^*)$. *Therefore,* $\mathrm{pcl}(\Gamma_r^*)$ *is a set–theoretic complete intersection of dimension* $d - m$.

Proof. By the Newton form of the polynomial interpolating $F(\boldsymbol{A_0}, T)$ at the points $T_1, \ldots, T_r$ we conclude that

$$F(\boldsymbol{A_0}, T_j) = \sum_{i=1}^{r} \Delta^{i-1}F(\boldsymbol{A_0}, T_1, \ldots, T_i)\,(T_j - T_1) \cdots (T_j - T_{i-1}).$$

Homogenizing both sides of this equality we deduce that $F(\boldsymbol{A_0}, T_j)^h$ belongs to the ideal of $\mathbb{F}_q[\boldsymbol{A_0}, T_0, \boldsymbol{T}]$ generated by $G_1^h, \ldots, G_m^h$ and $\Delta^{i-1}F(\boldsymbol{A_0}, T_1, \ldots, T_i)^h$. Denote by $V^h \subset \mathbb{P}^{d+r}$ the variety defined by all these polynomials. Any point of $V^h \cap \{T_0 = 0\} \subset \mathbb{P}^{d+r}$ satisfies

$$F(\boldsymbol{A_0}, T_i)^h|_{T_0=0} = T_i^d + A_{d-1}T_i^{d-1} = T_i^{d-1}(T_i + A_{d-1}) = 0 \quad (1 \leq i \leq r),$$

$$G_k^h(\boldsymbol{A}, T_0)^h|_{T_0=0} = G_k^{d_k}(\boldsymbol{A}) = 0 \quad (1 \leq k \leq m).$$

We deduce that $V^h \cap \{T_0 = 0\}$ is contained in the union

$$V^h \cap \{T_0 = 0\} \subset \bigcup_{\mathcal{I} \subset \{1,\ldots,r\}}^{r} V_{\mathcal{I}} \cap \{T_0 = 0\},$$

where $V_{\mathcal{I}} \subset \mathbb{P}^{d+r}$ is the variety defined by $T_i = 0$ $(i \in \mathcal{I})$, $T_j + A_{d-1} = 0$ $(j \in \{1,\ldots,r\} \setminus \mathcal{I})$ and $G_k^{d_k} = 0$ $(1 \le k \le m)$. As any $V_{\mathcal{I}} \cap \{T_0 = 0\}$ is of pure dimension $d - m - 1$, $V^h \cap \{T_0 = 0\}$ is of pure dimension $d - m - 1$.

Lemma 12 implies that V^h is of pure dimension $d - m$, and thus it has no irreducible component in the hyperplane at infinity. In particular, it agrees with the projective closure of its restriction to $\mathbb{A}^{d+r}$ (see, e.g.,[18, Proposition I.5.17]). As this restriction is Γ_r^*, we have $V^h = \mathrm{pcl}(\Gamma_r^*)$. $\square$

5.1. *The singular locus of* $\mathrm{pcl}(\Gamma_r^*)$

Next we study the singular locus of $\mathrm{pcl}(\Gamma_r^*)$. We start with the following characterization of the points of $\mathrm{pcl}(\Gamma_r^*)$ at infinity.

Lemma 13. $\mathrm{pcl}(\Gamma_r^*) \cap \{T_0 = 0\} \subset \mathbb{P}^{d+r-1}$ *is contained in a union of* $r+1$ *normal complete intersections defined over* $\mathbb{F}_q$, *each of pure dimension* $d - m - 1$ *and degree* $\prod_{i=1}^{m} d_i$.

Proof. We claim that $\Delta^1 F(\boldsymbol{A}_0, T_i, T_j)^h \in J^h$ for $1 \le i < j \le r$. Indeed, we have the identity $\Delta^1 F(\boldsymbol{A}_0, T_i, T_j)(T_i - T_j) = F(\boldsymbol{A}_0, T_i) - F(\boldsymbol{A}_0, T_j)$. Since $F(\boldsymbol{A}_0, T_k)$ vanishes in Γ_r^* for $1 \le k \le r$, we deduce that $\Delta^1 F(\boldsymbol{A}_0, T_i, T_j)$ vanishes on the nonempty Zariski open dense subset $\{T_i \ne T_j\} \cap \Gamma_r^*$ of Γ_r^*. This implies that $\Delta^1 F(\boldsymbol{A}_0, T_i, T_j)$ vanishes in Γ_r^*, which proves the claim.

Combining the claim with the fact that $F(\boldsymbol{A}_0, T_i)^h \in J^h$ for $1 \le i \le r$, we conclude that any $(\boldsymbol{a}_0, \boldsymbol{\alpha}) \in \mathrm{pcl}(\Gamma_r^*) \cap \{T_0 = 0\}$ satisfies the following identities for $1 \le i \le r$ and $1 \le i < j \le r$ respectively:

$$F(\boldsymbol{A}_0, T_i)^h|_{T_0=0} = T_i^d + A_{d-1} T_i^{d-1} = T_i^{d-1}(T_i + A_{d-1}) = 0, \quad (13)$$

$$\Delta^1 F(\boldsymbol{A}_0, T_i, T_j)^h|_{T_0=0} = \frac{T_i^d - T_j^d}{T_i - T_j} + A_{d-1} \frac{T_i^{d-1} - T_j^{d-1}}{T_i - T_j}$$

$$= \sum_{k=0}^{d-2} T_j^k T_i^{d-2-k}(T_i + A_{d-1}) + T_j^{d-1} = 0. \quad (14)$$

From (13)–(14) we deduce that $\mathrm{pcl}(\Gamma_r^*) \cap \{T_0 = 0\}$ is contained in a finite union of $r + 1$ normal complete intersections of $\mathbb{P}^{d+r-1}$ defined over $\mathbb{F}_q$ of pure dimension $d - m - 1$. More precisely, it can be seen that

$$\mathrm{pcl}(\Gamma_r^*) \cap \{T_0 = 0\} \subset \bigcup_{i=0}^{r} V_i \cap \{T_0 = 0\},$$

where V_0 is the variety defined by $T_i = 0$ $(1 \leq i \leq r)$ and $G_k^{d_k} = 0$ $(1 \leq k \leq m)$, and V_i $(1 \leq i \leq r)$ is defined as the set of solutions of

$$T_i + A_{d-1} = 0, \quad T_j = 0 \ (1 \leq j \leq r, i \neq j), \quad G_k^{d_k} = 0 \ (1 \leq k \leq m).$$

By Proposition 1 we have that $\mathrm{pcl}(\Gamma_r^*)$ is of pure dimension $d-m$. Then each irreducible component $\mathcal{C}$ of $\mathrm{pcl}(\Gamma_r^*) \cap \{T_0 = 0\}$ has dimension at least $d - m - 1$, and is contained in an irreducible component of a variety V_i for some $i \in \{0, \ldots, r\}$. By, e.g.,[24, §6.1, Theorem 1], $\mathcal{C}$ is an irreducible component of a variety V_i, finishing thus the proof of the lemma. $\qquad\square$

Now we are able to upper bound the dimension of the singular locus of $\mathrm{pcl}(\Gamma_r^*)$ at infinity.

Lemma 14. *The singular locus of* $\mathrm{pcl}(\Gamma_r^*)$ *at infinity has dimension at most* $d - m - 2$.

Proof. By [15, Lemma 1.1], the singular locus of $\mathrm{pcl}(\Gamma_r^*)$ at infinity is contained in the singular locus of $\mathrm{pcl}(\Gamma_r^*) \cap \{T_0 = 0\}$. Lemma 13 proves that $\mathrm{pcl}(\Gamma_r^*) \cap \{T_0 = 0\}$ has pure dimension $d-m-1$. Therefore, its singular locus has dimension at most $d - m - 2$. $\qquad\square$

Lemma 15. *The polynomials* $G_1^h, \ldots, G_m^h$ *and* $\Delta^{i-1} F(A_0, T_1, \ldots, T_i)^h$ $(1 \leq i \leq r)$ *generate* J^h. *Hence,* $\mathrm{pcl}(\Gamma_r^*)$ *is an ideal–theoretic complete intersection of dimension* $d-m$ *and multidegree* $(d_1, \ldots, d_m, d, \ldots, d-r+1)$.

Proof. According to [12, Theorem 18.15], it suffices to prove that of the set of points of $\mathrm{pcl}(\Gamma_r^*)$ for which the Jacobian of $G_1^h, \ldots, G_m^h$ and $\Delta^{i-1} F(A_0, T_1, \ldots, T_i)^h$ $(1 \leq i \leq r)$ does not have full rank, has codimension at least one in $\mathrm{pcl}(\Gamma_r^*)$. The set of points of $\{T_0 \neq 0\}$ satisfying this condition has codimension one, because $G_1, \ldots, G_m$ and $\Delta^{i-1} F(A_0, T_1, \ldots, T_i)$ $(1 \leq i \leq r)$ define a radical ideal. On the other hand, $\mathrm{pcl}(\Gamma_r^*) \cap \{T_0 = 0\}$ has codimension one in $\mathrm{pcl}(\Gamma_r^*)$. This proves the first assertion of the lemma.

We deduce that $\mathrm{pcl}(\Gamma_r^*)$ is an ideal–theoretic complete intersection of dimension $d - m$, and the Bézout theorem proves that $\deg \mathrm{pcl}(\Gamma_r^*) = \prod_{i=1}^m d_i \cdot d!/(d - r)!$. $\qquad\square$

Finally, we prove the main result of this section.

Theorem 3. *For* $q > d \geq m + 2$, $\mathrm{pcl}(\Gamma_r^*) \subset \mathbb{P}^{d+r}$ *is a normal ideal–theoretic complete intersection of dimension* $d - m$ *and multidegree* $(d_1, \ldots, d_m, d, \ldots, d - r + 1)$.

Proof. Lemma 15 shows that $\mathrm{pcl}(\Gamma_r^*)$ is an ideal–theoretic complete intersection of dimension $d - m$ and multidegree $(d_1, \ldots, d_m, d, \ldots, d - r + 1)$. On the other hand, Theorem 2 and Lemma 14 show that the singular locus of $\mathrm{pcl}(\Gamma_r^*)$ has codimension at least 2 in $\mathrm{pcl}(\Gamma_r^*)$. This implies $\mathrm{pcl}(\Gamma_r^*)$ is regular in codimension 1 and thus normal. $\square$

By Theorems 1 and 3 we conclude that $\mathrm{pcl}(\Gamma_r^*)$ is absolutely irreducible of dimension $d - m$, and the same holds for $\Gamma_r^* \subset \mathbb{A}^{d+r}$. Since Γ_r is a nonempty Zariski open subset of Γ_r^* of dimension $d - m$ and Γ_r^* is absolutely irreducible, the Zariski closure of Γ_r is Γ_r^*.

6. The number of $\mathbb{F}_q$–rational points of Γ_r

As before, let d, m be positive integers with $q > d \geq m+2$. Let $A_{d-1}, \ldots, A_1$ be indeterminates over $\mathbb{F}_q$, set $\boldsymbol{A} := (A_{d-1}, \ldots, A_1)$, and let $G_1, \ldots, G_m$ be polynomials of $\mathbb{F}_q[A_{d-1}, \ldots, A_1]$ satisfying hypotheses (H_1), (H_2), (H_3) and (H_4). Let $\boldsymbol{G} := (G_1, \ldots, G_m)$ and $\mathcal{A} := \mathcal{A}(\boldsymbol{G})$ be the family defined as

$$\mathcal{A} := \{T^d + a_{d-1}T^{d-1} + \cdots + a_1 T \in \mathbb{F}_q[T] : \boldsymbol{G}(a_{d-1}, \ldots, a_1) = 0\}.$$

In this section we determine the asymptotic behavior of the average value set $\mathcal{V}(\mathcal{A})$ of $\mathcal{A}$. By Lemma 1 we have

$$\mathcal{V}(\mathcal{A}) = \frac{1}{|\mathcal{A}|} \sum_{r=1}^{d} (-1)^{r-1} \sum_{\mathcal{X}_r \subset \mathbb{F}_q} |S_{\mathcal{X}_r}^{\mathcal{A}}|,$$

where the second sum runs through all the subsets $\mathcal{X}_r \subset \mathbb{F}_q$ of cardinality r and $S_{\mathcal{X}_r}^{\mathcal{A}}$ denotes the number of $f + a_0 \in \mathcal{A} + \mathbb{F}_q$ such that $(f + a_0)(\alpha) = 0$ for any $\alpha \in \mathcal{X}_r$.

Let $S_r^{\mathcal{A}} := \sum_{\mathcal{X}_r \subset \mathbb{F}_q} |S_{\mathcal{X}_r}^{\mathcal{A}}|$. According to Lemmas 2 and 3, for $1 \leq r \leq d$ we have

$$S_r^{\mathcal{A}} = \frac{|\Gamma_r(\mathbb{F}_q)|}{r!} = \frac{1}{r!} \left| \Gamma_r^*(\mathbb{F}_q) \setminus \bigcup_{i \neq j} \{T_i = T_j\} \right|.$$

We shall apply the results on the geometry of Γ_r^* of the previous section in order to estimate the number of $\mathbb{F}_q$–rational points of Γ_r^*.

6.1. An estimate for $S_r^{\mathcal{A}}$

We shall rely on the following estimate ([5, Theorem 1.3]; see also [15],[3] or [22] for similar estimates): if $W \subset \mathbb{P}^n$ is a normal complete intersection defined over $\mathbb{F}_q$ of dimension $l \geq 2$ and multidegree $(e_1, \ldots, e_{n-l})$, then

$$\big| |W(\mathbb{F}_q)| - p_l \big| \leq (\delta_W (D_W - 2) + 2) q^{l - \frac{1}{2}} + 14 D_W^2 \delta_W^2 q^{l-1}, \qquad (15)$$

where $p_l := q^l + q^{l-1} + \cdots + 1$, $\delta_W := \prod_{i=1}^{n-l} e_i$ and $D_W := \sum_{i=1}^{n-l}(e_i - 1)$.

In what follows, we shall use the following notations:

$$\delta_V := \prod_{i=1}^{m} d_i, \qquad \delta_{\Delta_r} := \prod_{i=1}^{r}(d-i+1) = \frac{d!}{(d-r)!}, \qquad \delta_r := \delta_V \delta_{\Delta_r},$$

$$D_V := \sum_{i=1}^{m}(d_i - 1), \; D_{\Delta_r} := \sum_{i=1}^{r}(d-i) = rd - \frac{r(r+1)}{2}, \; D_r := D_V + D_{\Delta_r}.$$

We start with an estimate on the number of elements of $\mathcal{A}$.

Lemma 16. *For* $q > 16(D_V \delta_V + 14D_V^2 \delta_V^2 q^{-\frac{1}{2}})^2$, *we have*

$$\frac{1}{2} q^{d-m-1} < |\mathcal{A}| \leq q^{d-m-1} + 2\big(\delta_V(D_V - 2) + 2 + 14D_V^2 \delta_V^2 q^{-\frac{1}{2}}\big) q^{d-m-\frac{3}{2}}.$$

Proof. Hypothesis (H$_1$), (H$_2$) and (H$_3$) imply that both the projective closure $\mathrm{pcl}(V) \subset \mathbb{P}^{d-1}$ of V and the set $\mathrm{pcl}(V)^{\infty} := \mathrm{pcl}(V) \cap \{T_0 = 0\}$ of points at infinity are normal complete intersections defined over $\mathbb{F}_q$, both of multidegree $(d_1, \ldots, d_m)$ in $\mathbb{P}^{d-1}$ and $\{T_0 = 0\} \cong \mathbb{P}^{d-2}$ respectively. Therefore, by (15) it follows that

$$\big||\mathcal{A}| - q^{d-m-1}\big| = \big||\mathrm{pcl}(V)(\mathbb{F}_q)| - |\mathrm{pcl}(V)^{\infty}(\mathbb{F}_q)| - p_{d-m-1} + p_{d-m-2}\big|$$

$$\leq \big||\mathrm{pcl}(V)(\mathbb{F}_q)| - p_{d-m-1}\big| + \big||\mathrm{pcl}(V(\mathbb{F}_q))^{\infty}| - p_{d-m-2}\big|$$

$$\leq \big(\delta_V(D_V - 2) + 2\big)(q+1)q^{d-m-\frac{5}{2}} + 14D_V^2 \delta_V^2 (q+1)q^{d-m-3}$$

$$\leq 2\big(\delta_V(D_V - 2) + 2 + 14D_V^2 \delta_V^2 q^{-\frac{1}{2}}\big) q^{d-m-\frac{3}{2}}.$$

By the hypothesis on q of the statement, the lemma readily follows. $\qquad\square$

By Theorem 3, $\mathrm{pcl}(\Gamma_r^*) \subset \mathbb{P}^{d+r}$ is a normal complete intersection defined over $\mathbb{F}_q$ of dimension $d - m$ and multidegree $(d_1, \ldots, d_m, d, \ldots, d - r + 1)$. Therefore, applying (15) we obtain

$$\big||\mathrm{pcl}(\Gamma_r^*)(\mathbb{F}_q)| - p_{d-m}\big| \leq (\delta_r(D_r - 2) + 2)q^{d-m-\frac{1}{2}} + 14D_r^2 \delta_r^2 q^{d-m-1}.$$

On the other hand, since $\mathrm{pcl}(\Gamma_r^*)^{\infty} := \mathrm{pcl}(\Gamma_r^*) \cap \{T_0 = 0\} \subset \mathbb{P}^{d+r-1}$ is a finite union of at most $r + 1$ varieties, each of pure dimension $d - m - 1$ and degree δ_V, by (7) we have $|\mathrm{pcl}(\Gamma_r^*)^{\infty}(\mathbb{F}_q)| \leq (r+1)\delta_V p_{d-m-1}$. Hence,

$$\big||\Gamma_r^*(\mathbb{F}_q)| - q^{d-m}\big| \leq \big||\mathrm{pcl}(\Gamma_r^*)(\mathbb{F}_q)| - p_{d-m}\big| + \big||\mathrm{pcl}(\Gamma_r^*(\mathbb{F}_q))^{\infty}| - p_{d-m-1}\big|$$

$$\leq (\delta_r(D_r - 2) + 2)q^{d-m-\frac{1}{2}} + 14D_r^2 \delta_r^2 q^{d-m-1} + (r+1)\delta_V p_{d-m-1}$$

$$\leq (\delta_r(D_r - 2) + 2)q^{d-m-\frac{1}{2}} + (14D_r^2 \delta_r^2 + 4r\delta_V)q^{d-m-1}. \qquad (16)$$

We also need an upper bound on the number $\mathbb{F}_q$–rational points of

$$\Gamma_r^{*,=} := \Gamma_r^* \cap \bigcup_{1 \leq i < j \leq r} \{T_i = T_j\}.$$

We observe that $\Gamma_r^{*,=} = \Gamma_r^* \cap \mathcal{H}_r$, where $\mathcal{H}_r \subset \mathbb{A}^{d+r}$ is the hypersurface defined by the polynomial $F_r := \prod_{1 \le i < j \le r}(T_i - T_j)$. From the Bézout inequality (5) it follows that

$$\deg \Gamma_r^{*,=} \le \delta_r \binom{r}{2}. \tag{17}$$

Furthermore, we claim that $\Gamma_r^{*,=}$ has dimension at most $d - m - 1$. Indeed, let $(\boldsymbol{a}_0, \boldsymbol{\alpha})$ be any point of $\Gamma_r^{*,=}$. Assume without loss of generality that $\alpha_1 = \alpha_2$. By the definition of divided differences we deduce that $f'_{\boldsymbol{a}_0}(\alpha_1) = 0$, which implies that $f_{\boldsymbol{a}_0}$ has multiple roots. By the proof of Corollary 2, the set of points $(\boldsymbol{a}_0, \boldsymbol{\alpha})$ of Γ_r^* for which $f_{\boldsymbol{a}_0}$ has multiple roots is contained in a subvariety of Γ_r^* of codimension 1, which proves the claim.

Combining the claim with (17) and (6), we obtain

$$\left|\Gamma_r^{*,=}(\mathbb{F}_q)\right| \le \delta_r \binom{r}{2} q^{d-m-1}. \tag{18}$$

Since $\Gamma_r(\mathbb{F}_q) = \Gamma_r^*(\mathbb{F}_q) \setminus \Gamma_r^{*,=}(\mathbb{F}_q)$, from (16) and (18) we deduce that

$$\left| |\Gamma_r(\mathbb{F}_q)| - q^{d-m} \right| \le \left| |\Gamma_r^*(\mathbb{F}_q)| - q^{d-m} \right| + |\Gamma_r^{*,=}(\mathbb{F}_q)|$$
$$\le (\delta_r(D_r - 2) + 2) \, q^{d-m-\frac{1}{2}} + \left(14 D_r^2 \delta_r^2 + \binom{r}{2} \delta_r + 4r\delta_V \right) q^{d-m-1}.$$

As a consequence, we obtain the following result.

Theorem 4. *Let $q > d \ge m + 2$. For any r with and $1 \le r \le d$, we have*

$$\left| S_r^{\mathcal{A}} - \frac{q^{d-m}}{r!} \right| \le \left(\frac{\delta_r(D_r - 2) + 2}{r!} q^{\frac{1}{2}} + 14 \frac{D_r^2 \delta_r^2}{r!} + \binom{r}{2} \frac{\delta_r}{r!} + \frac{4r}{r!} \delta_V \right) q^{d-m-1}.$$

6.2. An estimate for the average value set $\mathcal{V}(\mathcal{A})$

Theorem 4 is the critical step in our approach to estimate $\mathcal{V}(\mathcal{A})$.

Corollary 3. *With assumptions as in Lemma 16 and Theorem 4,*

$$|\mathcal{V}(\mathcal{A}) - \mu_d q| \le 2^d \delta_V (3D_V + d^2) q^{1/2} + \frac{7}{4} \delta_V^2 (D_V + 2)^2 d^4 \sum_{k=0}^{d-1} \binom{d}{k}^2 (d-k)!. \tag{19}$$

Proof. According to Lemma 1,

$$
\begin{aligned}
\mathcal{V}(\mathcal{A}) - \mu_d\, q &= \frac{1}{|\mathcal{A}|}\sum_{r=1}^{d}(-1)^{r-1}\left(S_r^{\mathcal{A}} - \frac{|\mathcal{A}|q}{r!}\right) \\
&= \frac{1}{|\mathcal{A}|}\sum_{r=1}^{d}(-1)^{r-1}\left(S_r^{\mathcal{A}} - \frac{q^{d-m}}{r!}\right) - \frac{1}{|\mathcal{A}|}\sum_{r=1}^{d}(-1)^{r-1}\left(\frac{|\mathcal{A}|q - q^{d-m}}{r!}\right) \\
&= \frac{1}{|\mathcal{A}|}\sum_{r=1}^{d}(-1)^{r-1}\left(S_r^{\mathcal{A}} - \frac{q^{d-m}}{r!}\right) + \mu_d\left(\frac{q^{d-m} - |\mathcal{A}|q}{|\mathcal{A}|}\right).
\end{aligned}
\tag{20}
$$

We consider the absolute value of the first sum in the right–hand side of (20). From Lemma 16 and Theorem 4 we have

$$
\frac{1}{|\mathcal{A}|}\sum_{r=1}^{d}\left|S_r^{\mathcal{A}} - \frac{q^{d-m}}{r!}\right| \le \frac{2}{q^{d-m-1}}\sum_{r=1}^{d}\left|S_r^{\mathcal{A}} - \frac{q^{d-m}}{r!}\right|
$$

$$
\le 2q^{\frac{1}{2}}\sum_{r=1}^{d}\frac{\delta_r(D_r - 2) + 2}{r!} + 28\sum_{r=1}^{d}\frac{D_r^2\delta_r^2}{r!} + 2\delta_V\sum_{r=1}^{d}\frac{\binom{r}{2}\delta_{\Delta_r} + 4r}{r!}.
$$

Concerning the first sum in the right–hand side, we see that

$$
\sum_{r=1}^{d}\frac{\delta_r(D_r - 2) + 2}{r!} \le \delta_V\left(D_V\sum_{r=1}^{d}\binom{d}{r} + \sum_{r=1}^{d}\frac{\delta_{\Delta_r}(D_{\Delta_r} - 2) + 2}{r!}\right)
$$

$$
\le \delta_V\left(D_V 2^d + d^2 2^{d-1}\right) = 2^{d-1}\delta_V(2D_V + d^2).
$$

On the other hand,

$$
\sum_{r=1}^{d}\frac{D_r^2\delta_r^2}{r!} = \delta_V^2\left(D_V^2\sum_{r=1}^{d}\frac{\delta_{\Delta_r}^2}{r!} + 2D_V\sum_{r=1}^{d}\frac{D_{\Delta_r}\delta_{\Delta_r}^2}{r!} + \sum_{r=1}^{d}\frac{D_{\Delta_r}^2\delta_{\Delta_r}^2}{r!}\right)
$$

$$
\le \delta_V^2\left(\frac{D_V^2}{4} + D_V + 1\right)\sum_{r=1}^{d}\frac{D_{\Delta_r}^2\delta_{\Delta_r}^2}{r!}
$$

$$
\le \delta_V^2\frac{(D_V + 2)^2}{4}\frac{1}{64}(2d - 1)^4\sum_{k=0}^{d-1}\binom{d}{k}^2(d - k)!.
$$

Finally, we consider the last sum

$$
\sum_{r=1}^{d}\frac{\delta_{\Delta_r}}{2(r-2)!} = \sum_{r=1}^{d}\binom{d}{r}\frac{r(r-1)}{2} = \sum_{k=0}^{d-1}\binom{d}{k}\frac{(d-k)!}{2(d-k-2)!}.
$$

As a consequence, we obtain

$$
\frac{1}{|\mathcal{A}|}\sum_{r=1}^{d}\left|S_r^{\mathcal{A}} - \frac{q^{d-m}}{r!}\right| \le q^{\frac{1}{2}}2^d\delta_V(2D_V + d^2) + \frac{7}{64}\delta_V^2(D_V + 2)^2(2d - 1)^4 \cdot
$$

$$
\cdot\sum_{k=0}^{d-1}\binom{d}{k}^2(d-k)! + \delta_V\sum_{k=0}^{d-1}\binom{d}{k}(d-k)! + 8\delta_V\sum_{k=0}^{d-1}\frac{1}{(d-k-1)!}.
$$

Concerning the second sum in the right–hand side of (20), by Lemma 16 it follows that

$$\left|\frac{q^{d-m} - |\mathcal{A}|q}{|\mathcal{A}|}\right| \leq 4\big(\delta_V(D_V - 2) + 2 + 14D_V^2\delta_V^2 q^{-\frac{1}{2}}\big)q^{\frac{1}{2}}.$$

The statement of the corollary follows by elementary calculations. □

Next we analyze the behavior of the right–hand side of (19). This analysis consists of elementary calculations, which are only sketched.

Fix k with $0 \leq k \leq d - 1$ and denote $h(k) := \binom{d}{k}^2(d - k)!$. From an analysis of the sign of the differences $h(k + 1) - h(k)$ for $0 \leq k \leq d - 1$ we deduce the following remark, which is stated without proof.

Remark 1. Let $k_0 := -1/2 + \sqrt{5 + 4d}/2$. Then h is either an increasing function or a unimodal function in the integer interval $[0, d - 1]$, which reaches its maximum at $\lfloor k_0 \rfloor$.

From Remark 1 we see that

$$\sum_{k=0}^{d-1}\binom{d}{k}^2(d - k)! \leq d\binom{d}{\lfloor k_0\rfloor}^2(d - \lfloor k_0\rfloor)! = \frac{d\,(d!)^2}{(d - \lfloor k_0\rfloor)!\,(\lfloor k_0\rfloor!)^2}. \quad (21)$$

Now we use the following version of the Stirling formula (see, e.g.,[13, p. 747]): for $m \in \mathbb{N}$, there exists θ with $0 \leq \theta < 1$ such that

$$m! = (m/e)^m\sqrt{2\pi m}\,e^{\theta/12m}.$$

By the Stirling formula there exist θ_i $(i = 1, 2, 3)$ with $0 \leq \theta_i < 1$ such that

$$C(d) := \frac{d\,(d!)^2}{(d - \lfloor k_0\rfloor)!\,(\lfloor k_0\rfloor!)^2} \leq \frac{d\,d^{2d+1}e^{-d+\lfloor k_0\rfloor}e^{\frac{\theta_1}{6d} - \frac{\theta_2}{12(d-\lfloor k_0\rfloor)} - \frac{\theta_3}{6\lfloor k_0\rfloor}}}{(d - \lfloor k_0\rfloor)^{d-\lfloor k_0\rfloor}\sqrt{2\pi(d - \lfloor k_0\rfloor)}\lfloor k_0\rfloor^{2\lfloor k_0\rfloor+1}}.$$

By elementary calculations we obtain

$$(d - \lfloor k_0\rfloor)^{-d+\lfloor k_0\rfloor} \leq d^{-d+\lfloor k_0\rfloor}e^{\frac{\lfloor k_0\rfloor(d-\lfloor k_0\rfloor)}{d}}, \qquad \frac{d^{\lfloor k_0\rfloor}}{\lfloor k_0\rfloor^{2\lfloor k_0\rfloor}} \leq e^{\frac{d-\lfloor k_0\rfloor^2}{\lfloor k_0\rfloor}}.$$

It follows that

$$C(d) \leq \frac{d^{d+2}e^{2\lfloor k_0\rfloor}e^{-\frac{\lfloor k_0\rfloor^2}{d} + \frac{1}{6d} + \frac{d-\lfloor k_0\rfloor^2}{\lfloor k_0\rfloor}}}{\sqrt{2\pi}e^d\sqrt{d - \lfloor k_0\rfloor}\lfloor k_0\rfloor}.$$

By the definition of $\lfloor k_0\rfloor$, it is easy to see that $d/\lfloor k_0\rfloor\sqrt{d - \lfloor k_0\rfloor} \leq 5/2$ and that $2\lfloor k_0\rfloor \leq -1 + \sqrt{5 + 4d} \leq -1/5 + 2\sqrt{d}$. Therefore, taking into account that $d \geq 2$, we conclude that

$$C(d) \leq \frac{5}{2}\frac{e^{\frac{109}{30}}d^{d+1}e^{2\sqrt{d}}}{\sqrt{2\pi}e^d}.$$

Combining this bound with Corollary 3 we obtain the following result.

Theorem 5. *For $q > \max\left\{d, 16(D_V \delta_V + 14 D_V^2 \delta_V^2 q^{-\frac{1}{2}})^2\right\}$ and $d \geq m + 2$, the following estimate holds:*

$$|\mathcal{V}(\mathcal{A}) - \mu_d\, q| \leq 2^d \delta_V (3 D_V + d^2) q^{\frac{1}{2}} + 67 \delta_V^2 (D_V + 2)^2 \, d^{d+5} e^{2\sqrt{d}-d}.$$

6.3. *Applications of our main result*

We discuss two families of examples where hypotheses (H_1), (H_2), (H_3) and (H_4) hold. Therefore, the estimate of Theorem 5 is valid for these families.

Our first example concerns linear families of polynomials. Let $L_1, \ldots, L_m \in \mathbb{F}_q[A_{d-1}, \ldots, A_2]$ be polynomials of degree 1. Assume without loss of generality that the Jacobian matrix of $L_1, \ldots, L_m$ with respect to $A_{d-1}, \ldots, A_2$ is of full rank $m \leq d - 2$. Consider the family $\mathcal{A}_{\mathcal{L}}$ defined as

$$\mathcal{A}_{\mathcal{L}} := \left\{ T^d + a_{d-1} T^{d-1} + \cdots + a_0 \in \mathbb{F}_q[T] : L_i(a_{d-1}, \ldots, a_2) = 0 \, (1 \leq i \leq m) \right\}.$$

It is clear that hypotheses (H_1), (H_2) and (H_3) hold. Now we analyze the validity of (H_4). Denote by $\mathcal{L} \subset \mathbb{A}^d$ the linear variety defined by $L_1, \ldots, L_m$, and let $\mathcal{D}(\mathcal{L}) \subset \mathbb{A}^d$ and $\mathcal{S}_1(\mathcal{L}) \subset \mathbb{A}^d$ be the discriminant locus and the first subdiscriminant locus of $\mathcal{L}$. Since the coordinate ring $\overline{\mathbb{F}}_q[\mathcal{L}]$ is a domain, hypothesis (H_4) holds if the coordinate function defined by the discriminant $\mathrm{Disc}(F(A_0, T))$ in $\overline{\mathbb{F}}_q[\mathcal{L}]$ is nonzero, and the class of the subdiscriminant $\mathrm{Subdisc}(F(A_0, T))$ in the quotient ring $\overline{\mathbb{F}}_q[\mathcal{L}]/\mathrm{Disc}(F(A_0, T))$ is not a zero divisor. For fields $\mathbb{F}_q$ of characteristic p not dividing $d(d-1)$, both assertions are consequences of [21, Theorem A.3]. Taking into account that $\delta_{\mathcal{L}} = 1$ and $D_{\mathcal{L}} = 0$, applying Theorem 5 we obtain the following result.

Theorem 6. *For $p := \mathrm{char}(\mathbb{F}_q)$ not dividing $d(d-1)$ and $q > d \geq m + 2$,*

$$|\mathcal{V}(\mathcal{A}_{\mathcal{L}}) - \mu_d\, q| \leq 2^d d^2 q^{\frac{1}{2}} + 268 d^{d+5} e^{2\sqrt{d}-d}.$$

Our second example consists of a nonlinear family of polynomials. Let s, m be positive integers with $m \leq s \leq d - m - 4$, let $\Pi_1, \ldots, \Pi_s$ be the first s elementary symmetric polynomials of $\mathbb{F}_q[A_{d-1}, \ldots, A_2]$ and let $G_1, \ldots, G_m \in \mathbb{F}_q[A_{d-1}, \ldots, A_2]$ be symmetric polynomials of the form $G_i := S_i(\Pi_1, \ldots, \Pi_s)$ $(1 \leq i \leq m)$. Consider the weight function $\mathsf{wt} : \mathbb{F}_q[Y_1, \ldots, Y_s] \to \mathbb{N}$ defined by setting $\mathsf{wt}(Y_i) := i$ $(1 \leq i \leq s)$ and denote by $S_1^{\mathsf{wt}}, \ldots, S_m^{\mathsf{wt}}$ the components of highest weight of $S_1, \ldots, S_m$. Assume that both $S_1, \ldots, S_m$ and $S_1^{\mathsf{wt}}, \ldots, S_m^{\mathsf{wt}}$ form regular sequences of $\mathbb{F}_q[Y_1, \ldots, Y_s]$, and the Jacobian matrices of $S_1, \ldots, S_m$ and $S_1^{\mathsf{wt}}, \ldots, S_m^{\mathsf{wt}}$ with respect to $Y_1, \ldots, Y_s$ have full rank in $\mathbb{A}^s$. We remark that varieties defined by polynomials of this type arise in several combinatorial problems over finite fields (see, e.g.,[4],[8],[21],[7] and [23]). Finally, let

$$\mathcal{A}_\mathcal{N} := \{T^d + a_{d-1}T^{d-1} + \cdots + a_0 \in \mathbb{F}_q[T] : G_i(a_{d-1}, \ldots, a_2) = 0 \, (1 \le i \le m)\}.$$

Hypotheses (H_1), (H_2) and (H_3) hold due to general facts of varieties defined by symmetric polynomials (see [23] for details). Further, it can be shown that (H_4) holds by a generalization of the arguments proving the validity of (H_4) for the linear family $\mathcal{A}_\mathcal{L}$. As a consequence, applying Theorem 5 we deduce the following result.

Theorem 7. *For* $p := \mathrm{char}(\mathbb{F}_q)$ *not dividing* $d(d-1)$, $m \le s \le d - m - 4$ *and* $q > \max\{d, 16(D_\mathcal{N}\delta_\mathcal{N} + 14D_\mathcal{N}^2\delta_\mathcal{N}^2 q^{-\frac{1}{2}})^2\}$, *where* $\delta_\mathcal{N} := \prod_{i=1}^m d_i$ *and* $D_\mathcal{N} := \sum_{i=1}^m (d_i - 1)$, *the following estimate holds:*

$$|\mathcal{V}(\mathcal{A}_\mathcal{N}) - \mu_d \, q| \le 2^d \delta_\mathcal{N}(3D_\mathcal{N} + d^2)q^{\frac{1}{2}} + 67\delta_\mathcal{N}^2(D_\mathcal{N} + 2)^2 \, d^{d+5} e^{2\sqrt{d}-d}.$$

References

[1] B. Birch and H. Swinnerton-Dyer, Note on a problem of Chowla, *Acta Arith.* **5**, 4, pp. 417–423 (1959).

[2] A. Cafure and G. Matera, Improved explicit estimates on the number of solutions of equations over a finite field, *Finite Fields Appl.* **12**, 2, pp. 155–185 (2006).

[3] A. Cafure and G. Matera, An effective Bertini theorem and the number of rational points of a normal complete intersection over a finite field, *Acta Arith.* **130**, 1, pp. 19–35 (2007).

[4] A. Cafure, G. Matera and M. Privitelli, Singularities of symmetric hypersurfaces and Reed-Solomon codes, *Adv. Math. Commun.* **6**, 1, pp. 69–94 (2012).

[5] A. Cafure, G. Matera and M. Privitelli, Polar varieties, Bertini's theorems and number of points of singular complete intersections over a finite field, *Finite Fields Appl.* **31**, pp. 42–83 (2015).

[6] L. Caniglia, A. Galligo and J. Heintz, Equations for the projective closure and effective Nullstellensatz, *Discrete Appl. Math.* **33**, pp. 11–23 (1991).

[7] E. Cesaratto, G. Matera and M. Pérez, The distribution of factorization patterns on linear families of polynomials over a finite field, Preprint arXiv:1408.7014 [math.NT], to appear in Combinatorica (2015).

[8] E. Cesaratto, G. Matera, M. Pérez and M. Privitelli, On the value set of small families of polynomials over a finite field, I, *J. Combin. Theory Ser. A* **124**, 4, pp. 203–227 (2014).

[9] S. Cohen, Uniform distribution of polynomials over finite fields, *J. Lond. Math. Soc. (2)* **6**, 1, pp. 93–102 (1972).

[10] S. Cohen, The values of a polynomial over a finite field, *Glasg. Math. J.* **14**, 2, pp. 205–208 (1973).

[11] V. Danilov, Algebraic varieties and schemes, in I. Shafarevich (ed.), *Algebraic Geometry I, Encyclopaedia of Mathematical Sciences*, Vol. 23. Springer, Berlin Heidelberg New York, pp. 167–307 (1994).

[12] D. Eisenbud, *Commutative Algebra with a View Toward Algebraic Geometry*, *Grad. Texts in Math.*, Vol. 150. Springer, New York (1995).

[13] P. Flajolet and R. Sedgewick, *Analytic combinatorics*. Cambridge Univ. Press, Cambridge (2009).

[14] W. Fulton, *Intersection Theory*. Springer, Berlin Heidelberg New York (1984).

[15] S. Ghorpade and G. Lachaud, Étale cohomology, Lefschetz theorems and number of points of singular varieties over finite fields, *Mosc. Math. J.* **2**, 3, pp. 589–631 (2002).

[16] J. Harris, *Algebraic Geometry: a first course*, *Grad. Texts in Math.*, Vol. 133. Springer, New York Berlin Heidelberg (1992).

[17] J. Heintz, Definability and fast quantifier elimination in algebraically closed fields, *Theoret. Comput. Sci.* **24**, 3, pp. 239–277 (1983).

[18] E. Kunz, *Introduction to Commutative Algebra and Algebraic Geometry*. Birkhäuser, Boston (1985).

[19] G. Lachaud and R. Rolland, On the number of points of algebraic sets over finite fields, *J. Pure Appl. Algebra* **219**, 11, pp. 5117–5136 (2015).

[20] R. Lidl and H. Niederreiter, *Finite fields*. Addison–Wesley, Reading, Massachusetts (1983).

[21] G. Matera, M. Pérez and M. Privitelli, On the value set of small families of polynomials over a finite field, II, *Acta Arith.* **165**, 2, pp. 141–179 (2014).

[22] G. Matera, M. Pérez and M. Privitelli, Explicit estimates for the number of rational points of singular complete intersections over a finite field, *J. Number Theory* **158**, 2, pp. 54–72 (2016).

[23] G. Matera, M. Pérez and M. Privitelli, Number of rational points of symmetric complete intersections over a finite field and applications, Preprint arXiv:1510.03721 [math.NT] (2015).

[24] I. Shafarevich, *Basic Algebraic Geometry: Varieties in Projective Space*. Springer, Berlin Heidelberg New York (1994).

[25] S. Uchiyama, Note on the mean value of $V(f)$, *Proc. Japan Acad.* **31**, 4, pp. 199–201 (1955).

[26] W. Vogel, *Results on Bézout's theorem, Tata Inst. Fundam. Res. Lect. Math.*, Vol. 74. Tata Inst. Fund. Res., Bombay (1984).

The density of unimodular matrices over integrally closed subrings of function fields

G. Micheli

Massachussets Institute of Technology, Cambridge, MA, USA
gmicheli@mit.edu

R. Schnyder

Universität Zürich, Zürich, Switzerland
reto.schnyder@math.uzh.ch

1. Introduction

Let q be a prime power and $F/\mathbb{F}_q$ an algebraic function field in one variable over $\mathbb{F}_q$, with genus g. We will mostly be using the notation of [8] in this paper. Let $\mathcal{P}$ be the set of places of F, and let $\emptyset \neq \mathcal{S} \subsetneq \mathcal{P}$ be a nonempty proper subset. Then the *holomorphy ring* of $\mathcal{S}$ is defined as

$$R := \bigcap_{P \in \mathcal{S}} \mathcal{O}_P.$$

As is well known, these rings are integrally closed and any integrally closed subring of F has this particular form.

In this paper, we wish to study the $k \times m$ matrices over the ring R which can be extended to $m \times m$ unimodular matrices, for some integers $k < m$. In particular, our goal is to compute the density of the set E of such matrices.

This question is an extension of an earlier result [5] about the density of coprime m-tuples over a holomorphy ring. That question, in turn, was inspired by well-known results about tuples of coprime integers: the density of coprime m-tuples of integers is given by $1/\zeta(m)$, where ζ is the Riemann zeta function. See for example [6, 9].

Intuitively, the density of E should give the probability that a randomly selected matrix of size $k \times m$ over R (with entries bounded in some way) lies in E. We define the density of a subset $M \subseteq R$ in analogy to the definition of natural density over the integers. For this, we first need to cover R with a suitable increasing sequence (or net) of finite subsets, for each of which the proportion of elements lying in M is well-defined. The overall density of M is then the limit of these proportions over the sequence of subsets. A straightforward choice for such a net are the Riemann-Roch spaces, as we describe now.

Let $\mathcal{D} := \{D \in \mathrm{Div}(F) \mid D \geq 0, \mathrm{supp}(D) \subseteq \mathcal{P} \setminus \mathcal{S}\}$ be the set of positive divisors of F supported away from $\mathcal{S}$. Let $\mathcal{L}(D) = \{f \in F \mid (f) + D \geq 0\} \cup \{0\}$ denote the Riemann-Roch space associated to a divisor D, and $\ell(D)$ its dimension over $\mathbb{F}_q$. Note that for $D \in \mathcal{D}$, the elements of $\mathcal{L}(D)$ have no poles in the set $\mathcal{S}$, whence $\mathcal{L}(D) \subseteq R$. In fact, R is the union of all $\mathcal{L}(D)$ for $D \in \mathcal{D}$. Since $\mathcal{D}$ is a directed set, these $\mathcal{L}(D)$ therefore form a net of increasing finite subsets covering R, as desired.

For any set T, we denote by $T^{k \times m}$ the set of $k \times m$ matrices having entries in T. For a subset $M \subseteq R^{k \times m}$, we can now define its *density* by setting

$$\overline{\mathbb{D}}(M) := \limsup_{D \in \mathcal{D}} \frac{|M \cap \mathcal{L}(D)^{k \times m}|}{|\mathcal{L}(D)^{k \times m}|},$$

$$\underline{\mathbb{D}}(M) := \liminf_{D \in \mathcal{D}} \frac{|M \cap \mathcal{L}(D)^{k \times m}|}{|\mathcal{L}(D)^{k \times m}|},$$

$$\mathbb{D}(M) := \lim_{D \in \mathcal{D}} \frac{|M \cap \mathcal{L}(D)^{k \times m}|}{|\mathcal{L}(D)^{k \times m}|},$$

the last being defined only when $\overline{\mathbb{D}}(M) = \underline{\mathbb{D}}(M)$. The limits are to be understood as limits of nets via Moore-Smith convergence [4, Chapter 2].

Our main result gives the density of $k \times m$ matrices over R which can be extended to unimodular matrices in the case that $2k - 1 \leq m$.

Theorem 1. *Let* $2k - 1 \leq m$, *then*

$$\mathbb{D}(E) = \prod_{j=m-k+1}^{m} \frac{1}{\zeta_R(j)},$$

where we define the zeta function of R *as*

$$\zeta_R(s) = \prod_{P \in \mathcal{S}} \left(1 - \frac{1}{q^{s \deg(P)}}\right)^{-1}.$$

The arguments we use for the proof of the theorem are similar to the ones in [5]. The basic ingredients are in fact the Riemann-Roch theorem and the Hasse-Weil bound. In any case, the main difficulty we will encounter is giving an estimate for the number of matrices having entries in $\mathcal{L}(D)$ whose reduction modulo a place P is not of full rank. This is provided in Proposition 2.

2. An unconditional upper bound

On the way to our proof of Theorem 1, we first give an upper bound to the superior density $\overline{\mathbb{D}}(E)$. This bound does not require the condition $2k - 1 \leq m$, but holds whenever $k \leq m$.

Note first that the set E of $k \times m$ matrices which can be extended to unimodular matrices is exactly

$$E = \{A \in R^{k \times m} \mid I_A = R\},$$

where I_A is the ideal generated by the $k \times k$ minors of A (see for example [3]). The condition $I_A = R$ holds if and only if $I_A + Q_R = R$ for every place $Q \in \mathcal{S}$, where $Q_R = Q \cap R$, since these Q_R range over all maximal ideals of R.

In order to study the density of E, we proceed by first restricting our attention to only finitely many places. Hence, for a finite subset $\mathcal{Q} \subseteq \mathcal{S}$, we define

$$E_{\mathcal{Q}} := \{A \in R^{k \times m} \mid I_A + Q_R = R \text{ for all } Q \in \mathcal{Q}\}.$$

As discussed above, E is the intersection of the $E_{\mathcal{Q}}$, with $\mathcal{Q}$ ranging over all finite subsets of $\mathcal{S}$. The following lemma gives the density of $E_{\mathcal{Q}}$. It is very similar to a part of the main proof in [5].

Lemma 1. *Let $\mathcal{Q} = \{Q_1, \ldots, Q_n\} \subseteq \mathcal{S}$ be a finite subset of places. The set $E_{\mathcal{Q}} \subseteq R^{k \times m}$ has density*

$$\mathbb{D}(E_{\mathcal{Q}}) = \prod_{i=1}^{n} \prod_{j=0}^{k-1} \left(1 - q^{(j-m)\deg Q_i}\right).$$

Proof. A matrix $A \in R^{k \times m}$ lies in $E_{\mathcal{Q}}$ if and only if for each $i \in \{1, \ldots, n\}$, the ideal I_A is not contained in $(Q_i)_R$, which is to say that the image of A in $(R/(Q_i)_R)^{k \times m} \cong \mathbb{F}_{q^{\deg Q_i}}^{k \times m}$ has nonzero $k \times k$ minors, i.e. has full rank.

Consider now the projection

$$\phi \colon R \to R/((Q_1)_R \cdots (Q_n)_R) \cong \prod_{i=1}^{n} R/(Q_i)_R \cong \prod_{i=1}^{n} \mathbb{F}_{q^{\deg Q_i}},$$

where the first isomorphism is by the Chinese remainder theorem. The number of matrices in $\prod_{i=1}^{n} \mathbb{F}_{q^{\deg Q_i}}^{k \times m}$ with full rank in each component is

$$\prod_{i=1}^{n} \prod_{j=0}^{k-1} (q^{m \deg Q_i} - q^{j \deg Q_i}) \tag{1}$$

by a simple counting argument. These form exactly the image of E_Q under $\phi^{k \times m}$.

Consider now a divisor $D \in \mathcal{D}$. We wish to count the number of matrices in $E_Q \cap \mathcal{L}(D)^{k \times m}$. First, we will show that ϕ maps $\mathcal{L}(D)$ surjectively onto $\prod_{i=1}^{n} \mathbb{F}_{q^{\deg Q_i}}$ if $\deg D$ is large enough.

For this, note that the image of $\mathcal{L}(D)$ under ϕ is $\mathcal{L}(D)/(\mathcal{L}(D) \cap ((Q_1)_R \cdots (Q_n)_R))$. The space

$$\mathcal{L}(D) \cap ((Q_1)_R \cdots (Q_n)_R) = \mathcal{L}(D) \cap Q_1 \cap \cdots \cap Q_n$$

consists of all elements in $\mathcal{L}(D)$ with a root at each Q_i, so it is equal to $\mathcal{L}(D - \sum_{i=1}^{n} Q_i)$. Hence, its dimension as an $\mathbb{F}_q$-vector space is $\ell(D - \sum_{i=1}^{n} Q_i)$, which by the Riemann-Roch theorem is equal to $\deg D - \sum_{i=1}^{n} \deg Q_i + 1 - g$ if $\deg D$ is large enough. On the other hand, the dimension of $\mathcal{L}(D)$ is then $\ell(D) = \deg D + 1 - g$, and so the image has dimension $\sum_{i=1}^{n} \deg Q_i$, the same as $\prod_{i=1}^{n} \mathbb{F}_{q^{\deg Q_i}}$.

This argument also gives us the dimension of the kernel of ϕ restricted to $\mathcal{L}(D)$, which is $\ell(D - \sum_{i=1}^{n} Q_i)$.

From this together with (1), we can count the number of matrices in $E_Q \cap \mathcal{L}(D)^{k \times m}$, and we get that

$$\frac{|E_Q \cap \mathcal{L}(D)^{k \times m}|}{|\mathcal{L}(D)^{k \times m}|} = q^{km(\ell(D - \sum_{i=1}^{n} Q_i) - \ell(D))} \cdot \prod_{i=1}^{n} \prod_{j=0}^{k-1} (q^{m \deg Q_i} - q^{j \deg Q_i})$$

$$= \prod_{i=1}^{n} \prod_{j=0}^{k-1} (1 - q^{(j-m) \deg Q_i})$$

if $\deg D$ is large enough. Taking the limit over all $D \in \mathcal{D}$, we get the claim. $\square$

Proposition 1. *We have an upper bound for the superior density of E given by*

$$\overline{\mathbb{D}}(E) \leq \prod_{j=0}^{k-1} \prod_{Q \in \mathcal{S}} \left(1 - q^{(j-m) \deg Q}\right).$$

Proof. For each finite subset $\mathcal{Q} \subseteq \mathcal{S}$, we have that $E \subseteq E_\mathcal{Q}$, and hence by Lemma 1

$$\overline{\mathbb{D}}(E) \le \mathbb{D}(E_\mathcal{Q}) = \prod_{j=0}^{k-1} \prod_{Q \in \mathcal{Q}} (1 - q^{(j-m)\deg Q}).$$

By taking the limit over all such finite sets $\mathcal{Q}$, we get our claim. $\qquad\square$

Remark 1. The possibly infinite product $\prod_{Q \in \mathcal{S}}(1 - q^{(j-m)\deg Q})$ in Theorem 1 converges absolutely if $j < m - 1$, since it is a subproduct of the reciprocal of the Zeta function of F. If $k = m$, we get the term $\prod_{Q \in \mathcal{S}}\left(1 - q^{-\deg Q}\right)$, which corresponds to the pole of the Zeta function. Nevertheless, since each term is strictly between 0 and 1, the product makes sense, though it may diverge to zero.

3. The exact density in the case $2k - 1 \le m$

In the case where $2k - 1 \le m$, we can prove that the upper bound of Proposition 1 is in fact exact. For this, we will need the following proposition.

Proposition 2. *Let n be an integer and S be a nonempty subset of $\mathbb{F}_{q^n}$. Let N be the set of $k \times m$ full rank matrices having entries in S. Then, independently of n, we have that*

$$|N| \ge \prod_{i=0}^{k-1} (|S|^m - |S|^i).$$

Under these assumptions, this bound is the best possible.

Proof. Let us bound $|N|$ via progressively counting the rows which can occur for a matrix in N. The first row $(v_{1,1}, \dots, v_{1,m})$ of a matrix A in N can be fixed in at least $|S|^m - 1$ ways: everything but the zero row can be extended to be invertible. The second row can be fixed in $|S|^m - |B_1|$ where B_1 are the $\mathbb{F}_{q^n}$-multiples of the first row which have all entries in S. Since the first row is non-zero, there exists $v_{1,j}$ different from zero. By the obvious inclusion of sets, we can now bound $|B_1|$ from above with the number of $\mathbb{F}_{q^n}$-multiples of the first row which have the j-th component in S. These are in bijection with the $a \in \mathbb{F}_{q^n}$ for which $av_{1,j} \in S$, of which there are exactly $|S|$.

We can iterate this procedure as follows: The i-th row $(v_{i,1}, \dots, v_{i,m})$ can be chosen in $|S|^m - |B_{i-1}|$ ways, where B_{i-1} are the rows in S^m which are $\mathbb{F}_{q^n}$-linearly dependent on the first $i - 1$ rows. By construction, the first $i - 1$ rows are linearly independent, which implies that there exists a full rank $(i - 1) \times (i - 1)$ submatrix K. Let $j_1, \ldots, j_{i-1} \in \{1, \ldots, m\}$ be the indices corresponding to the columns of the full rank submatrix. As in the simpler case before, B_{i-1} is contained in the set of rows which are $\mathbb{F}_{q^n}$-linearly dependent on the first $i - 1$ and for which the components $j_1, \ldots, j_{i-1} \in \{1, \ldots, m\}$ lie in S. It is easily seen that the rows satisfying this weaker condition are in bijection with the vectors $w \in \mathbb{F}_{q^n}^{i-1}$ such that $wK \in S^{i-1}$, so there are exactly $|K^{-1}S^{i-1}| = |S|^{i-1}$ of them.

The reader should now observe that this bound is uniform in n and also that it is the best possible general bound, since it is attained when S is a subfield of $\mathbb{F}_{q^n}$. $\qquad\square$

Proposition 3. *If $2k - 1 \leq m$, we have for the inferior density of E*

$$\mathbb{D}(E) = \prod_{j=0}^{k-1} \prod_{Q \in \mathcal{S}} \left(1 - q^{(j-m)\deg Q}\right).$$

Proof. Fix a finite set of places $\mathcal{Q} \subseteq \mathcal{S}$. Having $E \subseteq E_{\mathcal{Q}}$, it is easily seen that (see also [1, Lemma 4])

$$\mathbb{D}(E) = \mathbb{D}(E_{\mathcal{Q}}) - \overline{\mathbb{D}}(E_{\mathcal{Q}} \setminus E).$$

If we prove that $\overline{\mathbb{D}}(E_{\mathcal{Q}} \setminus E)$ tends to zero, the claim follows from Lemma 1 by taking the limit over all finite subsets $\mathcal{Q} \subseteq \mathcal{S}$.

To show this, we first fix the set of places $\mathcal{Q}$, then a divisor $D \in \mathcal{D}$ depending on $\mathcal{Q}$. We can now write:

$$(E_{\mathcal{Q}} \setminus E) \cap \mathcal{L}(D)^{k \times m} \subseteq \{A \in \mathcal{L}(D)^{k \times m} \mid I_A \subseteq P_R \text{ for some } P \in \mathcal{S} \setminus \mathcal{Q}\}.$$

$$= \bigcup_{P \in \mathcal{S} \setminus \mathcal{Q}} \{A \in \mathcal{L}(D)^{k \times m} \mid I_A \subseteq P_R\}.$$

In this union, we can limit ourselves to P of degree less than $k \deg D$: indeed, suppose we have $\deg P \geq k \deg D$ and $A \in \mathcal{L}(D)^{k \times m}$ with $I_A \subseteq P_R$. Since all entries of A are in $\mathcal{L}(D)$ and the $k \times k$ minors are degree k polynomials in these, we see that each such minor lies in $\mathcal{L}(kD)$. Because $I_A \subseteq P_R$, they furthermore lie in $\mathcal{L}(kD - P)$, which is trivial when $\deg(kD - P) = k \deg D - \deg P \leq 0$. The ideal I_A is generated by these minors and is hence (0). It follows that $I_A \subseteq P'_R$ for every P', and so A is already contained in the following restricted union (assuming D is chosen large enough that there is at least one $P' \in \mathcal{S} \setminus \mathcal{Q}$ of degree less than $k \deg D$):

$$(E_{\mathcal{Q}} \setminus E) \cap \mathcal{L}(D)^{k \times m} \subseteq \bigcup_{\substack{P \in \mathcal{S} \setminus \mathcal{Q} \\ \deg P < k \deg D}} \{A \in \mathcal{L}(D)^{k \times m} \mid I_A \subseteq P_R\}.$$

We write $S_P := \{A \in \mathcal{L}(D)^{k \times m} \mid I_A \subseteq P_R\}$.

With this, we can now give the following estimate:

$$\frac{|(E_Q \setminus E) \cap \mathcal{L}(D)^{k \times m}|}{|\mathcal{L}(D)^{k \times m}|} \leq \sum_{\substack{P \in \mathcal{S} \setminus \mathcal{Q} \\ \deg P < k \deg D}} \frac{|S_P|}{|\mathcal{L}(D)^{k \times m}|}.$$

We concentrate on the individual summands. Consider as in Lemma 1 the projection

$$\phi \colon R \to R/P_R \cong \mathbb{F}_{q^{\deg P}}$$

and its restriction to $\mathcal{L}(D)$, as an $\mathbb{F}_q$-linear map. Let d be the dimension of $\phi(\mathcal{L}(D))$. We easily see that $\phi^{k \times m}(S_P)$ is the set of matrices with entries in $\phi(\mathcal{L}(D))$ not of full rank, which by Proposition 2 has cardinality

$$|\phi^{k \times m}(S_P)| \leq q^{mkd} - \prod_{i=0}^{k-1}(q^{md} - q^{id}).$$

Furthermore, the kernel of $\phi|_{\mathcal{L}(D)}$ is $\mathcal{L}(D - P)$, so we get

$$|S_P| \leq q^{mk\ell(D-P)} \cdot \left(q^{mkd} - \prod_{i=0}^{k-1}(q^{md} - q^{id}) \right)$$

$$\leq q^{mk\ell(D-P)} \cdot C q^{(k-1)(m+1)d},$$

where C is a constant depending only on k ($C = 2^k$ works). We have $d = \ell(D) - \ell(D - P)$ by the rank-nullity theorem, so we can write

$$\frac{|S_P|}{|\mathcal{L}(D)^{k \times m}|} \leq q^{mk(\ell(D-P)-\ell(D))} \cdot C q^{(k-1)(m+1)d}$$

$$= q^{-mkd} \cdot C q^{(k-1)(m+1)d}$$

$$= C q^{(k-m-1)d}$$

$$\leq C q^{-kd}, \tag{2}$$

the last follows from the assumption $2k - 1 \leq m$.

We now look at two cases separately.

Case 1: $\deg D < \deg P < k \deg D$. In this case, we see that $\ell(D-P) = 0$ and $d = \ell(D)$. The Hasse-Weil bound says that the number of places of F of degree at most r is asymptotically equal to $\frac{q+1}{q} \frac{q^r}{r}$. With this and (2), we can estimate for large D the sum

$$\sum_{\substack{P \in \mathcal{S} \setminus \mathcal{Q} \\ \deg D < \deg P < k \deg D}} \frac{|S_P|}{|\mathcal{L}(D)^{k \times m}|} \leq C' \cdot \frac{q^{k \deg D}}{k \deg D} \cdot q^{-k\ell(D)},$$

for some new constant $C' = C \cdot (\frac{q+1}{q} + 1)$. Since $\ell(D) = \deg D + 1 - g$ for large D, we can rewrite the above to

$$\frac{C'}{k \deg D} \cdot q^{-k(1-g)}$$

which goes to zero upon taking the limit in D.

Case 2: $\deg P \leq \deg D$. In this case, we note that $\ell(D) \geq \deg D + 1 - g$ and $\ell(D-P) \leq \deg(D-P)$ by Riemann-Roch, and so $d = \ell(D) - \ell(D-P) \geq \deg P + 1 - g$. Using this with (2), we can estimate the sum

$$\sum_{\substack{P \in \mathcal{S} \backslash \mathcal{Q} \\ \deg P \leq \deg D}} \frac{|S_P|}{|\mathcal{L}(D)^{k \times m}|} \leq \sum_{\substack{P \in \mathcal{S} \backslash \mathcal{Q} \\ \deg P \leq \deg D}} C' \cdot q^{-k \deg P},$$

for $C' = Cq^{-k(1-g)}$. Taking the limit in D, we get

$$\sum_{P \in \mathcal{S} \backslash \mathcal{Q}} C' \cdot q^{-k \deg P}.$$

Up to the factor C', this is the tail of a subseries of the Zeta function of F evaluated at q^{-k}, which converges absolutely for $k > 1$. For $k = 1$, $m > 1$, note that the inequality $2k - 1 \leq m$ is not sharp, so the proof still works with a slight modificiation to the estimate (2).

Putting together all we have done so far, we see that

$$\overline{\mathbb{D}}(E_{\mathcal{Q}} \backslash E) = \limsup_{D \in \mathcal{D}} \frac{|(E_{\mathcal{Q}} \backslash E) \cap \mathcal{L}(D)^{k \times m}|}{|\mathcal{L}(D)^{k \times m}|}$$

$$\leq \limsup_{D \in \mathcal{D}} \left(\sum_{\substack{P \in \mathcal{S} \backslash \mathcal{Q} \\ \deg D < \deg P < k \deg D}} \frac{|S_P|}{|\mathcal{L}(D)^{k \times m}|} \right.$$

$$\left. + \sum_{\substack{P \in \mathcal{S} \backslash \mathcal{Q} \\ \deg P \leq \deg D}} \frac{|S_P|}{|\mathcal{L}(D)^{k \times m}|} \right)$$

$$\leq 0 + \sum_{P \in \mathcal{S} \backslash \mathcal{Q}} C' \cdot q^{-k \deg P}.$$

This final sum can be seen as the tail of an absolutely convergent series, and hence it converges to zero as $\mathcal{Q}$ grows. We conclude that

$$\lim_{\text{finite } \mathcal{Q} \subseteq \mathcal{S}} \overline{\mathbb{D}}(E_{\mathcal{Q}} \backslash E) = 0,$$

from which the proposition follows. $\square$

The main result of our paper, Theorem 1, now follows immediately from Propositions 1 and 3.

Remark 2. We tried to keep the tools used in the paper as elementary as possible. Nevertheless, the reader who is willing to pursue a more sophisticated approach (based on p-adic analysis) can refer to a function field version of [7, Lemma 20]. Still, a part of Lemma 1 will be needed and Proposition 2 will still be entirely necessary to satisfy the condition of [7, Lemma 20].

Remark 3. It is worth observing that in [2], this result has been presented for the special case of polynomial rings (which of course can be regarded as holomorphy rings of the rational function field). Unfortunately, the proof of [2] is not correct as it is presented (see the exchange of lim sup and an infinite sum in the proof of [2, Theorem 1]) and needs a fix, which we were able to perform in the case $2k - 1 \leq m$. It would be of great interest if one could adjust this proof to work in the most general case (i.e. $k < m$), which we conjecture to be true at least in the case of the polynomial ring.

Acknowledgements

We like to thank Pietro Speziali, Daniele Bartoli, Bence Csajbók, Jens-Dietrich Bauch for interesting discussions at Fq12 aimed at removing the bound in the main theorem.

The first author was supported in part by Swiss National Science Foundation grant numbers 149716 and 161757. The second author was supported in part by Swiss National Science Foundation grant number 149716 and *Armasuisse*.

References

[1] Dotti, E. and Micheli, G. (2015). Eisenstein polynomials over function fields, To appear on *Applicable Algebra in Engineering Communication and Computing*, http://arxiv.org/abs/1506.05380 doi: 10.1007/s00200-015-0275-2.

[2] Guo, X. and Yang, G. (2013). The probability of rectangular unimodular matrices over $\mathbb{F}_q[x]$, *Linear Algebra and its Applications* **438**, 6, pp. 2675–2682.

[3] Gustafson, W. H., Moore, M. E. and Reiner, I. (1981). Matrix completions over dedekind rings, *Linear and Multilinear Algebra* **10**, 2, pp. 141–144.

[4] Kelley, J. L. (1955). *General topology* (New York: Van Nostrand).

[5] Micheli, G. and Schnyder, R. (preprint). On the density of coprime m-tuples over holomorphy rings, *arXiv:1411.6876 [math.NT] (To appear in International journal of number theory)* doi: 10.1142/S1793042116500536 .

[6] Nymann J. E. (1972). On the probability that k positive integers are relatively prime, *Journal of Number Theory.* **4**, 5, pp. 469–473.

[7] Poonen, B. and Stoll, M. (1998). The Cassels-Tate pairing on polarized abelian varieties, *Annals of Mathematics* **150**, 3, pp. 1109–1149.

[8] Stichtenoth, H. (2009). *Algebraic function fields and codes*, Vol. 254 (Springer).

[9] Ferraguti, A. and Micheli, G. (2015). On Mertens-Cesaro Theorem for Number Fields *Bulletin of the Australian Mathematical Society*, doi: http://dx.doi.org/10.1017/S0004972715001288.

Some open problems arising from my recent finite field research

Gary L. Mullen

Department of Mathematics, The Pennsylvania State University,
University Park, PA 16802, USA `mullen@math.psu.edu`

We discuss a number of open problems and conjectures which have arisen in my recent research related to finite fields. These discussions will focus on a variety of areas including some theoretical topics as well as some topics from combinatorics and algebraic coding theory.

1. Introduction

The results in this paper correspond to my talk of the same title given at "The 12-th International Conference on Finite Fields and Their Applications" held at Skidmore College, in Saratoga Springs, NY, July 13-17, 2015. This is a personal journey through various topics which interest me because in recent years, I have worked on these finite field problems with various colleagues. While we made considerable progress on many of these problems, as will be seen below, unanswered questions still remian.

Throughout this paper q will represent a prime power and $\mathbb{F}_q$ will denote the finite field containing q elements. In an effort to keep this paper to a manageable length, we only provide some details concerning a variety of open problems; we do not provide any proofs. Further details concerning the problems discussed here, along with proofs, can be found in the references [1],[3],[5],[7],[8], and [9]. The volume [15] is a rather comprehensive treatise dealing with various theoretical as well as applied aspects of finite fields.

2. E-perfect codes

The classification of the parameters for all perfect codes is one of the most important and deepest results in algebraic coding theory. In this section, we discuss e-perfect codes and raise a conjecture concerning the classification of the parameters for all e-perfect codes. It will be seen that e-perfect codes are generalizations of perfect codes. Many more details related to e-perfect codes can be found in [1].

We denote a linear code C over $\mathbb{F}_q$ of length n, dimension k, and minimum distance d as an $[n, k, d]$ code. A linear code C is a k-dimensional subspace (and thus contains q^k codewords) of the n-dimensional vector space $\mathbb{F}_q^n$ over $\mathbb{F}_q$. If a code C has length n, has M codewords and minimum distance d, but may not be linear, we will denote C as an (n, M, d) code. It is known that a code C can correct t errors if $t = \lfloor \frac{d-1}{2} \rfloor$.

The following bound can be obtained by counting the number of distinct vectors in the spheres of radius t about each codeword.

Theorem 1. *(Hamming bound) Let C be a t-error-correcting code of length n over $\mathbb{F}_q$. Then*

$$|C| \left[1 + \binom{n}{1}(q-1) + \binom{n}{2}(q-1)^2 + \cdots + \binom{n}{t}(q-1)^t \right] \leq q^n.$$

A code C is **perfect** if the code's parameters yield an equality in the Hamming bound. Sections 8-10 of Chapter 6 in [14] contain a discussion of perfect codes.

The parameters of all perfect codes are known, and can be listed as follows:

The **trivial** perfect codes are:

(1) The zero vector $(0, \ldots, 0)$ of length n;
(2) The entire vector space $\mathbb{F}_q^n$;
(3) The binary repetition code of odd length n.

The non-trivial perfect codes must have the parameters of a Hamming or a Golay code whose parameters can be listed as follows:

(1) The Hamming code $\left[\frac{q^m - 1}{q - 1}, \frac{q^m - 1}{q - 1} - m, 3 \right]$ over $\mathbb{F}_q$, where $m \geq 2$ is a positive integer;
(2) The $[11, 6, 5]$ Golay code over $\mathbb{F}_3$;
(3) The $[23, 12, 7]$ Golay code over $\mathbb{F}_2$.

As defined in [1], a t-error correcting code C with parameters $(n, M, d), t = \lfloor \frac{d-1}{2} \rfloor$, is e-**perfect**, if in the above Hamming bound, equality is achieved when, on the right hand side, q^n is replaced by q^e. An n-perfect code is thus a perfect code.

Based upon considerable machine calculation, the authors of [1], made the following conjecture.

Conjecture 1. *Let C be an (n, M, d) non-trivial e-perfect code over $\mathbb{F}_q$. Then C must have one of the following sets of parameters:*

(1) $\left(\frac{q^m-1}{q-1}, q^{e-m}, 3 \right)$, *with q a prime power and $m < e \leq n$, where $m \geq 2$;*
(2) $(11, 3^{e-5}, 5)$, *with $q = 3$ and $5 < e \leq 11$;*
(3) $(23, 2^{e-11}, 7)$, *with $q = 2$ and $11 < e \leq 23$;*
(4) $(90, 2^{e-12}, 5)$, *with $q = 2$ and $12 < e \leq 89$.*

Problem 1. *Prove this conjecture regarding the classification of e-perfect codes.*

The authors of [1] indicate that they can construct e-perfect codes with each of the parameters listed above, except for the case when $n = 90$ and $e = 89$. This case is essentially open Research Problem 6.7 in [14]. As was the case for perfect codes, there could be many e-perfect codes with a given set of parameters.

3. R-closed subsets of Z_p

Results discussed in this section along with many other results and details as well as proofs related to this problem can be found in [9].

Let G be a finite abelian group with $|G| = g$ and let S be a subset of G with $|S| = s$. We begin with

Definition 1. *Let $0 \leq r \leq s^2$. A set S is r-**closed** if, among the s^2 ordered pairs (a, b) with $a, b \in S$, there are exactly r pairs such that $a + b \in S$. The r-value of the r-closed set S is denoted by $r(S)$.*

The function $r(S)$ can be viewed as a measure of how close the subset S is to being a subgroup of the group G. In particular, if S is a subgroup of G then S is s^2-closed. On the other hand, $r(S)$ can also be viewed as a measure of how close the set S is to being a sum-free set; i.e., if S is a sum-free set then S is 0-closed.

For a given G, what (if anything) can be said about the spectrum of r-values of the subsets of G?

We now state the classical Cauchy-Davenport theorem.

Theorem 2 (Cauchy-Davenport). *If A and B are non-empty subsets of Z_p then $|A + B| \geq \min(p, |A| + |B| - 1)$.*

Motivated by the Cauchy-Davenport Theorem, the authors of [9] were particularly interested in the case when $G = Z_p$ under the operation of addition modulo the prime p.

Later in this section we will make a conjecture (verified computationally in [9] for all primes $p \leq 23$) about the complete spectrum of r-values for any subset cardinality in Z_p. We observe that it has been proved for any prime p; that all conjectured r-values in the spectrum are attained when the subset cardinality s is suitably small ($s < \frac{2p+2}{7}$).

We begin with a result from [9] which relates the r value for a subset S of the group G and the r value of the complement subset T in the group G.

Theorem 3. *Let G be a finite abelian group of order g. Let s be a positive integer with $0 \leq s \leq g$, and let S be a subset of G of size s. Let T be the complement of S in G. Then*

$$r(S) + r(T) = g^2 - 3gs + 3s^2.$$

The function $k[p]$ defined in the next definition plays an important role in the determination of r values.

Definition 2. For p a prime, define

$$k[p] = \lfloor \frac{p+1}{3} \rfloor = \begin{cases} \frac{p-1}{3}, & p \equiv 1 \bmod 3 \\ \frac{p}{3}, & p \equiv 0 \bmod 3 \\ \frac{p+1}{3}, & p \equiv -1 \bmod 3. \end{cases}$$

The importance of the function $k[p]$ can be seen in the next result.

Proposition 1. *Let p be a prime. If $S \subseteq Z_p$ is 0-closed then $|S| \leq k[p]$.*

Our next definition gives lower and upper bounds on the range of possible r values of a set S of cardinality s.

Definition 3. Let p be an odd prime. For $0 \leq s \leq p$, define f_s and g_s as follows:

$$f_s = \begin{cases} 0 & s \le k[p] \\ \frac{(3s-p)^2-1}{4} & s > k[p] \text{ and } s \text{ even} \\ \frac{(3s-p)^2}{4} & s > k[p] \text{ and } s \text{ odd} \end{cases}$$

$$g_s = \begin{cases} \frac{3s^2}{4} & s \le p - k[p] \text{ and } s \text{ even} \\ \frac{3s^2+1}{4} & s \le p - k[p] \text{ and } s \text{ odd} \\ p^2 - 3sp + 3s^2 & s > p - k[p] \end{cases}$$

Note that $f_s + g_{p-s} = p^2 - 3sp + 3s^2$.

Combining several theorems from [9], the authors have established that, for prime $p(> 2)$, the smallest r for which a subset of $\mathbb{Z}_p$ of size s is r-closed is f_s and the largest r for which a subset of $\mathbb{Z}_p$ of size s is r-closed is g_s. We do not include details here as the results are quite technical in nature; instead we refer to the orignal paper [9].

The next result from [9] determines the spectrum for very small and very large subsets S of the additive group Z_p.

Proposition 2. *Let $p > 11$. For $1 \le s \le 3$ and $p - 3 \le s \le p$, the r-values for subsets of Z_p of size s are precisely the integers in the interval $[f_s, g_s]$ with the following exceptions:*

s	f_s	g_s	exceptions
1	0	1	—
2	0	3	2
3	0	7	4
p	p^2	p^2	—
$p-1$	$p^2 - 3p + 2$	$p^2 - 3p + 3$	—
$p-2$	$p^2 - 6p + 9$	$p^2 - 6p + 12$	$p^2 - 6p + 10$
$p-3$	$p^2 - 9p + 20$	$p^2 - 9p + 27$	$p^2 - 9p + 23$

For intermediate values of s, the next definition will be of great importance.

Definition 4. If $4 \le s \le p - 4$, define $V(s)$ by

$$V(s) = \begin{cases} 0 & \text{if } s \le k[p] \\ \lceil \frac{p-s-3}{4} \rceil & \text{if } s \ge \lfloor \frac{p+1}{2} \rfloor \\ \lceil \frac{3s-p-1}{4} \rceil & \text{otherwise} \end{cases}.$$

We are now able to state the main conjecture from [9].

Conjecture 2. *For a prime $p > 11$ and $4 \le s \le p - 4$, there are $V(s)$ exceptional values at the low end of the interval $[f_s, g_s]$ and $V(p-s)$ exceptional values at the high end of the interval $[f_s, g_s]$. All other values in the interval can be obtained as r-values. The exceptions are given by:*

$$f_s + 3i + 1 \text{ for } 0 \le i < V(s) \text{ if } s \equiv p \bmod 2$$

$$f_s + 3i + 2 \text{ for } 0 \le i < V(s) \text{ if } s \not\equiv p \bmod 2$$

$$g_s - 3i - 1 \text{ for } 0 \le i < V(p - s) \text{ if } s \text{ is even}$$

$$g_s - 3i - 2 \text{ for } 0 \le i < V(p - s) \text{ if } s \text{ is odd.}$$

Problem 2. *Prove Conjecture 1.2.*

This conjecture has been verified computationally for all primes $p \le 23$ and all corresponding s with $4 \le s \le p - 4$.

The next result from [9] proves that the conjecture is valid for small values of s, i.e., for $s < \frac{2p+2}{7}$.

Theorem 4. *For any prime p, all conjectured r-values in the spectrum are attained when the subset cardinality s is suitably small ($s < \frac{2p+2}{7}$).*

To obtain this result, the authors of [9] used techniques to actually construct sets S of cardinality s with the desired values of r. For example, they used intervals of consecutive integers and punctured intervals which are intervals with one or more points removed. For these kinds of sets, it is relatively easy to determine the corresponding r values. However in order to resolve Conjecture 1.2 regarding the full spectrum of r values, we believe that new techniques will be required.

We close this section by raising several additional problems.

Problem 3. *Determine the spectrum of r values when the additive group Z_p is replaced by the additive ring Z_n of integers modulo n.*

Even more generally one could ask

Problem 4. *Determine the spectrum of r values when the group Z_p is replaced by an arbitrary finite Abelian group.*

4. Subfield value sets

Let $\mathbb{F}_{q^d}$ be a subfield of $\mathbb{F}_{q^e}$ so that $d|e$. For $f \in \mathbb{F}_{q^e}[x]$, the **subfield value set** of the polynomial f is defined by
$$V_f(q^e; q^d) = \{f(c) \in \mathbb{F}_{q^d} | c \in \mathbb{F}_{q^e}\}.$$
Many details concerning subfield value sets of various kinds of polynomials over finite fields can be found in [3].

We begin by determining the subfield value set of the power polynomial x^n over the field $\mathbb{F}_{q^e}$, see [3] for details.

Theorem 5.
$$|V_{x^n}(q^e; q^d)| = \frac{(n(q^d - 1), q^e - 1)}{(n, q^e - 1)} + 1.$$

We note that when $d = e = 1$, this result reduces to the well known fact that the value set of the polynomial x^n over the finite field $\mathbb{F}_q$ has cardinality $\frac{q-1}{(n,q-1)} + 1$, where as usual, (a,b) denotes the greatest common divisor of the positive integers a and b.

In [3] the authors determined the cardinalities of the subfield value sets of, among various kinds of polynomials, Dickson polynomials over finite fields. The **Dickson polynomial of degree n and parameter $a \in \mathbb{F}_q$** is defined by

$$D_n(x,a) = \sum_{i=0}^{\lfloor n/2 \rfloor} \frac{n}{n-i}\binom{n-i}{i}(-a)^i x^{n-2i}.$$

When the parameter $a = 0$, we note the special case $D_n(x,0) = x^n$ so that Dickson polynomials can be viewed as generalizations of the power or cyclic polynomials x^n. Many algebraic and number theoretic properties of Dickson polynomials over various algebraic structures can be found in [12]. The reader should also consult Section 9.6 of [15] for a recent survey of Dickson polynomial results.

If $a \neq 0 \in \mathbb{F}_q$ it is known that the Dickson polynomial $D_n(x,a)$ induces a permutation on the field $\mathbb{F}_q$ if and only if $(n, q^2 - 1) = 1$; see [13] Theorem 7.16. Moreover in [2], the authors determined the candinality of the value set $V_{D_n(x,a)}$ of the Dickson polynomial $D_n(x,a)$ over the field $\mathbb{F}_q$. Their result can be stated as follows.

Theorem 6. *The cardinality of the value set of the Dickson polynomial of degree n over the field $\mathbb{F}_q$ satisfies*
$$|V_{D_n(x,a)}| = \frac{q-1}{2(n,q-1)} + \frac{q+1}{2(n,q+1)} + \alpha,$$

where α can be explicitly stated and is usually 0.

This result provides motivation for the next result concerning subfield value sets of Dickson polynomials as studied in [3].

In [3] the cardinality of the subfield value set of the Dickson polynomial $D_n(x, a)$ was determined if $a^n \in \mathbb{F}_{q^d}$. As an illustraion of these results we state Theorem 4.6, Part (a), from [3].

Theorem 7. *If q is odd and $a \neq 0 \in \mathbb{F}_{q^e}$ with $a^n \in \mathbb{F}_{q^d}$, $\eta_{q^e}(a) = 1$ and $\eta_{q^d}(a^n) = 1$, then*

$$|V_{D_n(x,a)}(q^e; q^d)| = \frac{(q^e - 1, n(q^d - 1)) + (q^e - 1, n(q^d + 1))}{2(q^e - 1, n)}$$

$$+ \frac{(q^e + 1, n(q^d - 1)) + (q^e + 1, n(q^d + 1))}{2(q^e + 1, n)} - \frac{3 + (-1)^{n+1}}{2}.$$

We note that the function η used in Theorem 1.7 is the quadratic character which is defined by $\eta(b) = 1$ if b is a non-zero square in the field and $\eta(b) = -1$ if b is a non-square in the field.

We also point out that when q is odd, there are three other analogous formulas for the cardinalities of subfield value sets of Dickson polynomials depending on values of the quadratic character η. There are also similar results when q is even which are given in [3] (see Theorem 4.9 of [3]).

In an effort to save space, we only give one of these formulas (Theorem 4.6, Part (a)) and refer to Theorem 4.6 and Theorem 4.9 of [3] for the others.

Problem 5. *Determine the cardinality $|V_{D_n(x,a)}(q^e; q^d)|$ of the subfield value set of the Dickson polynomial $D_n(x, a)$ when $a \in \mathbb{F}_{q^e}^*$ and $a^n \notin \mathbb{F}_{q^d}$.*

The following provides some motivation for this problem. The functional equation for Dickson polynomials states that if $x \in \mathbb{F}_q$ and $x = y + \frac{a}{y}$ with $y \in \mathbb{F}_{q^2}$, then $D_n(x, a) = y^n + \frac{a^n}{y^n}$.

In order to have $D_n(x, a) = y^n + \frac{a^n}{y^n} \in \mathbb{F}_{q^d}$ we require that

$$(y^n + \frac{a^n}{y^n})^{q^d} = y^n + \frac{a^n}{y^n}.$$

If $a^n \in \mathbb{F}_{q^d}$ we obtain the factorization

$$(y^{n(q^d - 1)} - 1)(y^{n(q^d + 1)} - a^n) = 0.$$

In this setting one can then count solutions to this equation to obtain our subfield value set results as illustrated in Theorem 1.7.

However if $a^n \notin \mathbb{F}_{q^d}$, the above equation does not seem to yield a useful factorization and thus the authors of [3] were unable to determine the cardinality of the value set in this case.

5. Hypercubes of class r

In the study of latin squares and hypercubes of higher dimension, one usually defines the order of the object to be the number of distinct symbols in the object. For example, a latin square of order n contains exactly n distinct symbols.

In [5] the authors allow the number of symbols to be a positive integer power of the order of the object under study. Kishen very briefly studied such objects in [10].

We begin this section with the definition of hypercubes of class $r \geq 1$.

Definition 5. Let d, n, r, t be integers, with $d > 0, n > 0, r > 0$ and $0 \leq t \leq d - r$. A (d, n, r, t)-**hypercube of dimension d, order n, class r and type t** is an $n \times \cdots \times n$ (d times) array on n^r distinct symbols such that in every t-subarray (that is, fix t coordinates of the array and allow the remaining $d - t$ coordinates to vary) each of the n^r distinct symbols appears exactly n^{d-t-r} times.

If $d \geq 2r$, two such hypercubes are **orthogonal**, if when superimposed, each of the n^{2r} possible distinct pairs occurs exactly n^{d-2r} times.

A set of such hypercubes is **mutually orthogonal** if any two distinct hypercubes in the set are orthogonal.

We note that a $(2, n, 1, 1)$ hypercube is a latin square of order n and if $r = 1$ we obtain the usual notion of a latin hypercube as studied, for example, in [11].

As an illustration of a cube of class $r = 2$, we consider the cube

$$0\,1\,2 \mid 4\,5\,3 \mid 8\,6\,7$$
$$3\,4\,5 \mid 7\,8\,6 \mid 2\,0\,1$$
$$6\,7\,8 \mid 1\,2\,0 \mid 5\,3\,4$$

which is a hypercube of dimension 3, order 3, class 2, and type 1. Note that the number of symbols in this example is $n^r = 3^2 = 9$.

The following result from [5] provides an upper bound on the maximum number of mutually orthogonal hypercubes.

Theorem 8. *The maximum number of mutually orthogonal hypercubes of dimension d, order n, type t and class r is bounded above by*

$$\frac{1}{n^r-1}\left(n^d-1-\binom{d}{1}(n-1)-\binom{d}{2}(n-1)^2-\cdots-\binom{d}{t}(n-1)^t\right).$$

Corollary 1. *There are at most $n-1$ mutually orthogonal latin squares of order n.*

The following result relates the number of hypercubes of class r with the number of **maximal distance separable** (MDS) codes (a code is an MDS code if its parameters yield an equality in the Singleton bound $d \le n-k+1$), for an $[n,k,d]$ code.

Theorem 9. *Let q be a prime power. The number of $(2r,q,r,r)$-hypercubes is at least the number of linear MDS codes over $\mathbb{F}_q$ of length $2r$ and dimension r.*

Also proved in [5] is the following result.

Theorem 10. *There are at most $(n-1)^r$, $(2r,n,r,r)$ mutually orthogonal hypercubes.*

By using polynomials over finite fields as illustrated in [5], one can construct sets of mutually orthogonal hypercubes. We now illustrate with several examples.

Theorem 11. *Let n be a prime power. For any integer $r < n$, there is a set of $n-1$ mutually orthogonal $(2r,n,r,r)$-hypercubes.*

Theorem 12. *Let $n = 2^{2k}$, $k \in \mathbb{N}$. Then there is a complete set of $(n-1)^2$ mutually orthogonal hypercubes of dimension 4, order n, and class 2.*

In [4] it is shown that if $r = 2$ and n is odd, there is a complete set of $(n-1)^2$ mutually orthogonal hypercubes.

We now provide several problems dealing with sets of mutually orthogonal hypercubes.

(1) Construct a complete set of mutually orthogonal $(4,n,2,2)$-hypercubes when $n = 2^{2k+1}$.
    In [4] it is shown that if $r = 2$ and $n = 2^{2k+1}$ there are $(n-1)(n-2)$ orthogonal hypercubes. Are there $(n-1)^2$ orthogonal hypercubes?

(2) Is the $(n-1)^r$ bound tight when $r > 2$? If so, construct a complete set
of mutually orthogonal $(2r, n, r, r)$-hypercubes of class $r > 2$. If not,
determine a tight upper bound and construct such a complete set.
In [4] it is shown that if $r \geq 1$ and $n \equiv 1 \pmod{r}$, there is a complete
set. This same paper also proves that if $n = p^r k$, there is a complete
set.
(3) Find constructions (other than the standard Kronecker product constructions) for sets of mutually orthogonal hypercubes when n is not a
prime power. Such constructions will require a new method not based
on the use of polynomials over finite fields.
(4) What happens for $d > 2r$? This is a big open problem as not much is
known in the case when $d > 2r$.

6. k-Normal elements

This section is devoted to a brief discussion of k-normal elements in finite
fields, i.e., to elements which generalize normal basis elements in finite
fields. We refer to [8] for many details and proofs.

An element $\alpha \in \mathbb{F}_{q^n}$ yields a **normal basis** for the field $\mathbb{F}_{q^n}$ over $\mathbb{F}_q$ if
the set $\{\alpha, \alpha^q, \dots, \alpha^{q^{n-1}}\}$ is a basis for $\mathbb{F}_{q^n}$ over $\mathbb{F}_q$; such an element α is a
normal element of $\mathbb{F}_{q^n}$ over $\mathbb{F}_q$.

The next theorem characterizes normal elements; see Theorem 2.39 of
[13].

Theorem 13. *For $\alpha \in \mathbb{F}_{q^n}$, the set $\{\alpha, \alpha^q, \dots, \alpha^{q^{n-1}}\}$ is a normal basis
for the field $\mathbb{F}_{q^n}$ over $\mathbb{F}_q$ if and only if the polynomials $x^n - 1$ and $\alpha x^{n-1} +
\alpha^q x^{n-2} + \cdots + \alpha^{q^{n-1}}$ in $\mathbb{F}_{q^n}[x]$ are relatively prime.*

Motivated by this result, we make the

Definition 6. Let $\alpha \in \mathbb{F}_{q^n}$. Denote by $g_\alpha(x)$ the polynomial
$\sum_{i=0}^{n-1} \alpha^{q^i} x^{n-1-i} \in \mathbb{F}_{q^n}[x]$. If $\gcd(x^n - 1, g_\alpha(x))$ over $\mathbb{F}_{q^n}$ has degree k
(where $0 \leq k \leq n - 1$), then α is a k-**normal** element of $\mathbb{F}_{q^n}$ over $\mathbb{F}_q$.

A normal element of $\mathbb{F}_{q^n}$ over $\mathbb{F}_q$ is thus a 0-normal element.

The following generalization of the classical Euler function from elementary number theory plays an important role in determining the number of
k-normal, and hence normal, elements in an extension field.

Definition 7. Let $f \in \mathbb{F}_q[x]$ be a monic polynomial. The *Euler Phi function for polynomials* is defined by

$$\Phi_q(f) = |(\mathbb{F}_q[x]/f\mathbb{F}_q[x])^*|.$$

The reader can find results concerning how to compute the number $\Phi_q(f)$ for a given polynomial f in Lemma 3.69 of [13].

Our next result determines the number of k-normal elements of $\mathbb{F}_{q^n}$ over $\mathbb{F}_q$.

Theorem 14. *The number of k-normal elements of $\mathbb{F}_{q^n}$ over $\mathbb{F}_q$ is given by*

$$\sum_{\substack{h|x^n-1, \\ \deg(h)=n-k}} \Phi_q(h), \tag{1}$$

where divisors are monic and polynomial division is over $\mathbb{F}_q$.

In [16] the authors showed that k-normal elements can be described as cyclic vectors of the Frobenius automorphism. This gives a very useful way to study k-normal elements. However this approach does not appear to be useful in determining the multiplicative order of k-normal elements.

An important extension of the **Normal Basis Theorem** (which states that every extension field has a normal basis over the base field), the **Primitive Normal Basis Theorem** establishes that, for all pairs (q,n), a normal basis $\{\alpha, \alpha^q, \ldots, \alpha^{q^{n-1}}\}$ for $\mathbb{F}_{q^n}$ over $\mathbb{F}_q$ exists with α a primitive element of $\mathbb{F}_{q^n}$.

We ask whether an analogous claim can be made regarding k-normal elements for certain non-zero values of k. In particular, when $k = 1$, does there always exist a primitive 1-normal element of $\mathbb{F}_{q^n}$ over $\mathbb{F}_q$? In [8] the authors proved

Theorem 15. *Let $q = p^e$ be a prime power and n be a positive integer with $p \nmid n$. Assume that $n \geq 6$ if $q \geq 11$, and that $n \geq 3$ if $3 \leq q \leq 9$. Then there exists a primitive 1-normal element of $\mathbb{F}_{q^n}$ over $\mathbb{F}_q$.*

Problem 6. *For which values of q, n, and k can explicit formulas be obtained for the number of k-normal primitive elements of $\mathbb{F}_{q^n}$ over $\mathbb{F}_q$?*

This question is closely related to being able to determine the factorization of the polynomial $x^n - 1$ into irreducibles over $\mathbb{F}_q$. Factoring $x^n - 1$ into a product of irreducibles over $\mathbb{F}_q$ is important in the study of cyclic vectors of the Frobenius automorphism; see [16].

Problem 7. *Obtain a complete existence result for primitive 1-normal elements of $\mathbb{F}_{q^n}$ over $\mathbb{F}_q$ (with or without a computer). We conjecture, based*

upon considerable computational work from [8], that such elements always exist.

Problem 8. *Determine the pairs (n, k) such that there exist primitive k-normal elements of $\mathbb{F}_{q^n}$ over $\mathbb{F}_q$.*

Problem 9. *Determine the existence of high order k-normal elements $\alpha \in \mathbb{F}_{q^n}$ over $\mathbb{F}_q$.*

The author and his undergraduate student (L. Anderson) who studied with him during the Fall Semester 2014 at Penn State, have made the

Conjecture 3. *Let $p \geq 5$ be a prime and let $n \geq 3$. Let (a, k) be one of the four ordered pairs $(a, k) = (1, 0)$ or $(1, 1)$ or $(2, 0)$ or $(2, 1)$. Then there is an element $\alpha \in \mathbb{F}_{p^n}$ of order $\frac{p^n - 1}{a}$ which is k-normal.*

The $a = 1, k = 0$ case of Conjecture 1.3 is really not a conjecture, it is the statement of the primitive normal basis theorem, see Theorem 2.40 of [13]. A proof of Conjecture 1.3 must thus contain a proof of the primitive normal basis theorem.

7. Reversed Dickson polynomials

In this final section we very briefly discuss reversed Dickson polynomials (RDP) as discussed in [7]. Such polynomials are obtained from Dickson polynomials $D_n(x, a)$ by interchanging the roles of x and a. When studying Dickson polynomials one normally considers the parameter a to be fixed, and the variable x then runs through the elements of the finite field $\mathbb{F}_q$. When studying RDPs, we hold the element x fixed and let the variable now be the value a.

As illustrated in Section 4 of this paper, the **Dickson polynomial of degree n and parameter $a \in \mathbb{F}_q$** is defined by

$$D_n(x, a) = \sum_{i=0}^{\lfloor n/2 \rfloor} \frac{n}{n - i} \binom{n - i}{i} (-a)^i x^{n - 2i}.$$

Fix $x \in \mathbb{F}_q$ and let a be the variable in the polynomial $D_n(x, a)$. Some basic results on reversed Dickson polynomials which induce permutations of the field $\mathbb{F}_q$ can be found in [7].

In [7] it was shown that the reversed Dickson polynomial $D_n(a, x)$ is a permutation polynomial on the field $\mathbb{F}_q$ if and only if the polynomial

$D_n(1, x)$ is a permutation polynomial on $\mathbb{F}_q$. It therefore suffices to consider the reversed Dickson polynomial $D_n(1, x)$. The reader should note that with this notation, the variable of the reversed Dickson polynomial is now considered to be x.

As indicated in the next result, reversed Dickson polynomials play an important role in the study of almost perfect nonlinear functions defined over finite fields. A function $f : \mathbb{F}_q \to \mathbb{F}_q$ is **almost perfect nonlinear (APN)** if for each $a \in \mathbb{F}_q^*$ and $b \in \mathbb{F}_q$, the equation $f(x + a) - f(x) = b$ has at most two solutions in $\mathbb{F}_q$. See Section 9.2 of [15] for a discussion of APN functions.

In [7] the following result was obtained.

Theorem 16. *For p an odd prime, if the function x^n is an APN function on the field $\mathbb{F}_{p^{2e}}$ then the polynomial $D_n(1, x)$ is a permutation polynomial on $\mathbb{F}_{p^e}$*
and this implies that x^n is an APN function on the field $\mathbb{F}_{p^e}$.

The following conjecture postulates the complete set of reversed Dickson polynomials which are permutations of the finite field $\mathbb{F}_p$ where $p > 3$ is a prime.

Conjecture 4. *Let $p > 3$ be a prime and let $1 \le n \le p^2 - 1$. Then the polynomial $D_n(1, x)$ induces a permutation polynomial on the field $\mathbb{F}_p$ if and only if*

$$
n = \begin{cases}
2, \ 2p, \ 3, \ 3p, \ p+1, \ p+2, \ 2p+1 & \text{if } p \equiv 1 \pmod{12}, \\
2, \ 2p, \ 3, \ 3p, \ p+1 & \text{if } p \equiv 5 \pmod{12}, \\
2, \ 2p, \ 3, \ 3p, \ p+2, \ 2p+1 & \text{if } p \equiv 7 \pmod{12}, \\
2, \ 2p, \ 3, \ 3p & \text{if } p \equiv 11 \pmod{12}.
\end{cases}
$$

This conjecture has been verified by machine calculation for each prime $p \le 200$.

Problem 10. *Prove Conjecture 1.4 for the permutation polynomial classification of RDPs over $\mathbb{F}_p$.*

Problem 11. *What happens over $\mathbb{F}_q$ when q is a prime power?*

Problem 12. *Determine the cardinalities of the value sets for RDPs over $\mathbb{F}_p$.*

Chapter 7 of [13] provides a detailed study of permutation polynomials over finite fields. The paper [6] and Section 8.1 of [15] also provide recent detailed discussions of permutation polynomials over finite fields.

Acknowledgment The author would like to sincerely thank his co-authors who over the past few years, have worked with him on the various projects discussed in this paper. These co-authors include, in alphabetical order, F. Castro, W.-S. Chou, J. Ethier, J. Gomez-Calderon, X.-D. Hou, S. Huczynska, H. Janwa, D. Panario, I. Rubio, J. Sellers, B. Stevens, D. Thomson, and J. Yucas.

References

[1] F. Castro, H. Janwa, G.L. Mullen, I. Rubio, E-perfect codes, Bull. Institute Combinatorics and Its Applications, 75(2015), 83-90.

[2] W.-S. Chou, J. Gomez-Calderon, and G.L. Mullen, Value sets of Dickson polynomials over finite fields, J. Number Thy. 30(1988), 334-344.

[3] W.-S. Chou, J. Gomez-Calderon, G.L. Mullen, D. Panario, and D. Thomson, Subfield value sets of polynomials over finite fields, Funct. Approx. Comment. Math. 48(2013), 147-165.

[4] D. Droz, Orthogonal Sets of Latin Squares and class r hypercubes generated by finite algebraic systems. Ph.D. thesis, Pennsylvania State University, 2016.

[5] J. Ethier, G.L. Mullen, D. Panario, B. Stevens, and D. Thomson, Sets of orthogonal hypercubes of class r, J. Combinatorial Theory, Ser. A 119(2011), 430-439.

[6] X.-D. Hou, A survey of permutation binomials and trinomials over finite fields, Topics in Finite Fields, Contemporary Math., Amer. Math. Soc., Vol. 632 (2015), 177-191.

[7] X.-D. Hou, G.L. Mullen, J.A. Sellers, and J.L. Yucas, Reversed Dickson polynomials over finite fields, Finite Fields Appl. 15(2009), 748-773.

[8] S. Huczynska, G.L. Mullen, D. Panario, and D. Thomson, Existence and properties of k-normal elements over finite fields, Finite Fields Appl. 24(2013), 170-183.

[9] S. Huczynska, G.L. Mullen, and J.L. Yucas, The extent to which subsets are additively closed, J. Combinatorial Theory, Ser. A., 116(2009), 831-843.

[10] K. Kishen, On the construction of Latin and hyper-Graeco-Latin cubes and hypercubes, J. Indian Society of Agricultural Statistics, Vol. 2 (1950), 20-48.

[11] C.F. Laywine and G.L. Mullen, Discrete Mathematics Using Latin Squares, Wiley, New York, 1998.

[12] R. Lidl, G.L. Mullen, and G. Turnwald, Dickson Polynomials, Longman Scientific and Technical, Pitman Mono. and Surveys in Pure and Appl. Math., Essex, United Kingdom, Vol. 65, 1993.

[13] R. Lidl and H. Niederreiter, <u>Finite Fields</u>, Sec. Ed., Cambridge University Press, Cambridge, 1997.

[14] F.J. MacWilliams and N.J.A. Sloane, <u>The Theory of Error-Correcting Codes</u>, North-Holland, Amsterdam, 1977.

[15] G.L. Mullen and D. Panario, <u>Handbook of Finite Fields</u>, CRC Press, Boca Raton, FL, 2013.

[16] D. Thomson and C. Weir, k-normal elements are cyclic vectors of Frobenius, preprint.

On coefficients of powers of polynomials and their compositions over finite fields

Gary L. Mullen

Department of Mathematics, The Pennsylvania State University,
University Park, PA 16802, USA, mullen@math.psu.edu

Amela Muratović-Ribić

University of Sarajevo, Department of Mathematics, Zmaja od Bosne
33-35, 71000 Sarajevo, Bosnia and Herzegovina, amela@pmf.unsa.ba

Qiang Wang

School of Mathematics and Statistics, Carleton University, 1125 Colonel
By Drive, Ottawa, Ontario, K1S 5B6, Canada,
wang@math.carleton.ca

For any given polynomial f over the finite field $\mathbb{F}_q$ with degree at most $q - 1$, we associate it with a $q \times q$ matrix $A(f) = (a_{ik})$ consisting of coefficients of its powers $(f(x))^k = \sum_{i=0}^{q-1} a_{ik}x^i$ modulo $x^q - x$ for $k = 0, 1, \ldots, q - 1$. This matrix has some interesting properties such as $A(g \circ f) = A(f)A(g)$ where $(g \circ f)(x) = g(f(x))$ is the composition of the polynomial g with the polynomial f. In particular, $A(f^{(k)}) = (A(f))^k$ for any k-th composition $f^{(k)}$ of f with $k \geq 0$. As a consequence, we prove that the rank of $A(f)$ gives the cardinality of the value set of f. Moreover, if f is a permutation polynomial then the matrix associated with its inverse $A(f^{(-1)}) = A(f)^{-1} = PA(f)P$ where P is an antidiagonal permutation matrix. As an application, we study the period of a nonlinear congruential pseudorandom sequence $\bar{a} = \{a_0, a_1, a_2, \ldots\}$ generated by $a_n = f^{(n)}(a_0)$ with initial value a_0, in terms of the order of the associated matrix. Finally we show that $A(f)$ is diagonalizable in some extension field of $\mathbb{F}_q$ when f is a permutation polynomial over $\mathbb{F}_q$.

1. Introduction

Let $\mathbb{F}_q$ be the finite field of order $q = p^n$ where p is a prime number and n is a positive integer. Let $f(x) = \sum_{i=0}^{q-1} a_i x^i$ be a polynomial over $\mathbb{F}_q$ with degree at most $q-1$. To compute its composition with another polynomial $g(x) = \sum_{i=0}^{q-1} b_i x^i$, we can either use interpolation to obtain its expression directly, or calculate all the powers $f(x)^i \pmod{x^q - x}$ in the expression $(g \circ f)(x) = \sum_{i=0}^{q-1} b_i (f(x))^i$.

Denote by

$$(f(x))^k = \sum_{i=0}^{q-1} a_{ik} x^i \mod (x^q - x)$$

the k-th power of the polynomial $f(x)$ for $k = 1, 2, \ldots, q-1$. Denote by f_0 the zero polynomial in $\mathbb{F}_q[x]$. If $f \neq f_0$ we will define $(f(x))^0 = 1$ and $f_0(x)^0 = 0$.

For any polynomial $f(x) = \sum_{i=0}^{q-1} a_i x^i$ we associate a coefficient vector v_f with it, namely,

$$v_f = (a_0, a_1, \ldots, a_{q-1})^T.$$

We define a $q \times q$ matrix associated with $f(x) \neq f_0(x)$ by

$$A(f) = \begin{bmatrix} 1 & a_{01} & a_{02} & \cdots & a_{0,q-2} & a_{0,q-1} \\ 0 & a_{11} & a_{12} & \cdots & a_{1,q-2} & a_{1,q-1} \\ \vdots & \vdots & & \ddots & \vdots & \vdots \\ 0 & a_{q-2,1} & a_{q-2,2} & \cdots & a_{q-2,q-2} & a_{q-2,q-1} \\ 0 & a_{q-1,1} & a_{q-1,2} & \cdots & a_{q-1,q-2} & a_{q-1,q-1} \end{bmatrix},$$

where the k-th column consists of the coefficients of the $(k-1)$-th power of $f(x)$. In particular, we define $A(f_0)$ to be the zero $q \times q$ matrix. We note that we can build the matrix $A(f)$ by directly computing each of the k-th powers of $f(x)$ modulo $x^q - x$. For example, finding each column of $A(f)$ takes $q^{1+o(1)}$ bit operations using the result of Kedlaya and Umans [9]. On the other hand, using Lagrange's interpolation $f(x)^k = \sum_{\alpha \in \mathbb{F}_q} f(\alpha)^k (1 - (x - \alpha)^{q-1})$, one can obtain the explicit expression for all the entries of $A(f)$. Namely, for all $1 \leq i, j \leq q-1$, we have $a_{ij} = -\sum_{\alpha \in \mathbb{F}_q} f(\alpha)^j \binom{q-1}{i} (-\alpha)^{q-1-i}$ and $a_{0j} = f(0)^j = \sum_{\alpha \in \mathbb{F}_q} f(\alpha)^j (1 - (-\alpha)^{q-1})$.

The *Bell matrix* of an analytic function f is an infinite matrix defined as

$$B[f]_{jk} = \frac{1}{j!} \left[\frac{d^j}{dx^j} (f(x))^k \right]_{x=0},$$

where $(f(x))^k = \sum_{j=0}^{\infty} B[f]_{jk} x^j$. It is sometimes called a *Jabotinsky matrix*. The transpose of a Bell matrix is called a *Carleman matrix*, which is often used in iteration theory to find the continuous iteration of a function [8].

In this paper we show that our matrix $A(f)$ of a polynomial f over $\mathbb{F}_q$ is indeed a finite field analogue of the Bell matrix. Some fundamental properties in terms of the composition of polynomials are proved similarly. Moreover, we derive a few results specifically related to finite field theory. In Section 2 we show that the matrix associated with the composition of two polynomials over a finite field is the product of two associated matrices. That is, $A(g \circ f) = A(f)A(g)$. As a corollary, we prove that the value set size of any polynomial f over $\mathbb{F}_q$ is the rank of its associated matrix $A(f)$, which is equivalent to an earlier result of Chou and Mullen [3], which deals with the transpose of the $(1,1)$-minor of $A(f)$. In Section 3, we concentrate on permutation polynomials over $\mathbb{F}_q$. In particular, we prove that the associated matrix for the compositional inverse $f^{(-1)}$ satisfies $A(f^{(-1)}) = PA(f)P$, where P is an antidiagonal permutation matrix defined by $P_{i,q-i} = 1$ for $i = 1, \ldots, q$ and zero otherwise. Moreover, we show $A(f)$ is diagonalizable in some extension field of $\mathbb{F}_q$. Throughout this paper, we note that $f^k(x)$ or $(f(x))^k$ denotes the k-th power of $f(x)$ modulo $x^q - x$, while $f^{(k)}(x)$ denotes the k-th composition of $f(x)$ modulo $x^q - x$.

2. The matrix of a composition of polynomials

First we derive the following obvious result.

Proposition 1. *Let* $f(x) = \sum_{i=0}^{q-1} a_i x^i \in \mathbb{F}_q[x]$ *and* $g(x) = \sum_{i=0}^{q-1} b_i x^i \in \mathbb{F}_q[x]$. *Then*

$$v_{g \circ f} = A(f)v_g.$$

Proof. The $(i+1)$-th coordinate of $A(f)v_g$ is given by $(A(f)v_g)_{i+1} = \sum_{k=0}^{q-1} a_{ik} b_k$. On the other hand, we obtain $g \circ f(x) = \sum_{k=0}^{q-1} b_k (f(x))^k = \sum_{k=0}^{q-1} b_k \sum_{i=0}^{q-1} a_{ik} x^i = \sum_{i=0}^{q-1} (\sum_{k=0}^{q-1} b_k a_{ik}) x^i$. Therefore we obtain $(v_{g \circ f})_{i+1} = \sum_{k=0}^{q-1} a_{ik} b_k = (A(f)v_g)_{i+1}$ for $i = 0, 1, \ldots q - 1$. $\square$

Theorem 1. *Let* $g(x) = \sum_{i=0}^{q-1} c_i x^i \in \mathbb{F}_q[x]$ *and* $f(x) = \sum_{j=0}^{q-1} a_j x^j \in \mathbb{F}_q[x]$. *Let* $(g \circ f)(x) = g(f(x))$ *be the composition of* g *with* f. *Then*

$$A(g \circ f) = A(f)A(g).$$

Proof. By Proposition 1, we see that $A(f)v_{g^k} = v_{g^k \circ f}$ for any k-th power of the polynomial g. Let $\sigma_k(x) = x^k$. Because the composition of polynomials is an associative operation, we have $g^k \circ f = (\sigma_k \circ g) \circ f = \sigma_k \circ (g \circ f) = (g \circ f)^k$. Therefore $A(f)v_{g^k} = v_{(g \circ f)^k}$ for all $k = 0, 1, 2 \ldots, q-1$. Partitioning the matrix $A(g)$ with columns $v_{g^0}, v_g, v_{g^2}, \ldots, v_{g^{q-1}}$, we derive

$$A(f)A(g) = \left(A(f)v_{g^0}, A(f)v_g, A(f)v_{g^2}, \ldots, A(f)v_{g^{q-1}} \right)$$

$$= \left(v_{(g \circ f)^0}, v_{(g \circ f)}, v_{(g \circ f)^2}, \ldots, v_{(g \circ f)^{q-1}} \right) = A(g \circ f).$$

$\square$

We recall that $f^k(x)$ denotes the k-th power of $f(x)$, while $f^{(k)}(x)$ denotes the k-th composition of $f(x)$.

Corollary 1. *For any given polynomial $f \in \mathbb{F}_q[x]$ we have that $A(f^{(k)}) = (A(f))^k$, for any $k = 1, 2, \ldots$.*

This provides an algebraic way to study the composition of polynomials in terms of multiplication of matrices. Although the matrices associated with polynomials are large and costly to build, this still gives us some interesting theoretical consequences. We note that the transpose of the $(1,1)$-minor of $A(f)$ was earlier studied by Chou and Mullen [3]. They gave a result on the size of the value set of f in terms of the rank of the $(1,1)$-minor of $A(f)$; see also page 234 in [13]. However, our proof is different.

Corollary 2. *Let f be a polynomial over a finite field $\mathbb{F}_q$ and $|V_f|$ be the size of the value set $V_f = \{f(a) \mid a \in \mathbb{F}_q\}$ of f. Then $|V_f| = rank(A(f))$.*

Proof. If $f(x) \in \mathbb{F}_q[x]$ is not a permutation polynomial then we define $D = V_f$ and let $g \in \mathbb{F}_q[x]$ be a nonzero polynomial of least degree m such that $g : D \to \{0\}$. Let $g(x) = b_m x^m + b_{m-1} x^{m-1} + \cdots + b_1 x + b_0$. Then we have $g \circ f(x) = 0$, and thus $A(f)v_g = 0$ by Proposition 1. This means that the first $m+1$ columns of $A(f)$ are linearly dependent and thus the coefficients of g determine a linear dependence among the polynomials $1, f(x), f^2(x), \ldots, f^m(x)$ in the sense that $\sum_{i=0}^m b_i(f(x))^i = 0$. Moreover, $(f(x))^0, f(x), \ldots, (f(x))^{m-1}$ are linearly independent because $g(x)$ is the lowest degree polynomial such that $g \circ f = 0$. Therefore, $rank(A(f)) = deg(g(x)) = |V_f|$.

If $f \in \mathbb{F}_q[x]$ is permutation polynomial, then all the powers of f and corresponding columns of $A(f)$ are linearly independent. $\square$

Corollary 2 states that the size of the value set of f is given by the rank of the matrix $A(f)$. One would also like to know which elements $c \in \mathbb{F}_q$ show up in the value set V_f of f, and if c shows up in the value set, how many times does it appear?

Again, we consider the polynomial $f(x) = a_0 + a_1 x + \cdots + a_{q-1} x^{q-1}$ over $\mathbb{F}_q$. First we want to find the number of nonzero solutions to $f(x) = c$. Let us consider the polynomial $h(x) = (a_0 + a_{q-1} - c) + a_1 x + \cdots + a_{q-2} x^{q-2}$. Then, by the König-Rados Theorem (Theorem 6.1 in [10]), the number of nonzero solutions to $f(x) = c$ is $q - 1 - r$, where r the rank of the $(q-1) \times (q-1)$ left circulant matrix

$$
C(h) := \begin{bmatrix}
a_0 + a_{q-1} - c & a_1 & \cdots & a_{q-2} \\
a_1 & a_2 & \cdots\ a_0 + a_{q-1} - c \\
a_2 & a_3 & \cdots & a_1 \\
\vdots & \vdots & \vdots & \vdots \\
a_{q-3} & a_{q-2} & \cdots & a_{q-4} \\
a_{q-2} & a_0 + a_{q-1} - c & \cdots & a_{q-3}
\end{bmatrix}.
$$

Therefore, if $c \neq f(0)$, then c appears in the value set V_f of f if and only if the rank of the matrix $C(h)$ is less than $q - 1$. And the number of times that c appears in the value set V_f of f is r if and only if the rank of the matrix $C(h)$ is $q - 1 - r$. If $c = f(0)$ then c appears in the value set $q - r$ times.

Let k be the largest positive integer such that $\{1, f, \ldots, f^k\}$ is linearly independent over $\mathbb{F}_q$. Then we have $rank(A(f)) \geq k + 1$. For example, let $f \in \mathbb{F}_q[x]$ be a polynomial of degree d, then it is clear that $1, f, \ldots, f^{\lfloor (q-1)/d \rfloor}$ are linearly independent and thus the value set V_f has size $|V_f| \geq \lfloor (q-1)/d \rfloor + 1$. We note that polynomials satisfying $|V_f| = \lfloor (q-1)/d \rfloor + 1$ are called *minimum value set polynomials*. The classification of minimum value set polynomials is the subject of several papers; see [1, 2, 6, 7, 11]. Using the discussion after Corollary 2, we have the following.

Corollary 3. *Let f be a polynomial of degree d over the finite field $\mathbb{F}_q$. Then f is a minimum value set polynomial if and only if $rank(A(f)) = \lfloor (q-1)/d \rfloor + 1$. Equivalently, $\{1, f, \ldots, f^{\lfloor (q-1)/d \rfloor}\}$ is a basis which spans the space of all nonnegative powers of f.*

Let us consider the $(1,1)$-minor $M(f)$ of $A(f)$. If the i-th row of $M(f)$ consists entirely of 0's or entirely of 1's, set $l_i = 0$. Otherwise for a nonzero i-th row of $M(f)$, arrange the entries in a circle and define l_i to be the maximum number of consecutive zeros appearing in this circular arrangement. Let L_f be the maximum of the values of l_i, where the maximum is taken over all of the $q-1$ rows of the matrix $M(f)$. Using the linear independence of these columns, we can derive a lower bound of the size of the value set V_f.

Corollary 4. *(Theorem 1,[4]) Let f be a polynomial over $\mathbb{F}_q$ and L_f be defined as above. Then $|V_f| \geq L_f + 2$.*

In [12], Mullen fully classified polynomials $f(x)$ over finite fields which commute with linear permutations, that is, $f(bx + a) = bf(x) + a$. We note that $A(bx + a)$ is an upper triangular matrix. Comparing the second column of $A(bx + a)A(f) = A(f)A(bx + a)$, one can derive the following corollary.

Corollary 5. *(Theorem 1,[12]) The polynomial $f(x) = b_0 + b_1x + \cdots + b_{q-1}x^{q-1}$ satisfies $f(bx + a) = bf(x) + a$ if and only if*

$$b_0(b - 1) = -a + \sum_{t=1}^{q-1} b_t a^t, \tag{1}$$

$$b_s(1 - b^{s-1}) = b^{s-1} \sum_{t=s+1}^{q-1} \binom{t}{s} a^{t-s} b_t, \quad (1 \leq s \leq q - 1)$$

3. Permutation polynomials

Permutation polynomials over the field $\mathbb{F}_q$ under the operation of functional composition form a group isomorphic to the symmetric group $(S_q, \circ)$ with $q!$ elements. There is a representation of the permutation polynomials in terms of circulant matrices such that its centralizer is commutative [14], but here we consider the representation of f in terms of the invertible matrix $A(f)$. We note that the mapping $f \to A(f)$ is one-to-one. Hermite's criterion (Theorem 7.4 in [10]) states that $f(x)$ is permutation polynomial if and only if the coefficient $a_{q-1,k}$ in the k-th power of $f(x)$ is 0 for all $k = 1, 2, \ldots, q - 2$ and $f(x)$ has exactly one root in $\mathbb{F}_q$, say $f(e) = 0$. This means that all entries of the last row of $A(f)$ are zero except $a_{q-1,q-1} = 1$. Indeed, $f(x)^{q-1} = 1$ if $x \neq e$ and $f(x)^{q-1} = 0$ if $x = e$. Hence
$$f(x)^{q-1} = \sum_{a \neq e}(1 - (x - a)^{q-1})$$
$$= x^{q-1} - \sum_{i=1}^{q-2}\sum_{a \neq e} \binom{q-1}{i}(-a)^{q-1-i}x^i + \sum_{a \neq e}(1 - (-a)^{q-1}).$$

Therefore $a_{q-1,q-1} = 1$. Moreover, $a_{0,q-1} = 1$ if $e \neq 0$, and $a_{0,q-1} = 0$ if $e = 0$.

We now consider the compositional inverse $f^{(-1)}$ of a permutation polynomial f with respect to composition. Since $A(g \circ f) = A(f)A(g)$ and the matrix associated with $f^{(0)}(x) = Id(x) = x$ is the identity matrix, it is easy to see that $A(f^{(-1)}) = A(f)^{-1}$. Moreover, we have

Theorem 2. *Let f be a permutation polynomial of $\mathbb{F}_q$. Let P be the antidiagonal permutation matrix, i.e. P is defined by $P_{i+1,q-i} = 1$ for $i = 0, 1, \ldots, q-1$ and zero otherwise. Then $A(f^{(-1)}) = (A(f))^{-1} = PA(f)P$.*

Proof. Obviously, $A(f^{(-1)}) = (A(f))^{-1}$. Denote the k-th power of f and the inverse polynomial $f^{(-1)}$ by $f^k(x) = \sum_{i=0}^{q-2} a_{ik}x^i$ and $(f^{(-1)}(x))^k = \sum_{i=0}^{q-2} b_{ik}x^i$ respectively, for $k = 1, 2, \ldots, q-1$. For any permutation polynomial $g(x) = \sum_{i=0}^{q-2} c_i x^i$, it is well known (see for example [15]) that its coefficients can be calculated by $c_i = -\sum_{s \in \mathbb{F}_q} s^{q-1-i}g(s)$, for $i = 0, 1, \ldots, q-2$.

For $1 \leq k \leq q-2$, by Hermite's criterion, the polynomial $f^k(x)$ must have degree at most $q-2$. Therefore we have for $0 \leq i \leq q-2$ and $1 \leq k \leq q-2$,

$$a_{ik} = -\sum_{s \in \mathbb{F}_q}(f(s))^k s^{q-1-i} = -\sum_{s \in \mathbb{F}_q} s^k (f^{(-1)}(s))^{q-1-i}) = b_{q-1-k,q-1-i},$$

i.e.

$$a_{ik} = b_{q-1-k,q-1-i}, \ for \ 0 \leq i \leq q-2 \ and \ 1 \leq k \leq q-2.$$

Moreover, $a_{q-1,k} = 0$ for $1 \leq k \leq q-2$ and $a_{q-1,q-1} = 1$. In addition, $a_{q-1,k} = b_{q-1-k,0}$ by the definition of $A(f^{(-1)})$.

On the other hand, for any polynomial $g(x) = \sum_{i=0}^{q-1} c_i x^i$, it is well known that its coefficients can be calculated by

$$c_i = -\sum_{s \in \mathbb{F}_q} s^{q-1-i}g(s),$$

for $i = 1, \ldots, q-2$ and

$$c_0 + c_{q-1} = -\sum_{s \in \mathbb{F}_q} g(s)^{q-1}.$$

Hence we can compute

$$a_{i,q-1} = -\sum_{s \in \mathbb{F}_q} s^{q-1-i}(f(s))^{q-1} = -\sum_{s \in \mathbb{F}_q}(f^{(-1)}(x))^{q-1-i}s^{q-1} = b_{0,q-1-i},$$

for $1 \leq i \leq q - 2$ and

$$a_{0,q-1} + a_{q-1,q-1} = - \sum_{s \in \mathbb{F}_q} f(s)^{q-1} = - \sum_{s \in \mathbb{F}_q} s^{q-1} = 1.$$

Because $a_{q-1,q-1} = 1$, we have $a_{0,q-1} = 0$, which is equal to $b_{q-1,0}$ by definition. Hence we have proven that

$$a_{ik} = b_{q-1-k,q-1-i},$$

for all $0 \leq i \leq q - 1$ and $1 \leq k \leq q - 1$. Since the multiplication by P on both sides reverses the order of rows and columns of $A(f)$, it follows that $A(f^{(-1)}) = PA(f)P$. □

Corollary 6. *Let f be a permutation polynomial and P be the antidiagonal permutation matrix as defined in Theorem 2. Then the matrix $PA(f)$ is the inverse of itself.*

Proof. By Theorem 2, we have $(A(f))^{-1} = A(f^{(-1)}) = PA(f)P$. Therefore $(P(A(f))^2 = I$. □

Corollary 7. *Let S be a group of invertible $q \times q$ matrices over $\mathbb{F}_q$ equipped with the operation $A * B = B \cdot A$ where $B \cdot A$ denotes the usual product of the matrices B and A. Denote by f_π the permutation polynomial of degree at most $q - 2$ induced by a permutation $\pi \in S_q$. Then the mapping $\psi : S_q \to S$ given by $\psi(\pi) = A(f_\pi)$ is a monomorphism and thus S_q is isomorphic to the subgroup $\mathcal{A} = \{A(f_\pi) | \pi \in S_q\}$ of the group S.*

Proof. It is easy to show that $(S, *)$ is a group and that the mapping ψ is injective. Now $\psi(\pi \circ \alpha) = A(f_\pi \circ f_\alpha) = A(f_\alpha) \cdot A(f_\pi) = \psi(\pi) * \psi(\alpha)$. □

Finally we comment on some potential applications of our results in sequence designs. For any permutation polynomial f, we define a nonlinear congruential pseudorandom sequence $\bar{a} = \{a_0, a_1, a_2, ...\}$ such that $a_n = f^{(n)}(a_0)$ and a_0 is the initial value. The period of $\bar{a}$ is equal to the smallest k such that $f^{(k+i)}(a_0) = f^{(i)}(a_0)$ for some i. Character sums of these sequences are studied in [5, 16–19]. For each initial value that is not fixed by f, we find the period of the nonlinear congruential pseudorandom sequence. If we take K as the least common multiple of all these periods, then we obtain $f^{(K)} = id$ and thus $A(f^{(K)}) = I$. Conversely, if $A(f)^K = I$ then the period of the nonlinear congruential sequence is a divisor of the order of the invertible matrix. Next we demonstrate the following two simple examples although they can be obtained easily without using these matrices.

Let $f(x) = x^m$ be a polynomial over $\mathbb{F}_q$ such that $(m, q-1) = 1$. Then $A(f)$ is a permutation array such that the only nonzero entry in column k is in $(km \pmod{q-1}, k)$ position where $1 \leq k \leq q - 1$. The period of $\bar{a}$ is well known, which is the order of m modulo $q - 1$.

Let $f(x) = ax + b \in \mathbb{F}_p[x]$, where a is a primitive element in $\mathbb{F}_p$ and $b \neq 0$. Then

$$A(f) = \begin{bmatrix} 1 & b & b^2 & \dots & b^{p-2} & 1 \\ 0 & a & 2ab & \dots & (p-2)ab^{p-3} & (p-1)ab^{p-2} \\ \vdots & \vdots & \ddots & \vdots & & \vdots \\ 0 & 0 & 0 & \dots & a^{p-2} & (p-1)a_{p-2}b \\ 0 & 0 & 0 & \dots & 0 & 1 \end{bmatrix}.$$

The matrix is an upper triangular matrix such that its eigenvalues are all the nonzero elements $(a^k, k = 1, \dots, p-1)$ in $\mathbb{F}_p$ and the multiplicity of 1 is 2. Hence the period of A is equal to $p - 1$, the least common multiple of orders of these eigenvalues.

Computing the order of the matrix $A(f)$ associated with a permutation polynomial f provides an algebraic way to find out the period of this kind of pseudorandom sequence, although the matrix $A(f)$ itself is costly to build. For example, finding each column of $A(f)$ takes $q^{1+o(1)}$ bit operations using the result of Kedlaya and Umans [9]. We wonder whether we could overcome this drawback by pre-computing the initial matrix, or taking a sparse matrix, or diagonalizing the matrix. As an attempt, we end our paper with a diagonalization result of $A(f)$ over some extension field of $\mathbb{F}_q$.

Theorem 3. *Let $f \in \mathbb{F}_q[x]$ be a permutation polynomial of $\mathbb{F}_q$ such that the disjoint cycles $C_1, C_2, \dots, C_k$ of f have lengths $L_1, L_2, \dots, L_k$ respectively. Let K be an extension field of $\mathbb{F}_q$ that contains all solutions of the equations $x^{L_i} - 1 = 0$ for $i = 1, 2, \dots, k$ and ψ_i be a fixed primitive L_i-th root of unity in K for each i. Then $A(f)$ is diagonalizable with eigenvalues ψ_i^j for $i = 1, \dots, k$ and $j = 0, \dots, L_i - 1$.*

Proof. For each cycle C_i we pick a starting point (arbitrarily) and denote it by b_0, so our cycle can be denoted by $(b_0, b_1, \dots, b_{L_i-1})$. For each j such that $0 \leq j \leq L_i - 1$, we can define

$$g_{i,j}(x) = \begin{cases} (\psi_i^j)^t & if \ x = b_t \in C_i; \\ 0 & if \ x \in \mathbb{F}_q \setminus C_i. \end{cases}$$

Obviously,

$$g_{i,j}(f(x)) = (\psi_i^j)g_{i,j}(x)$$

i.e., each $g_{i,j}$ produces an eigenvector of $A(f)$ with the corresponding eigenvalue (ψ_i^j). Indeed, if $x \notin C_i$, then $f(x) \notin C_i$ and so $g_{i,j}(x) = 0 = g_{i,j}(f(x))$. If $x \in C_i$ then $x = b_t$ for some $t = 0, 1, \ldots, L_i - 1$. Then $f(x) = b_{t+1 \pmod{L_i}}$. Thus $g_{i,j}(f(x)) = (\psi_i^j)^{t+1} = (\psi_i^j)(\psi_i^j)^t = (\psi_i^j)g_{i,j}(x)$. In this way we obtain a set $\{g_{i,j}(x) : i = 1, \ldots, k, j = 0, 1, \ldots, L_i - 1\}$ of q polynomials in $K[x]$. For each fixed i, it is easy to see that $\{g_{i,j}(x) : j = 0, \ldots, L_i - 1\}$ is linearly independent because ψ_i is a primitive L_i-th root of unity. Moreover, if $i \neq i'$ then $g_{i,j}(x)g_{i',j'}(x) = 0$. Therefore the set of q polynomials $\{g_{i,j}(x) : i = 1, \ldots, k, j = 0, 1, \ldots, L_i - 1\}$ is linearly independent. Because the size of $A(f)$ is q and these $g_{i,j}(x)$'s provide us q linearly independent eigenvectors corresponding to eigenvalue ψ_i^j, the proof is complete. $\qquad\square$

Remark 1. From the proof of Theorem 3, all polynomials $g(x) \in K[x]$ such that $g(f(x)) = \lambda g(x)$ for some λ satisfy

$$g(x) = \sum_{i=1}^{k} \sum_{j=0}^{L_i-1} a_{i,j}g_{i,j}(x).$$

Remark 2. Theorem 3 can be extended to non permutation polynomials such that either x or $f(x)$ is in a cycle of the functional graph of f, that is, the tail length of any element in the functional graph is at most one. For each such a leaf d in the functional graph of f, we define the function

$$g_{i,d}(x) = \begin{cases} 1 & if \ x = d; \\ 0 & if \ x \neq d. \end{cases}$$

Obviously, $d \notin V_f$. Hence $g_{i,d}(f(x)) = 0 = 0g_{i,d}(x)$ for all $x \in \mathbb{F}_q$ and thus $g_{i,d}(x)$ derives an eigenvector corresponding to the eigenvalue 0. Together with the eigenvectors corresponding to the nodes in the cycles, we have q linearly independent eigenvectors and thus $A(f)$ is diagonalizable. However, in general $A(f)$ is not necessarily diagonalizable in each of its extension fields. For example, let $f(x) = x^2 + x + 1 \in \mathbb{F}_5[x]$. Then

$$A(f) = \begin{pmatrix} 1 & 1 & 1 & 1 & 1 \\ 0 & 1 & 2 & 1 & 0 \\ 0 & 1 & 3 & 2 & 0 \\ 0 & 0 & 2 & 2 & 0 \\ 0 & 0 & 1 & 1 & 0 \end{pmatrix}$$

It is easy to check that the rank of $A(f)$ is 3 over $\mathbb{F}_5$. However, eigenvalues of $A(f)$ over $\mathbb{R}$ are $5, 1, 1, 0, 0$ and thus are $0, 1, 1, 0, 0$ over $\mathbb{F}_5$. Hence $A(f)$ cannot be diagonalizable over any extension field of $\mathbb{F}_5$.

Acknowledgment   We thank the anonymous referee for helpful suggestions. Research of Qiang Wang is partially supported by NSERC of Canada.

References

[1] H. Borges and R. Conceicao, *On the characterization of minimal value set polynomials*, J. Number Theory 133 (2013), 2021-2035.

[2] L. Carlitz, D. J. Lewis, W. H. Mills, and E. G. Straus, *Polynomials over finite fields with minimal value sets*, Mathematika 8 (1961), 121-130.

[3] W.-S. Chou and G. L. Mullen, *A note on value sets of polynomials over finite fields*, preprint, 2012.

[4] P. Das and G. L. Mullen, *Value sets of polynomials over finite fields*, Finite Fields with Applications to Coding Theory, Cryptography and Related Areas (Oaxaca, 2001), 80-85, Springer, Berlin, 2002.

[5] D. Gomez and A. Winterhof, *Character sums for sequences of iterations of Dickson polynomials*, Finite Fields and Applications, 147-151, Contemp. Math., 461, Amer. Math. Soc., Providence, RI, 2008.

[6] J. Gomez-Calderson, *A note on polynomials with minimal value set over finite fields*, Mathematika 35 (1988), 144-148.

[7] J. Gomez-Calderon and D. J. Madden, *Polynomials with small value set over finite fields*, J. Number Theory 28 (1988), no. 2, 167-188.

[8] P. Gralewicz and K. Kowalski, *Continuous time evolution from iterated maps and Carleman linearization*, Chaos Solitons Fractals 14 (2002), no. 4, 563-572.

[9] K. S. Kedlaya and C. Umans, *Fast polynomial factorization and modular composition*, SIAM J. Comput. 40 (2011), no. 6, 1767-1802.

[10] R. Lidl and H. Niederreiter, *Finite Fields*, Encyclopedia of Mathematics and Its Applications 20, Cambridge University Press, 1997.

[11] W. H. Mills, *Polynomials with minimal value sets*, Pacific J. Math 14 (1964), 225-241.

[12] G. L. Mullen, *Polynomials over finite fields which commute with linear permutations*, Proc. Amer. Math. Soc. 84 (1982), no. 3, 315-317.

[13] G. L. Mullen and D. Panario, *Handbook of Finite Fields*, CRC Press, Boca Raton, FL, 2013.

[14] A. Muratović-Ribić, *Representation of polynomials over finite fields with circulants*, Sarajevo J. Math. 1(13) (2005), no. 1, 21-26.

[15] A. Muratović-Ribić, *A note on the coefficients of inverse polynomials*, Finite Fields Appl. 13 (2007) 977-980.

[16] H. Niederreiter, *Design and analysis of nonlinear pseudorandom number generators*, Monte Carlo Simulation, A.A. Balkema Publishers, Rotterdam, 2001, 3-9.

[17] H. Niederreiter and I. E. Shparlinski, *On the distribution and lattice structure of nonlinear congruential pseudorandom numbers*, Finite Fields Appl. 5 (1999), 246-253.

[18] H. Niederreiter and I. E. Shparlinski, *Recent advances in the theory of nonlin-*

ear pseudorandom number generators, Monte Carlo and Quasi-Monte Carlo Methods, 2000 (Hong Kong), 86-102, Springer, Berlin, 2002.

[19] H. Niederreiter and A. Winterhof, *Multiplicative character sums for nonlinear recurring sequences*, Acta Arith. 111 (2004), 299-305 .

On the structure of certain reduced linear modular systems

E. Orozco

University of Puerto Rico at Rio Piedras, PR,
edusmildo.orozco1@upr.edu

1. Introduction

The discrete Fourier transform of a multidimensional array of real or complex numbers can be efficiently computed with a fast Fourier transform (FFT) algorithm. If the length of the array is a prime number, the FFT is known as a *multidimensional FFT of prime edge–length.*

Let p be any prime and $\mathbb{F}_q$ be a finite field of characteristic p. Given a nonsingular linear modular system (LMS) $L_S = (\mathbb{F}_q^n, S)$ over $\mathbb{F}_q$ and a nonsingular matrix M that commutes with S, a *reduced linear modular system* (RLMS) R_{MS}, associated with L_S, is the finite dynamical system that results from the action of M on the cycles of L_S. The structure of such systems resemble the cycle structure of LMSs. The study of RLMSs involves two problems. The first one is to determine the cycle structure of an RLMS R_{MS} when S and M are nonsingular. The second problem is, given a nonsingular LMS L_S, find an associated nonsingular RLMS R_{MS} with the least possible number of cycles. The solution to the last problem has implications in the efficient computation of multidimensional FFTs of prime edge–length with linear symmetries in its inputs [8]. When the characteristic polynomial of an LMS is irreducible over $\mathbb{F}_p$, it is shown in [9] that the structure of any associated RLMS is composed of cycles of equal length, besides the trivial zero–cycle of length one. In this case, an *optimal* RLMS is one where the characteristic polynomial of M is primitive and the cycle structure of the RLMS is composed of one nontrivial cycle with length depending on the periods of S and M.

A matrix is said to be non–derogatory if its minimal and characteristic polynomials are the same. In this work we present some results about the case when the minimal polynomials of S and M are nontrivial powers of irreducible polynomials and S and M are non–derogatory matrices over $\mathbb{F}_q$. We also present a connection with Lucas' theorem and the Chinese remainder theorem to solve a special matrix version of the discrete logarithm problem.

1.1. *Notation*

Throughout this document, p is prime and $\mathbb{F}_q$ denotes the finite field of size q of characteristic p. The multiplicative order of $\alpha \in \mathbb{F}_q^*$ is denoted by k_α and $\langle \alpha \rangle$ represents the cycle group generated by α. If $\alpha, \beta \in \mathbb{F}_q^*$, we define $\bar{k}_{\alpha,\beta} = \frac{k_\alpha}{\gcd(k_\alpha, k_\beta)}$. If g is a generator of $\mathbb{F}_q^*$, the *index* of α with respect to g, is denoted by $\operatorname{ind}_g(\alpha)$. Given a matrix S, ϕ_S and m_S represent the characteristic and the minimal polynomial of S, respectively. If n and x are positive integers and $\alpha \in \mathbb{F}_q$, $J_\alpha(n)$ represents a Jordan block matrix of size n with α in its main diagonal and $J_\alpha^x(n)$ denotes $J_\alpha(n)$ raised to the power x. If t is a positive integer, $r_t = \lfloor \log_p(t-1) \rfloor + 1$ represents the least positive integer for which $p^{r_t} \geq t$. We use LMS and RLMS to abbreviate *linear modular system* and *reduced linear modular system*, respectively.

2. System of binomial congruences

Lucas' theorem is a classical result that gives a straightforward method for computing a binomial coefficient modulo p [1].

Theorem 1 (Lucas 1878). *Let* $m = m_0 + m_1 p + m_2 p^2 + \cdots + m_d p^d$ *and* $n = n_0 + n_1 p + n_2 p^2 + \cdots + n_d p^d$ *be the p–ary expansions of the nonnegative integers* m *and* n. *Then*

$$\binom{m}{n} \equiv \binom{m_0}{n_0}\binom{m_1}{n_1} \cdots \binom{m_d}{n_d} \pmod{p}.$$

Many researchers have studied, generalized, and applied Lucas' theorem to new branches of mathematics. McIntosh applied it to integer functions [6]. Granville generalized it to compute binomial coefficients modulo prime powers [4]. Sun and Davis [5] applied it for combinatorial congruences modulo prime powers and Loveless [7] investigated certain products of binomial coefficients modulo a composite number. In this section we use Lucas' theorem to determine when certain system of congruence equations modulo p has a solution and provide a formula for it, if such a solution exists.

Theorem 2. *Let p be any prime, $c_1, c_2, \ldots, c_n$ be a sequence of integers with $0 \leq c_i \leq p - 1$, and $d = \lfloor \log_p n \rfloor$. Then, the system of congruences*

$$\binom{x}{k} \equiv c_k \pmod{p}, \ (1 \leq k \leq n), \tag{1}$$

in the unknown x, has a solution b if and only if

$$c_k \equiv \binom{b_0}{k_0}\binom{b_1}{k_1}\binom{b_2}{k_2}\cdots\binom{b_d}{k_d} \pmod{p}, \ (1 \leq k \leq n),$$

where

$$b = b_0 + b_1 p + b_2 p^2 + \cdots + b_d p^d, \ 0 \leq b_i \leq p - 1,$$
$$k = k_0 + k_1 p + k_2 p^2 + \cdots + k_d p^d, \ 0 \leq k_i \leq p - 1,$$

and in particular, $c_{p^i} = b_i$ for $0 \leq i \leq d$. The solution, if it exists, is unique modulo p^{d+1}.

Proof. The system of congruence equations (1) has a solution b if and only if

$$\binom{b}{k} \equiv c_k \pmod{p}, \ (1 \leq k \leq n),$$

and by Lucas' theorem this is true if and only if

$$c_k \equiv \binom{b_0}{k_0}\binom{b_1}{k_1}\binom{b_2}{k_2}\cdots\binom{b_d}{k_d} \pmod{p}, \ (1 \leq k \leq n).$$

In particular,

$$c_{p^i} \equiv \binom{b_0}{0}\binom{b_1}{0}\binom{b_2}{0}\cdots\binom{b_i}{1}\cdots\binom{b_d}{0} = b_i \pmod{p}, \ (1 \leq i \leq d).$$

Clearly, $b_0 + b_1 p + b_2 p^2 + \cdots + b_d p^d < p^{d+1}$. $\qquad\square$

Example 1. Find a solution, if it exists, of the system of congruences

$$\binom{x}{k} \equiv c_k \pmod{7}, \ (1 \leq k \leq 8), \text{ where } c = (3, 3, 1, 0, 0, 0, 1, 3) \tag{2}$$

Solution: Let $d = \lfloor \log_7 8 \rfloor = 1$. By Theorem 2, System (2) has the solution $b = c_1 + 7c_7 = 3 + 7 \cdot 1 = 10$ if and only if

$$c_k \equiv \binom{b_0}{k_0}\binom{b_1}{k_1} \equiv \binom{3}{k_0}\binom{1}{k_1} \pmod{7}, \ (1 \leq k \leq 8).$$

In this case, assuming the congruences are modulo 7,

$$c_1 \equiv \binom{3}{1}\binom{1}{0} \equiv 3, \ c_2 \equiv \binom{3}{2}\binom{1}{0} \equiv 3, \ c_3 \equiv \binom{3}{3}\binom{1}{0} \equiv 1,$$

$$c_4 \equiv \binom{3}{4}\binom{1}{0} \equiv 0, \ c_5 \equiv \binom{3}{5}\binom{1}{0} \equiv 0, \ c_6 \equiv \binom{3}{6}\binom{1}{0} \equiv 0,$$

$$c_7 \equiv \binom{3}{0}\binom{1}{1} \equiv 1, \ c_8 \equiv \binom{3}{1}\binom{1}{1} \equiv 3.$$

Therefore, $b = 10$ is the unique solution modulo 49 of System (2).

2.1. *Some basic results*

Lemmas 1 and 2 can be found in [9]. Lemmas 4 and 5 are the key ingredients to prove the main result of this section.

Lemma 1. *Let* $a, b, m > 0$ *be integers,* $d = \gcd(a, m)$, *with* a *and* b *not both zero and suppose that* $\gcd(b, d) = 1$. *Then, a solution* (u, v) *for*

$$ax \equiv by \pmod{m}$$

is (rb, d), *where* r *is a positive integer such that* $r\frac{a}{d} \equiv 1 \pmod{\frac{m}{d}}$. *Furthermore, the least positive integer* y_0 *such that* (x_0, y_0) *satisfies* $ax_0 \equiv by_0$ (mod m), *for some integer* x_0, *is* $y_0 = d$.

If G is a cyclic group and g is a generator of G, we say that the *index* of $a \in G$ with respect to g, written $\text{ind}_g(a)$, is the least positive integer t such that $a = g^t$.

Lemma 2. *Let* G *be a cyclic group of order* m *and* g *be a generator of* G. *Then, for any* $\alpha \in G$ *of order* k_α *there is an integer* $1 \leq r_\alpha \leq k_\alpha$ *with* $\gcd(r_\alpha, k_\alpha) = 1$ *such that* $\text{ind}_g(\alpha) = r_\alpha \frac{m}{k_\alpha}$.

Lemma 3 and its proof are given for the sake of completeness since they are used in several subsequent results. This lemma establishes a necessary and sufficient condition for an element to belong to a cyclic group. Let $\langle a \rangle = \{a, a^2, \dots\}$.

Lemma 3. *Let* $\alpha, \beta \in G$. *Then* $\beta \in \langle \alpha \rangle$ *if and only if* $k_\beta | k_\alpha$. *If this is the case, then there exists an integer* δ, $1 \leq \delta \leq k_\beta$, *such that* $(\alpha^{\frac{k_\alpha}{k_\beta}})^\delta = \beta$.

Proof. First, assume $\beta \in \langle \alpha \rangle$. Then, $\beta = \alpha^j$, for some integer $1 \leq j \leq k_\alpha$. Thus, $k_\beta = \frac{k_\alpha}{\gcd(k_\alpha, j)}$. Then, $k_\beta | k_\alpha$.

Now, assume $k_\beta | k_\alpha$ and let $t = \frac{k_\alpha}{k_\beta}$. Let r_α and r_β be the integers of Lemma 2 associated with α and β, respectively. Note that $r_\alpha^{-1} r_\alpha \equiv 1$ (mod k_α) implies that $r_\alpha^{-1} r_\alpha = 1 + k_\alpha s$, for some integer s. Also, note that $\alpha^{tr_\alpha^{-1} r_\beta} = g^{\text{ind}_g(\alpha) tr_\alpha^{-1} r_\beta}$, where g is a generator of G. Let $j = tr_\alpha^{-1} r_\beta$ (mod k_α). Thus, $\alpha^j = g^{r_\alpha \frac{m}{k_\alpha} \frac{k_\alpha}{k_\beta} r_\alpha^{-1} r_\beta} = g^{r_\beta \frac{m}{k_\beta} r_\alpha^{-1} r_\alpha} = g^{\text{ind}_g(\beta)(1 + k_\alpha s)} = \beta$, since $k_\alpha = k_\beta t$. Therefore, $\beta \in \langle \alpha \rangle$.

The computation of j is reduced to finding an integer δ, $1 \leq \delta \leq k_\beta$, such that $(\alpha^{\frac{k_\alpha}{k_\beta}})^\delta = \beta$. Thus, $j \equiv \frac{k_\alpha}{k_\beta} \delta$ (mod k_α). $\qquad\square$

In [9], the *reach* of β with respect to α, $\text{reach}_\alpha(\beta)$, is defined as the least positive integer ι for which $\alpha^j = \beta^\iota$, for some positive integer j. However, no explicit formulas for the computation of ι or its associated j are provided. Lemma 4 provides such formulas.

Lemma 4. *Let $\alpha, \beta \in \mathbb{F}_q^*$. Then,*

    i. $\text{reach}_\alpha(\beta) = \bar{k}_{\beta,\alpha}$ and

    ii. there exists a positive integer δ such that $\alpha^{k_{\alpha,\beta}\delta} = \beta^{\bar{k}_{\beta,\alpha}}$ and the least positive integer j satisfying $\alpha^j = \beta^{\bar{k}_{\beta,\alpha}}$ is $j = \bar{k}_{\alpha,\beta}\delta$.

Proof. Part i : By definition, $\text{reach}_\alpha(\beta)$ is the least positive integer ι for which $\alpha^j = \beta^\iota$, for some positive integer j. This equation, written in index calculus, is equivalent to $j\,\text{ind}_g(\alpha) \equiv \iota\,\text{ind}_g(\beta) \pmod{q-1}$. Using Lemma 2, this equation can be rewritten as $r_\alpha \bar{k}_{\beta,\alpha}\, j \equiv r_\beta \bar{k}_{\alpha,\beta}\, \iota \pmod{k_\alpha \bar{k}_{\beta,\alpha}}$, where $\bar{k}_{\alpha,\beta} = \frac{k_\alpha}{\gcd(k_\alpha, k_\beta)}$ and $\bar{k}_{\beta,\alpha} = \frac{k_\beta}{\gcd(k_\alpha, k_\beta)}$. Applying Lemma 1, $\iota = \bar{k}_{\beta,\alpha}$ is the $\text{reach}_\alpha(\beta)$, for some j. In other words, $\iota \equiv 0 \pmod{\bar{k}_{\beta,\alpha}}$.

Part ii : The order of $\beta^{\bar{k}_{\beta,\alpha}}$ is equal to $\gcd(k_\alpha, k_\beta)$ and, clearly, it divides k_α. Then, by Lemma 3, there exists a positive integer δ such that $\alpha^{k_{\alpha,\beta}\delta} = \beta^{\bar{k}_{\beta,\alpha}}$. It is easy to see that $j = \bar{k}_{\alpha,\beta}\delta$ is the least positive integer for which $\alpha^j = \beta^{\bar{k}_{\beta,\alpha}}$. $\square$

As we shall see in Section 4, in order to determine the cycle structure of a nonsingular RLMS $R_{MS} = (\mathcal{F}_S, M)$, associated with a nonsingular LMS $L_S = (\mathbb{F}_p^n, S)$, we have to solve the matrix equation

$$S^x = M^y, \tag{3}$$

in the unknowns x and y. The notion of the reach of β with respect to α also applies to nonsingular square matrices. The least positive integer y, for some positive integer x, is known as the *reach* of M with respect to S and is denoted by $\text{reach}_S(M)$. Since S and M are nonsingular, such a number always exists. In the case that S is non–derogatory (i.e., $\phi_S = m_S$) with ϕ_S equal to a nontrivial power of an irreducible polynomial over $\mathbb{F}_p$, equation (3) leads us to an equivalent equation

$$J_\alpha^x(n) = J_\beta^y(n), \tag{4}$$

where $\alpha, \beta \in \mathbb{F}_q^*$ are the roots of ϕ_S and ϕ_M, respectively.

Lemma 5. *Let $\alpha, \beta \in \mathbb{F}_q^*$ such that $\alpha\beta^{-1} \notin \mathbb{F}_p^*$. Then, the least positive integer y'', for some positive integer x'', that satisfies the system of congruences*

$$\binom{x}{k}\beta^k \equiv \binom{y}{k}\alpha^k \pmod{p}, \ (1 \leq k \leq n) \tag{5}$$

in the unknowns x and y, is $y'' = p^{d+1}$ and the corresponding $x'' = p^{d+1}$, where $d = \lfloor \log_p n \rfloor$.

Proof. It is clear that $(x'', y'') = (p^{d+1}, p^{d+1})$ satisfies Equation (5). Now, assume that (x', y') is also a solution, where x' and y' are positive integers. Then,

$$\binom{x'}{k} \equiv \binom{y'}{k} (\alpha\beta^{-1})^k \pmod{p}, \ (1 \le k \le n).$$

Let $x' = x'_0 + x'_1 p + \cdots + x'_s p^s$ and $y' = y'_0 + y'_1 p + \cdots + y'_r p^r$ be the base p representations of x' and y', respectively. Observe that the condition $\alpha\beta^{-1} \notin \mathbb{F}_p^*$ implies that either α or β must be in $\mathbb{F}_q^*$ but not in $\mathbb{F}_p^*$. Applying Lucas' theorem we see that $x'_0 \equiv y'_0 (\alpha\beta^{-1}) \pmod{p}$. Since $x'_0 \in \mathbb{F}_p$, it must be the case that $y'_0 = 0$ since $\alpha\beta^{-1}$ is not an integer modulo p. In what follows we prove that $y'_1 = \cdots = y'_d = 0$ by using the same idea.

Let $c_k \equiv \binom{y'}{k} (\alpha\beta^{-1})^k \pmod{p}, \ (1 \le k \le n)$. Then,

$$c_k \equiv \binom{y'_0}{k_0}\binom{y'_1}{k_1} \cdots \binom{y'_r}{k_r} (\alpha\beta^{-1})^k \pmod{p}, \ (1 \le k \le n).$$

In particular, for $k = p^i$,

$$c_{p^i} \equiv \binom{y'_0}{0}\binom{y'_1}{0} \cdots \binom{y'_i}{1} \cdots \binom{y'_r}{0} (\alpha\beta^{-1})^{p^i}$$

$$\equiv y'_i(\alpha\beta^{-1})^{p^i} \pmod{p}, \ (0 \le i \le d).$$

Then,

$$c_{p^i} - y'_i(\alpha\beta^{-1})^{p^i} \equiv (c_{p^i} - y_i(\alpha\beta^{-1}))^{p^i} \equiv 0 \pmod{p}$$

if and only if

$$c_{p^i} - y'_i(\alpha\beta^{-1}) \equiv 0 \pmod{p}$$

if and only if

$$c_{p^i} \equiv y'_i(\alpha\beta^{-1}) \pmod{p}$$

if and only if $y'_i = 0$, $(0 \le i \le d)$, since $\alpha\beta^{-1}$ cannot be an integer. Thus, $y' = y'_{d+1}p^{d+1} + \cdots + y'_r p^r \ge p^{d+1}$ and $c_{p^i} \equiv 0 \pmod{p}$.

On the other hand, if $k = k_0 + k_1 p + \cdots + k_d p^d$ is not a power of p, there is $0 \le t \le d$ for which $y'_t \ne 0$.

$$c_k \equiv \binom{y'_0}{k_0}\binom{y'_1}{k_1} \cdots \binom{y'_t}{k_t} \cdots \binom{y'_d}{k_d}\binom{y'_{d+1}}{0} \cdots \binom{y'_r}{0} (\alpha\beta^{-1})^k \pmod{p}$$

$$\equiv \binom{0}{k_0}\binom{0}{k_1} \cdots \binom{0}{k_t} \cdots \binom{0}{k_d} (\alpha\beta^{-1})^k \pmod{p}$$

$$\equiv 0 \pmod{p}$$

since $\binom{0}{k_t} = 0$. Therefore, $c_k \equiv 0 \pmod{p}$, $(1 \leq k \leq n)$. Finally, since (x', y') satisfies System (5), $\binom{x'}{k} \equiv 0 \pmod{p}$, $(1 \leq k \leq n)$. Therefore, by a straightforward application of Theorem 2, $x' \equiv 0 \pmod{p^{d+1}}$. □

The following theorem is crucial in the determination of the cycle structure of certain reduced linear modular systems that will be studied in Section 4.

Theorem 3. *Let $\alpha, \beta \in \mathbb{F}_q^*$ such that $\alpha\beta^{-1} \notin \mathbb{F}_p^*$. Then, the least positive integer y'', for some positive integer x'', that satisfies the system*

$$\alpha^x = \beta^y \tag{6}$$

$$\binom{x}{k}\alpha^k \equiv \binom{y}{k}\beta^k \pmod{p}, \ (1 \leq k \leq n) \tag{7}$$

in the unknowns x and y, is $y'' = \bar{k}_{\beta,\alpha}p^{d+1}$ and its corresponding $x'' = \bar{k}_{\alpha,\beta}\delta p^{d+1}$, where $d = \lfloor \log_p n \rfloor$, and δ is defined as in Lemma 3.

Proof. Clearly, $(x'', y'') = (\bar{k}_{\alpha,\beta}\delta p^{d+1}, \bar{k}_{\beta,\alpha}p^{d+1})$ is a simultaneous solution of Equations (6) and (7), since

$$\alpha^{x''} = \alpha^{\bar{k}_{\alpha,\beta}\delta p^{d+1}} = \beta^{\bar{k}_{\beta,\alpha}p^{d+1}} = \beta^{y''}$$

and, by applying Lucas' theorem,

$$\binom{x''}{k}\alpha^k \equiv 0 \equiv \binom{y''}{k}\beta^k \pmod{p}, (1 \leq k \leq n).$$

Let (x', y') be any other solution. Then, $y' \equiv 0 \pmod{k_\beta}$ and $y' \equiv 0 \pmod{p^{d+1}}$, from the proofs of Lemma 4 and Lemma 5, respectively. Thus, by a simple application of the Chinese remainder theorem, $y' \equiv 0 \pmod{\bar{k}_{\beta,\alpha}p^{d+1}}$, which implies that $y' \geq y''$. A similar argument yields that $x' \equiv 0 \pmod{\bar{k}_{\alpha,\beta}\delta p^{d+1}}$. □

3. Discrete logarithm problem on Jordan block matrices

Let $\alpha \in \mathbb{F}_q^*$, S be a nonsingular matrix similar to a Jordan block $J_\alpha(n)$, and M be a nonsingular matrix that commutes with S. It is important for the computation of the cycle structure of certain *reduced linear modular systems* [9] to determine whether M is a power of S or not. That is to say, we need to find out if

$$S^j = M, \text{ for some positive integer } j. \tag{8}$$

If this is the case, the RLMS R_{MS} associated with the LMS L_S, does not provide any advantage on the computation of a multidimensional FFT of prime edge–length with linear symmetries induced by S [8, 17], and should be ruled out in any algorithm that searches for matrices that optimize such a computation.

Equation (8) can be viewed as a special version of the *discrete logarithm problem* for matrices [14, 15]. In this section we solve this problem and present necessary and sufficient conditions to determine if this equation has a solution or not.

The following lemma is the result of a direct application of the binomial theorem and will be useful in the proof of Theorem 4.

Lemma 6. *Let* $\alpha \in \mathbb{F}_q$. *Then,*

$$J_\alpha^j(n) = \begin{pmatrix} \binom{j}{0}\alpha^j & \binom{j}{1}\alpha^{j-1} & \binom{j}{2}\alpha^{j-2} & \cdots & \binom{j}{n-1}\alpha^{j-n+1} \\ 0 & \binom{j}{0}\alpha^j & \binom{j}{1}\alpha^{j-1} & \cdots & \binom{j}{n-2}\alpha^{j-n+2} \\ & & & \vdots & \\ 0 & 0 & 0 & \cdots & \binom{j}{0}\alpha^j \end{pmatrix}$$

The next theorem solves the discrete logarithm problem over matrices that are similar to Jordan blocks over finite fields.

Theorem 4. *Let S be a matrix similar to $J_\alpha(n)$ and M be any nonsingular matrix that commutes with S. Then, $S^j = M$, for some positive integer j, if and only if the system of equations*

$$\alpha^j = \beta_0 \tag{9}$$

$$\binom{j}{k} \equiv c_k \pmod{p}, \ (1 \le k \le n-1), \tag{10}$$

has a solution in the unknown j, where $c_k = \beta_0^{-1}\beta_k\alpha^k$. Furthermore, if such j exists, it is unique modulo $k_\alpha p^{d+1}$, where $d = \lfloor \log_p(n-1) \rfloor$.

Proof. Let P be the nonsingular matrix for which $S = PJ_\alpha(n)P^{-1}$. Now, since S is non–derogatory and M commutes with S, M must be a polynomial combination of S. Let $Q(x)$ be the polynomial for which $M = Q(S)$. Thus, $M = Q(PJ_\alpha(n)P^{-1}) = PQ(J_\alpha(n))P^{-1}$. Hence, $P^{-1}MP = Q(J_\alpha(n))$.

Then, $S^j = M$ if and only if $J_\alpha^j(n) = Q(J_\alpha(n))$. Thus,

$$Q(J_\alpha(n)) = \begin{pmatrix} \beta_0 & \beta_1 & \beta_2 & \cdots & \beta_{n-1} \\ 0 & \beta_0 & \beta_1 & \cdots & \beta_{n-2} \\ & & \vdots & & \\ 0 & 0 & 0 & \cdots & \beta_0 \end{pmatrix}, \text{ for some } \beta_0 \neq 0, \beta_1, \ldots, \beta_{n-1} \in \mathbb{F}_q.$$

Then, by Lemma 6, $J_\alpha^j(n) = Q(J_\alpha(n))$ if and only if

$$\alpha^j = \beta_0 \tag{11}$$

$$\binom{j}{k} \equiv \beta_0^{-1}\beta_k\alpha^k \pmod{p}, \ (1 \le k \le n-1), \tag{12}$$

if and only if, by Lemma 3, $k_{\beta_0}|k_\alpha$ and, by Theorem 2, $c_k = \beta_0^{-1}\beta_k\alpha^k \in \mathbb{F}_p$ and $\binom{j}{k} \equiv c_k \pmod{p}$, $(1 \le k \le n-1)$. If all these conditions are met, a solution j_0 must satisfy

$$j_0 \equiv \frac{k_\alpha}{k_{\beta_0}}\delta \pmod{k_\alpha}$$

$$j_0 \equiv c_1 + c_p p + c_{p^2}p^2 + \cdots + c_{p^d}p^d \pmod{p^{d+1}},$$

where δ is the integer defined as in Lemma 3 for which $\alpha^{\frac{k_\alpha}{k_{\beta_0}}\delta} = \beta_0$ and $d = \lfloor \log_p(n-1) \rfloor$. Now, since $\gcd(k_\alpha, p^{d+1}) = 1$, the Chinese remainder theorem provides the least positive integer for j_0. $\qquad\square$

4. A special case of reduced linear modular systems

A *finite dynamical system* (FDS) is a pair $(\mathbb{S}, f)$, where $\mathbb{S}$ is a finite set and $f : \mathbb{S} \to \mathbb{S}$. The *functional graph* associated with $(\mathbb{S}, f)$ is the graph $\mathcal{G}_f = (V, E)$, where V is the set of elements of $\mathbb{S}$ and, for any $a, b \in \mathbb{S}$, $(a, b) \in E$ if and only if $f(a) = b$. It is well-known that if $\mathbb{S}$ is a finite field and f is an invertible linear function (i.e., f can be regarded as a nonsingular matrix or a linearized polynomial), $\mathcal{G}_f$ is composed of cycles (also called *orbits*) of different lengths depending on the number and periods of the distinct factors of the minimal polynomial of f [2, 3]. Current interest in the study of $\mathcal{G}_f$ over finite fields is motivated by a growing number of application in different domains [10–12]. In this work we continue the study of reduced linear modular systems [9].

Let us recall some definitions and previous results presented in [9]. Some of the notation is slightly changed to reflect the fact that, in this work, we are defining our matrices over finite fields, instead of restricting ourselves to the integers modulo a prime p. A *linear modular system* (LMS) is an FDS $L_S = (\mathbb{F}_q^n, S)$, where S is a nonsingular $n \times n$ matrix over $\mathbb{F}_q$. For any $u, v \in \mathbb{F}_q^n$, define $u \approx_S v$ if and only if $v = S^j u$, for some positive integer j. For any $u \in \mathbb{F}_q^n$, let $O_S(u) = \{S^j u | j \in \mathbb{Z}\}$ be the cycle of u. The relation $\approx_S$ is an equivalence relation on $\mathbb{F}_q^n$ and its equivalence classes are the cycles of $\mathcal{G}_S$. It is customary to use a $\dot{+}$ to express the formal sum of cycles of $\mathcal{G}_S$:

$$\sum\nolimits_S = 1 \dot{+} \mu_1(t_1) \dot{+} \mu_2(t_2) \dot{+} \cdots \dot{+} \mu_r(t_r),$$

where the 1 stands for the trivial 0–cycle of length one, and $\mu_i(t_i)$ indicates that $\mathcal{G}_S$ contains μ_i cycles of length t_i.

A set of representatives of $\approx_S$ is called a *fundamental set* of S, and is denoted by $\mathcal{F}_S$. Let M be a nonsingular matrix that commutes with S. The action of M on $\mathcal{F}_S$, defined by $O_S(u) \approx_{MS} O_S(v)$ if and only if $M^i u = S^j v$, for some positive integers i and j, induces an equivalence relation $\approx_{MS}$ on the cycles of $\mathcal{G}_S$. The resulting FDS, $R_{MS} = (\mathcal{F}_S, M)$, is what we call a *reduced linear modular system* (RLMS).

The functional graph associated with $(\mathcal{F}_S, M)$ is $\mathcal{G}_{MS} = (V, E)$, where $V = \mathcal{F}_S$ and for any $u, v \in \mathcal{F}_S$, $(u, v) \in E$ if and only if $Mu = S^j v$, for some positive integer j. The structure of $\mathcal{G}_{MS}$ is composed of cycles called MS–cycles (i.e., cycles of cycles of $\mathcal{G}_S$). The MS–cycle of $u \in \mathcal{F}_S$ is denoted by $O_{MS}(u)$. The length of $O_{MS}(u)$, denoted $|O_{MS}(u)|$, is the least positive integer ι for which $M^\iota u = S^j u$, for some positive integer j. The structure of $\mathcal{G}_{MS}$ is characterized by the formal sum

$$\textstyle\sum_{MS} = 1 \ddot{+} \eta_1(\iota_1) \ddot{+} \eta_2(\iota_2) \ddot{+} \cdots \ddot{+} \eta_r(\iota_r),$$

where the 1 stands for the trivial 0–cycle of length one, and $\eta_i(\iota_i)$ indicates that $\mathcal{G}_{MS}$ contains η_i MS–cycles of length ι_i. We use a $\ddot{+}$ to distinguish the cycle structure of $\mathcal{G}_{MS}$ from that of $\mathcal{G}_S$.

For a general nonsingular RLMS $R_{MS} = (\mathcal{F}_S, M)$, the determination of the cycle structure of $\mathcal{G}_{MS}$ is still an open problem. In the case that the characteristic polynomial of S is irreducible over $\mathbb{F}_p$, any nonsingular RLMS R_{MS} associated with $L_S = (\mathbb{F}_p^n, S)$, has the structure $1 \ddot{+} \eta[\iota]$, where $\iota = \text{reach}_S(M)$ and $\eta = \frac{p^n - 1}{\iota}$ (Theorem 2.5 in [9]). A straightforward application of Lemma 4 gives us $\iota = \frac{k_M}{\gcd(k_S, k_M)}$, where k_N denotes the period of a matrix N.

The following theorem solves the MS–structure problem when the LMS L_S is defined by certain Jordan block of size n.

Theorem 5. *Let* $L_S = (\mathbb{F}_q, S)$, $S = J_\alpha(n)$, *be an LMS with structure*

$$1 \ddot{+} \mu_1(k_\alpha) \ddot{+} \mu_2(k_\alpha p^{r_2}) \ddot{+} \cdots \ddot{+} \mu_n(k_\alpha p^{r_n}).$$

Also, let $R_{MS} = (\mathcal{F}_S, M)$ *be an RLMS with* $M = J_\beta(n)$ *and* $\alpha\beta^{-1} \notin \mathbb{F}_p$. *Then, the cycle structure of* R_{MS} *is*

$$1 \ddot{+} \eta_1[\bar{k}_{\beta,\alpha}] \ddot{+} \eta_2[\bar{k}_{\beta,\alpha} p^{r_2}] \ddot{+} \cdots \ddot{+} \eta_n[\bar{k}_{\beta,\alpha} p^{r_n}],$$

where $\bar{k}_{\beta,\alpha} = \frac{k_\beta}{\gcd(k_\beta, k_\alpha)}$ *and* $\eta_e = \frac{\mu_e}{\bar{k}_{\beta,\alpha} p^{r_e}}$.

Proof. $S = J_\alpha(n)$ partitions $\mathbb{F}_q^n$ into a sequence of nested subspaces

$$U_0 \subseteq U_1 \subseteq U_2 \subseteq \cdots \subseteq U_n,$$

where $U_e = \{v \in \mathbb{F}_q^n | P(S)^e v = 0\}$ and P is the minimal polynomial of α. The orbits structure of L_S in U_1 is $1 + \mu_1(k_\alpha)$, where k_α is the period of P (i.e., the multiplicative order of α in $\mathbb{F}_q^*$), d is the degree of P, and $\mu_1 = \frac{p^d-1}{k_\alpha}$. The structure of L_S in $U_e \backslash U_{e-1}$ is $\mu_e(k_\alpha p^{r_e})$, where $\mu_e = \frac{p^{d(e-1)}(p^d-1)}{k_\alpha p^{r_e}}$ and $r_e = \lfloor \log_p (e-1) \rfloor + 1$, $2 \le e \le n$ [2].

In [9] it is shown that the cycle structure of R_{MS} in U_1 is given by the formal sum $1 \ddot{+} \eta_1[\iota_1]$, where ι_1 is the reach of M with respect to S restricted to U_1, and $\eta_1 = \frac{\mu_1}{\iota_1}$. By Lemma 4, $\iota_1 = \frac{k_\beta}{\gcd(k_\alpha, k_\beta)} = \bar{k}_{\beta,\alpha}$. Thus, the structure of R_{MS} in U_1 is $1 \ddot{+} \eta_1[\bar{k}_{\beta,\alpha}]$.

If $v \in U_e \backslash U_{e-1}$, $P(S)^e v = 0$ and $P(S)^{e-1} v \ne 0$. Thus, $v = (v_1, \ldots, v_e, 0, \ldots, 0)$ and $v_e \ne 0$. The length of the cycle of R_{MS} containing v is the least positive integer ι_v for which $J_\alpha^{j_v} v = J_\beta^{\iota_v} v$, for some positive integer j_v. Applying Lemma 6, this last equation is equivalent to $J_\alpha^{j_v} = J_\beta^{\iota_v}$. That is, ι_v and j_v do not depend on v and, hence, all cycles of R_{MS} in $U_e \backslash U_{e-1}$ are of the same length, call it ι_e. So, our problem is to find the least positive integer ι_e for which $J_\alpha^{j_e} = J_\beta^{\iota_e}$, for some positive integer j_e. This is equivalent to solving the system

$$\alpha^{j_e} = \beta^{\iota_e} \tag{13}$$

$$\binom{j_e}{k} \equiv \binom{\iota_e}{k}(\alpha\beta^{-1})^k \pmod{p}, \ (1 \le k \le e-1). \tag{14}$$

Applying Theorem 3, $\iota_e = \bar{k}_{\beta,\alpha} p^{r_e}$, where $r_e = \lfloor \log_p(e-1) \rfloor + 1$. Now, R_{MS} has $\eta_e = \frac{\mu_e}{\iota_e}$ cycles of length ι_e, $(2 \le e \le n)$. Therefore, the structure of R_{MS} is $\sum_{MS} = 1 \ddot{+} \eta_1[\bar{k}_{\beta,\alpha}] \ddot{+} \eta_2[\bar{k}_{\beta,\alpha} p^{r_2}] \ddot{+} \cdots \ddot{+} \eta_n[\bar{k}_{\beta,\alpha} p^{r_n}]$. $\qquad \square$

It is known that a prime edge–length multidimensional fast Fourier transform with linear symmetries induced by a nonsingular LMS $L_S = (\mathbb{F}_q^n, S)$, can be efficiently computed via cyclic convolutions by determining a nonsingular RLMS $R_{MS} = (\mathcal{F}_S, M)$ that minimizes the number of cycles of $\mathcal{G}_{MS}$ (i.e., maximizes the length of its cycles) [8]. An *optimal* RLMS $R_{MS} = (\mathcal{F}_S, M)$ for a nonsingular LMS $L_S = (\mathbb{F}_q^n, S)$ is one that minimizes the number of MS–cycles (i.e., minimizes the sum $\eta_1 + \eta_2 + \cdots + \eta_r$).

The following corollary of Theorem 5, finds an optimal RLMS for an LMS L_S when it is defined by a Jordan block of size n.

Corollary 1. *Let γ be primitive of $\mathbb{F}_q^*$, $\alpha \in \mathbb{F}_q^*$, and $S = J_\alpha(n)$. Then, an optimal RLMS for $L_S = (\mathbb{F}_q^n, S)$ is $R_{MS} = (\mathcal{F}_S, M)$, where*

$$M = \begin{cases} J_{\gamma^p}(n), & \text{if } \alpha\gamma^{-1} \in \mathbb{F}_p^* \\ J_\gamma(n), & \text{otherwise} \end{cases}$$

and the structure of R_{MS} is given by

$$1 \dotplus 1[\mu_1] \dotplus p^{d-2r_2}[\mu_1 p] \dotplus \cdots \dotplus p^{d(n-1)-2r_n}[\mu_1 p^{r_n}],$$

where $\mu_1 = \frac{p^d-1}{k_\alpha}$ and d is the degree of the minimal polynomial of α.

Proof. Let γ be a primitive of $\mathbb{F}_q^*$. We will use the fact that $\gamma^{\frac{q-1}{p-1}s} \in \mathbb{F}_p^*$ for any integer s. Note that the order of $\gamma^{\frac{q-1}{p-1}}$ is $\frac{q-1}{\gcd(q-1,\frac{q-1}{p-1})} = p - 1$. Thus, $\gamma^{\frac{q-1}{p-1}}$ is a primitive in $\mathbb{F}_p$. Now, if $\alpha\gamma^{-1} \in \mathbb{F}_p^*$, then, $\alpha = \gamma^{\frac{q-1}{p-1}s+1}$ for some integer s. Let $M = J_{\gamma^p}(n)$. We will show that $\alpha\gamma^{-p} \notin \mathbb{F}_p^*$. Assume, by contradiction, that $\alpha\gamma^{-p} \in \mathbb{F}_p^*$. Then, $\alpha = \gamma^{\frac{q-1}{p-1}t+p}$, for some integer t. Thus,

$$\frac{q-1}{p-1}t + p \equiv \frac{q-1}{p-1}s + 1 \pmod{q-1}$$

if and only if

$$p \equiv 1 \pmod{\frac{q-1}{p-1}}$$

if and only if $p \equiv 1 \pmod{1 + p + p^2 + \cdots + p^{d-1}}$. Contradiction.

Now, set $\beta = \gamma^p$, if $\alpha\gamma^{-1} \in \mathbb{F}_p^*$; otherwise, set $\beta = \gamma$. Thus, $k_\beta = q - 1$. So, $\bar{k}_{\beta,\alpha} = \frac{q-1}{\gcd(q-1,k_\alpha)} = \mu_1$ and $\eta_1 = 1$. The largest MS–cycles are achieved precisely when $\bar{k}_{\beta,\alpha} = \mu_1$. Therefore, $(\mathcal{F}_S, M)$, where $M = J_\beta(n)$, is an optimal RLMS for $(\mathbb{F}_q^n, S)$. Finally, Theorem 5 gives us the desired MS–cycle structure: $\sum_{MS} = 1 \dotplus 1[\mu_1] \dotplus p^{d-2r_2}[\mu_1 p] \dotplus \cdots \dotplus p^{d(n-1)-2r_n}[\mu_1 p^{r_n}]$. $\square$

5. Conclusion and further work

In this paper we present partial results concerning the problem of determining the cycle structure of reduced linear modular systems. Specifically, we provide a method for the computation of the cycle structure of a nonsingular RLMS $R_{MS} = (\mathcal{F}_S, M)$, associated with a nonsingular LMS $L_S = (\mathbb{F}_q^n, S)$, when S and M are Jordan blocks of the form $J_\alpha(n)$ and $J_\beta(n)$, with the condition that $\alpha\beta^{-1}$ is not an integer modulo p. We also provide a method for computing an optimal RLMS for an LMS of this type. For a general nonsingular RLMS, the determination of its cycle structure remains an open problem and is the subject of further investigation. We conjecture that, based on the ideas presented in [11, 13], we only need to consider RLMSs that are non–equivalent as FDSs. Finally, given nonsingular commuting matrices S and M, we give necessary and sufficient conditions in order for M to be a power of S, using Lucas' theorem and the Chinese remainder theorem. This last result is useful for ruling out matrices M that do not provide any advantage on the computation of a multidimensional FFT of prime edge–length with linear symmetries induced by matrix S.

Acknowledgment We are grateful to Dr. Dorothy Bollman for her guidance and joyful approach to research and to Dr. Oscar Moreno for introducing us into the interesting area of finite fields. We thank Dr. David Thomson for sharing his perspective about the structure of linear modular systems in a visit he made to Puerto Rico. We also would like to thank the reviewers for their valuable suggestions and comments which helped to improve the quality of this manuscrit. This work was partially supported by NIH/NIGMS Award Number P20GM103475.

References

[1] E. Lucas. *Sur les congruences des nombres eulériens et des coefficients différentiels des fonctions trigonométriques, suivant un module premier.* Bull. Soc. Math. de France, 6, (1878), pp. 49–54.

[2] B. Elspas. *The theory of autonomous linear sequential networks, linear sequential switching circuits.* (eds) W. Kautz, Holden–day, Inc. (1965), pp. 21–61.

[3] R. Lidl and N. Niederreiter. *Finite fields.* Encyclopedia of Mathematics and its Applications, Vol. 20, (1997), Cambridge University Press.

[4] A. Granville. *Arithmetic properties of binomial coefficients. I. Binomial coefficients modulo prime powers.* Organic mathematics (Burnaby, BC, 1995) CMS Conf. Proc., vol. 20, Amer. Math. Soc., Providence, RI, (1997), pp. 253–276.

[5] Z. Sun and D. M. Davis. *Combinatorial congruences modulo prime powers.* Trans. Amer. Math. Vol. 359, No. 11, (2007), pp. 5525–5553.

[6] R. J. McIntosh. *A Generalization of a congruential property of Lucas.* The American Mathematical Monthly. Vol. 99, No. 3, (March 1992), pp. 231–238.

[7] A. D. Loveless. *A congruence for products of binomial coefficients modulo a composite.* INTEGERS: 7 (2007), # A44, pp. 1–9.

[8] J. Seguel, D. Bollman, and E. Orozco. *A new prime edge–length crystallographic FFT.* Lecture Notes in Computer Science, Part II (2002), pp. 548–557, Springer–Verlag.

[9] E. Orozco. *Reduced linear modular systems.* Contemporary Mathematics, Vol. 461, (2008), pp. 205–212.

[10] S. V. Konyagin, F. Luca, B. Mans, L. Mathieson, M. Sha, and I. E. Shparlinski. *Functional graphs of polynomials over finite fields.* J. Comb. Theory Ser. B 116, C (January 2016), pp. 87–122. DOI=http://dx.doi.org/10.1016/j.jctb.2015.07.003

[11] A. Ostafe and M. Sha. *Counting dynamical systems over finite fields.* Contemporary Mathematics (2015), Amer. Math. Soc., in press.

[12] N. Boston, A. Ostafe, I. Shparlinski, and M. Zieve. *The art of iterating rational functions over finite fields.* BIRS Workshop. Final report http://www.birs.ca/workshops/2013/13w5141/report13w5141.pdf

[13] E. Bach and A. Bridy. *On the number of distinct functional graphs of affine-linear transformations over finite fields.* Linear Algebra and its Applications 439.5 (2013), pp. 1312–1320.

[14] A. Mahalanobis. *The discrete logarithm problem in the group of non–singular circulant matrices.* Groups Complexity Cryptology. 2 (2010), pp. 83–89. DOI 10.1515/GCC.2010.006.

[15] A. Menezes and Y. Wu. *The discrete logarithm problem in $GL(n,q)$.* Ars Combinatoria, 47 (1997), pp. 22–32.

[16] G. L. Mullen and D. Panario. *Handbook of Finite Fields.* (2013), CRC Press.

[17] E. Orozco and D. Bollman. *Optimizing symmetric FFTs with prime edge-length.* Computational Science and Its Applications, Springer–Verlag, Part III, (2004), LNCS 3045, pp. 736–744.

[18] *Sage Mathematics Software (Version 5.9)*, The Sage Developers, http://www.sagemath.org.

Finding a Gröbner basis for the ideal of recurrence relations on m-dimensional periodic arrays

Ivelisse M. Rubio

Department of Computer Science, University of Puerto Rico, Box 70377, S.J., PR 00936-8377 iverubio@gmail.com

Moss Sweedler

Department of Mathematics, Cornell University, 310 Malott Hall, Ithaca, NY 14853 3.14159x2.71828@gmail.com

Chris Heegard

Native Intelligence, 17179 La Brisa Ct., Sugarloaf, FL 33042 heegard@nativei.com

Recent developments in applications of multidimensional periodic arrays [9] have drawn new attention to the computation of Gröbner bases for the ideal of linear recurrence relations on the arrays. An m-dimensional infinite array can be represented by a multivariate power series sitting within the ring of multivariate Laurent series. We reinterpret the problem of finding linear recurrence relations on m-dimensional periodic arrays as finding the kernel of a module map involving quotients of Laurent series and present an algorithm to compute a Gröbner basis for this kernel. The algorithm does not assume the knowledge of a generating set for the kernel of this ideal and it is based on linear algebra computations. Finding a generating set is one application of the algorithm.

1. Introduction

There are different algorithms to find Gröbner bases for the ideal of linear recurrence relations on multidimensional arrays (or equivalently, multivariate power series). One of the best known is Sakata's algorithm [14, 15]. This algorithm is an extension to m-dimensions of the Berlekamp-Massey algorithm and has been used for decoding multidimensional cyclic codes and algebraic geometric codes [12, 13, 16]. Recent developments in applications of multidimensional periodic arrays [9] have drawn new attention to the computation of Gröbner bases for the ideal of linear recurrence relations on the arrays. The approach that we present here can be implemented easily and is suitable for these new applications.

In [7], Faugere et al. presented an efficient algorithm to transform a Gröbner basis of a 0-dimensional ideal with respect to a given monomial order into a Gröbner basis with respect to another monomial order. Shortly after it appeared, Moss Sweedler and Lee Taylor showed in [17] that the ideas in [7] could be expanded and cast as: 1) an initial Gröbner basis allows one to effectively do linear algebra in the finite dimensional vector space quotient; 2) if one can effectively do linear algebra in a finite dimensional vector space quotient module, then one can find a reduced Gröbner basis for the kernel ideal, without having an initial generating set for this kernel.

Let $\mathcal{F}$ be a field and $\mathcal{F}[\mathbf{x}] := \mathcal{F}[x_1, \ldots, x_m]$ be the ring of polynomials in m variables and coefficients in $\mathcal{F}$. The Sweedler-Taylor algorithm in [17] computes reduced Gröbner bases for 0-dimensional ideals $I \subseteq \mathcal{F}[\mathbf{x}]$ and it is based upon considering ascending subspaces of $\mathcal{F}[\mathbf{x}]/I$. The ascending subspaces are spanned by the image of monomials which are ascending in the monomial order. An ascending chain of subspaces of a finite dimensional vector space must stabilize and this forces termination of the algorithm. A description of the Sweedler-Taylor algorithm and a complete proof of its validity was presented in [11]. Like Sakata's algorithm, this algorithm does not assume the knowledge of a basis for the ideal. Other algorithms to compute a basis for $\mathcal{F}[\mathbf{x}]/I$ have been presented in [1–4].

After reinterpreting the definition and basics of linear recurrence relations on multidimensional periodic arrays, we modify the Sweedler-Taylor algorithm to find a Gröbner basis that generates these relations. This analog to the Sweedler-Taylor algorithm computes what we call *lead monomial generating sets* and, under certain conditions, a lead monomial generating set turns out to be a Gröbner basis for the ideal of linear recurrence relations. We will see in Proposition 2 that twice the period minus the dimension of the array is an upper bound for the degree of the polynomials required to form a Gröbner basis for the ideal of linear recurrence relations valid on the array with respect to the graded lexicographic order.

The next example on a one dimensional array illustrates the first steps of the algorithm.

Example 1. Consider the one dimensional array $S = (3, 5, 9, 6, \ldots)$ over $\mathbb{F}_{11}$, the finite field with 11 elements. This sequence is defined by the recurrence $S_0 = 3$, and $S_{i+1} = 2^{i+1} + S_i$, for $i > 0$. One can write this sequence as the power series:

$$S(x) = S_0 + S_1 x + S_2 x^2 + S_3 x^3 + \cdots = 3 + 5x + 9x^2 + 6x^3 + \cdots.$$

A polynomial $C = \sum_{i=0}^{L} C_i x^i \in \mathbb{F}_{11}[x]$ defines a linear recurrence relation at a point S_u of the array S if $L \leq u$ and $\sum_{i=0}^{L} C_i S_{i+u-L} = 0$. We say that the polynomial is valid for the array S if it defines a linear recurrence relation for each $u \geq L$. The set of all linear recurrence relations valid on the array S form an ideal in $\mathbb{F}_{11}[x]$. To find a polynomial valid for the array one needs a polynomial $C_0 + C_1 x + C_2 x^2 + \cdots + C_L x^L$ such that

$$C_0 3 + C_1 5 + C_2 9 + \cdots + C_L S_L = 0, \quad C_0 5 + C_1 9 + \cdots + C_L S_{L+1} = 0,$$

$$C_0 9 + C_1 6 + \cdots + C_L S_{L+2} = 0, \ldots.$$

This would be the same as finding constants $C_0, C_1, \ldots, C_L$ such that, in the following table, when one sums the rows multiplied by their respective constant, and consider the reduction mod 11, one gets a sequence of 0's.

$$
\begin{array}{c|cccc}
C_0 & 3 & 5 & 9 & 6 & \cdots \\
C_1 & 5 & 9 & 6 & & \cdots \\
C_2 & 9 & 6 & & & \cdots \\
\vdots & \vdots & & & \vdots & \cdots \\
C_k & S_k & S_{k+1} & & & \cdots \\
\hline
 & 0 & 0 & & & \cdots
\end{array}
\tag{1}
$$

Note that each shift is equivalent to multiplying the power series $S(x)$ by $x^{-1}, x^{-2}, \ldots$ respectively and "mod out" (or ignore) the terms with negative exponents. In Section 3.1 we define these shift operations algebraically. In this example, if one consider $C_0 = 2, C_1 = 8, C_2 = 1, C_i = 0$ for $i > 2$, one gets the following operations modulo 11 on the power series $S(x)$:

$$
\begin{aligned}
& 2 \; (3 + 5x + 9x^2 + 6x^3 + \cdots) \\
+ \; & 8x^{-1} \; (3 + 5x + 9x^2 + 6x^3 + \cdots) \\
+ \; & x^{-2} \; (3 + 5x + 9x^2 + 6x^3 + \cdots).
\end{aligned}
\tag{2}
$$

Ignoring the terms with negative exponents one gets $S_2' x^2 + S_3' x^3 + \cdots$; hence, the polynomial $x^2 + 8x + 2$ gives a recurrence relation up to S_3. The theory and algorithm in this paper imply that this is actually the minimal polynomial valid for the array S.

For this particular example, since we have the explicit recurrence, $S_{i+1} = 2^{i+1} + S_i$, we may directly verify the relation $x^2 + 8x + 2$:

$$C_0 S_i + C_1 S_{i+1} + C_2 S_{i+2} = 2S_i + 8S_{i+1} + S_{i+2} = 2S_i + 8\left(2^{i+1} + S_i\right) + S_{i+2}$$

$$= 2S_i + 8\left(2^{i+1} + S_i\right) + 2^{i+2} + 2^{i+1} + S_i = 11S_i + 112^{i+1} = 0.$$

Since this holds for the recurrence, it tells us that $x^2 + 8x + 2$ holds not just up to S_3 but for the full array S. What if we did not have the recurrence but had a black-box that gave as much (but only a finite amount) of a sequence requested? How might one determine such an equation, as we did here, finding $x^2 + 8x + 2$? How might one determine if there even exists such an equation? Suppose you had the black-box and the information that there is an equation of degree less than some specific degree. How might one determine the equation?

Note that one could find the coefficients of C by performing row reduction in (1) and finding a linear dependency relation among the rows. Of course, the array is infinite and one might not be able to compute the dependency relations effectively. But the arrays that we will consider are periodic and we will see that it will be enough to compute the dependency relations in a finite subarray.

From (2) note that, intuitively, what one is looking is for a polynomial $C(x)$ such that in $C\left(x^{-1}\right) S(x)$ the only non-zero terms have negative exponents.

The method used here is an application of the analog to the Sweedler-Taylor algorithm that will be described in Section 4 after we introduce the concept of lead monomial generating sets. In Section 3 we present the basics of linear recurrence relations and reformulate the problem in order to use the algorithm to find a Gröbner basis for the ideal of linear recurrence relations on an m-dimensional periodic array.

2. Gröbner bases and the Sweedler-Taylor algorithm

The set of polynomials that define the linear recursion relations on an m-dimensional array form an ideal over $\mathcal{F}[\mathbf{x}]$. These ideals have a finite generating set and Gröbner bases are ideal generating sets with "nice properties".

Let $\mathbb{N}_0 := \{0, 1, 2, \ldots\}$ and consider the set of exponents $\alpha =$ $(\alpha_1, \ldots, \alpha_m) \in \mathbb{N}_0^m$. Set $\alpha \leq \beta$ if and only if $\alpha_i \leq \beta_i$ for $i = 1, \ldots, m$. This defines the partial order of divisibility where $\mathbf{x}^\alpha | \mathbf{x}^\beta$ if and only if $\alpha \leq \beta$. We say that a monomial $\mathbf{x}^\alpha = x_1^{\alpha_1} x_2^{\alpha_2} \cdots x_m^{\alpha_m}$ in a set of monomials is minimal with respect to $\leq$ if there is no other monomial $\mathbf{x}^\beta$ in the set with $\beta < \alpha$.

We say that $<_T$ is a **monomial order** if it is a well ordering in $\mathbb{N}_0^m$ such that $<_T$ is a total order, and $\alpha <_T \beta$ implies that $\alpha + \gamma <_T \beta + \gamma$ for $\alpha, \beta, \gamma \in \mathbb{N}_0^m$. Note that divisibility is not a total order and hence is not a monomial order, but it is compatible with any monomial order in the sense that $\mathbf{x}^\alpha | \mathbf{x}^\beta$ implies that $\mathbf{x}^\alpha \leq_T \mathbf{x}^\beta$.

Define $|\alpha| = \sum_{i=1}^m \alpha_i$. Two common examples of monomial orders are the **lexicographical order**, where $\alpha <_{lex} \beta$ if in $\beta - \alpha$ the left most non-zero entry is positive, and the **graded lexicographical order**, where $\alpha <_{grlex} \beta$ if $|\alpha| < |\beta|$ or if $|\alpha| = |\beta|$ and $\alpha <_{lex} \beta$. We will use the following notation:

Let $f = \sum_\alpha a_\alpha \mathbf{x}^\alpha$ be a nonzero polynomial with each $a_\alpha \neq 0$ and $I \subset \mathcal{F}[\mathbf{x}]$. Then,

(1) $LE(f) = leadexp(f)$ is the largest exponent vector α in f with respect to $<_T$.
(2) $LM(f)$ denotes the leading monomial of f and it equals $\mathbf{x}^{LE(f)}$.
(3) $LC(f)$ denotes the coefficient of $LM(f)$. In other words, the so called leading term of f is $LC(f)LM(f)$.
(4) $LE(I) := \{LE(f) \mid 0 \neq f \in I\} \subseteq \mathbb{N}_0^m$. (Note that if $I = \{0\}$, then $LE(I) = \{\}$.)
(5) $LM(I) := \{LM(f) \mid 0 \neq f \in I\} = \{\mathbf{x}^\alpha \mid \alpha \in LE(I)\}$. (If $I = \{0\}$, then $LM(I) = \{\}$.)

Definition 1. Let $G = \{g_1, \ldots, g_l\} \subset I$, I an ideal in $\mathcal{F}[\mathbf{x}]$. One says that G is a **Gröbner basis** for I with respect to $<_T$ if $\langle LM(g_1), \ldots, LM(g_l) \rangle = \langle LM(I) \rangle$. If $LC(g_i) = 1$ for $i = 1, \ldots, l$ and $LM(g_i)$ does not divide any term of g_j for $i \neq j$, then one says that G is a **reduced Gröbner basis** for I with respect to $<_T$.

It is a standard result that a Gröbner basis for an ideal generates the ideal. Also $G = \{g_1, \ldots, g_l\} \subset I$ is a Gröbner basis for I if and only if for any $p \in I$, $LM(g_i)|LM(p)$ for some $g_i \in G$ (see [6]).

An important concept that will be used to determine a finite subarray $S' \subset S$ that is sufficient to compute a Gröbner basis for the ideal of linear recurrence relations on S is the concept of a *delta set*. A set $\Delta \subset \mathbb{N}_0^m$ is called a **delta set** if it satisfies 1 in the following lemma:

Lemma 1. *Let* $\Delta, \Gamma \subset \mathbb{N}_0^m$ *be set theoretic complements. The following conditions are equivalent:*

(1) For $\beta \in \Delta, \alpha \in \mathbb{N}_0^m$, *if* $\alpha \leq \beta$ *then* $\alpha \in \Delta$.
(2) For $\alpha \in \Gamma, \beta \in \mathbb{N}_0^m$, *if* $\alpha \leq \beta$ *then* $\beta \in \Gamma$.

Obviously the set of exponents of all monomials which occur as leading monomials of an ideal I satisfies 2 of Lemma 1. Hence, the set of exponents of all monomials that do not occur as leading monomials of an ideal I is a delta set (these monomials are called *standard monomials* in [5]). We will denote *the delta set of the ideal I* as Δ_I. So, $\Delta_I = \mathbb{N}_0^m \backslash LE(I)$. Of course the delta set of an ideal depends on the specific monomial order chosen.

When we talk about the *dimension* of an ideal $I \subset R$ we mean the *Krull dimension* of the ring R/I. $I \subset \mathcal{F}[\mathbf{x}]$ is a 0-dimensional ideal if and only if $\mathcal{F}[\mathbf{x}]/I$ has finite dimension as a $\mathcal{F}$-vector space. The set of monomials which do not occur as leading monomials of elements in I map one to one to a basis of $\mathcal{F}[\mathbf{x}]/I$. Thus, one has that Δ_I corresponds to a basis of $\mathcal{F}[\mathbf{x}]/I$ as a $\mathcal{F}$ vector space and hence its size does not depend on the specific monomial order chosen. So, I is a 0-dimensional ideal if and only if Δ_I is a finite set.

Algorithms for computing Gröbner bases usually assume the knowledge of some basis for the ideal. We will describe an algorithm, due to Moss Sweedler and Lee Taylor [11, 17], that computes a reduced Gröbner basis for a 0-dimensional ideal without necessarily knowing a basis for the ideal. This algorithm is based on linear algebra and generalizes a common technique relating to the irreducible polynomial of an element in a field extension, the minimal polynomial of a matrix and more. Namely, suppose that B is a finite degree field extension of $\mathcal{F}$, or B is the $\mathcal{F}$ algebra of $n \times n$ matrices over $\mathcal{F}$, or B is any finite dimensional algebra over $\mathcal{F}$. Suppose that $b \in B$ and consider $1 = b^0, b^1, b^2, \ldots$. Since B is finite dimensional, there is a first t where b^t is linearly dependent upon $1, b^1, b^2, \ldots, b^{t-1}$. Thus there is a linear relation:

$$0 = \lambda_0 + \lambda_1 b + \cdots + \lambda_{t-1} b^{t-1} + \lambda_t b^t,$$

with $\lambda_i \in \mathcal{F}$ and $\lambda_t \neq 0$. This gives the following degree t polynomial in $\mathcal{F}[x]$ which b satisfies:

$$\lambda_0 + \lambda_1 x + \cdots + \lambda_{t-1} x^{t-1} + \lambda_t x^t.$$

By the minimality of t, this is the lowest degree non-zero polynomial satis
fied by b.

From a Gröbner basis viewpoint, here is what we have just done. Let
$\Pi : \mathcal{F}[x] \rightarrow B$ be the map defined by: $f(x) \mapsto f(b)$, for $f(x) \in \mathcal{F}[x]$.
This is an $\mathcal{F}$ algebra map and gives B an $\mathcal{F}[x]$-module structure. In the
general procedure, we shall assume and use that Π is a module map. Under
the *unique* monomial order on $\mathcal{F}[x]$, the ascending list of monomials is:
$1, x, x^2, x^3, \ldots$. Consider

$$\Pi(1), \ \Pi(x), \ \Pi(x^2), \ \Pi(x^3), \ \ldots \ = \ 1, b, b^2, b^3, \ldots.$$

Since B is finite dimensional, there is a first t where $\Pi(x^t)$ is linearly dependent upon

$$\Pi(1), \ \Pi(x), \ \Pi(x^2), \ \Pi(x^3), \ \ldots, \ \Pi(x^{t-1}).$$

Thus, there is a linear relation

$$0 = \lambda_0 \Pi(1) + \lambda_1 \Pi(x) + \cdots + \lambda_{t-1} \Pi(x^{t-1}) + \lambda_t \Pi(x^t)$$

or, equivalently,

$$0 = \Pi \left(\lambda_0 + \lambda_1 x + \cdots + \lambda_{t-1} x^{t-1} + \lambda_t x^t \right),$$

with $\lambda_i \in \mathcal{F}$ and $\lambda_t \neq 0$. This gives the following degree t polynomial in
$ker(\Pi)$:

$$\lambda_0 + \lambda_1 x + \cdots + \lambda_{t-1} x^{t-1} + \lambda_t x^t.$$

By the minimality of t, this polynomial is the lowest degree non-zero polynomial in $ker(\Pi)$. Hence it generates the principal ideal $ker(\Pi)$. In the
multivariate case which follows, $ker(\Pi)$ is still an ideal, but it no longer is
a principal ideal. One must iterate the preceding procedure. In order to
actually do what we just described, we must be able to effectively determine linear dependence in B. This could be given by a "linear dependence
oracle". The Sweedler-Taylor algorithm is not concerned with how linear
dependence is determined. Let us now present the general algorithm.

Let B be an $\mathcal{F}[\mathbf{x}]$-module, and consider an $\mathcal{F}[\mathbf{x}]$-module map $\Pi :$
$\mathcal{F}[\mathbf{x}] \longrightarrow B$, where $dim_{\mathcal{F}} (Im(\Pi)) = dim_{\mathcal{F}} (\mathcal{F}[\mathbf{x}]/ker(\Pi))$ is finite and one
can determine linear dependence among the elements of B. Under these
conditions, one can compute a reduced Gröbner basis for $I = ker(\Pi)$ with
respect to any monomial order using the following algorithm.

Suppose $<_T$ is a monomial order on $\mathcal{F}[\mathbf{x}]$. *Begin counting the monomi-*
als in $\mathcal{F}[\mathbf{x}]$ with respect to $<_T$, forming: $1 = t_{1,0} <_T t_{1,1} <_T t_{1,2} <_T \cdots$. *If*
$M \subset \mathcal{F}[\mathbf{x}]$, *begin counting the monomials in M with respect to $<_T$ means*

to form the sequence $(a <_T b <_T c <_T \cdots)$, where a is the smallest monomial in M with respect to $<_T$, b is the smallest monomial of $M\backslash\{a\}$ with respect to $<_T$, c is the smallest monomial of $M\backslash\{a,b\}$, etc. Since the set of monomials is well ordered with respect to a monomial order, there is always such a smallest monomial unless after a finite number of steps the set $M\backslash\{a,b,c,\ldots,e\}$ contains no more monomials. (This includes the possibility that M had no monomials in the first place, in which case the sequence is the empty sequence.) When M has but a finite number of monomials, they are eventually all selected for the sequence and no monomials remain to be selected. At this point the sequence simply stops.

Consider the subspaces $B_{1,i}$ of B, where $B_{1,i}$ is the $\mathcal{F}$ subspace spanned by $t_{1,0}, t_{1,1}, t_{1,2}, \ldots, t_{1,i}$. The spirit of the algorithm is better displayed if we define these sets recursively:

$$B_{1,-1} = \{0\} \subseteq B_{1,0} = \mathcal{F}\,\Pi\,(t_{1,0}) \subseteq B_{1,1} = B_{1,0} + \mathcal{F}\,\Pi\,(t_{1,1}) \subseteq$$

$$\cdots \subseteq B_{1,l} = B_{1,l-1} + \mathcal{F}\,\Pi\,(t_{1,l}) \subseteq \cdots \subseteq Im(\Pi).$$

Since $dim_{\mathcal{F}}\,(Im(\Pi)) < \infty$, it must happen that some $B_{1,i} = B_{1,i-1}$. This means that there will be a first linearly dependent element, call it $\Pi\,(t_{1,e_1})$, in $Im(\Pi)$. This is, $\Pi\,(t_{1,e_1}) \in B_{1,e_1-1}$. Thus,

$$\Pi\,(t_{1,e_1}) = \sum_{j=0}^{e_1-1} \lambda_{1,j}\Pi\,(t_{1,j}) \quad \text{or, equivalently,}$$

$$\Pi\left(t_{1,e_1} - \sum_{j=0}^{e_1-1} \lambda_{1,j}t_{1,j}\right) = 0. \tag{3}$$

Set $g_1 := t_{1,e_1} - \sum_{j=0}^{e_1-1} \lambda_{1,j}t_{1,j}$ and $G_1 := \{g_1\}$. Then, $G_1 \subset I = ker(\Pi)$ and $LM\,(g_1) = t_{1,e_1}$. Set $\Delta_{I,1} := \{LE\,(t_{1,j})\}_{j<e_1}$.

Now begin counting the set of monomials which are greater than t_{1,e_1} and not divisible by t_{1,e_1},

$t_{2,0} <_T t_{2,1} <_T t_{2,2} <_T \cdots$, and define

$$B_{2,-1} = B_{1,e_1-1} \subseteq B_{2,0} = B_{2,-1} + \mathcal{F}\,\Pi\,(t_{2,0}) \subseteq B_{2,1} = B_{2,0} + \mathcal{F}\,\Pi\,(t_{2,1})\cdots.$$

So one has

$$B_{1,-1} \subseteq B_{1,0} \subset B_{1,1} \subset \cdots \subset B_{1,e_1-1} = B_{1,e_1}$$

$$= B_{2,-1} \subseteq B_{2,0} \subseteq B_{2,1} \subseteq \cdots \subseteq Im(B),$$

and again, find the first t_{2,e_2} such that $\Pi\left(t_{2,e_2}\right) \in B_{2,e_2-1}.$

$$g_2 := t_{2,e_2} - \sum_{j=0}^{e_2-1} \lambda_{2,j} t_{2,j} - \sum_{i=0}^{e_1-1} \lambda_{1,i} t_{1,i}, \quad \text{and} \quad \Pi\left(g_2\right) = 0.$$

Then, $g_2 \in I = ker(\Pi)$, $LM\left(g_2\right) = t_{2,e_2}$. Set $\Delta_{I,2} := \Delta_{I,1} \cup \{LE\left(t_{2,j}\right)\}_{j<e_2}$, and $G_2 := G_1 \cup \{g_2\}.$

The general inductive step is the following. Suppose that $\Delta_{I,n-1}$ and G_{n-1} have been defined. Begin counting the set of monomials greater than $t_{n-1,e_{n-1}}$ and not divisible by

$$t_{1,e_1}, t_{2,e_2}, t_{3,e_3}, \ldots, t_{n-1,e_{n-1}},$$

the lead monomials of the elements in G_{n-1}. Suppose this counting yields:

$$t_{n,0} <_T t_{n,1} <_T t_{n,2} <_T \cdots.$$

Then define the subspaces $B_{n,i}$'s as:

$$B_{n,-1} = B_{n-1,e_{n-1}-1} \subseteq B_{n,0} = B_{n,-1} + \mathcal{F}\,\Pi\left(t_{n,0}\right) \subseteq B_{n,1}$$

$$= B_{n,0} + \mathcal{F}\,\Pi\left(t_{n,1}\right) \subseteq \cdots \subseteq Im(\Pi).$$

It can happen that there are only a finite number of monomials in the sequence

$$t_{n,0} <_T t_{n,1} <_T \cdots <_T t_{n,p}$$

and that all of them are linearly independent and one runs out of monomials. It can also happen that the sequence is empty, i.e. there are no monomials greater than $t_{n-1,e_{n-1}}$ and not divisible by a lead monomial in G_{n-1}. In these cases, set $\Delta_{I,n} := \Delta_{I,n-1} \cup \{LE(t_{n,j})\}_{j=0}^{p}$, and $G_n := G_{n-1}$. Note that in case the sequence was empty, $\Delta_{I,n} = \Delta_{I,n-1}$. The algorithm now terminates with $G_{n-1} = G_n$ a reduced Gröbner basis for $I = ker(\Pi)$ and $\Delta_{I,n}$ the set of lead exponents of the standard monomials. Set $G := G_n$ and $\Delta_I := \Delta_{I,n}.$

If the algorithm did not terminate, there will be a minimum e_n, where $B_{n,e_n-1} = B_{n,e_n}$, or equivalently,

$$\Pi\left(t_{n,e_n}\right) \in B_{n,e_n-1}$$

$$\Pi\left(t_{n,e_n}\right) = \sum_{j=0}^{e_n-1} \lambda_{n,j}\Pi\left(t_{n,j}\right) + \sum_{k=0}^{e_{n-1}-1} \lambda_{n-1,k}\Pi\left(t_{n-1,k}\right) + \cdots + \sum_{i=0}^{e_1-1} \lambda_{1,i}\Pi\left(t_{1,i}\right)$$

$$\Pi\left(t_{n,e_n} - \sum_{j=0}^{e_n-1} \lambda_{n,j}t_{n,j} - \sum_{k=0}^{e_{n-1}-1} \lambda_{n-1,k}t_{n-1,k} - \cdots - \sum_{i=0}^{e_1-1} \lambda_{1,i}t_{1,i}\right) = 0.$$

In this case, set

$$g_n := t_{n,e_n} - \sum_{j=0}^{e_n-1} \lambda_{n,j}t_{n,j} - \sum_{k=0}^{e_{n-1}-1} \lambda_{n-1,k}t_{n-1,k} - \cdots - \sum_{i=0}^{e_1-1} \lambda_{1,i}t_{1,i},$$

$$\Delta_{I,n} := \Delta_{I,n-1} \cup \{LE(t_{n,j})\}_{j<e_n}, \text{ and } G_n := G_{n-1} \cup \{g_n\}.$$

In short, what the algorithm does is to find $\mathbf{x}^\alpha$'s, minimal in terms of divisibility, such that $\Pi(\mathbf{x}^\alpha) \in Span_{\mathcal{F}}\left\{\Pi\left(\mathbf{x}^\beta\right)\right\}_{\beta<_T\alpha}$, where $Span_{\mathcal{F}}(V)$ is the $\mathcal{F}$ vector space spanned by the elements in V. The exponents β are exponents of independent monomials, and the Δ_I set is the set of exponents of all the independent monomials.

The above algorithm was presented in [17]. The algorithm together with a complete proof of its validity are included in [11]. The algorithm is summarized as follows:

Algorithm 1 (The Sweedler-Taylor Algorithm).

Input: 1) A $\mathcal{F}[\mathbf{x}]$-module B.
            2) A $\mathcal{F}[\mathbf{x}]$-module map $\Pi : \mathcal{F}[\mathbf{x}] \longrightarrow B$,
               where $dim_{\mathcal{F}}\left(Im(\Pi)\right) < \infty$,
               and we can compute linear dependence in B.
            3) A monomial order $<_T$ on $\mathcal{F}[\mathbf{x}]$.

Output: 1) A reduced Gröbner basis for $I = ker(\Pi)$ with respect to $<_T$.
             2) Δ_I.

1) BEGIN
2) Let A be the set of all monomials in $\mathcal{F}[\mathbf{x}]$, $l := 1$, $G_0 = \emptyset$, $\Delta_{I,0} = \emptyset$
3) Order all monomials in A with respect to $<_T$: $t_{l,0} <_T t_{l,1} <_T \cdots$

4) REPEAT
5) $i := -1$
6)   REPEAT
7) $i := i + 1$
8) $A := A - \{t_{l,i}\}$
9)   UNTIL $\Pi(t_{l,e_l}) = \sum_{h=1}^{l} \sum_{j=0}^{e_h-1} \lambda_{h,j} \Pi(t_{h,j})$ or $A = \emptyset$
10)   IF $\Pi(t_{l,e_l}) = \sum_{h=1}^{l} \sum_{j=0}^{e_h-1} \lambda_{h,j} \Pi(t_{h,j})$, THEN
11) $g_l := t_{l,e_l} - \sum_{h=1}^{l} \sum_{j=0}^{e_h-1} \lambda_{h,j} t_{h,j}$
12) $G_l := G_{l-1} \cup \{g_l\}$
13) $\Delta_{I,l} := \Delta_{I,l-1} \cup \{LE(t_{l,j})\}_{j<e_l}$
14)   Order the monomials in
           $A := A - \{t_{l,e_l} M \mid M$ a monomial in $\mathcal{F}[\mathbf{x}]\}$
           with respect to $<_T$
15) $l := l + 1$
16)   ELSE
17) $\Delta_{I,l} := \Delta_{I,l-1} \cup \{leadexp(t_{l,j})\}_{j \le i}$
18) UNTIL $A = \emptyset$
19) $G := G_{l-1}$, $\Delta_I := \Delta_{I,l-1}$
20) END

Theorem 2. *The above algorithm computes a reduced Gröbner basis* $G = \{g_1, \cdots, g_l\}$ *for* $ker(\Pi)$ *with respect to* $<_T$ *and* Δ_I *is the delta set of* $I = ker(\Pi)$.

3. Linear recurrence relations on m-dimensional arrays

Sakata [14] studied the relation between periodic arrays, linear recurrence relations and ideals. He proved that if an array S is m-dimensional periodic, then the ideal of linear recurrence relations valid on S is 0-dimensional, a fact that also follows from the paper of Gianni [8]. This will allow us to use the analog to the Sweedler-Taylor algorithm in Section 4 to find a Gröbner basis that generates the ideal of linear recurrences. In this section we present the basics from linear recurrence relations on m-dimensional arrays, and reformulate the problem in terms of Laurent series where the desired Gröbner basis will be the kernel of a module map.

Let $S \subset \mathcal{F}^{\mathbb{N}_0^m}$ be an m-dimensional infinite array. The arrays considered in this paper are m-dimensional periodic and this is key in order to effectively compute linear dependency in the algorithm that we will present.

Definition 2. An m-dimensional array S is said to be m-**dimensional periodic** if there is a m-tuple, that we call the **period vector**, $n = (n_1, \ldots, n_m) \in \mathbb{N}^m$, such that

$$S_{(\alpha_1, \ldots, \alpha_m)} = S_{(\alpha_1 + n_1 k_1, \ldots, \alpha_m + n_m k_m)}$$

for $k_i \in \mathbb{N}_0$ and all $(\alpha_1, \ldots, \alpha_m) \in \mathbb{N}_0^m$.

Let $C(\mathbf{x}) = \sum_{\alpha \in Supp(C)} C_\alpha \mathbf{x}^\alpha \in \mathcal{F}[\mathbf{x}]$, where $Supp(C) := \{\alpha \mid C_\alpha \text{ is a non-zero coefficient of } C\}$.

Definition 3. The polynomial $C(\mathbf{x})$ defines a linear recurrence relation at a point S_u of the array S if $LE(C) \leq u$ and

$$\sum_{\alpha \in Supp(C)} C_\alpha S_{\alpha + u - LE(C)} = 0. \tag{4}$$

In this case we say that C **is valid at the point** S_u. Also set C to be valid at S_u, if $LE(C) \not\leq u$.

Definition 4. A polynomial C **is valid for the array** S if the equation

$$\sum_{\alpha \in Supp(C)} C_\alpha S_{\alpha + \beta} = 0 \tag{5}$$

holds for all $\beta \in \mathbb{N}_0^m$. In this case we also say that S **satisfies the m-dimensional linear recurrence relation given by** C.

Note that C is a valid polynomial for S if and only if C is a valid polynomial at every point S_u such that $LE(C) \leq u$.

Let $Val(S)$ denote the set of all valid polynomials for the array S. The set $Val(S)$ is an ideal in $\mathcal{F}[\mathbf{x}]$ and our goal is to find a Gröbner basis for $Val(S)$. If S is an m-dimensional periodic array with period vector $n = (n_1, \ldots, n_m)$ then it is clear that $Val(S)$ contains the polynomials $x_1^{n_1} - 1, x_2^{n_2} - 1, \ldots, x_m^{n_m} - 1$. So, in this case, $Val(S)$ is a 0-dimensional ideal as it was proved in [8]. From now on, we only consider periodic arrays.

We now reformulate the definition of a valid polynomial in order to have $Val(S) = ker(\Pi)$, the kernel of a map $\Pi : \mathcal{F}[\mathbf{x}] \longrightarrow B$. By choosing Π as a $\mathcal{F}[\mathbf{x}]$-module map, we can then use the analog to the Sweedler-Taylor algorithm to find a Gröbner basis for the ideal of valid polynomials $Val(S)$. The approach that we use is a generalization of Example 1 to the multivariate case.

3.1. *Reformulation of the problem*

Let us look at Example 1 again. The algebraic way to view the power series operations in (2) is as follows: view the power series as sitting within the ring of Laurent series $\mathcal{F}\{\{x\}\}$ and multiply by x^{-1}. This is not quite correct because $x^{-1}\left(3 + 5x + 9x^2 + 6x^3 + \cdots\right) = 3x^{-1} + 5 + 9x + 6x^2 + \cdots$, and there is an "unwanted" $3x^{-1}$ term. Hence, factor out the subspace L^- of Laurent series spanned by $x^{-1}, x^{-2}, x^{-3}, \cdots$. Since $x^{-1}L^- \subseteq L^-$, multiplying by x^{-1} gives an operator on $\mathcal{F}\{\{x\}\}/L^-$. Now, $\mathcal{F}\{\{x\}\}/L^-$ has a complete basis $1, \overline{x}, \overline{x}^2, \ldots$, and the image of $S(x)$ in $\mathcal{F}\{\{x\}\}/L^-$ is:

$$3 + 5\overline{x} + 9\overline{x}^2 + 6\overline{x}^3 + \cdots, \text{ and}$$

$$x^{-1}\left(3 + 5\overline{x} + 9\overline{x}^2 + 6\overline{x}^3 + \cdots\right) = 5 + 9\overline{x} + 6\overline{x}^2 + \cdots,$$

since $3x^{-1} \equiv 0 \bmod L^-$. This algebraic treatment immediately generalizes to the multivariate case.

Denote the ring of Laurent series in several variables and coefficients in $\mathcal{F}$ by $\mathcal{F}\{\{\mathbf{x}\}\} = \mathcal{F}\{\{x_1, \ldots, x_m\}\}$. There are several definitions for Laurent series if $m > 1$ [10], the one considered here is not a field and can be defined as $\mathcal{F}\{\{\mathbf{x}\}\} =$

$$\left\{ \sum_{\alpha_1=-k_1}^{\infty} \sum_{\alpha_2=-k_2}^{\infty} \cdots \sum_{\alpha_m=-k_m}^{\infty} a_\alpha \mathbf{x}^\alpha \,|\, \alpha = (\alpha_1, \alpha_2, \ldots, \alpha_m)\ a_\alpha \in \mathcal{F}, k_i \in \mathbb{N} \right\}.$$

Here every $f(\mathbf{x}) \in \mathcal{F}\{\{\mathbf{x}\}\}$ has the property that there is a monomial $\mathbf{x}^\alpha \in \mathcal{F}[\mathbf{x}]$ where $\mathbf{x}^\alpha f(\mathbf{x}) \in \mathcal{F}[[\mathbf{x}]]$. An alternative way of defining $\mathcal{F}\{\{\mathbf{x}\}\}$ is as $\mathcal{F}[[\mathbf{x}]]$ localized at the multiplicative system consisting of the monomials of $\mathcal{F}[x]$. This gives the ring structure on $\mathcal{F}\{\{\mathbf{x}\}\}$.

One can associate the array S to a multivariate power series $S(\mathbf{x})$ by assigning the coefficient S_α to the monomial $\mathbf{x}^\alpha$. S is also a Laurent series with the coefficients of the terms with negative exponents equal to 0. For a polynomial $C(\mathbf{x})$ let $C(\mathbf{x}^{-1})$ denote $C\left(x_1^{-1}, x_2^{-1}, \ldots, x_m^{-1}\right)$. Since C is a polynomial, $C(\mathbf{x}^{-1})$ is a Laurent series in $\mathcal{F}\{\{\mathbf{x}\}\}$, and one can multiply S and $C(\mathbf{x}^{-1})$ as Laurent series. In this setting, we say that a polynomial C is **valid at a point** S_u if $u - LE(C) \not\geq 0$ or if $u - LE(C) \geq 0$, then in the product $C(\mathbf{x}^{-1})S(\mathbf{x})$ the term with exponent $u - LE(C)$ has coefficient 0. Similarly, the equation that defines a valid polynomial (5) can be "translated" to multiplication of Laurent series as follows: $\sum_{\alpha \in Supp(C)} C_\alpha S_{\alpha+\beta} = 0$ for all $\beta \in \mathbb{N}_0^m$ is equivalent to say that all the (non-zero) terms of the product $C(\mathbf{x}^{-1})S(\mathbf{x})$ have negative exponents, where we say that $ax_1^{e_1} \cdots x_m^{e_m}$ has negative exponents if at least one of the e_i's is negative. For example, $x_1^{5234}x_2^{-1}x_3^{749}$ has negative exponents!

Consider the 2-dimensional case and look at the product $C(\mathbf{x}^{-1})S(\mathbf{x})$ in the plane. The m-dimensional case is a straightforward generalization of this case. One has that C is a valid polynomial for the array S if the coefficients of the terms of $C(\mathbf{x}^{-1})S(\mathbf{x})$ outside the shaded area are zero:

Note that a term ax^iy^j lies in the shaded area if and only if $i < 0$ or $j < 0$. In the figure, N is the exponent of the least common multiple of all the monomials in $C(\mathbf{x})$.

To define $Val(S)$ as the kernel of a map one needs to be able to multiply by polynomials in x^{-1} and y^{-1}. Let $L := \mathcal{F}\{\{x,y\}\}$, and let L^- be the subspace of Laurent series all of whose non-zero terms have negative exponents.

Since L and L^- are $\mathcal{F}[x^{-1}, y^{-1}]$-modules, $L^- \subset L$, one can form the $\mathcal{F}[x^{-1}, y^{-1}]$-module $B := L/L^-$.

In pictures,

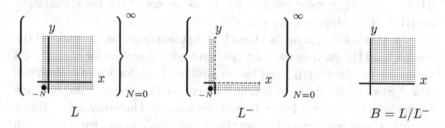

Note that the natural representation of B is by power series in x and y; i.e. B is isomorphic to $\mathcal{F}[[x,y]]$ as a vector space. The m-dimensional case is similar. From now on consider B to be represented in terms of $\mathcal{F}[[x_1,\ldots,x_m]]$ as a vector space.

Within $\mathcal{F}\{\{x_1,\ldots,x_m\}\}$, the elements $x_1^{-1},\ldots,x_m^{-1}$ are algebraically independent over $\mathcal{F}$. Let $\mathcal{F}[\mathbf{x}^{-1}] = \mathcal{F}[x_1^{-1},\ldots,x_m^{-1}]$ denote the subalgebra of $\mathcal{F}\{\{x_1,\ldots,x_m\}\}$ generated by $x_1^{-1},\ldots,x_m^{-1}$. By the algebraic independence of $x_1^{-1},\ldots,x_m^{-1}$, $\mathcal{F}[\mathbf{x}^{-1}]$ is (isomorphic to) the polynomial ring in $x_1^{-1},\ldots,x_m^{-1}$. Since $\mathcal{F}[\mathbf{x}^{-1}]L^- \subset L^-$, there is a natural multiplicative action of $\mathcal{F}[\mathbf{x}^{-1}]$ on $\mathcal{F}\{\{\mathbf{x}\}\}/L^- = B$. In (6), $x_1^{-1} = x^{-1}$ is the "shift left" operator and $x_2^{-1} = y^{-1}$ is the "shift down" operator.

In terms of the representation of B as power series in $x_1, \ldots, x_m$, for $p(\mathbf{x}) \in \mathcal{F}[[\mathbf{x}]]$:

$$\mathbf{x}^{-\alpha}p(\mathbf{x}) = \mathbf{x}^{-\alpha}p(\mathbf{x}) + L^-,$$

i.e. $\mathbf{x}^{-\alpha}p(\mathbf{x})$ is the coset of $\mathbf{x}^{-\alpha}p(\mathbf{x})$ in L/L^-.

The $\mathcal{F}\left[\mathbf{x}^{-1}\right]$-module action on B is also informally described as: when a polynomial $g(\mathbf{x}^{-1})$ in $\mathbf{x}^{-1}$ acts upon a power series $p(\mathbf{x})$ in $\mathbf{x}$, simply multiply $g(\mathbf{x}^{-1})$ by $p(\mathbf{x})$ as Laurent series and consider all resulting terms with exponents with negative entries to be zero.

Now, consider the coset of the periodic array S, $\overline{S(\mathbf{x})} \in B = L/L^-$. Then $\overline{\mathcal{F}[\mathbf{x}^{-1}]S(\mathbf{x})}$ is the cyclic submodule of L/L^- generated by $\overline{S(\mathbf{x})}$. Note that using our reformulation of the problem, a polynomial C is valid on the array S if and only if $\overline{C(\mathbf{x}^{-1})S(\mathbf{x})} = 0$. That is, all terms with exponents with non-negative coordinates are zero. So, if one defines the map

$$\Pi_S : \mathcal{F}[\mathbf{x}^{-1}] \longrightarrow \overline{\mathcal{F}[\mathbf{x}^{-1}]S(\mathbf{x})} \subseteq B = L/L^-,$$

$$C(\mathbf{x}^{-1}) \longmapsto \overline{C(\mathbf{x}^{-1})S(\mathbf{x}) + L^-},$$

one says that C is valid for the array S if and only if $C(\mathbf{x}^{-1}) \in ker(\Pi_S)$. That is, the set of valid polynomials is the kernel of the map, $Val(S) = ker(\Pi_S)$ (if we identify $C(\mathbf{x})$ with $C(\mathbf{x}^{-1})$).

Now that the 0-dimensional ideal $Val(S)$ is expressed as the kernel of the module map Π_S one may use the Sweedler-Taylor algorithm for computing a Gröbner basis for $ker(\Pi_S)$ and have a Gröbner basis for $Val(S)$. However, $\overline{\mathcal{F}[\mathbf{x}^{-1}]S(\mathbf{x})} \subseteq B$, B is not a finite dimensional $\mathcal{F}$-vector space and one cannot directly compute linear dependence as needed in the Sweedler-Taylor algorithm. But one can use the fact that S is periodic to see that it is enough to just consider a finite subarray of S to obtain $Val(S)$.

Although B is infinite dimensional, there are aspects of local finiteness about it. If $S(\mathbf{x})$ were a polynomial, then $\overline{\mathcal{F}[\mathbf{x}^{-1}]S(\mathbf{x})}$ is finite dimensional. There are also many power series $S(x)$ that are not polynomials but $\overline{\mathcal{F}[\mathbf{x}^{-1}]S(\mathbf{x})}$ is finite dimensional; these are precisely the periodic power series considered in this paper.

The analog to the Sweedler-Taylor algorithm was designed to deal with the situation of B being infinite dimensional. The linear dependency will be computed "modulo" multiples of certain monomials. With this we will be able to compute minimal polynomials that are valid "up to" a certain point in the array, this is, valid in a subarray. Because of the periodicity,

polynomials valid on certain subarray will be valid for the complete
array S.

In the next section we will introduce the concept of *being valid up to a
point S_u*. Then we define a set B_u such that B/B_u is finite dimensional.
Using the B_u's we modify the Sweedler-Taylor algorithm to compute what
we call lead monomial generating sets. The modified algorithm will produce
a reduced Gröbner basis for $Val(S)$.

4. Lead monomial generating sets for the set of recurrence relations valid on subarrays

Let $C\left(\mathbf{x}^{-1}\right) \in \mathcal{F}[\mathbf{x}^{-1}]$. Since $\mathcal{F}[\mathbf{x}^{-1}]$ is a polynomial ring in $x_1^{-1}, \ldots, x_m^{-1}$,
one may apply all the theory of Gröbner bases to it. For a monomial order
$<_T$ in $\mathcal{F}[\mathbf{x}]$, the monomial order in $\mathcal{F}[\mathbf{x}^{-1}]$ is given by $\mathbf{x}^{-\alpha} <_T \mathbf{x}^{-\beta}$ if
and only if $\mathbf{x}^\alpha <_T \mathbf{x}^\beta$. Then, for $C\left(\mathbf{x}^{-1}\right) = \sum_{\alpha \in Supp(C)} C_\alpha \mathbf{x}^{-\alpha}$, we set
$LE\left(C\left(\mathbf{x}^{-1}\right)\right) = LE\left(C(\mathbf{x})\right)$.

Let $u \in \mathbb{N}_0^m$ be such that $LE(C) \leq u$. Then, $\beta = u - LE(C) \geq 0$. From
(4) one has that C is valid at S_u if $\sum_{\alpha \in Supp(C)} C_\alpha S_{\beta+\alpha} = 0$. Using the
new notation, this is the same as saying that in $\overline{C(\mathbf{x}^{-1})S(\mathbf{x})}$ the term with
exponent β has coefficient zero.

Note that if $LE(C) \not\leq u$ then the m-tuple $\beta = u - LE(C)$ has at least
one negative entry and when one mods out by L^- the term with exponent
β will assuredly be zero. So, for checking the validity of C at a particular
entry S_u, one does not have to check whether $LE(C) \leq u$ or not.

We formalize this as:

Definition 5. The m-dimensional linear recurrence relation given by the
polynomial C **is valid at the point** S_u if in $\overline{C(\mathbf{x}^{-1})S(\mathbf{x})}$ the term with
exponent $\beta = u - LE(C)$ has coefficient zero. We say that **C is valid up
to** S_u if C is valid at each S_l with $l \leq_T u$.

For now, let's choose $<_T$ to be a monomial order such that for any
monomial M there are only a finite number of monomials $<_T M$. For
example, let $<_T$ be the graded lexicographic order (but not pure lex). Now
define

Definition 6.

$$B_u := \left\{ \sum_{\alpha >_T u} \lambda_\alpha \mathbf{x}^\alpha \right\} \subset B.$$

We have that B/B_u is a finite dimensional vector space. One can then define $Val_u(S)$, the set of polynomials that are "valid up to S_u", using the set B_u:

$$Val_u(S) := \Big\{0 \neq C(\mathbf{x}) \mid \text{ in } \overline{C(\mathbf{x}^{-1})S(\mathbf{x})}, \text{ for every } \beta \text{ with} \qquad (6)$$

$$\beta + LE(C) \leq_T u, \text{ the term with exponent } \beta \text{ has coefficient zero}\Big\} \cup \{0\}$$

$$= \Big\{0 \neq C(\mathbf{x}) \mid \text{ in } \overline{C(\mathbf{x}^{-1})S(\mathbf{x})} \text{ the only non-zero terms}$$

$$\text{are those with exponents } \beta \text{ where } \beta + LE(C) >_T u\Big\} \cup \{0\}.$$

It follows that

$$Val_u(S) = \Big\{0 \neq C(\mathbf{x}) \mid \Pi_S\left(C(\mathbf{x}^{-1})\right) \in \mathbf{x}^{-LE(C)}B_u\Big\} \cup \{0\}.$$

So, $Val_u(S)$ is composed by polynomials C that give linear dependency relations modulo $\mathbf{x}^{-LE(C)}B_u$. $Val_u(S)$ is not necessarily an ideal but it is easy to prove that the set $LM\left(Val_u(S)\right)$ is closed under monomial multiplication, as follows.

Lemma 2. $LM\left(Val_u(S)\right)$ *is a monomial ideal.*

Proof. Say $0 \neq C\left(\mathbf{x}^{-1}\right) \in Val_u(S)$. Since $LM\left(\mathbf{x}^{-\alpha}C\left(\mathbf{x}^{-1}\right)\right) = \mathbf{x}^{-\alpha}LM\left(C\left(\mathbf{x}^{-1}\right)\right)$, it suffices to show that $\mathbf{x}^{-\alpha}C\left(\mathbf{x}^{-1}\right) \in Val_u(S)$.

$$\Pi_S\left(\mathbf{x}^{-\alpha}C\left(\mathbf{x}^{-1}\right)\right) = \overline{\mathbf{x}^{-\alpha}C\left(\mathbf{x}^{-1}\right)S(\mathbf{x})}.$$

Since $\Pi_S\left(C\left(\mathbf{x}^{-1}\right)\right) \in \mathbf{x}^{-LE(C)}B_u$, we see that

$$\Pi_S\left(\mathbf{x}^{-\alpha}C\left(\mathbf{x}^{-1}\right)\right) \in \mathbf{x}^{-\alpha}\mathbf{x}^{-LE(C)}B_u = \mathbf{x}^{-LE\left(\mathbf{x}^{-\alpha}C(\mathbf{x}^{-1})\right)}B_u,$$

where the equality follows from the fact that $LE\left(\mathbf{x}^{-\alpha}C\left(\mathbf{x}^{-1}\right)\right) = \alpha + LE\left(C\left(\mathbf{x}^{-1}\right)\right)$. $\qquad\square$

Since $LM\left(Val_u(S)\right)$ is a monomial ideal, $\Gamma := LE\left(Val_u(S)\right)$ satisfies part 2 of Lemma 1. Therefore, its complement $\mathbb{N}_0^m \setminus \Gamma$ is a delta set, denoted by Δ_u.

Let u^+ be the m-tuple that immediately follows u with respect to $<_T$. Note that $B_{u^+} \subseteq B_u$ implies $Val_{u^+}(S) \subseteq Val_u(S)$. Therefore

$LM\left(Val_{u^+}(S)\right) \subseteq LM\left(Val_u(S)\right)$ and $\Delta_u \subseteq \Delta_{u^+}$. In particular $Val(S) \subseteq Val_u(S)$ and $\Delta_u \subseteq \Delta_{Val(S)}$ for all $u \in \mathbb{N}_0^m$.

Definition 7. Let $G = \{g_1, \ldots, g_l\} \subset \mathcal{F}[\mathbf{x}]$. The set G is called a **lead monomial generating set for the set** $A \subset \mathcal{F}[\mathbf{x}]$ with respect to $<_T$ if $\langle LM(G)\rangle = \langle LM(A)\rangle$. G is called a **reduced lead monomial generating set for** A if $LC(g_i) = 1, i = 1, \ldots, l$ and $LM(g_i)$ does not divide any term of g_j for every $j \neq i$.

Note that this concept is very similar to a reduced Gröbner basis for an ideal. In fact, if A is an ideal and $G \subseteq A$ then a lead monomial generating set for A is a Gröbner basis for A.

4.1. *Analog to the Sweedler-Taylor algorithm*

Building on the Sweedler-Taylor algorithm gives an algorithm to compute a reduced lead monomial generating set G_u for $Val_u(S)$. The set G_u is referred in other papers [14, 16] as a "minimal polynomial set" for $Val_u(S)$.

Unlike the Sweedler-Taylor algorithm which looks for linear dependency relations, one looks for "partial linear dependency" relations, where partial means modulo $\mathbf{x}^{-LE(C)}B_u$. In this setting, all the equations that give linear dependency relations, like equation (3), become:

$$\Pi_S\left((t_{1,e_1})^{-1}\right) \in B_{1,e_1-1} + (t_{1,e_1})^{-1} B_u$$

$$\Pi_S\left((t_{1,e_1})^{-1}\right) = \sum_{j=0}^{e_1-1} \lambda_j \Pi_S\left((t_{1,j})^{-1}\right) + (t_{1,e_1})^{-1} B_u \qquad (7)$$

$$\Pi_S\left((t_{1,e_1})^{-1} - \sum_{j=0}^{e_1-1} \lambda_j (t_{1,j})^{-1}\right) \in (t_{1,e_1})^{-1} B_u.$$

The lead monomial generating set for $Val_u(S)$ is $G_u = \{g_1, \ldots, g_n\}$, where $g_i := t_{i,e_i} - \sum_{j=0}^{e_i-1} \lambda_{i,j} t_{i,j} - \sum_{j=0}^{e_{i-1}-1} \lambda_{i-1,j} t_{i-1,j} - \cdots - \sum_{j=0}^{e_1-1} \lambda_{1,j} t_{1,j}$.

Hence, by changing Π to Π_S and lines 9) and 10) of Algorithm 1 to

9) UNTIL $\Pi_S\left((t_{l,e_l})^{-1}\right) = \sum_{h=1}^{l} \sum_{j=0}^{e_h-1} \lambda_{h,j} \Pi_S\left(((t_{h,j})^{-1}\right) + (t_{l,e_l})^{-1} B_u$ or $A = \emptyset$

10) IF $\Pi_S\left((t_{l,e_l})^{-1}\right) = \sum_{h=1}^{l} \sum_{j=0}^{e_h-1} \lambda_{h,j} \Pi_S\left((t_{h,j})^{-1}\right) + (t_{l,e_l})^{-1} B_u$, THEN,

we obtain an analog to the Sweedler-Taylor algorithm to compute lead monomial generating sets for $Val_u(S)$. A description of the analog to the Sweedler-Taylor algorithm in a more general setting can be found in [11].

4.2. *Finding u such that $Val_u(S) = Val(S)$*

The periodicity of the array S makes the ideal quotient $\mathcal{F}[\mathbf{x}]/Val(S)$ into a finite dimensional vector space, and, at some point S_u, $Val_u(S) = Val(S)$ and the partial dependencies (7) are in fact (complete) linear dependencies. This implies that there exists u such that one can use the analog to the Sweedler-Taylor algorithm described above to compute a reduced lead monomial generating set for $Val_u(S)$ and obtain a Gröbner basis for $Val(S)$. We now give a bound for u that guarantees that $Val_u(S) = Val(S)$.

Suppose that the array S has period $n = (n_1, \ldots, n_m)$. Then, $x_1^{n_1} - 1, \ldots, x_m^{n_m} - 1 \in Val(S)$, and, by Lemma 1, the following result is clear:

Lemma 3. *Let S be an array with period $n = (n_1, \ldots, n_m)$, $Val(S)$ be the ideal of polynomials valid in S and $\Delta_{Val(S)}$ be the delta set of $Val(S)$. Also let $C \in Val(S)$ with $LE(C) = (l_1, \ldots, l_m)$. Then*

(1) $\Delta_{Val(S)} \subseteq \{(\delta_1, \ldots, \delta_m) \in \mathbb{N}_0^m \mid (\delta_1, \ldots, \delta_m) \le (n_1 - 1, \ldots, n_m - 1)\}$,
(2) $|\Delta_{Val(S)}| \le n_1 \cdots n_m$.
(3) If C is minimal in $Val(S)$ with respect to divisibility of $LM(C)$, then $l_i \le n_i$ for $i = 1, \ldots, m$.

We now prove that if we have a minimal polynomial valid at all entries in a certain region of the array S then C is also valid outside that region and hence in the complete array.

Prop 1. Let S be an array with period $n = (n_1, \ldots, n_m)$. Suppose that C is valid at S_u for all $u \le (2n_1 - 1, \ldots, 2n_m - 1)$ and it is minimal with this property with respect to divisibility of $LM(C)$. If $v \not\le (2n_1 - 1, \ldots, 2n_m - 1)$, then C is valid at S_v, and hence $C \in Val(S)$.

Proof. If $v - LE(C) \not\ge 0$, then C is valid at S_v and there is nothing to prove. Suppose that $v - LE(C) \ge 0$. Then $v = LE(C) + \beta$, for some $\beta = (\beta_1, \ldots, \beta_m) \in \mathbb{N}_0^m$. We can rewrite $\beta_i = k_i n_i + r_i$, where $k_i, r_i \in \mathbb{N}_0$, $r_i < n_i$ and have
$$v = LE(C) + (r_1, \ldots, r_m) + (n_1 k_1, \ldots, n_m k_m).$$
Since C is $\le$ minimal, $LE(C) \le (n_1, \ldots, n_m)$, and since also $r_i < n_i$, we have that
$$\gamma = LE(C) + (r_1, \ldots, r_m) \le (2n_1 - 1, \ldots, 2n_m - 1),$$

and, by hypothesis, C is valid for S_γ. Therefore, by the periodicity of S,

$$0 = \sum_{\alpha \in Supp(C)} C_\alpha S_{\alpha+\gamma-LE(C)} = \sum_{\alpha \in Supp(C)} C_\alpha S_{\alpha+\gamma+(n_1 k_1, \ldots, n_m k_m)-LE(C)}$$

$$= \sum_{\alpha \in Supp(C)} C_\alpha S_{\alpha+v-LE(C)},$$

and C is valid at S_v.

$\square$

Any point S_u with $u \leq (2n_1 - 1, \ldots, 2n_m - 1)$ satisfies $|u| \leq 2|n| - m$. So if $S' = \{S_u \mid |u| \leq 2|n| - m\}$, then S' contains all S_u with $u \leq (2n_1 - 1, \ldots, 2n_m - 1)$. This selection for S' is convenient for computing a Gröbner basis with respect to the *grlex* order; other choices can be used for other monomial orders.

Corollary 1. *Let S be an array with period $n = (n_1, \ldots, n_m)$ and $S' \subset S$ be such that $S' = \{S_u\}_{|u| \leq 2|n|-m}$. If C is valid at every point $S_u \in S'$ and it is minimal with this property with respect to divisibility of $LM(C)$, then C is valid for the infinite array S. This is, $C \in Val(S)$.*

If we select a monomial order $<_T$ and order the elements of S such that all the elements in the subarray $S' = \{S_u\}_{|u| \leq 2|n|-m}$ are among the first elements of the ordered array S, then the minimal polynomials valid for S' are the minimal polynomials valid for S. If u^+ is the smallest with respect to $<_T$ such that $u^+ \geq_T u$ for all u with $|u| \leq 2|n| - m$, then a reduced lead monomial generating set for $Val_{u^+}(S)$ is a reduced Gröbner basis for $Val(S)$. In particular, if we consider the *graded lexicographical* order we get the next proposition.

Prop 2. *Let S be an array with period $n = (n_1, \ldots, n_m)$ and $S' \subset S$ be such that $S' = \{S_u\}_{|u| \leq 2|n|-m}$. If u^+ is the largest element in $\mathbb{N}_0^m$ with respect to $<_{grlex}$ such that $S_{u^+} \in S'$, then, the set of all minimal polynomials in $Val_{u^+}(S)$ forms a Gröbner basis for $Val(S)$ with respect to $<_{grlex}$. This is, $Val_{u^+}(S) = Val(S)$.*

We now illustrate how to use the analog to the Sweedler-Taylor algorithm discussed in Section 4.1 to compute a Gröbner basis for $Val(S)$ with respect to $<_{grlex}$, where S is a 2-dimensional array with period $n = (n_1, n_2)$. By Proposition 2, it is enough to use the points S_u with $u_1 + u_2 \leq 2(n_1 + n_2) - 2$. Following the idea of Example 1, to find the partial linear dependency relations we form a matrix where each row corresponds to $\mathbf{x}^{-\alpha}S(x)$, and there is a column for each monomial from 1 to $x^{2(n_1+n_2)-2}$. The matrix will have a form similar to the matrix below, where we used the *grlex* order with $y < x$.

$$1 \quad y \quad x \quad y^2 \quad xy \quad \cdots \quad x^{2(n_1+n_2)-2}$$

$$
\begin{array}{c}
\Pi_S(1) \\[2ex]
\Pi_S(y^{-1}) \\[2ex]
\Pi_S(x^{-1}) \\[2ex]
\vdots
\end{array}
\left(
\begin{array}{c}
S'(x,y) \\[2ex]
\overline{y^{-1}S'(x,y)} \\[2ex]
\overline{x^{-1}S'(x,y)} \\[2ex]
\vdots
\end{array}
\right)
$$

When a new row is adjoined to the matrix, it is reduced with the previous rows. We need to find relations that are valid up to $S_{(2(n_1+n_2)-2,0)}$. If the new row corresponds to the map $\Pi_S(\mathbf{x}^{\alpha}) = \Pi_S(x^{-\alpha_1}y^{-\alpha_2})$, we look for linear dependencies modulo $x^{-\alpha_1}y^{-\alpha_2}B_{(2(n_1+n_2)-2,0)}$, where $B_{(2(n_1+n_2)-2,0)} = \left\{ \sum_{\gamma >_{grlex}(2(n_1+n_2)-2,0)} \lambda_\gamma \mathbf{x}^\gamma \right\}$. That is, the row $\Pi_S(x^{-\alpha_1}y^{-\alpha_2})$ needs to be partial linearly dependent, which means that it should have 0 in every column $x^i y^j$, where $(\alpha_1 + i, \alpha_2 + j) \leq_T (2(n_1 + n_2) - 2, 0)$. Recall that we ignore any term that has negative exponents.

Example 2. Let S be the following 2-dimensional array with entries in $\mathbb{F}_{11}$ and period vector $n = (2, 2)$:

$$
S = \left(
\begin{array}{llll}
\vdots & \vdots & \vdots & \vdots \\
10 & 8 & 10 & 8 \cdots \\
3 & 1 & 3 & 1 \cdots \\
10 & 8 & 10 & 8 \cdots \\
3 & 1 & 3 & 1 \cdots
\end{array}
\right)
$$

To find a Gröbner basis for $Val(S)$ with respect to $<_{grlex}$ we use the analog to the Sweedler-Taylor algorithm to compute a lead monomial generating set for $Val_u(S)$, where $u = (6, 0)$. We order the monomials from 1 to x^6 with respect to $<_{grlex}$ and construct a matrix for reduction, where we include additional columns to keep track of the row operations and be able to obtain the coefficients of the Gröbner basis elements.

	1	y	x	y^2	yx	$\cdots$	x^5	y^6	$\cdots$	x^6	1	y
$\Pi_S(1)$	3	10	1	3	8	$\cdots$	1	3	$\cdots$	3	1	0
$\Pi_S(y^{-1})$	10	3	8	10	1	$\cdots$	8				0	1

	1	y	x	y^2	yx	$\cdots$	x^5	y^6	$\cdots$	x^6	1	y
$\Pi_S(1)$	3	10	1	3	8	$\cdots$	1	3	$\cdots$	3	1	0
$\Pi_S(y^{-1})$	0	10	1	0	0	$\cdots$	1				4	1

Since the row corresponding to $\Pi_S(y^{-1})$ is independent, we adjoin the row corresponding to $\Pi_S(x^{-1})$ and reduce it with the previous rows.

	1	y	x	y^2	$\cdots$	x^5	y^6	$\cdots$	x^6	1	y	x
$\Pi_S(1)$	3	10	1	3	$\cdots$	1	3	$\cdots$	3	1	0	0
$\Pi_S(y^{-1})$	0	10	1	0	$\cdots$	1				4	1	0
$\Pi_S(x^{-1})$	1	8	3	1	$\cdots$	3				0	0	1

Multiplying the first row by 10, the second and third rows by 3, adding the three rows and substituting the result in the third row we obtain:

	1	y	x	y^2	$\cdots$	x^5	y^6	$\cdots$	x^6	1	y	x
$\Pi_S(1)$	3	10	1	3	$\cdots$	1	3	$\cdots$	3	1	0	0
$\Pi_S(y^{-1})$	0	10	1	0	$\cdots$	1				4	1	0
$\Pi_S(x^{-1})$	0	0	0	0	$\cdots$	0				0	3	3

From this, one obtains that $3x^{-1} + 3y^{-1} \in ker(\Pi_S)$ and the monic polynomial $g_1 = x + y$ is in the reduced Gröbner basis for $Val(S)$. We

continue the algorithm considering the next monomial that is not a multiple
of x, and adjoin the row corresponding to $\Pi_S\left(y^{-2}\right)$:

	1 y x y^2 $\cdots$ x^5 y^6 $\cdots$ x^6	1 y x y^2
$\Pi_S(1)$	$3\ 10\ 1\ 3\ \cdots\ 1\ 3\ \cdots\ 3$	$1\ 0\ 0\ 0$
$\Pi_S(y^{-1})$	$0\ 10\ 1\ 0\ \cdots\ 1$	$4\ 1\ 0\ 0$
$\Pi_S(y^{-2})$	$3\ 10\ 1\ 3\ \cdots\ 3$	$0\ 0\ 0\ 1$

Multiplying the first row by 10 and adding the third row we obtain:

	1 y x y^2 $\cdots$ x^5 y^6 $\cdots$ x^6	1 y x y^2
$\Pi_S(1)$	$3\ 10\ 1\ 3\ \cdots\ 1\ 3\ \cdots\ 3$	$1\ 0\ 0\ 0$
$\Pi_S(y^{-1})$	$0\ 10\ 1\ 0\ \cdots\ 1$	$4\ 1\ 0\ 0$
$\Pi_S(y^{-2})$	$0\ 0\ 0\ 0\ \cdots\ 0$	$10\ 0\ 0\ 1$

This implies that the polynomial $y^2 + 10$ is in the reduced Gröbner
basis. Since there are no monomials left that are not multiples of the leading monomials in the Gröbner basis, $G = \left\{y^2 + 10, x + y\right\}$ is the reduced
Gröbner basis for $Val(S)$ with respect to $<_{grlex}$ and $y < x$.

4.3. *Conclusions*

We presented an algorithm for computing a Gröbner basis for the ideal
of linear recurrence relations on m-dimensional periodic arrays. The algorithm is based on linear algebra computations and can be implemented
easily. Since the algorithm does not assume that one already has a generating set for the ideal of linear recurrences, the fact that it gives a generating
set is one of the applications of the algorithm. By Proposition 2, twice
the period minus the dimension of the array is an upper bound for the
degree of the polynomials required to form a Gröbner basis for the ideal
of linear recurrence relations valid on the array with respect to the graded
lexicographic order. Recent applications for the algorithm include the computation of the linear complexity of m-dimensional periodic arrays [9].

References

[1] J. Abbott, A. Bigatti, M. Kreuzer, and L. Robbiano. Computing ideals of points. *J. Symbolic Comput.*, 30(4):341–356, 2000.

[2] J. Abbott, M. Kreuzer, and L. Robbiano. Computing zero-dimensional schemes. *J. Symbolic Comput.*, 39(1):31–49, 2005.

[3] M. Borges-Quintana, M. A. Borges-Trenard, and E. Martínez-Moro. A general framework for applying FGLM techniques to linear codes. In *Applied algebra, algebraic algorithms and error-correcting codes*, volume 3857 of *Lecture Notes in Comput. Sci.*, pages 76–86. Springer, Berlin, 2006.

[4] M. A. Borges-Trenard, M. Borges-Quintana, and T. Mora. Computing Gröbner bases by FGLM techniques in a non-commutative setting. *J. Symbolic Comput.*, 30(4):429–449, 2000.

[5] Bruno Buchberger. Bruno Buchbergers phd thesis 1965: An algorithm for finding the basis elements of the residue class ring of a zero dimensional polynomial ideal. *Journal of Symbolic Computation*, 41(34):475 – 511, 2006. Logic, Mathematics and Computer Science: Interactions in honor of Bruno Buchberger (60th birthday).

[6] David Cox, John Little, and Donal O'Shea. *Ideals, varieties, and algorithms.* Undergraduate Texts in Mathematics. Springer, New York, third edition, 2007. An introduction to computational algebraic geometry and commutative algebra.

[7] J. C. Faugère, P. Gianni, D. Lazard, and T. Mora. Efficient computation of zero-dimensional Gröbner bases by change of ordering. *J. Symbolic Comput.*, 16(4):329–344, 1993.

[8] Patrizia Gianni, Barry Trager, and Gail Zacharias. Gröbner bases and primary decomposition of polynomial ideals. *J. Symbolic Comput.*, 6(2-3):149–167, 1988. Computational aspects of commutative algebra.

[9] Domingo Gomez-Perez, Tom Hoholdt, Oscar Moreno, and Ivelisse Rubio. Linear complexity for multidimensional arrays - a numerical invariant. In *Information Theory (ISIT), 2015 IEEE International Symposium on*, pages 2697–2701, June 2015.

[10] Ainhoa Aparicio Monforte and Manuel Kauers. Formal Laurent series in several variables. *Expositiones Mathematicae*, 31(4):350 – 367, 2013.

[11] Ivelisse María Rubio. *Gröbner bases for 0-dimensional ideals and applications to decoding, PhD thesis.* Cornell University, Jan., 1998.

[12] Keith Saints and Chris Heegard. On hyperbolic cascaded Reed-Solomon codes. In *Applied algebra, algebraic algorithms and error-correcting codes (San Juan, PR, 1993)*, volume 673 of *Lecture Notes in Comput. Sci.*, pages 291–303. Springer, Berlin, 1993.

[13] Keith Saints and Chris Heegard. Algebraic-geometric codes and multidimensional cyclic codes: a unified theory and algorithms for decoding using Gröbner bases. *IEEE Trans. Inform. Theory*, 41(6, part 1):1733–1751, 1995. Special issue on algebraic geometry codes.

[14] Shojiro Sakata. Finding a minimal set of linear recurring relations capable of generating a given finite two-dimensional array. *J. Symbolic Comput.*,

5(3):321–337, 1988.

[15] Shojiro Sakata. The BMS algorithm. In *Gröbner Bases, Coding, and Cryptography*, pages 143–163. Springer, 2009.

[16] Shojiro Sakata, Helge Elbrønd Jensen, and Tom Høholdt. Generalized Berlekamp-Massey decoding of algebraic-geometric codes up to half the Feng-Rao bound. *IEEE Trans. Inform. Theory*, 41(6, part 1):1762–1768, 1995. Special issue on algebraic geometry codes.

[17] Moss Sweedler and Lee Taylor. An algorithm for finding reduced Gröbner bases for 0-dimensional ideals (preprint). 1996.

An introduction to hyperelliptic curve arithmetic

R. Scheidler

Department of Mathematics and Statistics, University of Calgary,
2500 University Drive NW, Calgary, Alberta T2N 1N4, Canada
`rscheidl@ucalgary.ca`

1. Introduction and Motivation

Secure authentication across insecure communication channels is crucial in today's digital world. When conducting an online shopping or banking transaction, for example, a customer needs to be assured that she is communicating with the intended retailer or bank, rather than falling victim to a phishing attack. Similarly, the retailer or bank must be sure that the transaction originated with a legitimate client as opposed to an impostor. Secure authentication between two or more entities is usually effected through the use of a *cryptographic key*, i.e. a shared secret that is only known to the participating communicants. Only parties with knowledge of this secret are legitimate, and a secure authentication system ensures that it is infeasible for impersonators to obtain the secret.

Geographic reality requires that shared cryptographic keys must themselves be established across insecure communication channels in such a way that eavesdroppers cannot discover them. One of the most common and efficient means by which this can be accomplished is the Diffie-Hellman key agreement protocol [12]. In this protocol's most general form, two communicating parties *Alice* and *Bob* first agree on a finite cyclic group G, written additively, and a generator g of G. Both G and g can be publicly known. To establish a shared cryptographic key, Alice and Bob proceed as follows:

- Alice generates a secret random integer a and sends $A = ag$ to Bob.

321

- Bob generates a secret random integer h and sends $B = bg$ to Alice.
- Upon receipt of B, Alice computes $K = aB$.
- Upon receipt of A, Bob computes $K = bA$.

The shared key is $K = aB = bA = (ab)g$. An eavesdropper, armed with knowledge of g and the ability to intercept A and B, must obtain K in order to successfully impersonate Alice or Bob. In all practical applications, the only know way to accomplish this is to solve an instance of the *discrete logarithm problem*[1] (DLP) in G: given g and a scalar multiple $xg \in G$, find x.

For reasons of practicality, the elements of the underlying group G should have a compact representation, and addition in G should be efficient. To ensure that the key K is protected from discovery by an adversary, the DLP in G must be intractable. In addition, the group order should satisfy certain properties. Most obviously, it should be sufficiently large to foil successful guesses at K. To thwart a Pohlig-Hellman attack on the DLP [35], the group order should additionally have at least one large prime factor; ideally G is of prime order. There may be additional requirements depending on the specific group in question.

The fastest algorithms for solving the DLP in a *generic* finite cyclic group G are Shanks' deterministic baby step giant step technique [37] and Pollard's randomized rho method [36]. Both require on the order of $\sqrt{|G|}$ group operations, which agrees asymptotically with the proved lower bound for solving the DLP in generic groups [38]; the baby step giant step algorithm additionally requires storage of approximately $\sqrt{|G|}$ group elements. To maximize security, DLP-based cryptography therefore seeks to employ group settings where this "square root" performance for discrete logarithm extraction is believed to be best possible.

Many groups have been proposed for discrete log based cryptography, but clearly not all groups are suitable (the reader readily convinces herself that the DLP in the additive group of integers modulo any $n \in \mathbb{N}$ is trivially solvable even for very large n). Exploiting structural properties inherent in certain specific groups makes it possible to achieve an asymptotic complexity for discrete log extraction that is significantly below the square root bound. For example, the fastest DLP algorithm in the multiplicative group of a large prime field, which is the setting originally proposed by Diffie and Hellman, is the Number Field Sieve [20] which is subexponential. For finite

---

[1] In a multiplicatively written group G, the DLP asks to obtain x from g^x; hence the name discrete *logarithm* problem.

fields of small characteristic, better subexponential [26] or even close to polynomial [2] performance is possible.

In 1985, Koblitz [28] and Miller [31] independently proposed the group of points on an elliptic curve over a finite field for discrete log based cryptography. Four years later, Koblitz suggested to use the Jacobian of a hyperelliptic curve in this context. DLP computation in this setting was subsequently shown to be subexponential for sufficiently large genus [1, 32] and faster than square root performance (though still exponential) for genus 3 and higher [16, 39]. The natural generalization to Jacobians of other types of curves of genus at least 3 was similarly established to achieve below square root complexity [11]. This left only the cases of genus 1 and 2 curves — which are elliptic and hyperelliptic, respectively — as suitable settings for discrete log based cryptography. To date, the fastest known DLP algorithms for these two scenarios are the aforementioned methods of square root complexity. As a result, genus 1 and 2 curves represent the most secure settings for discrete log based cryptography. Computing the corresponding group orders is possible [6, 18, 27, 30] but not easy. In order to avoid group order computation, one can instead construct a curve whose associated group has a prescribed group order. For elliptic curves, this is feasible, and the approach first presented in [8] has since undergone significant improvements. However, in genus 2, this is a much harder problem [7].

Genus 1 and 2 curves are also highly practical, particularly for cryptography on devices with constrained computing power and storage such as smart cards or smart phones. Elliptic curve cryptography enjoys commercial deployment in the Blackberry smart phone and Bluray technology, to name just two examples. Hyperelliptic curves have not seen such use, but are therefore also not subject to licensing fees. Their arithmetic is more complicated than that of elliptic curves, but has the potential to outperform it due to the following phenomenon. The order of the Jacobian of a curve of genus g over a finite field $\mathbb{F}_q$ lies in the *Hasse-Weil interval* $[(\sqrt{q} - 1)^{2g}, (\sqrt{q} + 1)^{2g}]$; for large q, it is thus very close to q^g. The *security level* is the computational effort of the fastest successful attack, i.e. in essence the asymptotic complexity of the fastest known algorithm for solving the DLP. In our context, this is $\sqrt{q^g}$. To achieve a fixed security level $\sqrt{q^g} \approx 2^n$ (see [3] for recommended values of n), a genus 1 curve needs to be defined over a field of size $q \approx 2^{2n}$, whereas the field of definition of a genus 2 curve need only have order $q \approx 2^n$. The group law operates on $2g$-tuples of field elements; in essence, points on the curve for genus 1 and pairs of points for genus 2. So genus 2 arithmetic employs quadruples of field elements, as opposed to pairs of field elements in genus 1, but in genus 2 the elements belong to a field of half the size. This can lead to overall faster performance in genus 2. See [5] for the race between genus 1 and 2 arithmetic.

This article provides a gentle introduction to arithmetic in the groups associated with elliptic and hyperelliptic curves. For the latter, we will focus on the genus 2 scenario, but present arithmetic for arbitrary genus as well. We chose this approach since in addition to being an essential ingredient in Diffie-Hellman key agreement and other curve based cryptographic protocols, hyperelliptic curve arithmetic can be used for determining the order and the group structure of the Jacobian, extracting discrete logarithms in this setting, and tackling other problems arising in computational number theory that are of interest for higher genus.

The amount of literature on elliptic and hyperelliptic curves, their arithmetic, and their uses in cryptography is too vast to cite and review here. Instead, we refer the reader to the comprehensive source [10] and the references cited therein. Throughout, let $\mathbb{K}$ be any field and $\overline{\mathbb{K}}$ some fixed algebraic closure of $\mathbb{K}$.

2. Elliptic Curves and Their Arithmetic

In order to understand hyperelliptic curve arithmetic, it is useful to first become familiar with the considerably simpler point arithmetic on elliptic curves. Formally, an *elliptic curve* over $\mathbb{K}$ is a pair consisting of a smooth (projective) curve of genus one and a distinguished point on the curve. For our purposes, we will think of an elliptic curve over $\mathbb{K}$ as given by a *Weierstraß equation*

$$E \ : \ y^2 + a_1 xy + a_3 y = x^3 + a_2 x^2 + a_4 x + a_6 \tag{1}$$

with $a_1, a_2, a_3, a_4, a_6 \in \mathbb{K}$. In addition, E must be *non-singular* (or *smooth*), i.e. there are no simultaneous solutions $(x, y) \in \overline{\mathbb{K}} \times \overline{\mathbb{K}}$ of (1) and its two partial derivatives with respect to x and y:

$$a_1 y = 3x^2 + 2a_2 x + a_4 \ ,$$

$$2y + a_1 x + a_3 = 0 \ .$$

The non-singularity condition guarantees that there is a unique tangent line to E at every point on E. It is equivalent to requiring the *discriminant* of E to be non-zero.

For any field $\mathbb{L}$ with $\mathbb{K} \subseteq \mathbb{L} \subseteq \overline{\mathbb{K}}$, the set of $\mathbb{L}$-*rational points* on E is

$$E(\mathbb{L}) = \{(x_0, y_0) \in \mathbb{L} \times \mathbb{L} \mid y_0^2 + a_1 x_0 y_0 + a_3 y_0 = x_0^3 + a_2 x_0^2 + a_4 x_0 + a_6\} \cup \{\infty\} \ .$$

Here, ∞ is the aforementioned distinguished point on E, also referred to as the *point at infinity*, and all other $\mathbb{L}$-rational points one E are said to be *affine* (or *finite*). The point at infinity arises as follows. The *homogenization* of a bivariate polynomial $F(x, y)$ of total degree $d \in \mathbb{N}$ is the homogeneous polynomial $F_{\text{hom}}(x, y, z) = z^d F(x/z, y/z)$ of degree d in three variables x, y, z. The equation $F_{\text{hom}}(x, y, z) = 0$ defines a *projective curve* whose (projective) points are equivalence classes on the space $\overline{\mathbb{K}}^d \setminus \{\mathbf{0}\}$ where two d-tuples in this space are equivalent if they are $\overline{\mathbb{K}}^*$-multiples of each other. Every projective point has a unique representation $[x_0 : y_0 : z_0]$, normalized so that the last non-zero entry is 1. Applying this procedure to E shows that the points $(x_0, y_0) \in \mathbb{L} \times \mathbb{L}$ are in one-to-one correspondence with the projective points $[x_0 : y_0 : 1]$ on the homogenization E_{hom} of E, and the point ∞ corresponds to the unique projective point on E_{hom} with $z = 0$, namely $[0 : 1 : 0]$.

If $\mathbb{K}$ has characteristic different from 2, then completing the square in y, i.e. replacing y by $y - (a_1 x + a_3)/2$ in (1), yields a curve

$$y^2 = x^3 + b_2 x^2 + b_4 x + b_6 \qquad (b_2, b_4, b_6 \in \mathbb{K}) \qquad (2)$$

that is $\mathbb{K}$-isomorphic[2] to (1). If, in addition, $\mathbb{K}$ has characteristic different from 3, then substituting x by $x - b_2/3$ in (2) yields a curve that is $\mathbb{K}$-isomorphic to (2) (and hence to (1)), and is given by a *short* Weierstraß equation

$$E \;:\; y^2 = x^3 + Ax + B \qquad (A, B \in \mathbb{K}) \,. \qquad (3)$$

Here, the non-singularity condition on E is easily seen to hold if and only if the cubic polynomial $x^3 + Ax + B$ has distinct roots, or equivalently, its discriminant $-(4A^3 + 27B^3)$ does not vanish. In characteristic 2 and 3, there are analogous shorter forms for Weierstraß equations; see Table 13.2, p. 274, of [10].

An abelian group structure can be imposed on $E(\mathbb{L})$ via the motto "any three collinear points on E sum to zero", where the point at infinity functions as the identity element (zero). To determine inverses, this is best considered projectively: for any affine point $P = (x_0, y_0) \in E(\mathbb{L})$, the line through the projective points $[x_0 : y_0 : 1]$ and $[0 : 1 : 0]$ is $x = x_0 z$ which intersects E_{hom} uniquely in the third point $[x_0 : -y_0 - a_1 x_0 - a_3 : 1]$. Thus, the inverse of P is the affine point $\overline{P} = (x_0, -y_0 - a_1 x_0 - a_3) \in E(\mathbb{L})$, and

---

[2] An *isomorphism* between two curves is a bijective rational map between the sets of points of the two curves. If such an isomorphism is defined over $\mathbb{K}$, then the two curves are said to be $\mathbb{K}$-*isomorphic*.

the line through P, $\overline{P}$ and ∞ is the line $x = x_0$. If E is in short Weierstraß form, then $\overline{P} = (x_0, -y_0)$ and inversion is geometrically simply reflection of a point on the x-axis; see Figure 1.

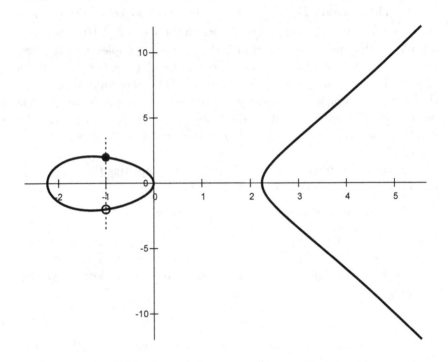

Fig. 1.   Point inversion on $E : y^2 = x^3 - 5x$ over $\mathbb{Q}$. The inverse of $P = (-1, -2)$ (white circle) is $\overline{P} = (-1, 2)$ (black circle), obtained by reflecting P on the x-axis.

To add two affine points $P = (x_1, y_1), Q = (x_2, y_2) \in E(\mathbb{L})$ with $Q \neq \overline{P}$, let $L : y = ax + b$ be the line through P and Q if $P \neq Q$ and the unique tangent line to E at P if $P = Q$. Substituting L into E yields a cubic polynomial in x with roots x_1 and x_2. Let x_3 be the third root of this polynomial and put $R = (x_3, y_3)$ with $y_3 = ax_3 + b$. Then $R \in E(\mathbb{L})$, and since P, Q and R are collinear, we see that $P + Q = \overline{R}$; see Figure 2. Because of this construction, elliptic curve point addition is also said to follow the *chord and tangent law*.

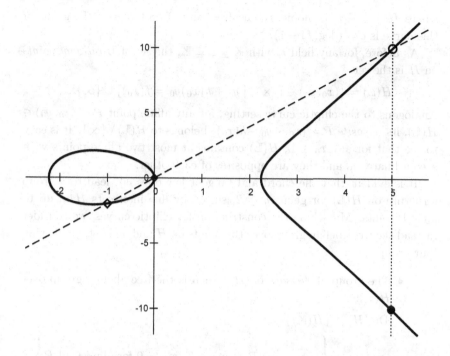

Fig. 2. Point addition on $E : y^2 = x^3 - 5x$ over $\mathbb{Q}$. The line $y = 2x$ (dashed line) through $P = (-1, -2)$ (white lozenge) and $Q = (0, 0)$ (black lozenge) intersects E in the third point $R = (5, 10)$ (white circle), so $P + Q = \overline{R} = (5, -10)$ (black circle).

3. Hyperelliptic Curves

By (1) and (2), an elliptic curve over $\mathbb{K}$ is given by a non-singular equation $y^2 + h(x)y = f(x)$ where $f(x), h(x) \in \mathbb{K}[x]$, $f(x)$ is monic, $\deg(f) = 3$, $\deg(h) \leq 1$ if $\mathbb{K}$ has characteristic 2, and $h(x) = 0$ otherwise. For our purposes, a *hyperelliptic curve* of *genus* $g \in \mathbb{N}$ over $\mathbb{K}$ is a curve given by a *generalized Weierstraß equation*, which is a non-singular equation of the form

$$H : y^2 + h(x)y = f(x) , \qquad (4)$$

where $f(x), h(x) \in \mathbb{K}[x]$, $f(x)$ is monic, $\deg(f) = 2g + 1$, $\deg(h) \leq g$ if $\mathbb{K}$ has characteristic 2, and $h(x) = 0$ otherwise. Thus, elliptic curves can be viewed as genus 1 hyperelliptic curves. In characteristic different from 2, a hyperelliptic curve is simply given by an equation $y^2 = f(x)$

where $f(x) \in \mathbb{K}[x]$ is monic, square-free, and of odd degree; the genus of this curve is $g = (\deg(f) - 1)/2$.

As before, for any field $\mathbb{L}$ with $\mathbb{K} \subseteq \mathbb{L} \subseteq \overline{\mathbb{K}}$, the set of $\mathbb{L}$-*rational points* on H is the set

$$H(\mathbb{L}) = \{(x_0, y_0) \in \mathbb{L} \times \mathbb{L} \mid y_0^2 + h(x_0)y_0 = f(x_0)\} \cup \{\infty\} .$$

Analogous to the elliptic curve setting, for any affine point $P = (x_0, y_0) \in H(\mathbb{L})$, its *opposite* $\overline{P} = (x_0, -y_0 - h(x_0))$ belongs to $H(\mathbb{L}) \setminus \{\infty\}$. It is easy to see that for any $x_0 \in \mathbb{L}$, $H(\mathbb{L})$ contains at most two affine points with x-coordinate x_0, and they are opposites of each other.

It is evident that the chord and tangent law does not lead to a group structure on $H(\mathbb{L})$ for genus $g \geq 2$, since any line intersects H in up to $2g + 1$ points. Mimicking the construction for elliptic curves, we consider instead the free abelian group over the points on H and define the following four groups:

- The group of *divisors* on H, which is the free abelian group over $H(\overline{\mathbb{K}})$:

$$\mathrm{Div}(H) = \langle H(\overline{\mathbb{K}}) \rangle$$

$$= \left\{ \sum_{P \in H(\overline{\mathbb{K}})} m_P P \mid m_P \in \mathbb{Z}, m_P = 0 \text{ for almost all } P \right\} .$$

- The subgroup of $\mathrm{Div}(H)$ of *degree zero* divisors on H:

$$\mathrm{Div}^0(H) = \langle [P] \mid P \in H(\overline{\mathbb{K}}) \rangle$$

$$= \left\{ \sum_{P \in H(\overline{\mathbb{K}})} m_P [P] \mid m_P \in \mathbb{Z}, m_P = 0 \text{ for almost all } P \right\} ,$$

where $[P] = P - \infty$.

- The subgroup of $\mathrm{Div}^0(H)$ of *principal* divisors on H:

$$\mathrm{Prin}(H) = \left\{ \sum_{P \in H(\overline{\mathbb{K}})} v_P(\alpha)[P] \mid \alpha \in \overline{\mathbb{K}}(H) \right\} ,$$

where $\overline{\mathbb{K}}(H) = \overline{\mathbb{K}}(x, y) = \{r(x) + s(x)y \mid r(x), s(x) \in \overline{\mathbb{K}}(x)\}$ is the *function field* of H, and for any function $\alpha \in \overline{\mathbb{K}}(H)$, $v_P(\alpha)$ is the *multiplicity* of the point $P \in H(\overline{\mathbb{K}})$ at α. That is, $v_P(\alpha) = 0$ if P is neither a zero nor a pole of α; if P is a zero of α, then $v_P(\alpha)$ is the multiplicity of this zero; if P is a pole of α, then $-v_P(\alpha)$ is the multiplicity of this pole.

- The *degree zero class group* or *Jacobian*[3] of H:

$$\text{Jac}(H) = \text{Div}^0(H) / \text{Prin}(H) .$$

The Jacobian is the appropriate hyperelliptic curve generalization of the group of points on an elliptic curve. The set of points on H embeds into the Jacobian of H by assigning each $P \in H(\overline{\mathbb{K}})$ the coset of $[P]$ in $\text{Jac}(H)$. For elliptic curves, this embedding is in fact a group isomorphism, so the point addition as defined in Section 2 is compatible with the group law on the Jacobian of an elliptic curve. However, for hyperelliptic curves of genus 2 and higher, this embedding is no longer surjective.

The identity in $\text{Jac}(H)$ is the coset of $[\infty]$. The generalization of the elliptic curve motto that any three collinear points on E sum to zero is "all the points on any *function* on H sum to zero". That is, if $P_1, P_2, \ldots, P_r$ is the complete collection of intersection points of H with some function $\alpha \in \mathbb{K}(H)$, with respective multiplicities $v_{P_i}(\alpha)$ for $1 \le i \le r$, then the divisor $D = \sum_{i=1}^{r} v_{P_i}(\alpha)[P_i]$ is principal. Since for any affine point $P = (x_0, y_0) \in H(\overline{\mathbb{K}})$, the line $x = x_0$ intersects H only in P, $\overline{P}$ and ∞, we see that the inverse of the class of $[P]$ is the class of $-[P] = [\overline{P}]$. More generally, the inverse of the class of a divisor $D = \sum m_P[P]$ in $\text{Jac}(H)$ is the class of $\overline{D} = \sum m_P[\overline{P}]$.

The *(affine) support* of a degree zero divisor $D = \sum m_P[P] \in \text{Div}^0(H)$, denoted $\text{supp}(D)$, is the set of points $P \in H(\mathbb{L}) \setminus \{\infty\}$ for which $m_P \ne 0$. The divisor D is *semi-reduced* if the following conditions are satisfied.

- $m_P > 0$ for all $P \in \text{supp}(D)$;
- If $P \in \text{supp}(D)$ with $P \ne \overline{P}$, then $\overline{P} \notin \text{supp}(D)$;
- If $P \in \text{supp}(D)$ with $P = \overline{P}$, then $m_P = 1$.

It is not hard to see that every class in $\text{Jac}(H)$ contains a semi-reduced divisor: simply replace every summand $-[P]$ by $[\overline{P}]$, and subsequently remove all combinations of the form $[P] + [\overline{P}] = [\infty]$ from D, noting that $2[P] = [\infty]$ when $P = \overline{P}$.

A semi-reduced divisor $D \sum m_P[P] \in \text{Div}^0(H)$ is *reduced* if $\sum m_P \le g$, where g is the genus of H. In particular, the support of a reduced divisor contains at most g points. For example, the reduced divisors on a genus 2 hyperelliptic curve H are exactly the divisors of the form $[P]$ and $[P] + [Q]$ with affine points $P, Q \in H(\overline{\mathbb{K}})$ such that $Q \ne \overline{P}$. In fact, divisors of the

---

[3]The Jacobian of an algebraic curve of genus g leads a double life as an abelian group and a principally polarized abelian variety of dimension g called the *Jacobian variety* of the curve.

form $[P]$ arising from affine points P on any hyperelliptic curve are always reduced. The key enabler for efficient Jacobian arithmetic is the following theorem, provable via Riemann-Roch theory:

Theorem 1. *Every class in* $Jac(H)$ *contains a unique reduced divisor.*

The above theorem distills Jacobian arithmetic down to the following question: given two reduced divisors D_1 and D_2, determine (efficiently) the reduced representative of the class of $D_1 + D_1$ in $Jac(H)$. We denote this reduced representative by $D_1 \oplus D_2$.

4. Arithmetic on Reduced Divisors

To avoid clutter, we will frequently omit the square brackets in the degree zero divisor notation and simply write semi-reduced divisors as sums of affine points. We begin this section with an illustration of Jacobian arithmetic via reduced divisors on a genus 2 example.

Example 1. Consider the genus 2 hyperelliptic curve $H : y^2 = f(x)$ over $\mathbb{Q}$ with $f(x) = x^5 - 5x^3 + 4x - 1$. We wish to find $D_1 \oplus D_2$ for the two reduced divisors $D_1 = P_1 + P_2$ and $D_2 = Q_1 + Q_2$ on H, where $P_1 = (-2, 1)$, $P_2 = (0, 1)$, $Q_1 = (2, 1)$ and $Q_2 = (3, -11)$. These four points all lie on the unique degree 3 function $y - v(x) \in \overline{\mathbb{Q}}(H)$ where $v(x) = -(4/5)x^3 + (16/5)x + 1$. The curve $y = v(x)$ intersects H in two more points R_1 and R_2. The x-coordinates of all six intersection points are the roots of the equation

$$0 = f(x) - v(x)^2 = -\left(x - (-2)\right)\left(x - 0\right)\left(x - 2\right)\left(x - 3\right)u(x)$$

with $u(x) = 16x^2 + 23x + 5$. Thus, the x-coordinates of R_1 and R_2 are the zeros of $u(x)$, and their y-coordinates are obtained by substituting their respective x-coordinates into $y = v(x)$. Since the six points $P_1, P_2, Q_1, Q_2, R_1, R_2$ form the complete intersection of H with $y = v(x)$, they sum to zero, so $D_1 \oplus D_2 = \overline{R}_1 + \overline{R}_2$, which is the divisor

$$\left(\frac{-23 + \sqrt{209}}{32}, \frac{1333 - 115\sqrt{209}}{2048}\right) + \left(\frac{-23 - \sqrt{209}}{32}, \frac{1333 + 115\sqrt{209}}{2048}\right) \quad (5)$$

This process is illustrated in Figure 3.

In general, to find the reduced sum $D_1 \oplus D_2$ of two divisors D_1 and D_2 on a hyperelliptic curve (4), first determine the semi-reduced sum D of D_1 and D_2; this is equal to the actual sum $D_1 + D_2$ unless $\overline{P} \in \text{supp}(D_2)$

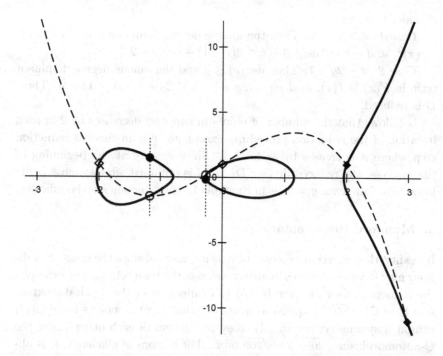

Fig. 3. Divisor addition on $H : y^2 = x^5 - 5x^3 + 4x - 1$ over $\mathbb{Q}$ (solid curve). The function $y = -(4/5)x^3 + (16/5)x + 1$ (dashed curve) through $D_1 = (-2, 1) + (0, 1)$ (white lozenges) and $D_2 = (2, 1) + (3, -11)$ (black lozenges) intersects H in the third reduced divisor $D = R_1 + R_2$ (white circles) whose points have respective x-coordinates $(-23 \pm \sqrt{209})/32$ and y-coordinates $(-1333 \pm 115\sqrt{209})/2048$. Hence $D_1 \oplus D_2 = \overline{D}$ (black circles).

for some $P \in \operatorname{supp}(D_1)$. Normally, D will not be reduced, and in fact $r = |\operatorname{supp}(D)|$ can be as large as $2g$. So assume that $r \geq g + 1$, and iterate over D as follows.

The r points in $\operatorname{supp}(D)$ all lie on an interpolation curve $y = v(x)$ with $\deg(v) \leq r - 1$. Substitute this curve into (4)) to obtain the polynomial $F(x) = f(x) - v(x)^2 - h(x)v(x)$. Put $d = \deg(F)$. Among the d zeros of $F(x)$ (counted with multiplicities), r are the x-coordinates of the points in $\operatorname{supp}(D)$. Determine the remaining $d - r$ zeros x_i ($1 \leq i \leq d - r$) of $F(x)$ and substitute them into $y = v(x)$ to define $d - r$ new points $(x_i, v(x_i))$ on H. Replace the r points in $\operatorname{supp}(D)$ by the opposites $(x_i, -v(x_i) - h(x_i))$ of these $d - r$ new points. This defines a new semi-reduced divisor in the divisor class of $D_1 + D_2$, again called D, with $|\operatorname{supp}(D)| = d - r$. Now consider two cases:

   Case 1: $d \geq 2g + 2$. Then the unique degree-dominant term in $F(x)$ is $-v(x)^2$, so $d - r = 2 \deg(v) - r \leq 2(r - 1) - r = r - 2$.

   Case 2: $d \leq 2g + 1$. Then $\deg(v) \leq g$ and the unique degree-dominant term in $F(x)$ is $f(x)$, so $d - r = 2g + 1 - r \leq 2g + 1 - (g + 1) = g$. Thus, D is reduced.

   It follows that the number of points in $\mathrm{supp}(D)$ decreases by 2 in each iteration of the reduction procedure, except possibly in the last reduction step where it decreases by at least 1. Since $r \leq 2g$ at the beginning of this process, the reduced divisor $D_1 \oplus D_2$ is obtained after at most $\lceil g/2 \rceil$ iterations. For genus $g = 2$, as in Example 1, one reduction step is sufficient.

5. Mumford Representations

In reality, the reduction process above is impractical since the points of a divisor may have coordinates in an extension of the base field $\mathbb{K}$. For example, the reduced divisor D given by (5) has support over the quadratic extension $\mathbb{L} = \mathbb{Q}(\sqrt{209})$ of $\mathbb{Q}$. Note, however, that the two points in $\mathrm{supp}(D)$ exhibit a symmetry; specifically, they are images of each other under the $\mathbb{Q}$-automorphism $\sqrt{209} \mapsto -\sqrt{209}$ on $\mathbb{L}$. For reasons of efficiency, it is obviously desirable to carry out Jacobian arithmetic exclusively in the base field $\mathbb{K}$.

Definition 1. Let $D = m_1[P_1] + \cdots + m_r[P_r]$ be a semi-reduced divisor on a hyperelliptic curve H as given in (4), and write $P_i = (x_i, y_i)$ with $x_i, y_i \in \overline{\mathbb{K}}$ for $1 \leq i \leq r$. The *Mumford representation* of D is a pair of polynomials $u(x), v(x) \in \overline{\mathbb{K}}[x]$ defined as follows:

$$u(x) = \prod_{i=1}^{r} (x - x_i)^{m_i} \, ,$$

$$\left(\frac{d}{dx} \right)^j \left[f(x) - v(x)^2 - v(x)h(x) \right]_{x=x_i} = 0 \tag{6}$$

$$(0 \leq j \leq m_i - 1, \quad 1 \leq i \leq r) \, .$$

Write $D = (u, v)$.

   Taking into account the appropriate multiplicities, the zeros of $u(x)$ are exactly the x-coordinates of the points in the support of D, and $v(x)$ is an interpolation polynomial through these points; in particular, $v(x_i) = y_i$ for $1 \leq i \leq r$. Note that $u(x)$ is monic and divides $f(x) - v(x)^2 - h(x)v(x)$. The divisor D uniquely determines $u(x)$ and $v(x)$ mod $u(x)$; conversely, any pair of polynomials $u(x), v(x)$ as defined in (6) defines a semi-reduced divisor $D = \sum_{i=1}^{r} m_i[P_i]$, with $P_i = (x_i, v(x_i))$ for $1 \leq i \leq r$, on the hyperelliptic curve (4). To ensure uniqueness, we will always choose $v(x)$ to be of least non-negative degree in its congruence class modulo $u(x)$. This means in particular that if $D = (u, v)$ is reduced, then $\deg(v) < \deg(u) \leq g$.

Example 2. Let H be a hyperelliptic curve as given in (4).

(1) For an affine point $P = (x_0, y_0)$ on H, the Mumford representation of the corresponding point divisor is $[P] = (x - x_0, y_0)$.
(2) If $D = (u, v)$, then $\overline{D} = (u, -(v + h) \bmod u)$.
(3) If $D_1 = (u_1, v_1)$ and $D_2 = (u_2, v_2)$ are semi-reduced divisors on H whose sum is semi-reduced, then $D_1 + D_2 = (u, v)$ where

$$u = u_1 u_2 , \quad v \equiv \begin{cases} v_1 \bmod u_1 , \\ v_2 \bmod u_2 . \end{cases}$$

The case when the sum $D_1 + D_2$ is not semi-reduced arises when $\overline{P} \in \mathrm{supp}(D_2)$ for some $P \in \mathrm{supp}(D_1)$. Every point $P = (x_0, y_0) \in \mathrm{supp}(D_1) \cap \mathrm{supp}(\overline{D}_2)$ satisfies $u_1(x_0) = u_2(x_0) = 0$ and $v_1(x_0) = -v_2(x_0) - h(x_0) = y_0$. It hence contributes a common factor $x - x_0$ to $u_1(x)$, $u_2(x)$ and $v_1(x) + v_2(x) + h(x)$. However, even in this general situation, the Mumford representation of the semi-reduced sum $D = (u, v)$ of D_1 and D_2 can be obtained efficiently through simple polynomial arithmetic, including an extended gcd computation:

$$d = \gcd(u_1, u_2, v_1 + v_2 + h) = s_1 u_1 + s_2 u_2 + s_3 (v_1 + v_2 + h) ,$$
$$u = u_1 u_2 / d^2 ,$$
$$v \equiv \frac{1}{d}(s_1 u_1 v_2 + s_2 u_2 v_1 + s_3 (v_1 v_2 + f)) \bmod u ,$$

with $s_1, s_2, s_3 \in \overline{\mathbb{K}}[x]$. If $D_1 + D_2$ is semi-reduced, we have $d = 1$ and $s_3 = 0$.

The iterative reduction process described above can also be effected easily via Mumford representations. If $D = (u, v)$ is a semi-reduced divisor, loop over the following to steps until $\deg(u) \leq g$:

$$u \leftarrow (f - vh - v^2)/u , \quad v \leftarrow -(v + h) \bmod u . \tag{7}$$

Updating $u(x)$ as above eliminates all the roots of $f(x) - v(x)h(x) - v(x)^2$ that are the x-coordinates of the points in $\mathrm{supp}(D)$, leaving only the x-coordinates of the remaining intersection points of H with $y = v(x)$. The formula for $v(x)$ in (7) replaces these intersections points by their opposites. The above process of divisor addition with subsequent reduction is due to Cantor [9].

Example 3. The respective Mumford representations of the reduced divisors D_1 and D_2 of Example 1) are

$$D_1 = (x^2 + 2x, 1), \quad D_2 = (x^2 - 5x + 6, -12x + 25) .$$

Thus, $D_1 + D_2 = (u, v)$ where

$$u(x) = x^4 - 3x^3 - 4x^2 + 12x , \quad v(x) = -\frac{4}{5}x^3 + \frac{16}{5}x + 1 .$$

The reader will recognize the polynomial $v(x)$ from Example 1. One reduction step (7) produces $u(x) = 16x^2 + 23x + 5$ (which the reader will again recognize from Example 1) and $v(x) = (16x - 23)/320$, yielding $D_1 \oplus D_2 = (16x^2 + 23x + 5, (16x - 23)/320)$.

Note that the Mumford polynomials $u(x)$ and $v(x)$ of the divisor $D_1 \oplus D_2$ of Example 3 have coefficients in $\mathbb{Q}$, whereas the points in its support have coordinates in $\mathbb{Q}(\sqrt{209})$. This is no accident. For any hyperelliptic curve H over $\mathbb{K}$, every $\mathbb{K}$-automorphism of the Galois group $\mathrm{Gal}(\overline{\mathbb{K}}/\mathbb{K})$ acts on points on H coordinatewise (leaving ∞ fixed), and on divisors on H pointwise. A divisor on H is *defined over* $\mathbb{K}$ if it is $\mathrm{Gal}(\overline{\mathbb{K}}/\mathbb{K})$-invariant. In other words, any Galois automorphism of $\overline{\mathbb{K}}/\mathbb{K}$ may permute the points in the support of a divisor that is defined over $\mathbb{K}$, but must leave the entire divisor fixed. For example, the reduced divisor D given in (5) is defined over $\mathbb{Q}$, since every $\mathbb{Q}$-automorphism of $\overline{\mathbb{Q}}$ is either the identity or the involution $\sqrt{209} \mapsto -\sqrt{209}$ when restricted to $\mathbb{L} = \mathbb{Q}(\sqrt{209})$. This can also be seen from the following key theorem that can easily be proved using Galois theory:

Theorem 2. *A semi-reduced divisor $D = (u, v)$ on a hyperelliptic curve H over a field $\mathbb{K}$ is defined over $\mathbb{K}$ if and only if $u(x)$ and $v(x)$ have coefficients in $\mathbb{K}$.*

Let $\mathrm{Jac}_{\mathbb{K}}(H)$ denote the subgroup of $\mathrm{Jac}(H)$ of divisor classes represented by reduced divisors defined over $\mathbb{K}$. Theorem 2, combined with the algorithms presented above, guarantees that arithmetic in $\mathrm{Jac}_{\mathbb{K}}(H)$ is entirely effected through simple and efficient polynomial arithmetic in $\mathbb{K}[x]$. Moreover, Theorems 1 and 2 establish that for a finite field $\mathbb{K} = \mathbb{F}_q$, $\mathrm{Jac}_{\mathbb{F}_q}(H)$ is finite. This group therefore satisfies all the efficiency requirements for cryptographic applications.

6. Beyond Weierstraß Models

Many elliptic curves can be described by equations other than (1) that may support faster point addition at the expense of the most general possible form. *Edwards curves*, for example, allow for very efficient point addition formulas which in addition are *complete*, i.e. exceptional cases such as adding two opposite points or adding the infinite point to some point are included in the formulas and need not be considered separately. *Hessian* models can also be very effective for point arithmetic. For a comprehensive overview of other elliptic curve models and their arithmetic, the reader is referred to the *Explicit-Formulas Database* [4]. For some operations on points, using other coordinate systems such as projective coordinates can be advantageous; sometimes *mixed* coordinates, where the two input points are given in different coordinate systems, are best.

The two aforementioned special models for elliptic curves unfortunately do not extend to higher genus. However, appropriate variable transformations applied to (4) can remove some terms. The simplest of these eliminates the coefficient c of x^{2g} in $f(x)$ when the characteristic of $\mathbb{K}$ does not divide $2g + 1$ by mapping x to $x - c/(2g + 1)$; for elliptic curves, this is precisely the isomorphism from (1) to (2). In a completely different vein, there is a highly efficient arithmetic framework for genus 2 curves that uses the *Kummer surface* associated to the curve, rather than its Jacobian [17, 19].

There are also even degree models for both elliptic and hyperelliptic curves. They take the same form as (4), except that $\deg(f) = 2g + 2$. Moreover, when $\mathbb{K}$ has characteristic 2, then $\deg(h) = g + 1$, $h(x)$ is monic, and the leading coefficient of $f(x)$ has the form $s^2 + s$ for some $s \in \mathbb{K} \setminus \{0, 1\}$. Even degree curves are more general than their odd degree counterparts, since every odd degree model can be converted to a $\overline{\mathbb{K}}$-isomorphic even degree model. However, the reverse transformation requires a zero of $f(x)$, and in fact a common zero of $f(x)$ and $h(x)$ in characteristic 2, and thus may only be defined over an extension field of $\mathbb{K}$ of degree up to $2g + 2$; see Theorem 12.4.12, p. 448, of [33].

Investigating the homogenization of even degree hyperelliptic curve models reveals that the projective point $[0 : 1 : 0]$ is singular; the observant reader will note that this is in fact also the case for odd degree hyperelliptic curves of genus $g \geq 2$. To ascertain the behaviour at infinity on these curves, put $F(x, y) = y^2 + h(x)y - f(x)$ and substitute $x = 0$ into the isomorphic curve $x^{2g+2}F(x^{-1}, yx^{-g-1}) = 0$. For odd genus (including genus 1), this yields $y^2 = 0$ and thus the unique point $(0, 0)$. For even degree and characteristic different from 2, we obtain $y^2 = 1$ and thus two distinct points $(0, 1)$ and $(0, -1)$. Finally, for even degree and characteristic 2, we get $y^2 + y = s^2 + s$, producing the two distinct points $(0, s)$ and $(0, s + 1)$. Thus, odd degree models have one infinite point, whereas even degree models have two infinite points that are opposites of each other. This extra degree of freedom at infinity leads to complications for arithmetic on even degree models. As a result, research on this subject is far less advanced than investigations into the more traditional odd degree models.

Paulus and Rück [34] found a unique representation of degree zero divisor classes on even degree hyperelliptic curves via reduced divisors with a very small contribution (usually none) at one of the infinite points. Unfortunately, the group operation on these representatives is considerably slower than that on odd degree models. Jacobian arithmetic can be sped up considerably, to the point where it is close to competitive with odd degree model arithmetic, by instead prescribing approximately equal contributions at the two infinite points (so-called *balanced* divisors) [15].

Another natural approach is to define reduced divisors on an even degree hyperelliptic curve H completely analogous to the odd degree scenario, except that the unique infinite point on an odd degree model is replaced by one of the two infinite points on H (with the other infinite point not appearing in the divisor). The divisors thus obtained represent almost all divisor classes — leaving out only a heuristically expected proportion of $1/q$ of them [21] — and form the *infrastructure* of H. Addition on the infrastructure can be defined completely analogous to Jacobian arithmetic on odd degree models by applying Cantor's algorithm to the affine parts of any two infrastructure divisors; note that this is is different from Jacobian arithmetic on even degree models. Under this operation, the infrastructure is closed but not necessarily associative. The heuristically expected proportion of infrastructure divisors that violate associativity is approximately $1/q$, the same as that of "missing" divisor classes. As a result, infrastructures over large finite fields behave "almost" like abelian groups, and can in fact serve as a suitable setting for discrete logarithm based cryptography [23, 24], providing the same degree of security as the Jacobian setting. Moreover, the infrastructure of H can be embedded into the cyclic subgroup of $\mathrm{Jac}_{\mathbb{K}}(H)$ generated by the divisor class of $\infty - \overline{\infty}$, where ∞ and $\overline{\infty}$ are the two infinite points on H [14].

For hyperelliptic curves of small genus, the algorithms on Mumford polynomials can be realized symbolically as arithmetic on their coefficients in the base field. Such *explicit formulas* were first presented for odd degree models of genus 2 in [29], and the effort to optimize explicit field arithmetic in this setting has spawned a considerable volume of literature too extensive to cite here. Explicit formulas on odd degree models of genus 3 and 4 exist as well; first presented in [41], the genus 3 formulas in particular have undergone much refinement. Explicit formulas for even degree models of genus 2 can be found in [13]; work on even degree genus 3 curves is currently in progress. For Jacobian arithmetic on (even and odd degree) hyperelliptic curves of higher genus, the NUCOMP algorithm described in [25, 40] significantly outperforms Cantor's algorithm, especially for large genus and/or a large finite base field [22]. Efficient realization of the Jacobian group law on a number of families of non-hyperelliptic curves has also been investigated; in the interest of space, we forego citing any sources here. All told, explicit arithmetic and algorithms in Jacobians of curves and more generally, on *abelian varieties* — which represent higher dimensional analogues of curves — are the subject of intense ongoing research.

Acknowledgment The author is supported by a Discovery Grant from the *National Science and Engineering Research Council* (NSERC) of Canada. Thanks go to Gove Effinger, host and organizer of the Fq12 conference, for inviting me to present this work at Fq12 and encouraging me to write this article. Finally, the helpful comments of an anonymous reviewer are appreciated.

References

[1] L. M. Adleman, J. DeMarrais amd M.-D. Huang, A subexponential algorithm for discrete logarithms over hyperelliptic curves of large genus over GF(q). *Theoret. Comput. Sci.* **226** (1999), no. 1-2, 7–18.

[2] R. Barbulescu, P. Gaudry, A. Joux and E. Thomé, A heuristic quasi-polynomial algorithm for discrete logarithm in finite fields of small characteristic. *Advances in Cryptology — EUROCRYPT 2014*, 1–16, Lecture Notes in Comput. Sci., 8441, Springer, Heidelberg, 2014.

[3] E. Barker, W. Barker, W. Polk, W. Burr and M. Smid, Recommendation for key management – part 1: general (revision 3), NIST Special Publication 800-57, July 2012.

[4] D. J. Bernstein and T. Lange, Explicit-Formulas Database, http:// hyperelliptic.org/EFD/.

[5] J. W. Bos, C. Costello, H. Hisil, K. Lauter, Fast Cryptography in Genus 2, *Advances in Cryptology — EUROCRYPT 2013*, 194–210, Lecture Notes in Comp. Sci., 7881, Springer, Heidelberg, 2014.

[6] A. Bostan, P. Gaudry and É. Schost, Linear recurrences with polynomial

coefficients and application to integer factorization and Cartier-Manin operator. *SIAM J. Comput.* **36** (2007), no. 6, 1777–1806.

[7] R. Bröker, E. W. Howe, K. E. Lauter and P. Stevenhagen, Genus-2 curves and Jacobians with a given number of points. *LMS J. Comput. Math.* **18** (2015), no. 1, 170–197

[8] R. Bröker and P. Stevenhagen, Elliptic curves with a given number of points. *Algorithmic Number Theory*, 117–131, Lecture Notes in Comput. Sci., 3076, Springer, Berlin, 2004.

[9] D. G. Cantor, Computing in the Jacobian of a hyperelliptic curve. *Math. Comp.* **48** (1987), no. 177, 95–101.

[10] H. Cohen, G. Frey, R. Avanzi, C. Doche, T. Lange, K. Nguyen, F. Vercauteren, *Handbook of Elliptic and Hyperelliptic Curve Cryptography*, Chapman & Hall/CRC, Boca Raton, Florida, 2006.

[11] C. Diem, An index calculus algorithm for plane curves of small degree. *Algorithmic Number Theory*, 543–557, Lecture Notes in Comput. Sci., 4076, Springer, Berlin, 2006.

[12] W. Diffie and M. E. Hellman, New directions in cryptography. *IEEE Trans. Information Theory* **IT-22** (1976), no. 6, 644–654.

[13] S. Erickson, M.J. Jacobson, and A. Stein. Explicit formulas for real hyperelliptic curves of genus 2 in affine representation. *Adv. Math. Communication* **5** (2011), no. 4, 623–666.

[14] F. Fontein, The infrastructure of a global field of arbitrary unit rank. *Math. Comp.* **80** (2011), no. 276, 2325–2357.

[15] S. D. Galbraith, M. Harrison and D. J. Mireles Morales, Efficient hyperelliptic arithmetic using balanced representation for divisors. *Algorithmic Number Theory*, 342–356, Lecture Notes in Comput. Sci., 5011, Springer, Berlin, 2008.

[16] P. Gaudry, An algorithm for solving the discrete log problem on hyperelliptic curves. *Advances in Cryptology — EUROCRYPT 2000 (Bruges)*, 19–34, Lecture Notes in Comput. Sci., 1807, Springer, Berlin, 2000.

[17] P. Gaudry, Fast genus 2 arithmetic based on theta functions. *J. Math. Cryptol.* **1** (2007), no. 3, 243–265.

[18] P. Gaudry and R. Harley, Counting points on hyperelliptic curves over finite fields. *Algorithmic number theory (Leiden, 2000)*, 313–332, Lecture Notes in Comput. Sci., 1838, Springer, Berlin, 2000.

[19] P. Gaudry and D. Lubicz, The arithmetic of characteristic 2 Kummer surfaces and of elliptic Kummer lines. *Finite Fields Appl.* **15** (2009), no. 2, 246–260.

[20] D. M. Gordon, Discrete logarithms in GF(p) using the number field sieve. *SIAM J. Discrete Math.* **6** (1993), no. 1, 124–138.

[21] M. J. Jacobson, Jr., M. Rezai Rad and R. Scheidler, Comparison of scalar multiplication on real hyperelliptic curves. *Adv. Math. Communications* **8** (2014), no. 4, 389–406.

[22] M. J. Jacobson, Jr., R. Scheidler and A. Stein, Fast arithmetic on hyperelliptic curves via continued fraction expansions. *Advances in Coding Theory and Cryptology*, 200–243, Ser. Coding Theory Cryptol., 3, World Scientific

Publishing Co. Pte. Ltd., Hackensack, New Jersey 2007.

[23] M. J. Jacobson, Jr., R. Scheidler and A. Stein, Cryptographic protocols on real hyperelliptic curves. *Adv. Math. Communications* **1** (2007), no. 2, 197–221.

[24] M. J. Jacobson, Jr., R. Scheidler and A. Stein, Cryptographic aspects of real hyperelliptic curves. *Tatra Mountains Math. Pub.* **47** (2010), no. 1, 31–65.

[25] M. J. Jacobson, Jr. and A. J. van der Poorten, Computational aspects of NUCOMP, *Algorithmic Number Theory (Sydney, 2002)*, 120–133, Lecture Notes in Comput. Sci., 2369, Springer, Berlin, 2002.

[26] A. Joux, A new index-calculus algorithm with complexity $L(1/4 + o(1))$ in very small characteristic. *Selected Areas in Cryptography — SAC 2013*, 355–379, Lecture Notes in Comp. Sci., 8282, Springer, Berlin 2014.

[27] K. S. Kedlaya, Counting points on hyperelliptic curves using Monsky-Washnitzer cohomology. *J. Ramanujan Math. Soc.* **16** (2001), no. 4, 323–338.

[28] N. Koblitz, Elliptic curve cryptosystems. *Math. Comp.* **48** (1987), no. 177, 203–209.

[29] T. Lange, Formulae for Arithmetic on Genus 2 Hyperelliptic Curves. *Applicable Algebra in Engineering, Communication and Computing* **15** (2005), no. 5, 295–328.

[30] A. Lauder and D. Wan, Computing zeta functions of Artin-Schreier curves over finite fields. *LMS J. Comput. Math.* **5** (2002), 34–55.

[31] V. S. Miller, Use of elliptic curves in cryptography. *Advances in Cryptology — CRYPTO '85 (Santa Barbara, Calif., 1985)*, 417–426, Lecture Notes in Comput. Sci., 218, Springer, Berlin, 1986.

[32] V. Müller, A. Stein and C. Thiel, Computing discrete logarithms in real quadratic congruence function fields of large genus. *Math. Comp.* **68** (1999), no. 226, 807–822.

[33] G. L. Mullen and D. Panario (eds.), *Handbook of Finite Fields.*. Discrete Mathematics and its Applications, CRC Press, Boca Raton, Florida, 2013.

[34] S. Paulus and H.-G. Rück, Real and imaginary quadratic representations of hyperelliptic function fields. *Math. Comp.* **68** (1999), no. 227, 1233–1241.

[35] S. C. Pohlig and M. Hellman, An improved algorithm for computing logarithms over $GF(p)$ and its cryptographic significance. *IEEE Trans. Information Theory* **IT-24** (1978), no. 1, 106–110.

[36] J. M. Pollard, A Monte Carlo method for factorization. *Nordisk Tidskr. Informationsbehandlung (BIT)* **5** (1975), no. 3, 331–334.

[37] D. Shanks, Class number, a theory of factorization, and genera. *1969 Number Theory Institute (Proc. Sympos. Pure Math., Vol. XX, State Univ. New York, Stony Brook, N.Y., 1969)*, pp. 415–440. Amer. Math. Soc., Providence, R.I., 1971.

[38] V. Shoup, Lower bounds for discrete logarithms and related problems. *Advances in Cryptology — EUROCRYPT '97 (Konstanz)*, 256–266, Lecture Notes in Comput. Sci., 1233, Springer, Berlin, 1997.

[39] N. Thériault, Index calculus attack for hyperelliptic curves of small genus. *Advances in Cryptology — ASIACRYPT 2003*, 75–92, Lecture Notes in Comput. Sci., 2894, Springer, Berlin, 2003.

[40] A. van der Poortem, A note om NUCOMP, *Math. Comp.* **72** (2003), no. 244, 1935–1946

[41] T. Wollinger, Software and Hardware Implementation of Hyperelliptic Curve Cryptosystems. Doctoral Dissertation, Ruhr-Universität Bochum (Germany), 2004.

On the existence of aperiodic complementary hexagonal lattice arrays

Yin Tan, Guang Gong

Department of Electrical and Computer Engineering, University of Waterloo, Canada {yin.tan, ggong}@uwaterloo.ca

Binary aperiodic and periodic complementary set of sequences have been studied extensively due to their wide range of applications in engineering, for example in optics, radar and communications. They have been generalized either by being defined over larger alphabets or by being defined from one dimension to multi-dimensions. Recently, Ding, Noshad and Tarokh introduced and constructed the aperiodic complementary two-dimensional arrays over the alphabet $\mathfrak{A}_p^* = \{\zeta_p^i : 0 \leq i \leq p-1, p \text{ is a prime number}\}$, whose support set is a subset of the hexagonal lattice (mostly is a set of ℓ-layer consecutive hexagons). Despite that several constructions of such complementary arrays have been discovered, the conditions for which they exist are open. In this paper, we show that aperiodic complementary hexagonal lattice arrays over the alphabet $\mathfrak{A}_p^*$ can be mapped to a particular type of two-dimensional rectangular lattice array; and it leads to aperiodic (hence periodic) complementary sequences over the alphabet $\mathfrak{A}_p = \mathfrak{A}_p^* \cup \{0\}$. Then we make use of group ring equations to characterize periodic complementary sequences over alphabet $\mathfrak{A}_p$. As of independent interest, we show that, if the alphabet of the periodic complementary sequences is $\mathfrak{A}_p^*$, the notion of periodic complementary sequences is equivalent to the notion of certain relative difference family. This generalizes the results by Bömer and Antweiler on the characterization of periodic complementary binary sequences with difference families. The conditions for the existence of periodic complementary sequences over alphabet $\mathfrak{A}_p$ are derived. As the applications of our characterization, we determine the existence of a pair (triple) aperiodic complementary binary hexagonal lattice arrays whose support set is an ℓ-layer consecutive hexagons. A table listing the existence status for a pair of complementary hexagonal lattice arrays with $1 \leq \ell \leq 20$ as well as some open problems are presented.

1. Introduction

Let $S = (s_0, \ldots, s_{n-1})$ be a binary sequence over alphabet $\{\pm 1\}$. The aperiodic autocorrelation function of S is defined by $A^S(t) = \sum_{i=0}^{n-1-t} s_i s_{i+t}$; and the periodic autocorrelation function of S is defined by $P^S(t) = \sum_{i=0}^{n-1} s_i s_{i+t}$, where the addition of the indices is performed modulo n and $0 \leq t \leq n-1$. It is well known that the aperiodic and periodic autocorrelation functions are related by $P^S(t) = A^S(t) + A^S(n-1-t)$. By convention, we call the values $A^S(t), P^S(t)$ *out-of-phase* if $(t \bmod n) \not\equiv 0$, and *in-phase* otherwise. More generally, we call S a *p-phase sequence* if S is over the alphabet $\mathfrak{A}_p = \{0, \zeta_p^i : 0 \leq i \leq p-1\}$ (one may wonder why we include $0 \ldots$ in the alphabet: this is for the convenience to state the results in Section 3), where p is a prime and ζ_p is a primitive p-th root of unity.

For applications in engineering, the sequences with small out-of-phase (periodic) aperiodic autocorrelation values are highly preferred. From this point of view, the *Barker sequences*, whose all out-of-phase aperiodic autocorrelation values equal either 0 or 1, are the most ideal objects. Unfortunately very few Barker sequences are known so far. In [16], Golay proposed the notion *Golay pair*, which is a pair of binary sequences of the same length such that the sum of the out-of-phase aperiodic autocorrelation values are 0. Subsequently, in [24], Tseng and Liu further studied a *set of complementary binary sequences* such that the set may contain more than two sequences and the sum of all out-of-phase aperiodic autocorrelation values are 0. Another two widely adopted approaches to generalize complementary sequences are to define the sequences over a larger alphabet, or to consider complementary d-dimensional arrays, see for example [17, 18]. In the rest of the paper, we refer complementary to aperiodic complementary unless explicitly stated. Note that *periodic complementary sequences* are sequences with the sum of all out-of-phase periodic autocorrelation values is zero. They have been studied in [1, 2, 14].

Recently, in [10], Ding, Noshad and Tarokh studied the complementary two-dimensional arrays over the alphabet $\mathfrak{A}_p^* = \mathfrak{A}_p \setminus \{0\}$, whose support set can be an arbitrary lattice (Defined in Section 3.1). This generalizes the complementary two-dimensional arrays considered in [18] in the sense that the support sets of the arrays in [18] are square lattices. In particular, Ding, Noshad and Tarokh demonstrated that, if the support set is a particular subset of the hexagonal lattice, the corresponding complementary arrays may be used in coded aperture imaging with ideal performances. For brevity, we call an array whose support set is a subset of the hexagonal lattice a *hexagonal lattice array*. Several constructions of complementary hexagonal lattice arrays were provided in [10]. However, it is not clear the conditions for the support sets, the number of arrays in the set and the alphabet $\mathfrak{A}_p^*$, for which this complementary set exits. It is the aim of this paper to develop the conditions for the existence of such a set.

In Section 3, we define a set of points of the hexagonal lattice, which is denoted by Ω_ℓ and is called ℓ-*layer consecutive hexagons* (Definition 1). Such sets are the most commonly chosen support sets for hexagonal lattice arrays in the application of coded aperture imaging described in [10]. Through explicitly determining the coordinates of the points in Ω_ℓ, we prove that a hexagonal lattice array whose support set is Ω_ℓ is a particular type of two-dimensional rectangular lattice array. More precisely, we express a hexagonal lattice array with support set Ω_ℓ by a $(2l-1) \times (2\ell-1)$ matrix with entries in $\mathfrak{A}_p$ and determine certain entries in the matrix are zero. This result provides the connection between hexagonal lattice arrays and the commonly studied rectangular lattice arrays. Furthermore, according to this expression and by making use of the techniques in [18], a set of complementary sequences of length $4\ell^2 - 6\ell + 3$ over alphabet $\mathfrak{A}_p$ exist, if a complementary set of hexagonal lattice arrays over alphabet $\mathfrak{A}_p^*$ and with Ω_ℓ as the support set exits (Proposition 2).

In Section 4, we focus on the existence of the periodic complementary sequences with length n and alphabet $\mathfrak{A}_p$ (here we use the well-known fact that aperiodic complementary sequences must be periodic complementary). To explore the existence conditions, we establish an important connection between the sequences with length n and alphabet $\mathfrak{A}_p$ and the group ring elements of $\mathbb{Z}[\mathbb{Z}_n \times \mathbb{Z}_p]$. We provide a characterization of the periodic complementary sequences by the group ring equation, from which yields several conditions for the existence of periodic complementary sequences (Corollaries 3,4). As of independent interest, we show that, if the alphabet of the sequences is $\mathfrak{A}_p^*$, then a set of periodic complementary sequences is equivalent to a relative difference family in $\mathbb{Z}_p \times \mathbb{Z}_n$ relative to $\mathbb{Z}_p \times \{1\}$ (Theorem 2). This relationship generalizes the result in [2], which points out that a set of periodic complementary binary sequences is equivalent to a difference family. To the best of our knowledge, this is the first combinatorial characterization of the periodic complementary p-phase sequences for any prime p.

In the light of the results in Sections 3 and 4, in Section 5 we focus on determining the existence of complementary binary hexagonal lattice arrays $\mathcal{C}_\ell^t = \{C_\ell^1, \ldots, C_\ell^t\}$ whose support set is Ω_ℓ for $t = 2$ and 3 (Theorems 3 and 4). We provide necessary conditions on ℓ such that $\mathcal{C}_\ell^2, \mathcal{C}_\ell^3$ exist. We give a table listing the existence of $\mathcal{C}_\ell^2$ for $1 \le \ell \le 20$. Some open problems are proposed as well. We should mention that in [10] the authors provided examples of $\mathcal{C}_\ell^t$ for $t = 4$.

The rest of the paper is organized as follows. In Section 2, we present necessary definitions and results used throughout the paper. In Section 3, we will introduce the arrays with support set is a hexagonal lattice array; and show the relationship between complementary hexagonal lattice arrays and complementary sequences. We provide the characterization of periodic complementary sequences through group ring equations in Section 4. The connection between periodic complementary sequences over alphabet $\{\zeta_p^i : 0 \le i \le p-1\}$ and certain relative difference family is presented as well. In Section 5, we study the existence of complementary binary hexagonal lattice arrays. Some concluding remarks are given in Section 6.

2. Preliminaries

In this section, we present necessary definitions and results that will be used throughout the paper.

2.1. *Group rings and character theory*

Character theory is one of the most important tools for applying group rings to combinatorial objects. In this section we only review the characters of the group ring $\mathbb{C}[G]$, where G is an Abelian group. For the theory of the representation of a general group ring, please refer to [21].

Let $\mathbb{F}$ be an arbitrary field (usually we take the complex field $\mathbb{C}$), and let G be a multiplicative written group. The group algebra $\mathbb{F}[G]$ consists of all formal sums

$$\sum_{g \in G} a_g g, \quad a_g \in \mathbb{F}.$$

The addition and multiplication for elements in $\mathbb{F}[G]$ are defined as follows:

$$\sum_{g \in G} a_g g + \sum_{g \in G} b_g g = \sum_{g \in G} (a_g + b_g) g,$$

$$\sum_{g \in G} a_g g \cdot \sum_{g \in G} b_g g = \sum_{g \in G} \left(\sum_{h \in G} a_h \cdot b_{gh^{-1}} \right) g.$$

Moreover, we have a scalar multiplication

$$\lambda \sum_{g \in G} a_g g = \sum_{g \in G} (\lambda a_g) g, \quad \lambda \in \mathbb{F}.$$

It is easy to verify that with these operations, $\mathbb{F}[G]$ is indeed an algebra over $\mathbb{F}$. In the language of group rings, we identify a subset S of G with the group ring element $\sum_{s \in S} s$, which will also be denoted by S. For an element $A = \sum_{g \in G} a_g g \in \mathbb{F}[G]$ and an integer t, we define $A^{(t)} = \sum_{g \in G} a_g g^t$.

A character χ of a finite Abelian group G is a homomorphism from G to $\mathbb{C}^* = \mathbb{C} \setminus \{0\}$. A character χ is called *principal* if $\chi(c) = 1$ for all $c \in G$; otherwise it is called *non-principal*. A principal character is usually denoted by χ_0. All characters form a group denoted by $\widehat{G}$, which is isomorphic to G. The following result states the well-known *orthogonal relations* of characters.

Result 1 (Orthogonal relations of characters). *Let G be an Abelian group. Then the following equations hold:*

$$\sum_{g \in G} \chi(g) = \begin{cases} 0 & \text{if } \chi \neq \chi_0, \\ |G| & \text{if } \chi = \chi_0; \end{cases}$$

and

$$\sum_{\chi \in \widehat{G}} \chi(g) = \begin{cases} 0 & \text{if } g \neq 1_G, \\ |G| & \text{if } g = 1_G. \end{cases}$$

By linearity, we may extend each character $\chi \in \widehat{G}$ to a ring homomorphism from $\mathbb{C}[G]$ to $\mathbb{C}$, and we denote this homomorphism by χ, again. In particular, if G is the additive group of the finite field $\mathbb{F}_{p^n}$, all characters of G can be represented as follows. Define $\chi_1 : \mathbb{F}_{p^n} \to \mathbb{C}$ as $\chi_1(x) = \zeta_p^{\text{Tr}(x)}$ for all $x \in \mathbb{F}_{p^n}$, where ζ_p is a primitive p-th root of unity and $\text{Tr}(x)$ is the absolute trace function defined as $\text{Tr}(x) = \sum_{i=0}^{n-1} x^{p^i}$. Then χ_1 is an additive character of $\mathbb{F}_{p^n}$ (i.e. χ_1 is a character of the additive group of $\mathbb{F}_{p^n}$). Moreover, every additive character χ is of the form χ_b ($b \in \mathbb{F}_{p^n}$), where χ_b is defined by $\chi_b(x) = \chi_1(bx)$ for all $x \in \mathbb{F}_{p^n}$. Furthermore, if $G = \mathbb{F}_{p^n} \times \mathbb{F}_{p^n}$, all characters of G can be represented by $\chi_{u,v}$, where $\chi_{u,v}(a,b) = \zeta_p^{\text{Tr}(au+bv)}$ for any $(a,b) \in G$. The following results are important properties of group rings.

Result 2 (Inversion Formula). *Let $D = \sum_{g \in G} a_g g \in \mathbb{C}[G]$. Then the following hold:*

$$a_g = \frac{1}{|G|} \sum_{\chi \in \widehat{G}} \chi(D) \chi(g^{-1}), \tag{1}$$

$$\sum_{g \in G} |a_g|^2 = \frac{1}{|G|} \sum_{\chi \in \widehat{G}} |\chi(D)|^2. \tag{2}$$

The above Inversion Formula provides a useful method for showing two group ring elements are equal.

Corollary 1. *Let $A = \sum_{g \in G} a_g g$ and $B = \sum_{g \in G} b_g g$ be two group ring elements of $\mathbb{C}[G]$. Then $A = B$ if and only if $\chi(A) = \chi(B)$ for all $\chi \in \widehat{G}$.*

2.2. Relative difference family

Let G be an Abelian group of order v and let N be a subgroup of order n. A $(v, n, K, \lambda; u)$ *relative difference family* (RDF for short) in G relative to N is a collection of subsets of G, $\mathcal{D} = \{D_1, D_2, \ldots, D_u\}$, such that the following equation holds for the multiset union

$$\sum_{i=1}^{u} \left\{ xy^{-1} : x, y \in D_i, x \neq y \right\} = \lambda(G - N),$$

where K is the set of cardinality of all the base blocks D_i. In the case of $K = \{k\}$, it is called a $(v, n, k, \lambda; u)$ relative difference family for brevity. If a $(v, n, k, \lambda; u)$ relative difference family exists, we have

$$uk(k - 1) = \lambda(v - n). \tag{3}$$

According to the Inversion formula (Corollary 1), the set $\mathcal{D}$ is a $(v, n, K, \lambda; u)$ if and only if the following equation holds:

$$\sum_{i=1}^{u} \chi(D_i D_i^{(-1)}) = \begin{cases} \sum_{i=1}^{u} k_i + \lambda(v - n), & \text{if } \chi = \chi_0, \\ \sum_{i=1}^{u} k_i - \lambda n, & \text{if } \chi|_N = \chi_0, \chi \neq \chi_0 \\ \sum_{i=1}^{u} k_i, & \text{if } \chi|_N \neq \chi_0. \end{cases}$$

The notion of relative difference family generalizes previously well-studied objects in combinatorics. For example, $\mathcal{D}$ is called a *difference family* if $N = \{1\}$; it is called a *difference set* if $u = 1$ and $N = \{1\}$; it is called a relative difference set if $u = 1$. Relative difference families and its variants are well studied [3, 4, 6, 9, 11, 12, 15, 20, 25]. Difference families have applications in coding theory and cryptography, see for instance [5, 22].

3. Complementary Hexagonal Lattice Arrays

In this section, we will first briefly review the definitions of the lattices and the (periodic) aperiodic complementary hexagonal lattice arrays. Then we show the relationship between them and (periodic) aperiodic complementary sequences.

3.1. *Hexagonal lattice array*

A *lattice* in $\mathbb{R}^n$ is a subgroup of $\mathbb{R}^n$ which is generated by forming all linear combinations with integer coefficients of the elements in a basis. In other words, a lattice $\mathfrak{L}$ in $\mathbb{R}^n$ has the form

$$\mathfrak{L} = \left\{ \sum_{i=1}^{n} c_i \mathbf{e}_i \mid c_i \in \mathbb{Z} \right\},$$

where $\{\mathbf{e}_i\}_{i=1}^{n}$ forms a basis of $\mathbb{R}^n$. In the following we only introduce the arrays whose support set is a subset of hexagonal lattices. Interested readers may refer to [10] for the definition of the array whose support set is an arbitrary lattice. When $n = 2$, the lattice $\mathbb{A}_2$ is called *hexagonal* if $\mathbf{e}_1 = (1,0), \mathbf{e}_2 = (1/2, \sqrt{3}/2)$. Note that some authors choose the basis vectors of $\mathbb{A}_2$ as $\mathbf{e}_1' = (1,0), \mathbf{e}_2' = (-1/2, \sqrt{3}/2)$, but it is not difficult to see that $\{\mathbf{e}_1, \mathbf{e}_2\}$ and $\{\mathbf{e}_1', \mathbf{e}_2'\}$ generate the same lattice. A *hexagonal lattice array* (HLA for short) over an alphabet $\mathfrak{A}$ and with a support set Ω, denoted by $C^{\Omega, \mathfrak{A}}$, is a mapping $C[\cdot] : \mathbb{A}_2 \to \mathfrak{A}$ such that $C[\mathbf{a}] = 0$ for all $\mathbf{a} \notin \Omega$ and $C[\mathbf{a}] \in \mathfrak{A}$ for all $\mathbf{a} \in \Omega$. For a hexagonal lattice array $C^{\Omega, \mathfrak{A}}$, the aperiodic autocorrelation function $A^C(\cdot)$ is defined by

$$A^C(\mathbf{u}) = \sum_{\mathbf{s} \in \Omega} C[\mathbf{s}]\overline{C[\mathbf{s} + \mathbf{u}]}, \quad \mathbf{u} \in \mathbb{A}_2, \tag{4}$$

where $\overline{x}$ denotes the complex conjugate of $x \in \mathbb{C}$. A set of hexagonal lattice arrays $\mathcal{C} = \{C_1^{\Omega, \mathfrak{A}}, \ldots, C_t^{\Omega, \mathfrak{A}}\}$ with the same alphabet and support set is called *aperiodic complementary* if

$$A^{C_1}(\mathbf{u}) + \cdots + A^{C_t}(\mathbf{u}) = 0, \quad \forall \mathbf{u} \neq \mathbf{0}.$$

3.2. *ℓ-layer Hexagonal lattice arrays*

In the application of complementary hexagonal lattice arrays to the coded aperture imagining described in [10], the support set of a hexagonal lattice array is often chosen as ℓ consecutive layers of hexagons (defined below). We call such special hexagonal lattice arrays *ℓ-layer hexagonal lattice arrays*; their definition is given below.

Definition 1. *Let Ω be a set of points of the hexagonal lattice $\mathbb{A}_2$. We call Ω a 1-layer consecutive hexagon if Ω consists of only one hexagon. For $\ell > 1$, if Ω is the union of the $(\ell - 1)$-layer consecutive hexagons and the hexagons adjacent with the $(\ell - 1)$-layer consecutive hexagons, then Ω is called ℓ-layer consecutive hexagons and denoted it by Ω_ℓ. We call a hexagonal lattice array with support set as Ω_ℓ an ℓ-layer hexagonal lattice array (ℓ-HLA for short).*

We give below the 1-, 2- and 3-HLAs to illustrate Definition 1.

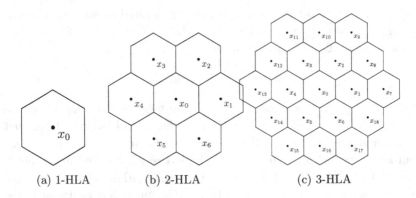

(a) 1-HLA   (b) 2-HLA   (c) 3-HLA

To compute the aperiodic autocorrelation function (defined in (4)) for an ℓ-HLA, in the following result we give the coordinates of the points (using the chosen basis vectors $\mathbf{e}_1 = (1,0), \mathbf{e}_2 = (1/2, \sqrt{3}/2)$ of $\mathbb{A}_2$) in the support set Ω_ℓ (defined in Definition 1). Furthermore, we make use of a $(2\ell - 1) \times (2\ell - 1)$ matrix to represent an ℓ-HLA. The proof of the following result may be given by induction on ℓ, we omit the details here due to its simplicity.

Proposition 1. *Let Ω_ℓ be the ℓ-layer consecutive hexagons. The set of the coordinates of the points in Ω_ℓ (by abusing our notations, we still use Ω_ℓ to denote the set of coordinates) is*

$$
\begin{aligned}
\Omega_\ell = \{(i,j) : &1 \le i \le \ell - 1, -(\ell - 1) \le j \le \ell - i - 1\} \\
&\cup \{(i,j) : -(\ell - 1) \le i \le -1, -\ell - i + 1 \le j \le \ell - 1\} \qquad (5) \\
&\cup \{(0,j) : -(\ell - 1) \le j \le (\ell - 1)\}.
\end{aligned}
$$

An ℓ-layer hexagonal lattice array can be represented by a $2\ell - 1$ by $2\ell - 1$ matrix $\mathcal{T}_\ell$ defined as

$$
\begin{array}{c}
\begin{array}{ccccccc}
-(\ell-1) & -(\ell-2) & \cdots & -1 & 0 & 1 & \cdots & \ell-1
\end{array} \\
\begin{array}{c}
-(\ell-1) \\ -(\ell-2) \\ \cdots \\ -1 \\ 0 \\ 1 \\ \cdots \\ \ell-1
\end{array}
\left(
\begin{array}{cccccccc}
0 & 0 & \cdots & 0 & a_{-(\ell-1),0} & a_{-(\ell-1),1} & \cdots & a_{-(\ell-1),\ell-1} \\
0 & 0 & \cdots & a_{-(\ell-2),-1} & a_{-(\ell-2),0} & a_{-(\ell-2),1} & \cdots & a_{-(\ell-2),\ell-1} \\
\cdots & \cdots & \cdots & \cdots & \cdots & \cdots & \cdots & \cdots \\
0 & a_{-1,-(\ell-2)} & \cdots & \cdots & \cdots & a_{-1,1} & \cdots & a_{-1,\ell-1} \\
a_{0,-(\ell-1)} & \cdots & \cdots & \cdots & \cdots & a_{0,1} & \cdots & a_{0,\ell-1} \\
a_{1,-(\ell-1)} & \cdots & \cdots & \cdots & \cdots & a_{1,1} & \cdots & 0 \\
\cdots & \cdots & \cdots & \cdots & \cdots & \cdots & \cdots & \cdots \\
a_{\ell-1,-(\ell-1)} & \cdots & \cdots & a_{\ell-1,-1} & a_{\ell-1,0} & 0 & \cdots & 0
\end{array}
\right),
\end{array}
$$

$$(6)$$

where the rows (resp. columns) are indexed from $-(\ell - 1)$ to $\ell - 1$ from top to bottom (resp. left to right).

By Proposition 1, the aperiodic autocorrelation for an ℓ-HLA C_ℓ can be computed as

$$A^{C_\ell}(\mathbf{u}) = \sum_{\mathbf{s} \in \Omega_\ell} C[\mathbf{s}]\overline{C[\mathbf{s}+\mathbf{u}]}, \quad \mathbf{u} \in \mathbb{A}_2, \tag{7}$$

where Ω_ℓ is defined in (5). Let us use a 2-HLA to illustrate Proposition 1.

Example 1. *Given a 2-layer hexagonal lattice array below, the set of the coordinates of the points* $\{x_0, x_1, x_2, x_3, x_4, x_5, x_6\}$ *is* $\Omega_2 = \{(0,0), (1,0), (0,1), (-1,1), (-1,0), (0,-1), (1,-1)\}$. *Now we define a 3 by*

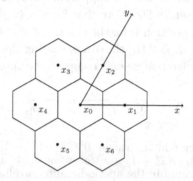

3 matrix $\mathcal{T}_2$ *as described in (6). If the point with coordinate* (i,j) *belongs to* Ω_2*, say* x_k*, the element* $\mathcal{T}_2(i,j)$ *is then* x_k*. Thus, we get*

$$\mathcal{T}_2 = \begin{array}{c} \\ -1 \\ 0 \\ 1 \end{array} \begin{array}{c} -1 \quad\; 0 \quad\;\; 1 \\ \begin{pmatrix} 0 & x_4 & x_3 \\ x_5 & x_0 & x_2 \\ x_6 & x_1 & 0 \end{pmatrix} \end{array}.$$

3.3. *Complementary ℓ-HLAs and complementary sequences*

In [17, 18], Jedwab and Parker constructed complementary s-dimensional arrays from complementary $(s+1)$-dimensional arrays. The following result is one of the key steps to apply for their technique.

Proposition 2. *Let* $\mathcal{C} = \{C_1^{\Omega_\ell,\mathfrak{A}}, \ldots, C_t^{\Omega_\ell,\mathfrak{A}}\}$ *be a set of ℓ-layer hexagonal lattice arrays over the alphabet* $\mathfrak{A}$*. For each i with $1 \le i \le t$, define a sequence S_i of length $(2\ell-1)^2$ as*

$$S_i(u(2\ell-1)+v) = C_i^{\Omega_\ell,\mathfrak{A}}(u,v), \quad 0 \le u,v \le 2\ell-2. \tag{8}$$

Then, for any integer α, β, we have the following relationship between the aperiodic autocorrelations

$$A^{S_i}(\alpha(2\ell - 1) + \beta) = A^{C_i^{\Omega_\ell,\mathfrak{A}}}(\alpha, \beta) + A^{C_i^{\Omega_\ell,\mathfrak{A}}}(\alpha + 1, \beta - (2\ell - 1)).$$

Therefore, the sequence set $\mathcal{S} = \{S_1, \ldots, S_t\}$ is aperiodic complementary if $\mathcal{C}$ is aperiodic complementary.

Proof. For simplicity, we denote $C_i^{\Omega_\ell,\mathfrak{A}}$ by C_i for $1 \leq i \leq t$. The aperiodic autocorrelation function of C_i is computed as equation (7). Since shifting the support set of C_i will not change the autocorrelation function values (see [10, Lemma 1]), we may assume the support set of C_i to be $\Omega_\ell + (\ell - 1, \ell - 1)$, and still denote this set by Ω_ℓ. Note that for $\mathbf{s} = (i, j) \in \Omega_\ell$, we have $0 \leq i, j \leq 2(\ell - 1)$. By Proposition 1, for the points with coordinates (i, j), where $0 \leq i, j \leq 2(\ell - 1)$ but $(i, j) \notin \Omega_\ell$, the corresponding element in the matrix $\mathcal{T}_\ell$ (defined in (6)) is 0, therefore we may compute the autocorrelation function for C_i as

$$A^{C_i}(a, b) = \sum_{0 \leq i, j \leq 2\ell - 2} C(i, j)\overline{C[i + a, j + b]}, \quad \mathbf{u} = (a, b) \in \mathbb{A}_2.$$

Note that any integer t in the range $[0, (2\ell - 1)^2 - 1]$ can be represented uniquely in the form $u(2\ell - 1) + v$ for some $0 \leq u, v \leq 2\ell - 2$. Also note that when we compute the aperiodic autocorrelation function for the sequence S_i, we take $S_i(t) = 0$ for $t \notin [0, (2\ell - 1)^2 - 1]$. Now for each $t = \alpha(2\ell - 1) + \beta, 0 \leq \alpha, \beta \leq 2\ell - 2$, we have

$$
\begin{aligned}
A^{S_i}(\alpha(2\ell - 1) + \beta) &= \sum_{j=0}^{(2\ell-1)^2-1} S_i(j)\overline{S_i(j + \alpha(2\ell - 1) + \beta)} \\
&= \sum_{u,v=0}^{2\ell-2} S_i(u(2\ell - 1) + v)\overline{S_i(u(2\ell - 1) + v + \alpha(2\ell - 1) + \beta)} \\
&= \sum_{u,v} S_i(u(2\ell - 1) + v)\overline{S_i((u + \alpha)(2\ell - 1) + v + \beta)} \\
&= \sum_u \left(\sum_{v=0}^{(2\ell-1)-\beta-1} S_i(u(2\ell - 1) + v)\overline{S_i((u + \alpha)(2\ell - 1) + v + \beta)} \right. \\
&\qquad + \sum_{v=(2\ell-1)-\beta}^{2\ell-2} S_i(u(2\ell - 1) + v) \cdot \\
&\qquad\qquad \left. \overline{S_i((u + \alpha + 1)(2\ell - 1) + v + \beta - (2\ell - 1))} \right) \\
&= \sum_u \left(\sum_{v=0}^{(2\ell-1)-\beta-1} C_i(u, v)\overline{C_i(u + \alpha, v + \beta)} \right. \\
&\qquad + \sum_{v=(2\ell-1)-\beta}^{2\ell-2} C_i(u, v)\overline{C_i(u + \alpha + 1, v + \beta - (2\ell - 1))} \right) \\
&= A^{C_i}(\alpha, \beta) + A^{C_i}(\alpha + 1, \beta - (2\ell - 1)).
\end{aligned}
$$

The last statement of the theorem is clear from the above equation. $\qquad\square$

Remark 1. *If the alphabet of an ℓ-HLA is $\mathfrak{A}$, then the sequence obtained from an ℓ-HLA described in Proposition 2 has the alphabet $\mathfrak{A} \cup \{0\}$.*

Let us use the 2-HLA stated in Example 1 to illustrate how to obtain a sequence from an HLA as described in Proposition 1.

Example 2. *Let C_2 be a 2-HLA stated in Example 1. The sequence defined in 8 from C_2 is $S = (0, x_4, x_3, x_5, x_0, x_2, x_6, x_1, 0)$.*

One may see from the matrix representation of an ℓ-HLA (6) and the definition of the sequence S from an ℓ-HLA (8) that the first and the last $(\ell-1)$ elements of the sequence S are 0s. Let us define a *shortened* sequence of length $(2\ell - 1)^2 - 2(\ell - 1) = 4\ell^2 - 6\ell + 3$ by removing the first and last consecutive $(\ell - 1)$ zeros. The following result shows that the set of sequences derived from ℓ-HLAs is aperiodic complementary if and only if the set of the shortened sequences is aperiodic complementary.

Proposition 3. *Let $\mathbf{0}$ be an all-zero sequence of length a and $\mathbf{S}$ be a sequence of length b. Let $\mathbf{T} = \mathbf{0} \parallel S \parallel \mathbf{0}$ be the sequence of length $2a + b$ defined by concatenating the sequences $\mathbf{0}$ and $\mathbf{S}$. Then we have $A^{\mathbf{T}}(\alpha) = A^{\mathbf{S}}(\alpha)$ for $1 \leq \alpha \leq b - 1$ and $A^{\mathbf{T}}(\alpha) = 0$ for $b \leq \alpha \leq 2a + b - 1$. Furthermore, let $\mathcal{S} = \{\mathbf{S}_1, \ldots, \mathbf{S}_t\}$ be a set of sequences of length b and $\mathcal{T} = \{\mathbf{0} \parallel \mathbf{S}_1 \parallel \mathbf{0}, \ldots, \mathbf{0} \parallel \mathbf{S}_t \parallel \mathbf{0}\}$ be the set of sequences of length $2a + b$. Then $\mathcal{T}$ is aperiodic complementary if and only if $\mathcal{S}$ if aperiodic complementary.*

Proof. Note that for the sequence $\mathbf{T}$ we have

$$\mathbf{T}_i = \begin{cases} 0, & \text{if } 1 \leq i \leq a, \\ S_i, & \text{if } a + 1 \leq i \leq a + b, \\ 0, & \text{if } a + b + 1 \leq i \leq 2a + b. \end{cases}$$

For α with $1 \leq \alpha \leq b - 1$, we have $A^{\mathbf{T}}(\alpha) = \sum_{i=1}^{2a+b-\alpha} T_i \overline{T_{i+\alpha}} = \sum_{i=1}^{a} T_i \overline{T_{i+\alpha}} + \sum_{i=a+1}^{a+b-\alpha} T_i \overline{T_{i+\alpha}} + \sum_{i=a+b-\alpha+1}^{2a+b-\alpha} T_i \overline{T_{i+\alpha}} = \sum_{i=1}^{a} 0 \cdot \overline{T_{i+\alpha}} + \sum_{j=1}^{b-\alpha} T_{a+j} \overline{T_{a+j+\alpha}} + \sum_{j=1}^{a} T_{(a+b-\alpha)+j} \overline{T_{(a+b)+j}} = \sum_{j=1}^{b-\alpha} S_j \overline{S_{j+\alpha}} = A^{\mathbf{S}}(\alpha)$. Similarly, one may show that $A^{\mathbf{T}}(\alpha) = 0$ for $b \leq \alpha \leq 2a + b - 1$.

Now assume $\mathcal{S}$ is a set of aperiodic complementary sequences, then

$$\sum_{i=1}^{t} A^{\mathbf{T}_i}(\alpha) = \begin{cases} \sum_{i=1}^{t} A^{\mathbf{S}_i}(\alpha) = 0, & \text{if } 1 \leq \alpha \leq a, \\ 0, & \text{if } a + 1 \leq \alpha \leq 2a + b - 1, \end{cases}$$

which follows that $\mathcal{T}$ is a set of aperiodic complementary sequences. The converse part is similar and we omit it here. $\square$

Thanks to Proposition 3, we will consider the *shortened* sequence S of length $(2\ell - 1)^2 - 2(\ell - 1) = 4\ell^2 - 6\ell + 3$ obtained by removing the first and last consecutive $2(\ell - 1)$ zeros in the rest of the paper. By abusing the notations, we still use S to denote this shortened sequence.

The following result will be used later to show the non-existence of complementary ℓ-HLAs.

Corollary 2. *If there exists a set of t aperiodic complementary ℓ-HLAs over alphabet $\mathfrak{A}$, then a set of aperiodic complementary sequences of length $4\ell^2 - 6\ell + 3$ over alphabet $\mathfrak{A} \cup \{0\}$ also exists. Moreover, there exists a set of t periodic complementary sequences of period $4\ell^2 - 6\ell + 3$ over alphabet $\mathfrak{A} \cup \{0\}$.*

By Corollary 2, one can see clearly that if a set of t periodic complementary sequences of period $4\ell^2 - 6\ell + 3$ over alphabet $\mathfrak{A} \cup \{0\}$ does not exist, then a set of t aperiodic complementary ℓ-HLAs over alphabet $\mathfrak{A}$ does not exist either.

4. Characterization of Periodic Complementary Sequences by Group Rings

First we fix some notation to be used in remaining sections. Let p be a prime and $\mathfrak{A}_p = \{0, \zeta_p^i : 0 \le i \le p - 1\}$, where ζ_p is a primitive p-th root of unity. Let $\mathfrak{A}_p^* = \mathfrak{A}_p \setminus \{0\}$. As mentioned in the previous section, the non-existence of periodic complementary sequences with period $4\ell^2 - 6\ell + 3$ and over alphabet $\mathfrak{A}_p$ implies the non-existence of aperiodic complementary ℓ-HLAs over alphabet $\mathfrak{A}_p^*$. In this section, we will make use of group rings to study periodic complementary sequences with period n and alphabet $\mathfrak{A}_p$. In particular, we show that a set of periodic complementary sequences with period n and over alphabet $\mathfrak{A}_p^*$ is equivalent to a relative difference family in the group $\mathbb{Z}_p \times \mathbb{Z}_n$ relative to $\mathbb{Z}_p \times \{1\}$.

4.1. *Periodic complementary sequences*

Let $\mathcal{S} = \{S_1, S_2, \ldots, S_t\}$ be a set of sequences. All sequences S_i have the period n and alphabet $\mathfrak{A}_p$. We may write each sequence S_i in the following form

$$S_i = \left\{ a_{i,0}\zeta_p^{b_{i,0}}, a_{i,1}\zeta_p^{b_{i,1}}, \ldots, a_{i,n-1}\zeta_p^{b_{i,n-1}} \right\},$$

where $a_{i,j} \in \{0,1\}, b_{i,j} \in \{0, \ldots, p-1\}$ for $1 \le i \le t$ and $0 \le j \le n - 1$. If $a_{i,j} = 0$, we may choose any $b_{i,j} \in \{0, \ldots, p-1\}$ which gives the same value of $a_{i,j}\zeta_p^{b_{ij}}$. Without loss of generality, we assume $b_{i,j} = 0$ when $a_{i,j} = 0$. For each S_i, we define an associated sequence $T_i = \{a_{i,0}, a_{i,1}, \ldots, a_{i,n-1}\}$ over alphabet $\{0, 1\}$. Now, let us define a group $G = \mathbb{Z}_p \times \mathbb{Z}_n$, where $\mathbb{Z}_p = \langle g \rangle$ and $\mathbb{Z}_n = \langle h \rangle$. For each S_i, we introduce a group ring element $D_i \in \mathbb{Z}[G]$ defined by

$$D_i = \sum_{j=0}^{n-1} a_{ij} g^{b_{ij}} h^j. \tag{9}$$

Given a character θ_α of the group $\mathbb{Z}_p$ defined by $\theta_\alpha(g) = \zeta_p^\alpha$ ($\alpha \in \{0, \ldots, p-1\}$), it is easy to see that $\theta_\alpha(D_i) = \sum_{j=0}^{n-1} a_{ij} \zeta_p^{\alpha b_{ij}} h^j$. The above arguments relate a set of sequences $\{S_1, \ldots, S_t\}$ to a set of group ring elements $\{D_1, \ldots, D_t\}$ in $\mathbb{Z}[G]$, which enables us to make use of the group rings to characterize the periodic complementary sequences. To present the main result of this section, we need the following two lemmas.

Lemma 1. *Let $X \in \mathbb{Z}[\mathbb{Z}_p]$, where p is a prime. Then $\chi(X) = 0$ for any non-principal character χ of $\mathbb{Z}_p$ if and only if there exists an integer n such that $X = n\mathbb{Z}_p$.*

Proof. If $X = n\mathbb{Z}_p$, then by Result 1, clearly $\chi(X) = 0$ for all non-principal characters χ. Conversely, assume that $X = \sum_{j=0}^{p-1} a_j g^j \in \mathbb{Z}[\mathbb{Z}_p]$. Let χ_i be a non-principal character defined by $\chi_i(g) = g^i$. Then we have $\chi_i(X) = \sum_{j=0}^{p-1} a_j \zeta_p^{ij} = \sigma_i(\sum_{j=0}^{p-1} a_j \zeta_p^j) = 0$, where $\sigma_i \in \mathrm{Gal}(\mathbb{Q}(\zeta_p)/\mathbb{Q})$ defined as $\sigma_i(\zeta_p) = \zeta_p^i$. Since $\{1, \zeta_p, \ldots, \zeta_p^{p-1}\}$ forms the integral basis of the algebraic integer ring $\mathbb{Z}[\zeta_p]$, we have $a_0 = a_1 = \cdots = a_{p-1}$ and therefore $X = a_0\mathbb{Z}_p$, which completes the proof. $\qquad\square$

Lemma 2. *Let $S = (s_0, s_1, \ldots, s_{n-1})$ be a sequence over the alphabet $\{0, 1\}$ and with length n. Assume the size of its support set is k. Then we have $\sum_{t=0}^{n-1} P^S(t) = k^2$, where P^S is the periodic autocorrelation function of the sequence S.*

Proof. We have that

$$\sum_{t=0}^{n-1} P^S(t) = \sum_{t=0}^{n-1} \sum_{j=0}^{n-1} s_j s_{j+t} = \left(\sum_{j=0}^{n-1} a_j \right) \left(\sum_{t=0}^{n-1} s_{j+t} \right) = k^2.$$

$\qquad\square$

Now we are ready to present the theorem.

Theorem 1. *Using the same notation as above and letting the size of the support set of each sequence S_i be k_i for $1 \leq i \leq t$, then S is a set of periodic complementary sequences if and only if the set of group ring elements defined in (9) $\mathcal{D} = \{D_1, \ldots, D_t\}$ satisfies the following group ring equation*

$$D_1 D_1^{(-1)} + \cdots + D_t D_t^{(-1)} = \sum_{i=1}^{t} k_i + \sum_{u=1}^{n-1} \left(\sum_{i=1}^{t} \frac{P^{T_i}(u)}{p} \mathbb{Z}_p \right) h^u, \quad (10)$$

where $P^{T_i}(u)$ is the periodic autocorrelation function of the sequence T_i.

Proof. For simplicity, we denote $\sum_{i=1}^{t} \frac{P^{T_i}(u)}{p} \mathbb{Z}_p$ by X_u. By Corollary 1, we need to show that for any character χ of G, the character values of both sides of Eq. (10) are equal. First, if χ is principal, for the left hand side of Eq. (10), it follows immediately that $\chi(\text{LHS}) = \sum_{i=1}^{t} k_i^2$. For the right hand side of Eq. (10), by noting that $\chi(X_u) = \sum_{i=1}^{t} P^{T_i}(u)$, we get:

$$\chi(\text{RHS}) = \sum_{i=1}^{t} k_i + \sum_{u=1}^{n-1} \sum_{i=1}^{t} P^{T_i}(u) = \sum_{i=1}^{t} \left(\sum_{u=0}^{n-1} P^{T_i}(u) - P^{T_i}(0) + k_i \right)$$

$$= \sum_{i=1}^{t} (k_i^2 - k_i + k_i) = \sum_{i=1}^{t} k_i^2,$$

where the second last equality uses Lemma 2 and $P^{T_i}(0) = k_i$.

Next, if χ is non-principal but is principal on $\mathbb{Z}_p$, namely $\chi = \theta_0 \eta_\beta$, then

$$\chi(D_i D_i^{(-1)}) = \chi(D_i) \overline{\chi(D_i)} = \sum_{j_1, j_2} a_{i,j_1} a_{i,j_2} \zeta_n^{\beta(j_1 - j_2)}$$

$$= \sum_v \left(\sum_{j_2} a_{i,j_2+v} a_{i,j_2} \right) \zeta_n^{\beta v} = \sum_v P^{T_i}(v) \zeta_n^{\beta t}.$$

Therefore we have $\chi(\text{LHS}) = \sum_{i=1}^{t} \chi(D_i D_i^{(-1)}) = \sum_{i=1}^{t} \sum_v P^{T_i}(v) \zeta_n^{\beta v}$. On the other hand, $\chi(\text{RHS}) = \sum_{i=1}^{t} k_i + \sum_{u=1}^{n-1} \sum_{i=1}^{t} P^{T_i}(u) \zeta_n^{\beta u} = \sum_{i=1}^{t} \left(k_i + \sum_{u=0}^{n-1} P^{T_i}(u) \zeta_n^{\beta u} - P^{T_i}(0) \right) = \sum_{i=1}^{t} \sum_{u=0}^{n-1} P^{T_i}(u) \zeta_n^{\beta u} = \chi(\text{LHS})$.

Finally, if χ is not principal on $\mathbb{Z}_p$, namely $\chi = \theta_\alpha \eta_\beta$ with $\alpha \neq 0$, we have that

$$
\chi(\text{LHS}) = \sum_{i=1}^{t} \theta_\alpha \eta_\beta (D_i D_i^{(-1)}) = \sum_{i=1}^{t} \eta_\beta \left(\theta_\alpha(D_i) \theta_\alpha(D_i^{(-1)}) \right)
$$

$$
= \sum_{i=1}^{t} \eta_\beta \left(\left(\sum_{j_1=0}^{n-1} a_{i,j_1} \zeta_p^{\alpha b_{i,j_1}} h^{j_1} \right) \cdot \left(\sum_{j_2=0}^{n-1} a_{i,j_2} \zeta_p^{-\alpha b_{i,j_2}} h^{-j_2} \right) \right)
$$

$$
= \sum_{i=1}^{t} \eta_\beta \left(\sum_{j_1,j_2=0}^{n-1} a_{i,j_1} a_{i,j_2} \zeta_p^{\alpha(b_{i,j_1}-b_{i,j_2})} h^{j_1-j_2} \right)
$$

$$
= \sum_{i=1}^{t} \eta_\beta \left(\sum_{u=0}^{n-1} \left(\sum_{j_1} a_{i,j_1} a_{i,j_1+u} \zeta_p^{\alpha(b_{i,j_1}-b_{i,j_1+u})} \right) h^u \right)
$$

$$
= \sum_{i=1}^{t} \eta_\beta \left(\sum_{u} \left(\sum_{j_1} (a_{i,j_1} \zeta_p^{\alpha b_{i,j_1}}) \overline{(a_{i,j_1+u} \zeta_p^{\alpha(b_{i,j_1+u})})} \right) h^u \right)
$$

$$
= \sum_{i=1}^{t} \eta_\beta \left(\sum_{u} P^{S_i}(u)^{\sigma_\alpha} h^u \right) = \sum_{u} \left(\sum_{i=1}^{t} P^{S_i}(u)^{\sigma_\alpha} \right) \zeta_n^{\beta u}
$$

$$
= \sum_{i=1}^{t} k_i,
$$

where $\sigma_\alpha \in \text{Gal}(\mathbb{Q}(\zeta_p)/\mathbb{Q})$ is defined by $\sigma_\alpha(\zeta_p) = \zeta_p^\alpha$ and the last equality uses the property that $\mathcal{S}$ is a complementary set of sequences. On the other hand, it is easy to check that $\chi(\text{RHS}) = \sum_{i=1}^{t} k_i$. This completes the proof. $\qquad\square$

Theorem 1 characterizes the periodic complementary sequences by a group ring equation. By exploring this equation using number theory techniques, we obtain the condition for the existence of complementary sequences over the alphabet $\mathfrak{A}_p$. The following results are obtained from Theorem 1. We will make use of them to provide the conditions for the existence of complementary ℓ-HLAs in the next section.

Corollary 3. *Assume* $\mathcal{S} = \{S_1, \ldots, S_t\}$ *is a set of periodic complementary sequences, where each* S_i *has period* n *and alphabet* $\mathfrak{A}_p$. *Then the following holds:*

(1) p *must be a divisor of* $\sum_{i=1}^{t} P^{T_i}(u)$ *for each* $1 \leq u \leq n-1$.
(2) *if* $p = 2$, *the sum of the sizes of the support sets* $\sum_{i=1}^{t} k_i$ *must be the sum of* t *squares.*

Proof. The first result follows directly from the Theorem 1 since both sides of Eq. (10) belong to $\mathbb{Z}[G]$ and hence the coefficients are integers. For the second result, when $p = 2$, for any character χ of G, we have $\chi(D_i) \in \mathbb{Z}$ and $\chi(D_i^{-1}) = \chi(D_i)$. In particular, if χ is non-principal on $\mathbb{Z}_p$, we get $\sum_{i=1}^{t} k_i$ on the right hand side of Eq. (10) and the left hand side is the summation of t squares. $\qquad\square$

We give several remarks on Theorem 1 and Corollary 3 below.

Remark 2. *In Theorem 1, the condition that p is prime is necessary. For instance, we computed the pair of complementary quaternion sequences listed in [21, page 76] and found that none of them satisfies the group equation (10).*

Let us give an application of Corollary 3 to show that there does not exist a pair of aperiodic complementary 2-HLAs over alphabet $\mathfrak{A}_p^*$ for any prime p. This result was also given in [10] but with a more tricky proof, while our proof is much shorter thanks to Corollary 3.

Proposition 4. *There does not exist a pair of aperiodic complementary 2-HLAs over alphabet $\mathfrak{A}_p^*$ for any prime p.*

Proof. By Corollary 2, a pair of aperiodic complementary 2-HLAs over $\mathfrak{A}_p^*$ will lead to a pair of periodic complementary sequences of length 7 over $\mathfrak{A}_p^*$ (note that the alphabet is not $\mathfrak{A}_p$ since from the matrix $\mathcal{T}_2$ in Example 1 we may see the shortened sequence from a 2-HLA is $(x_4, x_3, x_5, x_0, x_2, x_6, x_1)$ with alphabet the same as the 2-HLA). By Corollary 3(1) we have $p \mid 2 \cdot 7$ since $P^{T_i}(u) = 7$ for all $1 \le u \le 6$ and $i = 1, 2$. Thus there does not exist a pair of periodic complementary sequences with length 7 unless either $p = 7$ or $p = 2$. By Corollary 3(2), a pair of binary periodic complementary sequences does not exist since $7 + 7 = 14$ is not the sum of two squares. Furthermore, using MAGMA we performed an exhaustive search on a pair of sequences with period 7 and alphabet $\mathfrak{A}_7^*$; and we found that such a pair of periodic complementary sequences does not exist. $\qquad\square$

When the alphabet of each sequence S_i is $\mathfrak{A}_p^*$, we have the following result which is of independent interest.

Theorem 2. *Assume $\mathcal{S} = \{S_1, \ldots, S_t\}$ is a set of sequences with length n and alphabet $\mathfrak{A}_p^*$, where p is a prime. Then $\mathcal{S}$ is a set of periodic complementary sequences if and only if $\mathcal{D} = \{D_1, \ldots, D_t\}$ is a $\left(pn, p, n, \frac{tn}{p}; t\right)$*

relative difference family in $G = \mathbb{Z}_p \times \mathbb{Z}_n$ relative to $\mathbb{Z}_p \times \{1\}$, where D_i is defined as in (9). Equivalently, the following group ring equation holds

$$D_1 D_1^{(-1)} + \cdots + D_t D_t^{(-1)} = tn + \frac{tn}{p}\left(G - \mathbb{Z}_p \times \{1\}\right). \qquad (11)$$

Proof. It is easy to see that if the alphabet is $\mathfrak{A}_p^*$, then: (i) $k_i = n$; (ii) $T_i = (1, \ldots, 1)$ and $P^{T_i}(u) = n$ for $1 \le i \le t, 1 \le u \le n-1$. Therefore, by Theorem 1, we have that

$$
\begin{aligned}
D_1 D_1^{(-1)} + \cdots + D_t D_t^{(-1)} &= tn + \sum_{u=1}^{n-1} \frac{tn}{p}\mathbb{Z}_p h^u = tn + \frac{tn}{p}\mathbb{Z}_p \sum_{u=0}^{n-1} h^u \\
&= tn + \frac{tn}{p}\mathbb{Z}_p(\mathbb{Z}_n - 1) \\
&= tn + \frac{tn}{p}\left(G - \mathbb{Z}_p \times \{1\}\right).
\end{aligned}
$$

The proof is completed. $\qquad\square$

Remark 3. *When $p = 2$, the characterization of periodic complementary sequences using difference family was given by Bömer and Antweiler in [2]. Therefore we can regard Theorem 2 as a generalization of their result.*

We have the following immediate result.

Corollary 4. *Assume a set of t periodic complementary sequences of length n and alphabet $\mathfrak{A}_p^*$ exits, then $p \mid tn$.*

5. Existence of Aperiodic Complementary Binary Hexagonal Lattice Arrays

In this section we will provide the conditions for the existence of aperiodic complementary binary ℓ-HLAs $\mathcal{C}_\ell^t = \{C_\ell^1, \ldots, C_\ell^t\}$. It is worth to mention that Ding and Tarokh in [10] found an example when $\ell = 2$ and $t = 4$. In this section we provide the conditions for the existence of $\mathcal{C}_\ell^t$ when $t = 2$ and 3.

5.1. *Pair of complementary binary hexagonal lattice arrays*

First we present two useful lemmas. For a prime divisor p of an integer x, we use the notation $o_p(x)$ to denote the highest exponent of p that divides x, i.e. $p^{o_p(x)} \mid x$ but $p^{o_p(x)+1} \nmid x$.

Lemma 3 (Fermat). *An integer x can be represented as the sum of two squares if and only if $o_p(x)$ is even for all prime divisors p of x with $p \equiv 3$ mod 4.*

Lemma 4. *Let C_ℓ be an ℓ-layer hexagonal lattice array, then there are $3\ell^2 - 3\ell + 1$ elements in C_ℓ.*

Proof. Recall that the coordinates of the points in an ℓ-HLA are determined in (5). Then the number of points in the support set Ω_ℓ is

$$2 \cdot \sum_{i=1}^{\ell-1} (\ell - i - 1 + (\ell - 1) + 1) + 2(\ell - 1) + 1 = 2 \sum_{i=1}^{\ell-1} (2\ell - i - 1) + 2\ell - 1$$
$$= 3\ell^2 - 3\ell + 1.$$

This completes the proof. $\qquad\square$

Now we can state the following theorem.

Theorem 3. *There does not exist a pair of aperiodic complementary binary ℓ-HLAs when $3\ell(\ell - 1) + 1$ has a prime divisor p with $p \equiv 3$ mod 4 and $o_p(3\ell(\ell - 1) + 1)$ is odd.*

Proof. By Corollary 2 a pair of aperiodic complementary binary ℓ-HLAs will lead to a pair of periodic complementary sequences over the alphabet $\mathfrak{A}_2$. Further by Corollary 3(2) such a pair of periodic complementary sequences exist only if the sum of their support sets, $2(3\ell^2 - 3\ell + 1)$ (Lemma 4) is the sum of two integers. It follows from Lemma 3 that $2(3\ell^2 - 3\ell + 1)$ is the sum of two integers if and only if for all of its prime divisors p with $p \equiv 3$ mod 4, the order $o_p(2(3\ell^2 - 3\ell + 1))$ is even. Note that $3\ell(\ell - 1) + 1$ is odd (since $\ell(\ell - 1)$ is always even), hence the above statement is equivalent to $3\ell(\ell-1)+1$ cannot have a prime divisor $p \equiv 3$ mod 4 and $ord_p(3\ell(\ell-1)+1)$ is odd. This completes the proof. $\qquad\square$

The following corollary follows immediately.

Corollary 5. *There does not exist a pair of aperiodic complementary binary ℓ-layer hexagonal lattice arrays if $\ell \equiv 2, 3$ mod 4.*

Proof. By Theorem 3, the existence of a pair of aperiodic complementary binary ℓ-HLAs implies that $3\ell(\ell - 1) + 1 \equiv 1$ mod 4 (since for all prime divisors $p \equiv 3$ mod 4, its order is even). Now assume $3\ell(\ell - 1) + 1 \equiv 3$ mod 4 (note that $3\ell(\ell - 1) + 1$ is odd and then $3\ell(\ell - 1) + 1 \not\equiv 0, 2$ mod 4). This can only happen when $3\ell(\ell - 1) \equiv 2$ mod 4, or $\ell(\ell - 1) \equiv 2$ mod 4, which is equivalent to $\ell \equiv 2, 3$ mod 4. This completes the proof. $\qquad\square$

By Theorem 3 and Corollary 5, we use the following table to give the existence status for a pair of aperiodic complementary binary ℓ-HLAs for ℓ up to 20. The question mark "?" denotes the existence status is undetermined. For an ℓ-HLA, we use $|C_\ell|$ to denote the number of non-zero elements in it.

Table 1. Existence of pair of AC binary ℓ-HLAs for $1 \leq \ell \leq 20$

ℓ	Existence	Reference	ℓ	Existence	Reference
1	No	Trivial	2	No	Corollary 5
3	No	Corollary 5	4	?	
5	?		6	No	Corollary 5
7	No	Corollary 5	8	?	
9	No	Theorem 3 $|C_9| = 7 \cdot 31$	10	No	Corollary 5
11	No	Corollary 5	12	?	
13	No	Theorem 3 $|C_{13}| = 7 \cdot 67$	14	No	Corollary 5
15	No	Corollary 5	16	No	Theorem 3 $|C_{16}| = 7 \cdot 103$
17	No	Theorem 3 $|C_{17}| = 19 \cdot 43$	18	No	Corollary 5
19	No	Corollary 5	20	No	Theorem 3 $|C_{20}| = 7 \cdot 163$

One may see from Table 1 that the smallest undermined case is $\ell = 4$ for a pair of aperiodic complementary ℓ-HLAs. By Proposition 4, there are $3 \cdot 4^2 - 3 \cdot 4 + 1 = 37$ elements in a 4-HLA. Therefore the complexity for an exhaustive search for a pair of aperiodic complementary 4-HLAs is 2^{74}. We leave the following open problem for the future research.

Problem 1. *Determine the existence of a pair of aperiodic complementary binary ℓ-HLAs for the values ℓ which are not excluded in Theorem 3. In particular, determine the existence for the cases where $\ell = 4, 5, 8, 12$ as listed in Table 1.*

5.2. *Triple of complementary binary hexagonal lattice arrays*

Now we consider the existence of a triple of aperiodic complementary binary ℓ-HLAs.

Lemma 5 (Legendre theorem). *An integer x can be represented as the sum of three squares if and only if x is not of the form $4^a(8b+7)$, where a, b are integers.*

Theorem 4. *There do not exist aperiodic complementary binary ℓ-layer hexagonal lattice arrays $C_\ell^3 = \{C_\ell^1, C_\ell^2, C_\ell^3\}$ when $\ell \equiv 4, 5 \mod 8$.*

Proof. By Lemma 5, it is clear that $3(3\ell(\ell-1)+1)$ is not the sum of three squares if $3(3\ell(\ell-1)+1)$ is of the form $4^a(8b+7)$. In this case it follows from Corollary 3(2) that there does not exist a set of 3 periodic complementary binary sequences which the sum of the sizes of the support sets equals $3(3\ell(\ell-1)+1)$. This will further show the non-existence of a set of 3 aperiodic complementary binary ℓ-HLAs by Corollary 2. Now assume $3(3\ell(\ell-1)+1) = 4^a(8b+7)$ for some a, b. Since $3(3\ell(\ell-1)+1)$ is odd, the assumption is equivalent to $3(3\ell(\ell-1)+1) = 8b+7$. After simplification we get $9\ell(\ell-1)-4 = 8b$, hence $\ell(\ell-1) \equiv 4 \mod 8$. The rest of the proof is routine by checking only when $\ell \equiv 4, 5$ one may get $\ell(\ell-1) \equiv 4 \mod 8$. $\square$

We run a computer experiment to search for an aperiodic complementary triple set $C_\ell^3 = \{C_\ell^1, C_\ell^2, C_\ell^3\}$ for $\ell = 2$ and found no such a triple set exists. The complexity for an exhaustive search for the case $\ell = 3$ is 2^{57} (since by Proposition 4 each 3-HLA has 19 elements). We leave the following open problem.

Problem 2. *Determine the existence of complementary triple set of binary ℓ-HLAs for ℓ with $\ell \not\equiv 4, 5 \mod 8$.*

6. Conclusions

In this paper we explore the conditions for the existence of aperiodic complementary hexagonal lattice arrays over the alphabet $\mathfrak{A}_p^* = \{\zeta_p^i : 0 \leq i \leq p-1\}$, whose support set is a set of ℓ-layer consecutive hexagons. We call such a hexagonal lattice array an ℓ-HLAs for short. We first show that aperiodic complementary ℓ-HLAs lead to aperiodic complementary (and hence periodic complementary) sequences over the alphabet $\mathfrak{A}_p = \mathfrak{A}_p^* \cup \{0\}$. Through relating the sequences of period n and the alphabet $\mathfrak{A}_p$ to group ring elements in $\mathbb{Z}[\mathbb{Z}_p \times \mathbb{Z}_n]$, we provide a characterization of periodic complementary sequences by group rings. As of independent interest, we show that set of t periodic complementary sequences with period n and the alphabet $\mathfrak{A}_p^*$ is conceptually equivalent to a $(np, p, n, \frac{tn}{p}; t)$ relative difference family in $\mathbb{Z}_p \times \mathbb{Z}_n$ relative to $\mathbb{Z}_p \times \{1\}$. Let $\mathcal{C}_\ell^t = \{C_\ell^1, \ldots, C_\ell^t\}$ be a set of aperiodic complementary binary ℓ-HLAs. Thanks to the aforementioned characterization of periodic complementary sequences, we are able to give conditions for the existence of $\mathcal{C}_\ell^t$ when $t = 2, 3$. Note that Ding and Tarokh in [10] provided examples of aperiodic complementary binary 2-HLAs when $t = 4$.

There are some cases of ℓ that we cannot determine the existence of a pair or a triple of aperiodic complementary binary ℓ-HLAs. For a pair (resp. a triple) of aperiodic complementary binary ℓ-HLAs, the smallest undermined case of its existence is $\ell = 4$ (resp. $\ell = 3$). It will be interesting to develop more conditions to exclude, or to provide constructions of them for these cases. We leave two open problems (Problems 1 and 2) for the future research.

Acknowledgement

We thank Jie Ding and Vahid Tarokh for sending us their interesting manuscript [10] on hexagonal lattice complementary arrays.

References

[1] K.T. Arasu and Q. Xiang, On the existence of periodic complementary binary sequences, Designs, Codes and Cryptography 2, 257–262, (1992).

[2] L. Bömer and M. Antweiler, Periodic complementary binary sequences, IEEE Transaction on Information Theory 36(6), 1487–1494, (1990).

[3] M. Buratti, Constructions of $(q, k, 1)$ difference families with q a prime power and $k = 4, 5$, Discrete Mathematics 138, 169–175, (1995).

[4] M. Buratti, Recursive constructions for difference matrices and relative difference families, Journal of Combinatorial Designs 6(3), 165–182, (1998).

[5] M. Buratti, Y. Wei, D. Wu, P. Fan, M. Chen, Relative difference families with variable block sizes and their related OOCs, IEEE Transactions on Information Theory 57(11), 7489–7947, (2011).

[6] Y. Chang and C. Ding, Constructions of external difference families and disjoint difference families, Designs, Codes and Cryptography 40(2), 167–185, (2006).

[7] R. Craigen, W. Holzmann and H. Kharaghani, Complex Golay sequences: structure and applications, Discrete Mathematics 252, 73–89, (2002).

[8] J.A. Davis, J. Jedwab, and K.W. Smith, Proof of the Barker array conjecture, Proceedings of American Mathematical Society 135, 20112018, (2007).

[9] C. Ding, Two constructions of $(v,(v-1)/2,(v-3)/2)$ difference families, Journal of Combinatorial Designs 16(2), 164–171, (2008).

[10] J. Ding, M. Noshad and V. Tarokh, Complementary lattice arrays for coded aperture imaging, arXiv:1506.02160v2, (2015).

[11] J. H. Dinitz and P. Rbmodney, Disjoint difference families with block size 3, Utilitas Math. 52, 153–160, (1997).

[12] J. H. Dinitz and N. Shalaby, Block disjoint difference families for Steiner triple systems: $v \equiv 1$ (bmod 6), Journal of Statistical and Planning Inference 106, 77–86, (2002).

[13] S. Eliahou, M. Kervaire and B. Saffari, On Golay polynomial pairs, Advances in Applied Mathematics 12, 235–292, (1991).

[14] K. Feng, P. J. Shiue and Q. Xiang, On aperiodic and periodic complementary binary sequences, IEEE Transaction on Information Theory 45, 296-303, (1999).

[15] R. Fuji-Hara, Y. Miao and S. Shinohara, Complete sets of of disjoint difference families and their applications, Journal of Statistical and Planning Inference 106, 87–103, (2002).

[16] M.J.E. Golay, Complementary series, IRE Transaction on Information Theory, IT-7:8287, 1961.

[17] J. Jedwab and M.G. Parker, There are no Barker arrays having more than two dimensions, Designs, Codes and Cryptography 43, 79–84, (2007).

[18] J. Jedwab and M.G. Parker, Golay complementary array pairs, Designs, Codes and Cryptography 44, 209216, (2007).

[19] S.L. Ma, Polynomial addition sets, Ph.D. Thesis, University of Hong Kong, (1985).

[20] L. Martinez, D. Z. Dokovic, A. Vera-Lopez, Existence question for difference families and construction of some new families, Journal of Combinatorial Designs 12(4), 256–270, (2004).

[21] C. Milies and S. Sehgal, An Introduction To Group Rings, Algebras and applications, Volume 1. Springer, (2002).

[22] W. Ogata, K. Kurosawa, D. R. Stinson, H. Saido, New combinatorial designs and their applications to authentication codes and secret sharing schemes, Discrete Mathematics 279, 383–405, (2004).

[23] B. Schmidt, Cyclotomic integers and finite geometry, Journal of American Mathematical Society 12(4), 929-952, (1999).

[24] C.C. Tseng and C. Liu, Complementary sets of sequences, IEEE Transaction on Information Theory 18(5), 644-652, (2003).

[25] R. M. Wilson, Cyclotomy and difference families in elementary Abelian groups, Journal of Number Theory 4, 17–42, (1972).

Printed in the United States
By Bookmasters

Printed in the United States
By Bookmasters